Applied Stream Sanitation

ENVIRONMENTAL SCIENCE AND TECHNOLOGY

A Wiley-Interscience Series of Texts and Monographs

Edited by ROBERT L. METCALF, *University of Illinois*
JAMES N. PITTS, Jr., *University of California*

PRINCIPLES AND PRACTICES OF INCINERATION
Richard C. Corey

AN INTRODUCTION TO EXPERIMENTAL AEROBIOLOGY
Robert L. Dimmick

AIR POLLUTION CONTROL, Part I
Werner Strauss

APPLIED STREAM SANITATION
Clarence J. Velz

APPLIED STREAM SANITATION

BY

CLARENCE J. VELZ

Professor Emeritus of
Public Health Engineering
The University of Michigan

Wiley-Interscience,

A DIVISION OF JOHN WILEY & SONS

NEW YORK • LONDON • SYDNEY • TORONTO

Library of Congress Catalogue Card Number: 71-120710

ISBN 0 471 90525 9

Printed in the United States of America

10 9 8 7 6 5 4 3

SERIES PREFACE

Environmental Sciences and Technology

The Environmental Science and Technology Series of Monographs, Textbooks, and Advances is devoted to the study of the quality of the environment and to the technology of its conservation. Environmental science therefore relates to the chemical, physical, and biological changes in the environment through contamination or modification, to the physical nature and biological behavior of air, water, soil, food, and waste as they are affected by man's agricultural, industrial, and social activities, and to the application of science and technology to the control and improvement of environmental quality.

The deterioration of environmental quality, which began when man first collected into villages and utilized fire, has existed as a serious problem since the industrial revolution. In the last half of the twentieth century, under the ever-increasing impacts of exponentially increasing population and of industrializing society, environmental contamination of air, water, soil, and food has become a threat to the continued existence of many plant and animal communities of the ecosystem and may ultimately threaten the very survival of the human race.

It seems clear that if we are to preserve for future generations some semblance of the biological order of the world of the past and hope to improve on the deteriorating standards of urban public health environmental science and technology must quickly come to play a dominant role in designing our social and industrial structure for tomorrow. Scientifically rigorous criteria of environmental quality must be developed. Based in part on these criteria, realistic standards must be established and our technological progress must be tailored to meet them. It is obvious that civilization will continue to require increasing amounts of fuel, transportation, industrial chemicals, fertilizers, pesticides, and countless other products and that it will continue to produce waste products of all descriptions. What is urgently needed is a total systems approach to modern civilization through which the pooled talents of scientists and engineers, in cooperation with social scientists and the medical profession, can be focused on the development of order and equilibrium to the presently disparate segments of the human environment. Most of the skills and tools that are needed are already in existence. Surely a technology that has created such manifold environmental problems is also capable of solving them. It is our hope that this Series in Environmental Science and Technology will not only serve to make this challenge more explicit to the established professional but that it also will help to stimulate the student toward the career opportunities in this vital area.

Robert L. Metcalf
James N. Pitts, Jr.

PREFACE

Applied Stream Sanitation as developed in this text is intended as a complementary contribution to that well-known classic, *Stream Sanitation,* by the late Earle B. Phelps.* Much that we know today about the the fundamentals of stream sanitation and the capacity of streams to assimilate natural and man-made wastes stems from the early work of Phelps. His persistent interest in the stream as a living thing extended beyond scientific observation and description to a canny insight and capacity to synthesize its complex behavior into quantitative formulations. These are the foundation of modern stream sanitation, or more broadly the field of *potomology.*

As defined by Phelps in *Stream Sanitation,* potomology applies knowledge from many areas of the physical and biological sciences and mathematics; the potomologist need not qualify as a specialist in each of these areas, but he must be able to integrate knowledge from all of them as he pursues his own speciality, the *science of rivers.* The definition of potomology in the present volume is extended to include also knowledge in areas of the social sciences and engineering, essential elements in effective pollution control and wise use of water resources. Furthermore, in application potomology becomes as much an *art* as a *science,* tempered by experience and professional judgment.

In terms of practical usefulness the waste assimilation capacity of streams as a water resource has its basis in the complex phenomenon termed stream self-purification. This is a dynamic phenomenon reflecting hydrologic and biologic variations, and the interrelations are not yet fully understood in precise terms. However, this does not preclude applying what is known. Sufficient knowledge is available to permit quantitative definition of resultant stream conditions under expected ranges of variation to serve as practical guides in decisions dealing with water resource use, development, and management.

* Phelps, E. B., *Stream Sanitation,* Wiley, New York, 1944.

The orientation of this text is therefore toward practical application, in terms useful not only to the professional potomologist, be he scientist or engineer, but also to the conservationist, economist, lawyer, and administrator. In addition to presenting useful tools for evaluating solutions to water pollution, it is hoped it will assist in developing an informed, intelligent appreciation of the complex problem of waste disposal and pollution control in a technological society.

Over the years I have been fortunate in my associations with government and industry in studies of many rivers. From this work I have drawn liberally, including the practical illustrations. I acknowledge gratefully the cooperation and extensive data supplied by associates in government and industry, and the contributions by colleagues and my many students through academic years. I wish particularly to acknowledge the research support and the survey and river sampling data provided by National Council for Stream Improvement, Inc., which have made possible the detailed studies of many of these rivers.

Here I pay special tribute to the memory of my wife, Harriet O'Brien Velz, who participated in all of my river studies and without whose painstaking researches and constant assistance many could not have been undertaken and accomplished.

Finally, a tribute to the memory of Earle B. Phelps. It has been my great privilege to have known Professor Phelps intimately as his student, associate, and friend during his fruitful years. The potomologist will long be in his debt, and to me it is especially gratifying to extend the usefulness of his work in applying to practical problems the fundamentals of stream sanitation he so brilliantly conceived and formulated.

C. J. Velz

Longboat Key, Florida
February, 1969

CONTENTS

Applied Stream Sanitation

1

INTRODUCTION

THE WASTE PRODUCTS OF MAN'S ACTIVITIES

In the natural cycle there is no waste in the sense of loss; each living thing borrows its substance briefly only to return to the inanimate storehouse and itself be reused by countless generations. Man makes very inefficient use of the natural resources that he borrows and in the process creates a colossal amount and variety of waste products, *solid, gaseous,* and *liquid.* Raw material is seldom suitable for direct use. In the process of refining, fashioning, and converting a large fraction is discarded. Some is reused for other purposes, but inevitably there are unwanted, unprofitable waste products. Some are dumped in spoil banks; some are burned, with end products exhausted to the atmosphere; some are buried; some are promiscuously scattered in cities and over the countryside; and an increasing amount is discharged into the intricate water-carriage systems we call sewers and ultimately reaches streams, rivers, lakes, and estuaries.

In a simple agrarian society the waste products are few and are either plowed into the soil or are so widely distributed as to be unnoticeable as they return to the land. Not so in the modern urban-industrial concentrations. Technology has reached a stage at which it is relatively simple to extract and fashion the useful fraction of raw materials—the real challenge is how to dispose of the ever increasing quantities of unwanted solid, gaseous, and liquid waste products without befouling the environment in which we live and work.

The *solid wastes,* garbage and trash, collected by organized American community refuse systems in 1960 averaged over 4 lb per capita per day, or 0.73 ton per year; and it is estimated that by 1980 refuse production will approach 1 ton per capita per year. With more prepared packaged products on the market the trend is not only toward increase in weight but also larger bulk, estimated to increase from 4 cubic yards per capita per year in 1960 to about 6.5 cubic yards per year by 1980.

In addition, large quantities of refuse and other solid wastes are produced by industries that are not served by municipal collection systems, and there is considerable disposal also through home incinerators and kitchen garbage grinders.

There are few quantitative measures of the amount of *gaseous waste* products, but the pall over cities testifies to increasing magnitude. It is estimated that the burning of coal, oil, and natural gas adds six billion tons of carbon dioxide to the earth's atmosphere yearly. At this rate by the year 2000 there will be approximately 25 percent more carbon dioxide in the atmosphere than at present [1]. In addition, industrial processing releases a variety of other gaseous end products that usually contain large quantities of particulate matter settling thinly over wide areas, but massive in the aggregate.

The *liquid waste,* our primary concern in this book, comprises the *spent water supply* of industries, communities, and individual households, to which has been added in the process of use almost an infinite variety of unwanted waste products. In addition, agricultural drainage, urban storm drainage, and natural landwash carry large quantities of waste products to the streams. We shall deal with these in more detail later.

Interlinkage of Solid, Gaseous, and Liquid Waste Products

Although our focus in this text is on waste products that ultimately reach a watercourse, rationally the waste disposal problem must be considered as a unit in toto, the solids, the gases, and the liquids. The three states in which man produces his waste products are interlinked. Simply to transform a waste product from one state to another does not necessarily constitute satisfactory disposal; in fact it can aggravate the problem of ultimate disposal. Incineration of solid wastes may add to already serious air pollution. The products of evaporation and burning may present more difficulty in ultimate disposal in the atmosphere than treatment in liquid form with the residual effluent assimilated in the stream. On the other hand, condensates from evaporation, burning, scrubbing, and industrial recovery processing may produce liquid wastes greatly adding to stream pollution. Drainage from solid waste landfill disposal systems may degrade nearby streams. The contributions of ground garbage and other solids to the sewers may increase the residual effluent load to the watercourse already burdened beyond its waste assimilation capacity. The ease of water-carriage transport leads to an ever increasing wastewater burden, residuals of which ultimately reach the stream.

Overall waste disposal is a complex problem for which no simple solution can be applied wholesale. Regardless of how we treat the wastes, an inevitable *residual* remains for ultimate disposal on land, in the atmosphere, and in the stream. Each must be assigned a share of the burden, depending on relative assimilation capacities and technical and economic feasibility.

In a sense civilization is measured by how it deals with its waste products and the kind of environment in which it is content to live and work. Primitive man has no system; waste disposal is wholly promiscuous. When the environment becomes too befouled, he moves. Modern man, who cannot escape by moving, has to a degree applied science and engineering principles to waste disposal, but as yet he can claim only limited success. The magnitude is overwhelming, unpopular, and costly. There is, however, increasing evidence that modern man in his affluence is not satisfied with a deteriorated environment in which he merely can survive and that he is willing to work and pay for quality environment in which he can thrive.

In setting levels of quality environment it is well to keep in mind that it is the public who pays—that the cost is passed on either directly in taxes or indirectly in increased prices of products of industry or increased rates of utility services. Prudence dictates that careful evaluation be given to alternative methods.

Definition of Pollution

"Pollution" is a very general term and is defined in many ways. In its broadest sense as conceived by the layman it is the befouling of the environment by man's activities, particularly by the disposal of solid, gaseous, and liquid waste products. It also carries a relative connotation as to degree of befouling (impurity) in relation to interference with intended uses of land, air, and water. This leads to many conflicting interpretations of what constitutes pollution and the necessity for more precise definitions suitable for administrative management and legal control. These phases are considered further in Chapter 12.

By virtue of the physiological necessity of drinking water for man and other animals, water pollution falls into two rather sharply defined classifications—namely (a) pathogenic organisms or toxic substances harmful and hazardous and (b) substances that are deleterious, offensive to sight, taste, and smell, or detrimental to the usefulness of natural waters. "Pollution" is the inclusive term; "contamination" is the distinguishing term that is applied to harmful and hazardous waste constituents.

With the increasing variety and quantity of waste products and the interlinkage among solid, gaseous, and liquid states, the hazard of stream pollution is great. Gross pollution that seriously interferes with the beneficial uses of water resources is unnecessary, controllable, and manageable by scientific and engineering methods. However, complete elimination is *impossible;* some degree of pollution is *inevitable.* Man should not delude himself about this; he must accept and learn to accommodate himself to it. It is a matter of degree, not a matter of elimination. The degree of pollution that is acceptable depends on what we are willing to work for and pay for.

Waste disposal and pollution control are an integral part of the broader problem of the development, use, and management of the total water resource. The socioeconomic importance of water resources therefore forms a basis for more detailed consideration.

SOCIOECONOMIC IMPORTANCE OF WATER RESOURCES

The socioeconomic structure of entire regions is shaped to a large degree by the water resource available and how it is used, developed, and managed. Paradoxically, in areas blessed with abundant water resources the relation to growth and prosperity is not always adequately appreciated. Too frequently water is taken for granted, often with waste and abuse. In arid regions, when wells fail and rivers dry up, bitter experience teaches the meaning of water resources. Then comes fuller appreciation of the need for control, development of the little available, and careful use and management.

A glance at a map quickly reveals that all large cities straddle a river or border a large lake or an estuary; the same applies to the distribution of industry. In the arid regions population is sparse and cities are few, with but a scattering of industry. Water is the lifeblood of community and industrial activity. The regions that plan ahead and exercise intelligent management of their water resources will thrive; those that drift and do not foresee needs and problems, even if endowed with good water resources, will decline.

The world is in a technological race. Industrial output is estimated to double in the United States in intervals of 20 to 25 years. Paralleling industrial expansion is the current unprecedented population growth, some predictions exceeding 400 million for the United States by the year 2000. Not only is there growth in numbers, but the shift is strong from dispersed rural to high concentrations in the urban areas spilling over political boundary lines into complex urban-industrial corridors. This two-way trend toward urban-industrial concentration places tremendous demands on water resources.

It is not surprising that these new patterns of growth and increasing demands on limited water resources give rise to water use conflicts and rivalries, in which the complex problem of waste disposal and pollution control intensifies. An overview of the multiple use character of water resources and the impact on the waste assimilation capacity of streams provides an orientation to the role of the science and art of *applied stream sanitation.* A brief critical review of past piecemeal attack on water-resource problems also provides a background for understanding both why we are where we are and the need for a more rational approach in meeting future needs.

Rational Water Resource Development and Use

A rational approach to water resource development and use implies that first we be realistic about our concept of conservation. Interest in water resources is not for the sake of water but for the sake of people who must use it. Irrationally some hold that conservation implies forbidding use. *Nonuse is total waste* and is as much a loss as destructive overuse. (This does not imply that some areas should not be preserved in their natural state, as contemplated in the Wild Rivers program.) Irrationally others hold that "first come, first served," and would appropriate water resources for their exclusive use to the detriment of other users. This is *abuse.* The rational concept is balanced multiple use with the objective of fullest use over the years without waste or abuse.

Two unique characteristics of water resources should be appreciated if a rational balance among uses is to be attained: first, that water is a dynamic resource; second, that it is inherently a multiple use resource.

In addition to a very unequal geographical distribution, water in the natural setting is a highly *dynamic* resource. Unlike a fixed mineral deposit, water varies in the amount that is available at any location from day to day and season to season. Seen from aloft a river appears as a fixed entity whose size is measured by the number of square miles of tributary drainage area, but seen from a location on its bank a river becomes dynamic, its size varying with the amount of water flowing by in a unit of time. What appears to be a large river during the spring flood may become a small creek during the summer drought. It is this highly dynamic character of streams that adds to the difficulties of dependable use and control of quality.

Nevertheless, "dynamic" does not mean "chaotic"; observed over a period of years, stream runoff, although variable, is orderly, following the laws of chance occurrence—unless the natural setting has been tampered with. Statistical analyses of the runoff records determine the *range* of streamflow available and the drought severities that can be expected. Man must either design his activities within this range of variation and decide on a level of risk of water shortage during droughts or *develop* the water resource by such practice as drought control to regulate variations between the extremes of flood and drought. In turn the natural waste assimilation capacity of the stream and the associated water quality are not fixed quantities but rather ranges in potential paralleling the natural variations in streamflow.

When man enters the scene, nature's orderly system is altered. Depending on how they are planned and operated, man-made river developments and water uses can have an effect that is either beneficial or detrimental to the waste assimilation capacity and associated water quality. Thus a second complicating factor is the multiple use character of water resources.

Inherently water is a *multiple use* resource. Moreover seldom is a city alone on a river; others are either above or below, and what they or some private or governmental agency may do with water has an *impact* on other users. As the urban-industrial complex grows, competition for the limited water resources intensifies among potential uses:

1. Community and industrial water supply.
2. Electric power generation—hydro, fossil fueled, and atomic fueled.
3. Recreation—bathing, boating, and water sports.
4. Irrigation.
5. Navigation.
6. Fish, shellfish, and wildlife.
7. The inevitable necessity of ultimately disposing of the residual of wastes from communities, industries, agriculture, and natural sources.

Each of these individual water uses is important; often to the specific user his use is the most important. Herein lies the basis for difficulties and conflicts of interest. Incompatibility among uses is possible, but uses need not be mutually exclusive. The matter of priority and degree of use among users may change in place and time. The shift from an agricultural to an urban-industrial society places higher priorities on community and industrial water supply, tremendous needs of water for steam-electric generation, and water based recreation.

Although not generally recognized as such, a key water resource use in a technological society is satisfactory ultimate disposal of the residual of the waste products of man's activities. The waste assimilation capacity of the stream becomes a *prime* water resource asset to be wisely used and developed. Thus it is crucially important to carefully evaluate the impact, beneficial or detrimental, of other water resource developments and uses on the waste assimilation capacity of the stream.

As the pressures of urban-industrial growth continue, multiple water uses become more and more inextricably interrelated. No longer can each use be viewed in isolation; a *rational, integrated* approach is essential.

A brief review of past practice may provide a background illuminating the need for a more rational approach.

In the past water-resource development and use, with notable exceptions, have taken place largely on a piecemeal basis, each water use considered much as a problem independent of all others. In this practice pollution control did not keep pace with national growth.

In some instances a disaster, such as a flood, has dictated a one-sided view; development centered on flood control, with inadequate consideration of the overall water needs, including satisfactory waste disposal. In other instances some special public or private interest dominated, and a single-minded development, such as large-scale navigation works, proceeded on a narrow basis, often to the detriment of the waste assimilation capacity of the stream and the associated water quality. Another example is hydroelectric power, which often has been oriented to serving the peak of the electrical power demand without concern for the desirability of continuity in streamflow; if a power plant reduces streamflow to a trickle on weekends and at nonpeak hours, the effect on the waste assimilation capacity of the stream can be disastrous.

Public agencies assigned powers and responsibilities to deal with community and industrial water supply and waste disposal have, possibly unwittingly, fostered further fractionalization, piecemeal development, and inadequate stream pollution control. The two interrelated phases of the water contact cycle were split as though they were independent; incoming water supply was dealt with largely as a problem separate from that of the outgoing wastewater return. Incoming supply received the major attention, whereas the satisfactory disposal of the spent water supply, the wastewater return to the stream, received scant or, at best, belated attention.

In turn each of these two phases was further fractionalized. In many instances community water supply was planned and developed on a narrow basis, leaving industries and rapidly expanding fringes outside the central political unit to fend for themselves. Hodgepodge water-supply systems resulted, to the detriment of community and industrial growth.

Similarly the disposal of wastewater was fractionalized. In many instances industrial waste was considered as a problem separate from that of community sewage disposal and deliberately excluded from the municipal sewers and treatment works. Pollution control was disease oriented, centered on sewage, with divided and belated attention to industrial waste disposal problems.

Furthermore, regulatory agencies, with but few exceptions, depended almost exclusively on treatment as the cure-all for pollution prevention and abatement. Treatment is an essential line of defense, but, since it is a percentage removal, there always remains a residual to be handled by self-purification of the stream. When design emphasized details of treatment works rather than quantitative determination of the waste assimilation capacity of the receiving stream, there was a tendency to add costly features that did not contribute to the removal of the waste load. The expenditure could more profitably have been devoted to other means of enhancing the waste assimilation capacity and water quality.

Much is to be learned from our past practices, but clearly, if we are to meet society's current and future needs, much more rigorous scientific and engineering analysis must form the basis for integrated water resource development and use, including a more comprehensive approach to stream sanitation.

RATIONAL STREAM SANITATION

The Nature of Stream Self-Purification

Any system of stream sanitation is dependent on natural self-purification, the ability of the stream to assimilate wastes and restore its own quality. Without this capacity the world long ago would have been buried in its own wastes, and water resources would not be fit for any use. As a gift of nature, each watercourse has its individual self-purification capacity, and this valuable asset must be used. A brief sketch of the nature of this phenomenon will indicate the multiple avenues for a rational system of stream sanitation.

Self-purification is, to an extent, different for each type of waste, but it is sufficient at this stage to illustrate briefly three types: chemical, bacterial, and organic.

Stable chemical wastes undergo little or no change in their journey down a river. Here the primary factor is dilution, and self-purification is almost wholly dependent on streamflow along the course.

In a stream receiving bacterial wastes from sewage three factors aid in self-purification. Dilution takes place here as it does with chemical wastes, and bacteria are destroyed by the unfavorable conditions in the stream environment. The decline in the number of organisms is a function of water temperature and time; the warmer the water, the higher the death rate. Thus streamflow (dilution), time of exposure (measured in hours or days a particle takes to travel down the river from the sewer outlet), and water temperature are the three major determinants of self-purification in the case of bacterial wastes.

The third and most general type of waste is unstable organic matter from various sources, principally communities and industries. Here self-purification is a biochemical process of decay. Oxygen dissolved in the river water is utilized in the stabilization of the organic matter which is carried out by a chain gang of biological life. The stabilization is a time–temperature function, and utilization of the dissolved oxygen increases as the temperature rises. Nature replenishes the depleted oxygen through the complex phenomenon of reaeration from the atmosphere, but unfortunately the oxygen saturation capacity of water declines as the temperature rises. Thus stream runoff, time of passage down the river, water temperature, and reaeration are the four major factors that govern organic stream self-purification.

In the natural setting the waste assimilation potential, controlled by self-purification, depends on the size of the particular river and on how that river behaves. As already mentioned, rivers are dynamic because runoff is dependent on a broad range of variables in the hydrologic setting. Water temperature likewise varies, depending on the climatic setting.

The behavior of a river depends on the configuration of the drainage area and the physical characteristics of the channel along its course. Where the channel is shallow and steep, the time of passage is rapid and short, with good reaeration; where the river is deep and meandering, with natural pools, the time of passage is slow and long, with poor reaeration. Also, the time of passage down the river varies inversely with runoff, being slowest and longest during the drought season.

Thus in the natural setting waste assimilation capacity is not a fixed quantity but rather a range in potential that depends on variations in size and behavior through each reach of the river.

From this thumbnail sketch of the nature of self-purification it is apparent that man can exercise a very great influence on the self-purification (and therefore on the waste assimilation capacity and associated water quality) of a stream. This potential for influence offers an important opportunity for devising a rational system of stream sanitation.

THE DUAL OBJECTIVES OF STREAM SANITATION PRACTICE

There are two primary objectives in the practice of stream sanitation: (a) the satisfactory disposal of the water-carriage waste products of man's activities; (b) the enhancement of satisfactory stream water quantity and quality for intended multiple uses. These two objectives are interrelated, but in a rational system of stream sanitation they follow quite different lines of attack, the former being primarily *defensive* and the latter primarily *offensive.*

In an expanding technological society both lines of defense and lines of offense are essential to attain the dual objectives. Because of its setting and its pattern of variations, each river is unique; hence rational stream sanitation practice must be specifically designed for each particular river, using that combination of defense and offense that ensures maximal efficiency at minimal cost.

For convenience we shall consider lines of defense and lines of offense separately, but it should be emphasized that efficient combination of both is the essence of good practice.

Multiple Lines of Defense

The satisfactory disposal of the water-carriage waste products of man's activities is oriented to the reduction of waste loads and hence protection of the stream against overburden of its self-purification capacity (or the enhanced capacity provided by lines of offensive action) to maintain desired water quality. Effective defense is comprehensive; it is directed against all sources of man-made wastes—urban, industrial, and agricultural—and involves several lines of action.

Integrated Service Areas

In dealing with the urban-industrial concentration effective lines of defense are best developed for an integrated service area encompassing, if necessary, several political units, with services extending to the urban fringes. Likewise service encompasses the industries as well as the commercial and residential areas.

Segregation for Treatment or Special Disposal

Though the defense should be arranged for an integrated service area, it is necessary to segregate the wastewater in a *separate sewerage system* independent of the community storm water drainage system. Rigid control and inspection are required to ensure wastewater connections to the sewerage system, and roof, courtyard, and other rainwater collection to the storm drainage system. Since some form of wastewater treatment is usually essential before discharge of effluent to the stream, segregation is necessary, since diluting the waste prior to treatment complicates the problem. Furthermore, where *combined sewers* (wastewater and storm runoff collected in the same system) are provided, treatment is bypassed with almost every rain and the combined flow of raw sewage and rainwater is discharged directly to the river through many overflows.

Industrial wastewater should also be segregated prior to treatment to avoid dilution from rainwater or from relatively clean plant cooling water. In some instances extremely concentrated wastes are segregated for disposal by evaporation and burning or by pumping into deep disposal wells.

In-Plant Practices

Certain wet process industries can institute a number of in-plant lines of defense to reduce waste load at the source. More efficient use of raw materials, including recovery systems, reduces waste load. Equalization of the load by limited storage reduces slugs and peak discharges. Similarly neutralization of acids and alkalis can be effected prior to release. As industrial technology advances process changes can be arranged to reduce waste load substantially. In the critical summer–fall drought season industrial production can be tailored to waste assimilation variations in the receiving stream. In all industries wastewater surveillance, with control by competent staff for day-to-day operation and investigation, is a vital defense measure.

Planning and Zoning

As we have seen from the nature of self-purification, waste assimilation capacity is subject to two-way variation—in time and in place from reach to reach along the course of the river. Where waste effluent outlets are placed is as much of concern as the amount of waste load. Many industrial waste pollution problems stem from initial mislocation in relation to the waste assimilation capacity available and in relation to other water uses. Plant site location involves not only concern with water supply but, more important, the satisfactory disposal of wastewater effluent residuals.

The location of effluent outlets of community wastewater-treatment works is likewise important—though of course flexibility is more limited. In some situations advantages can be gained by joint sewers intercepting wastes of several communities for treatment at a joint plant with the effluent strategically located for favorable assimilation of the residual waste and without adverse effect on other water uses.

Community land use planning, including detailed analysis of interrelation of land and water, can protect against mislocation. Good zoning practice gives guidance to direction of growth and avoids conflicting uses.

Improved Agricultural Practices

Agricultural practice—especially the production of crops, livestock, and poultry—is a significant source of stream pollution. Unlike in the case of the urban-industrial concentration, the distribution is usually widespread. Organic matter, residual fertilizers, pesticides, and herbicides reach the stream through land drain systems and by landwash and erosion. The lines of defense currently are few; reduction in these wastes rests with improvements in such agricultural practices as erosion control and efficient use of agricultural chemicals and irrigation to reduce loss to the land drain system.

To a limited extent agricultural land serves as a means of ultimate disposal of wastewater from other sources. Irrigation of pasture and crop lands with wastewater effluent from small communities and industries has its best application in arid regions. Sewage effluent is not applied to vegetable crops that are consumed raw. The requirement for large land areas limits this method of disposal.

Control of the Environment

A significant source of pollution arises from urban wash flushed to storm water drains and natural drainage channels. The only effective line of defense against this source is a comprehensive system of control of the community environment. Important elements are efficient refuse collection and disposal, street cleaning, and public and private sanitation practice. A clean city is readily distinguishable from a littered, dirty city in the reflected quality of the stream that it borders.

Wastewater Treatment

The treatment of wastewater to remove a portion of the waste load before discharge to the stream has long been recognized as a major line of defense against excessive pollution from urban and industrial sources. There is, however, a popular misconception that treatment is all that is involved or necessary in satisfactory waste disposal and water quality control. It is a mistake to promote such a misconception of the problem. Wastewater treatment alone is not the solution; a much more sophisticated system is required if we are to meet today's and tomorrow's needs.

A major difficulty in treating water-carriage waste is its very dilute character. For example, the concentration of organic waste, expressed in terms of the biochemical oxygen demand (BOD), of the average American community is approximately 200 parts per million (ppm), or one part waste to 5000 parts water, or 0.02 percent waste to 99.98 percent water. Thus the treatment works not only must handle relatively large volumes of water but also must deal with the difficult task of removing a fraction of what already is very small. A treatment efficiency of 95 percent would require a reduction of BOD to 10 ppm, or a plant effluent purity of 99.999 percent water, 0.001 percent residual BOD. In dealing with such dilute wastes treatment becomes more and more difficult and costly as the standards for effluent purity are raised.

Treatment is a percentage removal; some residual fraction always remains—the law of diminishing returns is always in evidence. The natural law of geometric decrease indicates that it is as easy to produce the first 90 percent of improvement as the next 9; 100-percent efficiency has not been, nor is it likely to be, achieved in handling the tremendous volumes (100–200 gallons per capita per day) of wastewater produced by an urban-industrial society. Obviously then, although treatment will always play a primary role, there are technical and economic limits to its application in a multiple-line defense system. A brief discussion of efficiency and limitation of various practical treatment systems in general application will indicate its proper role.

Principles. Wastewater treatment systems employ principles aimed at the removal of waste matter in a particular state, *suspended, colloidal, dissolved,* or *living.* In the current art of treatment six general categories of fundamental principles are identified:

1. Mechanical separation.
2. Hydraulic separation.
3. Chemical coagulation.
4. Chemical and physical reaction.
5. Disinfection.
6. Biological reaction.

Mechanical principles are limited to the removal of the coarser suspended matter by purely mechanical means, such as screening.

Hydraulic separation is effective only in the removal of suspended matter, heavier or lighter than water, which will deposit on the bottom of a tank or rise to the surface where it can be removed.

Chemical coagulation is limited to the removal of colloidal matter and very fine, nonsettleable suspended matter bordering on the colloidal state. Dissolved matter in true solution is unaffected, and chemical coagulation is inefficient in removing settleable suspended matter.

The *chemical* principles, as distinct from chemical coagulation, bring about reactions and chemical changes in the dissolved salts and gases, with removal by precipitation. They also include physical reactions (aeration, adsorption) associated with, and related to, some of the chemical reactions.

Disinfection is aimed at the destruction of living organisms, chiefly the bacteria of sewage origin, by means of substances inimical to their life processes.

The *biological* principles, adaptations of the natural processes of decay, are capable of effectively removing waste matter in all four states and are the only principles now in general use capable of removing dissolved organic matter. These natural processes are *biochemical,* as distinguished from chemical and physical. They involve living systems—feeding, breathing, growth, reproduction, and death in a complex complementary as well as competitive chain of living forms, supported in large measure by the organic content of the wastewater. They also include aquatic forms capable of utilizing inorganic compounds, such as the nutrients nitrogen and phosphorus, contained in the wastewater or residues of the organic stabilization processes.

Contact between the waste matter and the biomass is usually effected either by passing the wastewater through a stationary biological bed (trickling filter) or by sweeping the biomass through the wastewater (activated sludge).

General Classification of Conventional Treatment Systems. Over the years conventional treatment systems have evolved as combinations of basic principles that can be classified on the basis of the physical state of the waste matter with which the system is particularly competent to deal. This leads to three general classifications:

1. Primary treatment, principally for the removal of suspended matter.
2. Intermediate treatment, principally for the removal of suspended and colloidal matter.
3. Complete treatment, for the removal of suspended, colloidal, and dissolved matter.

All three systems usually incorporate the principle of disinfection for the destruction of bacteria of sewage origin.

The predominant waste is organic matter, which in the wastewater of the average community is about equally divided among suspended, colloidal, and dissolved states. The general classification of treatment systems therefore affords a basis for the classification of effluent quality. Thus the quality of the effluent will contain residuals of roughly two-thirds for primary, one-third for intermediate, and a small residual for complete treatment. Again it should be emphasized that complete treatment refers to the fact that all three states—suspended, colloidal, and dissolved—are attacked; it does not mean, as is a common misconception, that wastes are completely removed.

With classification of treatment based on end results rather than means, widely varying means that accomplish the same end result should be classified as the same treatment.

More recently a further generalization of treatment systems has evolved on the basis of degree or stage of treatment:

1. Primary. The first stage, for removal of suspended matter.
2. Secondary. So-called complete treatment, generally implying a combination of primary and some form of biological principle.
3. Tertiary. Some form of advanced treatment following conventional complete treatment.

Although these classifications serve most purposes, it is recognized that full evaluation of any particular treatment system can be gained only by a breakdown of the combination of the basic principles of treatment employed. It follows also that the design of treatment systems should not be generalized; each treatment installation should employ the combination of principles that most efficiently removes the required portion of the waste load at the least cost. Since the required removal will vary depending on each specific stream requirement and defense and offense system, our general categorical classification will not hold, and treatment systems will be classified more appropriately on a continuous scale of efficiency.

Efficiency and Limitations. The efficiency of *primary treatment* in the removal of suspended matter ranges between 50 and 70 percent, depending on the concentration of suspended matter and the detention time provided for sedimentation. The corresponding removal of BOD ranges between 25 and 50 percent, depending on the proportion of BOD in the form of settleable solids. The removal of the settleable and floating solids prevents sludge deposits in the receiving stream and unsightly surface sleeks; where the dissolved oxygen assets of the stream are high, primary treatment may provide adequate removal of BOD.

The advantages of the primary treatment system are its simplicity of operation and relatively low capital and operating cost. Since it is limited to the removal of waste in the suspended state, the colloidal and dissolved matter remains in the effluent; primary treatment is definitely limited in flexibility and efficiency.

Intermediate treatment, aimed at the removal of waste matter in both the suspended and colloidal states, has a range in efficiency in BOD removal of 50 to 75 percent. A distinct advantage of intermediate treatment employing the principle of coagulation is its high degree of flexibility. It is responsive to change in the dosage of coagulant and can be readily discontinued and again resumed. During periods of high streamflow coagulation can be discontinued and the system can function as a primary plant; during periods of drought, when the waste assimilation capacity of the stream is reduced, coagulant dosage can be tailored to the required BOD removal, attaining efficiency as high as 75 percent. The system has the additional advantages of relatively small area for plant site, low capital cost, and freedom from odor and nuisance. The major disadvantages are its sensitivity to changes in raw waste and shock loads, and high operating costs for chemical coagulants and handling of sludge.

Intermediate systems using biological principles serve best in efficiencies above 60 percent and offer considerable flexibility, with ability to attain 75 percent or better removal of BOD. The biological units must remain in operation once they are matured, and hence cannot be bypassed, as in chemical coagulation. Areas required for plant site and capital costs are greater than those for systems using chemical coagulation, but operating costs are lower, although loss of head through trickling filters adds to pumping cost.

Complete treatment systems are complete only in the sense that waste matter in all of its states is subjected to attack. Efficiency in terms of BOD removal ranges from 80 to 95 (average 85) percent for trickling filters; 85 to 95 (average 90) percent for activated sludge, and as high as 98 percent for intermittent sand filters. Capital costs are highest for complete treatment systems. Operating costs are also highest for activated sludge processes because of the large volume of air that must be supplied and the expense of disposing of the large volume of sludge produced. The activated sludge process is sensitive to variations in waste loading and requires skilled supervision.

Trickling filter operation is less costly than the activated sludge process, capable of handling considerable variation in waste load, and almost automatic in operation. Trickling filter operations, however, are subject to odor and fly nuisance, whereas the activated sludge process is strictly aerobic and less subject to odors. Trickling filters require relatively large area for the stone beds housing the biological mass, whereas activated sludge plants require relatively little space.

Intermittent sand filtration is in the nature of a finishing treatment employed where highly purified effluent is required. Overall efficiency in average operation is enhanced, with up to 98 percent removal of BOD. The system is not applicable to large cities because of the very large areas needed for the sand beds, but it is practicable for small towns and isolated institutions where a high degree of removal is needed to protect the stream.

Chlorination is practiced in conjunction with conventional treatment systems where high efficiency in the destruction of bacteria of sewage origin is required. Bacterial reduction without chlorination roughly parallels the efficiency of BOD removal of the various systems of treatment. The numbers of bacteria are so great, however, that effective reduction requires the incorporation of the principles of disinfection. Chlorination functions poorly in application to raw sewage, since solid particles prevent proper penetration and organic matter increases the chlorine demand. Sedimentation therefore constitutes a minimum prerequisite for efficient chlorination. Where higher degrees of treatment are provided chlorine demand is reduced and the cost of disinfection is correspondingly lowered.

Disinfection by chlorination is highly sensitive to variations in chlorine demand, and effectiveness necessitates maintainng a free residual of 0.5 ppm. Practice includes postchlorination (following conventional treatment), prechlorination (preceding treatment), and split chlorination (pre and post). Prechlorination also provides an excellent protection against odors. Almost any degree of bacterial destruction can be attained, depending on the residual concentration of chlorine that is maintained. It is feasible to achieve and maintain a residual coliform bacteria count of 500 per 100 ml, representing an efficiency on the order of 99.995 percent.

Capital investment for chlorination practice is modest in relation to the total expenditure, and cost of operation is likewise reasonable, particularly if the chlorine demand is reduced by good BOD removal.

Advanced Treatment Methods and Nutrient Control. Since a major focus of advanced treatment is on the development of methods of removing nutrients from municipal and industrial wastewater, the pertinent facts concerning nutrients and their role in stream sanitation should be briefly set out.

Nutrient Control. It is well established that aquatic plants respond to the same nutritional stimuli as land plants. The aquatic plants of principal concern are the chlorophyll-bearing algae and rooted aquatics. The algae can proliferate in all parts of a lake where light intensities are adequate; the rooted aquatic plants are confined to the littoral areas. When the primary productivity of a lake is overstimulated, eutrophication or aging is accelerated and algal blooms occur. Many species impart taste and odor to water supplies. Large floating masses are concentrated by wind action and interfere with water-based recreational activities.

As these masses die, decomposition releases foul odors and may deplete dissolved oxygen to levels so low as to cause fish kills. These nuisances occur principally in lakes and reservoirs, which act as accumulators, and also may occur in some streams, particularly during the drought season.

The primary nutrients essential to growth are nitrogen and phosphorus; with macronutrients of calcium, magnesium, potassium, and sulfur; and micronutrients of iron, manganese, copper, zinc, molybdenum, vanadium, boron, chloride, cobalt, and silicon.

Aquatic plant growth, like land plant growth, can be inhibited by limiting one or more of the critical elements even though other nutritional material is adequate. It is on this principle, referred to as Liebig's law of the minimum, that efforts to control algal blooms and eutrophication hinge.

Although it is generally recognized that a substantial contribution of nutrient is contained in the effluent of conventional wastewater treatment systems, particularly since the advent of synthetic detergents, a significant contribution also derives from uncontrollable urban and land drainage, and from agricultural sources, especially with the extensive use of chemical fertilizers (see Chapter 2).

The question is which of the primary, macro, or micro elements is the critical nutrient and what is the critical level that will effect significant control of aquatic productivity.

The macro and micro elements are so commonly present in most rivers and lakes from natural drainage, and the required quantities are so small, that it is difficult to see how control could be effected by attempting to remove them from wastewater.

Of the primary nutrients nitrogen would appear to be the most logical critical element because of the relatively high content found in chemical analysis of algal composition. Sawyer [2] points out three reasons why control of nitrogen is not considered effective in most situations:

1. It occurs in most wastewaters largely in the form of ammonium and organic compounds. In a few wastewaters and some biologically treated effluents, part may be in the form of nitrites and nitrates. The removal of all forms by a single treatment method is a physical impossibility.

2. There is a significant leakage of combined nitrogen through soils, and thus appreciable amounts reach groundwater. Although nitrogen in organic forms does not penetrate the soil and the ammonium ion is retained effectively by adsorption or ion exchange, both forms gradually are converted to nitrate in aerobic soils and, since nitrates are not retained by any special mechanism, they are carried along in the groundwater unless removed by growing crops.

3. In the absence of adequate sources of inorganic and organic nitrogen, certain blue-green algae are capable of fixing atmospheric nitrogen, provided that other

nutrients are present in sufficient supply. Although there is good reason to believe that the amount of green algae and some forms of blue-green algae which will grow in water can be curtailed severely by reduction of the nitrogen budget, there is no assurance that it will limit the growth of the nitrogen-fixing blue-greens Anabaena, Gloetrichia, Nostoc, etc.

This then points to phosphorus as the most likely critical nutrient, and it is assumed by many that wholesale application of advanced treatment methods to reduce the phosphorus content of urban-industrial wastewaters will constitute effective control. This assumption appears to be an oversimplification. The critical concentration level of phosphorus to inhibit algal growth also remains in question. It has been considered as 0.1 mg/l, yet growth in lakes has persisted with no change in the amount of bloom when concentration in the receiving waters was reduced from 0.5 to 0.07 mg/l [3].

Sawyer [2] also questions this oversimplified assumption:

Whether phosphorus removal alone will accomplish what is expected of it is a big question. Much will depend on the effectiveness of removal at points where removal is practiced and on the amount of uncontrolled leakage from farmlands.

Nutrient control is highly complex. To some authorities it is conceivable that the uncontrollable sources from the urban, agricultural, and natural environments in general may maintain nutrients above critical levels even though all nutrients were to be removed from wastewater effluents. If this be true, any arbitrary program of advanced treatment applied to all urban wastewater would be disappointingly ineffective and an inefficient use of public funds.

A rational approach is to be more selective and analytical; to concentrate on those water bodies in which the existing or potential need for nutrient control is manifest. As Sawyer [2] suggests, this entails a thorough survey for at least a calendar year through the dormant and active growing seasons, employing modern methods of analysis for all conceivable critical nutrients to ascertain which are actually critical for the specific body of water. With this survey, and the measurement of the amounts of the critical elements contributed by the controllable sources relative to the amounts derived from uncontrollable sources, an estimate can be made of the probable effectiveness of the nutrient control of aquatic productivity. At present this estimate must be made on the basis of experienced professional judgment, as unfortunately at this stage of knowledge no quantitative methods of projecting an outcome are available, such as are presented in this text for other types of wastes.

Advanced Treatment. Extensive research is being pursued seeking practical advanced methods beyond conventional complete treatment to extend efficiencies and to remove chemical residuals and nutrients from urban-industrial wastewaters. Pilot and small scale plants have indicated that by further treatment of effluents of conventional systems, employing combinations of advanced biological methods and physicochemical principles, additional reductions in BOD, suspended matter, and persistent nutrients are possible, but costly.

Improvements in biological systems, principally increased aeration by various means (aerated lagoons and activated algae photosynthetic processes) show possibilities of further reduction of BOD, nitrogen, and phosphorus.

Chemical precipitation and coagulation of effluents of conventional biological systems with lime, alum, and synthetic polymeric flocculants is effective in removing residual suspended solids, colloids, color, and phosphates, but this adds significantly to the sludge disposal problem. If ammonia nitrogen remains, it can be effectively stripped with air in cooling tower equipment at high pH (10.8) and high air–liquid ratio (450–750 ft^3 of air per gallon).

Removal by activated carbon adsorption of non-BOD organics, which cause taste and odor and interfere with further treatment (such as ion exchange), is possible, but cost is excessive unless carbon can be reclaimed and successfully reactivated.

The emphasis given to development of desalination systems in producing fresh water from brines and seawater has opened the way to demineralization of wastewater for reuse. These methods include electrodialysis, in which electrical energy transports dissolved ionic impurities across membranes, and reverse osmosis, which transports pure water across membranes. Both systems encounter difficulties with residual organics and require high removal before demineralization.

Distillation can accomplish a high degree of purification but involves heating to induce rapid evaporation or vaporization through pressure reduction, followed by condensation of the vapor as the purified product. The most common types are multistage flash evaporators, multiple-effect evaporators, and vapor recompression evaporators. These systems have not as yet been effective in wastewater application. Scaling, corrosion, and organic fouling pose technical limitations, and the high cost of energy for heating precludes large scale application.

High-rate sand filtration serves as a finishing treatment and requires less area than conventional intermittent sand filtration. Such highly treated effluents can be returned to the groundwater rather than the stream providing the geologic formation permits infiltration of large volumes and there is no danger of well contamination.

A problem common to most advanced treatment systems stems from the fact that the additional impurities removed are simply concentrated in a smaller volume or are added to the sludge and remain as a difficult ultimate disposal problem.

In all the advanced treatment systems overall costs are high, ranging from two to four times that of conventional treatment. Large scale applications to the tremendous volumes of wastewater from an urban-industrial complex as yet have not proved to be technically or economically feasible. At the current state of the science and art the wholesale application of such refinements in treatment requirement relative to investment in other lines of defense and offense should be carefully evaluated.

This is not to say that advanced treatment methods do not show promise nor have a role in the overall problem. Many critical conditions exist from such causes as mislocation on small streams or lakes where no other, more efficient, alternative is available. Also the protection of unique natural resources, such as Lake Tahoe, may be justified. However, the promotion of wholesale application of tertiary treatment to all wastes without foreknowledge of the potential benefit or regardless of need is certainly open to question.

It is perhaps appropriate here also to bring into question the policy of applying complete treatment to all wastes before considering enhancement of stream quality by low flow augmentation and other offensive strategies. With the wide diversity in natural capacities of rivers throughout the country and the widely unequal distribution of the waste burden, a uniform treatment requirement with emphasis on effluent concentration regardless of load size or river capacity is irrational and wasteful. Hazen [4] has made the following suggestion:

> We should relax complete treatment requirements in coastal cities and along major rivers where there is ample (streamflow), and natural recovery can do much of the work. We should insist upon planning that provides for more treatment if and when it is needed; but nothing is gained by paying for it 30 or 40 years before that time.
>
> We should adopt a zoning policy that will encourage industries producing large volumes of waste to locate where simple treatment will suffice, and that will discourage them from areas where water resources already are limited.

It follows that treatment, conventional or advanced, should be adapted to the needs of the receiving stream and in balance with a rational system of multiple lines of defense and offense.

Multiple Lines of Offense

The enhancement of satisfactory stream water quantity and quality for intended uses constitutes the second part of the dual objective of stream sanitation. Lines of offensive action fall into two types: (a) positive developments to permit fuller utilization of the water resource, enhancing stream self-purification capacity; (b) protective or corrective measures against detrimental impact of other water resource uses and developments on the waste assimilation capacity of the stream. These are dealt with in more detail in the Chapters 9 and 11.

Drought Control, Low Flow Augmentation

As we have seen, the size of a river from the standpoint of quantity of streamflow may vary over a wide range, from flood to drought. Waste assimilation capacity and water quality fluctuate with these variations in quantity. If the total flow of a river could be equalized, a much higher assured flow would be continuously available and waste assimilation capacity and associated water quality would be greatly enhanced. Actually man has made very inefficient use of the total available, and in exposing himself to the risks of uncontrolled drought flow he frequently suffers drastic decline in quantity and degradation in quality.

Drought control is the protection against extremes of minimal streamflow and is a logical complement to protection against the extreme of flood. The floodwaters are decapitated and retained for release to augment the dry-weather streamflow. Various types of river developments are adaptable to the practice of drought control:

1. Headwaters storage reservoirs and regulated release.
2. Incremental storage reservoirs and regulated release.
3. Multipurpose river developments, including augmentation of low flow.
4. Adaptation of pumped storage for low flow augmentation.

The first three are developments of natural reservoir sites with sufficient tributary drainage area to provide the required water by storage of natural runoff. The pumped storage developments can be strategically located along the main stem of the river near points of need. They do not require tributary drainage area, since the water is pumped from the main channel to the off-channel reservoir. A portion of the water is recycled daily from the off-channel reservoir to the in-channel pool to generate electrical energy for peak power demands and repumping during off-peak power demand periods. The excess storage for flow augmentation is also pumped during off-peak demand in periods of high streamflow. This system provides two important ingredients for urban-industrial growth: water and electrical energy. The differential in value of energy between that produced for peak demand and that used during off-peak periods makes the system economically feasible.

It is seldom economically feasible to completely equalize the streamflow, but the practice of drought control ensures continuous flow several times greater than that available under the natural fluctuations of runoff. It not only provides enhanced waste assimilation capacity and water quality but also benefits all water users along the stream.

Storage of Wastes and Regulated Release

Where it is not feasible to store water to augment the low streamflow, a higher proportion of the total annual waste assimilation capacity can be made available by storing wastewater in lagoons and regulating the release of effluent in accord with the natural variation in streamflow. This is particularly adaptable to large, isolated industries where land area is available for construction of the required lagoon capacity. The lagoons serve a dual purpose as waste storage and as a treatment system. Reductions in BOD of 50 to 90 percent are attainable during the storage period.

Impact of River Developments and Uses on Waste Assimilation Capacity

The second kind of offensive action is effort to keep other users of the water resource developments from adversely affecting waste assimilation. Almost every river development has the potential to damage or to enhance waste assimilation through alteration of the paramenters that affect self-purification. The principal impacts are associated with the following:

1. Flood control works: dams, reservoirs, and alterations of channel.
2. Hydroelectric plants: affecting streamflow adversely by peaking practice but capable of enhancing flow by inclusion of reregulation to equalize the flow.
3. Navigation works: greatly increasing channel depth by locks, dams, and dredging with adverse effect or increasing flow from upland storage with beneficial effect.
4. Thermal electric plants: affecting the temperature of the river water, thereby stimulating the demand on dissolved oxygen and decreasing the oxygen saturation capacity of the water.
5. Diversion works: consumptive use of water by irrigation with adverse effect during the critical dry weather period or compensating for diversion from upland storage.

THE SCIENCE AND ART OF APPLIED STREAM SANITATION

Maturity in human endeavor is measured by the success with which we are able to anticipate or forecast the outcome or consequences of any plan of action. Through the application of scientific and engineering intelligence a high level of maturity has been attained in several areas of human endeavor. Large investments are made in almost complete confidence that products of predetermined quantity and quality will result. Bridges are crossed without fear or the slightest doubt that they will stand and carry the load. Such confidence stems from the fact that the approach is rational. Factors are quantitatively defined. Rigorous analyses precede design, construction, operation, and management.

The Rational Method of Stream Analysis

The science and art of applied stream sanitation has also attained a stage at which it is possible to forecast with reasonable reliability the waste assimilation capacity and resultant water quality of the stream for any combination of residual waste effluent loadings under the range in hydrologic variations expected and under the impact of man-made river developments and uses. This approach, as developed throughout the text, is termed the *rational method of stream analysis*.

Rigorous stream analysis is the basis of design of the lines of defense and offense in satisfactory waste disposal and in assurance of satisfactory quantity and quality of streamflow for intended uses. As we have seen, stream sanitation has as its foundation the efficient use of the waste assimilation capacity of the stream without waste or abuse. The best combination of defensive and offensive action to employ for a specific river is derived from the quantitative evaluation of the effectiveness and cost of each of the elements of the system. For example, the design of wastewater treatment systems is based on the determination of the allowable load placed on self-purification to ensure the desired water quality under prescribed probability of natural drought or regulated streamflow. Urban-industrial expansion is guided by comprehensive stream analysis prior to selection of sites and land uses. Decisions concerning the adoption of stream classifications and water quality criteria are based on thorough stream analysis to determine in advance the consequences of any proposal to be placed in effect. Consideration of the impact of water resource developments and uses contemplated by others is most convincingly argued by quantitative facts supported by rigorous analysis to show the consequences of any proposal.

Major Phases of Stream Analysis

Five major phases are usually entailed in the evaluation of waste assimilation capacity:

1. Determination of waste loadings.
2. Definition of hydrologic and climatologic factors.

3. Adoption and verification of self-purification factors.
4. Forecasts of expected stream conditions.
5. Evaluation of the impact of river developments.

The application of these phases in a practical river study entails care with a vast variety of detail, professional judgment in identification of relevant data, and extensive computations (none of which, however, need be excessively complicated).

Waste Loadings. Location, magnitude, types, and characteristics of existing municipal and industrial waste sources are best obtained by an onshore survey of the tributary wastewater service area. This entails measurements of volume and concentration in order to determine the residual daily load discharged to the stream as effluent from treatment systems and from all sources not receiving treatment. In addition to field measurement of existing loadings, future waste loads are projected from detailed population growth and economic base studies. Methods are discussed in Chapter 2.

Hydrologic and Climatologic Factors. The degree to which waste assimilation capacity can be quantitatively defined is dependent to a large extent on the detail with which the hydrology and climatology of the specific drainage basin are defined. Since these phenomena are highly variable in time and place, many of the parameters are defined in statistical terms from analyses of long-term records. Three variables are particularly pertinent: climate, stream runoff, and the physical characteristics of the river channel.

The *climatic setting* establishes the special differences among drainage basins and subdivisions. Physiographic land forms, geology, topography, vegetative cover, land use, rainfall, and temperature all play roles.

Stream runoff, the most important asset in waste assimilation, is difficult to define as a precise quantity available at a given time and place. The average flow of a river may appear good, but it is not always available. Depending on the climatic setting and the size, shape, and character of the tributary drainage area, the patterns of deviation about the average are quite different among rivers and even among the reaches of a given river. Intimate knowledge of the drainage area and extensive statistical analysis of long-term stream-gaging records define the probabilities of levels of streamflow available. Quantitative characterizations are also reflected in monthly, daily, and instantaneous hydrographs. Statistical analysis of recorded minimum flows, employing the theory of extreme values, defines the probability of occurrence of levels of severity.

An aspect of hydrology that is not usually fully appreciated is the nature of the stream channel. Channel characteristics determine how a river behaves, whether it is deep and slow moving or shallow and rapid. Occupied channel volumes, widths, depths, surface and cross-sectional areas, and hydraulic factors, particularly the important element of time of passage from reach to reach, are essential to quantitative determination of waste assimilation capacity along the course of the river. If such data are not available, they can be obtained readily in the field with echo sounding equipment by cross-section sounding at frequent intervals along the channel. Each runoff regime produces its corresponding channel characteristics; from measurements at one runoff condition adjustments can be made in channel parameters for other probabilities of streamflow. (Chapter 3 treats the relationship in detail.)

Adoption and Verification of Self-Purification Factors. The third phase in the evaluation of waste assimilation capacity is the adoption of the self-purification factors specific to the stream and type of waste. For example, in dealing with organic wastes deoxygenation and reaeration are determined separately; the liabilities and assets are then combined to give the resultant dissolved oxygen profile along the river. The liability, deoxygenation, is determined from the integration of the waste loads down the river at the predetermined rate of amortization and at the time of passage computed for the specific runoff regime, taking into account any abnormalities, such as sludge deposits. The assets, reaeration and increments of runoff, are developed from occupied channel volume, depth, temperature, and stream turnover as determined from the physical and hydraulic characteristics of each specific reach of the river, taking into account the initial dissolved oxygen in the streamflow and that contributed from tributaries along the course.

Verification of the self-purification factors adopted is afforded by comparison of the independently computed stream conditions with those observed in river sampling. For a verification to be significant it is axiomatic that conditions be comparable and that observed watercourse sampling reliably reflect a known waste loading and steady runoff regime. Self-purification is dealt with in Chapters 4 through 7 for freshwater streams and in Chapter 8 for estuaries; the technique of stream survey is discussed in Chapter 10.

Forecasts of Expected Stream Conditions. With self-purification characteristics fundamentally defined and verified, it is possible to forecast the stream conditions expected for various waste loadings over the whole range of natural or regulated stream runoff. These forecasts define the range of waste assimilation capacity available, from which the relation between stream runoff and allowable waste loading can be established to maintain any prescribed level of water quality. Such a reference frame affords the basis for designing an effective offensive and defensive system for satisfactory waste disposal and maintenance of satisfactory water quality. Applications of this phase are presented in Chapters 11 and 12.

Evaluation of the Impact of River Developments. There remains the important phase of evaluating the impact of river developments and other water uses on the waste disposal problem and stream quality control. Any existing or proposed river development or use that alters one or more of the basic parameters that govern self-purification can be quantitatively evaluated as a beneficial or detrimental effect on waste assimilation capacity and water quality. Chapter 9 considers this phase for the major types of river developments and uses.

A Suggestion for Teaching

Those who regard this book as a classroom text may feel the absence of problems for students to solve at the end of each chapter. They have been omitted quite deliberately as inimical to the author's theses. Applied stream sanitation works from the premise that solutions to stream pollution problems must always be multiple since the causes of pollution problems are always multiple. To ask students to work fragmentary problems would be to encourage in them that traditional fragmented approach that this volume seeks to supersede.

In the author's experience the rational method of stream analysis is best taught by assigning to each student a comprehensive basin study of an actual river system. The student will preferably undertake a river basin of which he has intimate knowledge or prior work experience; if not, a cooperative study of a basin might be arranged with a public or private agency willing to supply relevant data. A specific focus may be on an existing pollution problem or on prevention by evaluation of specific site locations as a guide to industrial growth.

The student will assemble the necessary raw data from library research and from public and private agencies, augmented where necessary by field survey. Since data are seldom entirely adequate, he must learn what is relevant and what may be employed as valid substitutions essential for meaningful quantitative analysis. The student will carry out the voluminous quantitative computations (where appropriate with the aid of a digital computer) make projections of stream conditions expected for various probabilities of natural or regulated streamflow under existing and proposed systems of river development, evaluate combinations of lines of defense and offense, reach specific practical conclusions and recommendations for a program of action, and write a comprehensive river basin report useful to the cooperating agency.

The teaching is of necessity informal, tutorial in nature, and extends beyond the classroom in the student's contact with professional men in applied fields. This approach is time consuming, but it presents a real challenge to the student, affords opportunity for research and the exercise of ingenuity, develops professional judgment, and, most important, obligates him to devise practical answers to pressing problems, thereby narrowing the gap between what is known and what is applied.

REFERENCES

[1] *Restoring the Quality of Our Environment,* The President's Advisory Committee, 1965.
[2] Sawyer, C. N., *J. Water Poll. Control Fed.,* **40,** No. 3, 363 (1968).
[3] Fitzgerald, G. P., Progress Report, USPH Project WP297, University of Wisconsin, 1964.
[4] Hazen, R., *Civil Engineering,* **37,** No. 12, 38 (1967).

2

SOURCES OF POLLUTION AND TYPES OF WASTES

In this chapter we deal with the four primary sources of pollution: urban, industrial, agricultural, and natural; the types of waste products originating from these sources; the methods of location and measurement of waste loads; and the projection of expected future loads.

SOURCES OF STREAM POLLUTION

The Urban Source

Urban concentration of population constitutes the major source of stream pollution and is the source that is most difficult to manage. The inevitable proximity of a city to a watercourse leads to two types of urban stream pollution sources, the *controllable* and the *uncontrollable*. Herein lies a major limitation to management.

Controllable Source

The water-carriage system of sewers leading to the treatment works collects the larger portion of the "liquid" waste products produced in the city. Normally this system serves the business and commercial areas, the residential districts, and increasingly much of the industrial area. The extent to which this system services the city and its surrounding suburban and industrial fringe is a measure of the controllable source. Seldom does the system serve the entire urban boundary, particularly in the rapidly expanding areas. Political boundaries seldom coincide with natural drainage service areas. The fringes outside the political boundary of the municipal system may be without sewerage service for years and usually depend on septic tanks or private systems. Many of the larger industries, particularly those located along the banks of rivers, have their own systems, which are independent of the municipal one. The objective from the standpoint of pollution control is to include, as far as feasible, *all* water-carriage waste products, excluding only those that can be handled more satisfactorily by other means.

A high degree of control is possible where the sewerage system and the treatment facilities are developed on a *service area* basis, independent of political boundary lines. The controllable source in an efficient service area system does not imply its elimination as a source of stream pollution. As was pointed out in Chapter 1, treatment of wastewater has technical and economic limitations. Some percentage of the waste load always remains in the effluent returned to the watercourse and ultimately must be handled by the waste assimilation capacity of the stream.

Uncontrollable Source

All urban wastes that reach the stream other than through the organized sewerage system and treatment works are the uncontrollable source. This constitutes a greater contribution to stream pollution than is generally recognized. The uncontrollable source is usually intermittent, associated with the occurrence of rainfall, reaching the river by two routes—namely, the *storm-water drain system* and *area wide urban wash to* small channels.

Many of the older cities are served by *combined sewers,* large drains designed to function in a dual capacity as a sewerage system to collect the water-carriage wastes of the community and as storm water drains. Inherently such a system requires many outlets to the river, since the storm-water volumes are too large to permit collection from extensive areas. During dry weather periods the wastewater is diverted from the combined sewer to *interceptor sewers,* usually constructed parallel to the river and leading to a central wastewater-treatment works. When rainfall and storms occur, normal wastewater is mixed with storm drainage, and, as the capacity of the dry weather interceptor is exceeded, the combined mixture is discharged directly to the river at the many combined sewer outlets. In addition to the current wastewater flow, the mixture usually contains a substantial quantity of previously accumulated sludge, which is flushed by the large volume storm flow. Thus the river is subjected to heavy *slugs* of pollution, many times during the year, in some cities almost with each slight shower.

The anachronism of combined sewers is incompatible with treatment of wastewaters before discharge and efficient stream pollution control. Improvements in interception can reduce the frequency and magnitude of untreated slugs that reach the river, but complete control is practically unattainable, short of conversion to absolutely separate systems. In large cities such reconstruction is prohibitively costly; even if feasible, this is not a complete solution. As a partial protection to adjacent and downstream watercourse areas of high use, such as ill-placed water supply and bathing beaches, diversion of storm overflows to large settling basins and heavy chlorination of the effluent serve as an expedient.

Current standard practice is the provision of two independent systems—that is, sewers to collect wastewater and storm drains to carry off the rainwater. In theory this should provide complete separation, but in practice many illegal wastewater connections are made to the storm drains with many roof and courtyard rainwater connections to the sewers. In addition, the storm drain system receives a multitude of waste products flushed to catchbasins during street washing and at every rainfall.

A comparative study of bacterial pollution discharged from a typical combined sewer system with that from a separate storm drain system reported by the U.S. Public Health Service [1,2] verifies the visual evidence with statistics. In these studies the combined sewer serves approximately 22,000 acres of northeastern Detroit, comprising a typical urban development (commercial, industrial, and residential), with two overflow outlets into the Detroit River. The separate storm drain system selected serves approximately 3800 acres of the city of Ann Arbor, Michigan, comprising a principally residential, and commercial development, with some light industry. The Ann Arbor area is served by a separate sanitary sewerage system and treatment works, with well organized supervision to control illegal connections. The Detroit system during dry weather is intercepted and conveyed to a central treatment works; during rainfall periods aggregating 0.2 to 0.3 in. the combined sewer overflows to the Detroit River. A systematic program of sampling of the overflow of the Detroit combined system was carried out during 1963 and of the outlet of the Ann Arbor separate storm-water system during 1964.

Since bacterial concentration is known to vary seasonally and is influenced by the intensity and duration of rainfall, for homogeneity and comparability in statistical analysis the seasonal peak period July–August is selected and data on the first storm of a day and reasonably similar cumulative rainfall are employed. The probability distributions of composited total coliform and fecal coliform bacteria of these samples are logarithmically normal, from which a graphical most probable mean,

Table 2–1 Bacterial Concentration of Discharges of the Detroit Combined Sewer Overflow and of the Ann Arbor Separate Storm Drain

	Detroit Combined Sewer Overflow	Ann Arbor Storm Drain
Total coliform bacteria (membrane filter):		
Number of samples	72	38
Mean per 100 ml	22,500,000	2,200,000
95% $\leqq$ per 100 ml	125,000,000	22,500,000
σ_{log}	0.45	0.65
$S.E._{log}$	0.053	0.105
Mean ± 2 S.E.	17,600,000	1,350,000
	28,700,000	3,570,000
Fecal coliform bacteria (membrane filter):		
Number of samples	64	38
Mean per 100 ml	8,200,000	270,000
95% $\leqq$ per 100 ml	49,000,000	3,800,000
σ_{log}	0.48	0.66
$S.E._{log}$	0.060	0.107
Mean ± 2 S.E.	6,200,000	165,000
	10,800,000	440,000

standard deviation, and standard error of mean are readily determined,* as summarized in Table 2–1.

A comparison of the total coliform bacteria (fecal and nonfecal) shows that the most probable mean concentration of the Detroit combined sewer overflow is 22.5 million per 100 ml, or approximately 10 times greater than the concentration in the Ann Arbor storm drain (i.e., 2.2 million), and the difference is statistically significant, with a high degree of confidence. The Detroit combined sewer overflow has a lower variation among sample results, with a standard deviation (σ_{log}) of 0.45 as compared with the Ann Arbor storm drain with a standard deviation (σ_{log}) of 0.65.

A comparison of the concentration of fecal coliform bacteria (originating from the intestines of man and other warm-blooded animals) reported in Table 2–1 shows that the most probable mean values are lower for both systems, with the concentration in the combined sewer overflow being approximately 30 times that of the separate storm drain. Variation

* For definitions and methods of analysis see Appendices A and B.

in individual sample results of fecal coliform bacteria as reflected by the standard deviations is similar to that for total coliform bacteria.

Two major facts are confirmed by these intensive studies: (a) combined sewers, though dry weather interceptors are provided, overflow with the occurrence of each slight rainfall, discharging heavy slugs of pollution; (b), although storm-water drains that are separate from sanitary sewers reduce the pollution load, a significant amount remains in the storm-water drain discharge.

In a similar manner uncontrollable organic oxygen demanding waste reaches the stream through the overflow of combined sewers, storm-water drains, and small drainage channels. Such organic loads may vary in magnitude from 5 to 10 percent of that of the controllable source collected by the sewerage system. Weibel and co-workers [3], in studies of a 27-acre residential, light commercial section of Cincinnati (37 percent impervious, 63 percent lawns, gardens, and parks) served by separate sewers and storm-water drains, found that the organic load, measured as BOD, of the storm-drain discharge averaged 6 percent of that carried in the separate sewerage system.

Thundershowers have a high probability of occurrence coincident with the drought season, and hence urban wash can constitute a critical pollution load. For example, in an urban area of 100,000 population with separate sewerage system and storm water drains, if it were possible to remove by treatment before discharge the entire waste load of the population connected to the sanitary sewers, there would remain in the storm-water discharge as the uncontrollable source from urban wash a load equivalent to the untreated wastewater of a city of 5000 to 10,000 population.

During rainfall *area-wide urban wash* also carries waste products to small natural drainage creeks and ditches from sections not served by sewers and storm drain systems. The larger the urban-industrial complex, the greater the accumulation of area-wide land pollution to be flushed; rubbish, garbage, deposited particulate matter from air pollution, dust, dirt, and filth from pet animals and promiscuous man. The magnitude of this source depends on community environmental control, but even in the cleanest cities it is surprisingly significant.

Thus the urban-industrial way of life contributes a controllable residual and an uncontrollable raw waste load to the river along its course through the city. Inevitably, therefore, a degree of water quality degeneration occurs in the city reach and downstream, even if it were possible to divert the entire controllable wastewater effluent to some other watercourse. This concomitant of urban-industrial concentration must be recognized in arriving at rational water and land use policies.

The Industrial Source

Although the variety and quantity of industrial water-carriage waste products are increasing, there is opportunity to exercise a high degree of control and management. The well organized industry usually takes an integrated view of waste disposal.

Enlightened corporations exert considerable internal management over product processes, segregation, collection, and treatment before ultimate disposal of solid, gaseous, and liquid residual on land, to the atmosphere, and to the watercourse. Since control of the industrial source may be practically complete, the defense against stream pollution depends to a great extent on the degree to which efficient management is exercised. As with the urban source, however, complete elimination is never attained, and some degree of stream pollution from the industrial source is inevitable.

Since our interest is in the residual that reaches the watercourse, the industrial source logically subdivides into (a) the part that is connected to *community systems* and (b) the part that is dealt with by independent *private systems*. Most small industries connect to the nearby urban sewerage system; many large wet process industries located along the river within the urban area provide independent collection and treatment; isolated industries beyond access to an urban system have no alternative but to provide their own facilities. Proliferation of independent private systems, particularly of relatively small capacity, leads to operational problems and difficulties in treatment, with poor efficiency unless provision is made for adequately trained operating staff. Where feasible and compatible, industrial and urban sources should be integrated into a common system.

Some industrial sources, such as waste heat from the cooling water of power plants, cannot be handled in the urban system because of the large volume and incompatibility with conventional treatment. As the size of electric power generating plants increases, the heat load concentrated at a single site may result in superelevation of the stream temperature to levels requiring cooling ponds, large reservoirs, or cooling towers to dissipate part of the heat load.

Although the industrial source is amenable to a high degree of control, the magnitude and the wide variety of wastes produced constitute a great challenge to satisfactory waste disposal and stream pollution management.

The Agricultural Source

Ordinarily one does not think of agricultural practices, such as crop and livestock production, as sources of pollution. However, as essential tools of modern agricultural practice, chemical fertilizers, pesticides, and herbicides are of increasing significance as a source of pollution. In the United States livestock and poultry wastes (excreta, bedding, and litter) are estimated to be in excess of 1.5 billion cubic yards per year, with obvious concentration in feedlots.

It also is common practice to spread manures on pasture and crop lands. If allowed to stand, much of the nutrients are leached by rainfall and thaw and are carried by landwash and drainage to watercourses. In cold climates where ground is frozen the loss of leached nutrients is correspondingly high. The magnitude of organic and nutrient loss is significant when it is recognized that the wastes of one dairy cow are equivalent to that of 17 humans.

The amount of wastes reaching the stream from agricultural sources has not been systematically measured except in isolated studies, but estimates indicate that it is substantially greater than is usually appreciated. Sawyer [4], in his studies of fertilization of lakes at Madison, Wisconsin, found the contribution from highly developed agricultural areas to range from 3800 to 5200 lb of nitrogen and 235 to 262 lb of phosphorus per year per square mile of tributary area. These wastes reach the stream primarily by two routes, through the land drain systems, and by landwash and erosion.

Drainage systems that serve cultivated lands collect seepage water containing residual applied chemicals leached from the topsoil. These land drainage wastewaters flow to nautral creeks, streams, and rivers without treatment. If the system covers extensive areas, natural groundwater seepage to the river is reduced and the severity of drought streamflow is aggravated. Where irrigation is practiced in conjunction with chemical fertilization, a portion of the applied water reaches the drainage system, and it usually contains a substantial amount of residual chemicals. Of the primary plant nutrients applied as fertilizer, the nitrogen moves readily with the drainage water, whereas the phosphorus tends to be fixed in the soil and reaches streams principally through erosion.

In a recent study [5] of tile-drainage effluents from systems on irrigated land in the San Joaquin Valley of California where crop fertilization and irrigation are intensively practiced, Johnston and associates found that in tilefields most suited to intercept irrigation drainage, 70 percent of the nitrogen applied was found in the drainage effluent, but only 3 percent of the applied phosphorus was recovered. The range in concentration of nitrogen in the effluent varied from 1.8 to 62.4 ppm, or a weighted average (in accordance with discharge) of 25.1 ppm, and the range of phosphorus was from 0.053 to 0.23 ppm, or a weighted average of 0.079 ppm. These and other studies point to the inefficient use of chemical fertilizers and the enormous waste discharge from agricultural sources.

Pesticides and herbicides applied to soil or plant surfaces are generally fixed by soils and are not usually found in measurable concentrations in the effluents of land drainage systems. However, measurable quantities reach streams as soil is eroded, in the process of spray application, from spillage, or from manufacture of the product.

Agricultural cultivation and overgrazing, in upsetting nature's balance of water, land, and vegetative cover, expose large areas of the earth's surface to *landwash* and *erosion*. Streams draining these exposed areas are subject to surface runoff after rainfall periods, and again where agricultural chemicals are employed residual slugs are washed into the natural watercourse. Studies carried out by the U.S. Public Health Service [6] confirm that landwash runoff of agricultural areas in which pesticides are used can result in widespread low level pollution of streams by these toxic chemicals. In addition, tons of topsoil are eroded and washed into streams and rivers. Clear water streams become muddy, silt laden watercourses, with the natural streambed blanketed with sediment. These sediments retain and accumulate agricultural chemicals fixed in the eroded soil, principally the phosphorus and the resistant pesticides and herbicides. The U.S. Department of Agriculture [7] estimates that the rivers of the United States carry approximately 1 billion tons of sediment to the oceans each year, most of which are eroded from poorly managed crop and range lands.

In current practice there is little or no effort devoted to control agricultural sources of stream pollution, except soil conservation to prevent erosion. Better recognition of the problem and exercise of lines of defense and offense as required can lead to improved management. Controlled application of agricultural chemicals and related irrigation practices make feasible a substantial reduction of the chemical losses to the drain system; wider extension of contour plowing, terracing, cover crops, and upland surface retarding works can further lessen the losses by surface wash and erosion. In instances it may be necessary to restrict the use of certain hazardous nondegradable pesticides and herbicides. The segregation and treatment of drainage from large, centralized livestock feedlots may be necessary in instances to reduce the residual organic waste load reaching streams.

The relative amounts of nitrogen and phosphorus that reach streams and lakes from agricultural, as distinct from urban and industrial, sources are unknown. Enough evidence is available concerning agricultural sources, however, to make it apparent that an attack on nutrient waste problems by treatment of only the controllable sources in urban-industrial effluents would be incorrect. The agricultural sources must be considered also.

However, with widespread and diverse agricultural practices complete control of agricultural wastes is not possible; inevitably a large fraction will continue to reach streams, rivers, and lakes.

The Natural Source

In the strict sense, no watercourse is without some degree of pollution, even in virgin territory. This pollution arises from stormwash, seepage from groundwater, swamp drainage, and aquatic life of the stream.

Stormwash from time to time carries to streams large quantities of organic matter from decaying flora and fauna and inorganic silt from soil erosion and bank scour. Some streams remain highly colored and muddy even during dry weather periods, depending on the character of the drainage area and its natural vegetative cover. Others are usually clear and silt laden only during the flood season.

Seepage from groundwater may contribute a variety of chemical compounds dissolved from the soil and geologic formations through which the seepage water passes in reaching the watercourse. The stream may appear clear and sparkling but during periods of low flow the concentration of dissolved salts may be detrimental to many water uses and even in instances contain material toxic to animals and man.

Swamp drainage contains high concentration of color, substantial organic and inorganic material, and is usually low in pH and dissolved oxygen. When flushed by a sudden rainstorm, the swamp waters drastically alter the normal quality of the streams and may result in a heavy fish kill. Also during periods of sudden high stage the river overflows its banks into swamp areas, and the back drainage on the succeeding decline carries large quantities of poor quality swamp waters to the main stream. This is illustrated in Figure 2–1, which shows the relation between runoff and dissolved oxygen for the lower Altamaha River. At low runoff stages, when the river is within its banks and swamp drainage is minimal, dissolved oxygen is 85 to 90 percent of saturation (and pH is about 8.0). At high runoff, particularily on the receding hydrograph, dissolved oxygen declines sharply, approaching 65 percent of saturation (the pH likewise sharply declines, approaching 6.5).

In the Chowan River basin tributary streams that drain extensive swamp areas normally have a dissolved oxygen content that is less than 20 percent of saturation, pH of 6.3 or lower, and color exceeding 1700 ppm. A dramatic fish kill occurred in the Chowan and other Atlantic coastal streams associated with the passage of the 1955 hurricane as reported by the North Carolina Wildlife Resources Commission [8] at

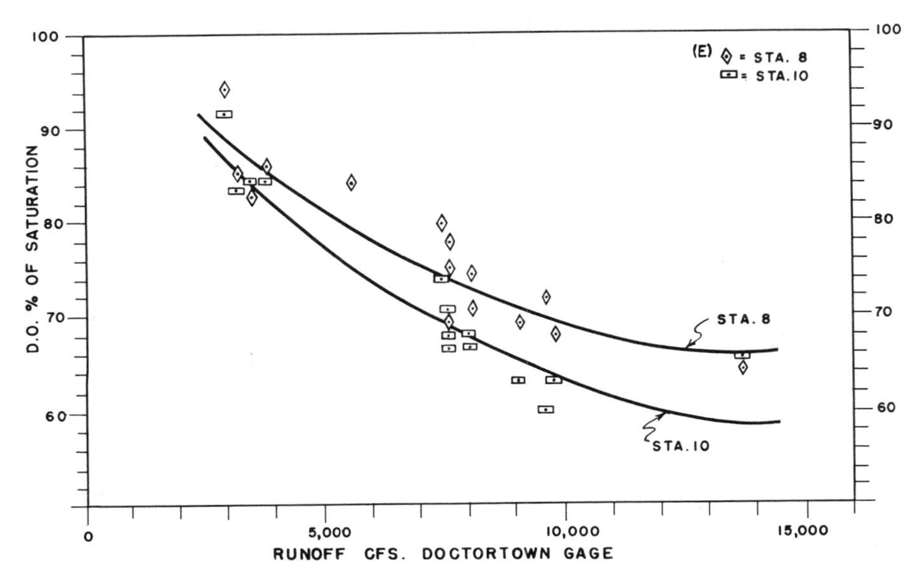

Figure 2–1 Altamaha River. Relation between receding runoff and dissolved oxygen contents, showing the influence of swamp drainage.

the hearings before the Stream Sanitation Committee:

> Not all of the streams were investigated but the major thoroughfares were checked. The fish in those streams were killed mainly by a decrease in oxygen content of the water. This, while not basically attributed to pollution, was a form of natural pollution. The water in these rivers overflowed the swamps, the organic material in the swamps began decomposing, with the resultant oxygen depletion which is a form of normal or natural pollution and was the cause of the fish kill. These streams showed a very high content of carbon dioxide, which in the presence of a very low oxygen content is toxic to fish, and also a high concentration of hydrogen sulfide. This gas is also toxic to fish. The fish die-off took place in a number of streams; in fact, it was a general die-off throughout the northeast coast in about every stream observed.

The *aquatic life* of the stream in its life cycle also constitutes a natural source of pollution. The normal flora and fauna depend on nutrients reaching the watercourse. When the latter is naturally well fertilized, a large and diverse population is maintained; in streams with poor nutrient content aquatic life is sparse. Excessive aquatic growths that die and decompose add significantly to the organic pollution load and deplete dissolved oxygen, particularly if residues deposit and accumulate in quiescent reaches, pools, and lakes. Some forms are toxic or produce toxic end products.

The chlorophyll forms, with light, produce oxygen and can result in supersaturation during the daylight hours. However, respiration during the night utilizes oxygen and may result in substantial depression of dissolved oxygen before sunrise.

The chain of biological life of the stream provides the "labor force" essential in the process we define as organic *self-purification* and constitutes one of the most valuable elements of water resources.

The combination of natural sources, of wastes from stormwash, seepage, and swamp drainage invariably results in a residual pollution in *all* streams and a degree of water quality degeneration. Normally streams that drain virgin territory carry a residual organic load of 0.5 to 1.0 ppm of 5-day standard BOD during the dry season; during high runoff this may increase to 1.0 to 2.0 ppm or higher. Seldom is a natural stream fully saturated with dissolved oxygen; during low flow periods dissolved oxygen is usually 85 to 90 percent of saturation; during high runoff dissolved oxygen may decline to 75 to 80 percent, and under some conditions where heavy drainage from swamps occurs it may decline to 20 percent of saturation or less. Thus it is well to recognize that in the natural setting, uninfluenced by man, there is pollution of streams. Nature has learned to live with it, and increasingly man must learn to manage his contributions by rational lines of defense and offense, and also must accept the fact that *complete elimination is not attainable.*

TYPES OF WASTE PRODUCTS

After discharge to the stream liquid waste products—urban, industrial, agricultural, and natural—to a large extent lose their identity and are blended into a heterogeneous mixture. However, the river in its self-purification processes reclassifies the waste products as to *type,* principally into five broad classifications:

1. Organic
2. Microbial
3. Radioactive
4. Inorganic
5. Thermal

Such a reclassification of sources as to type is essential in defining the waste assimilation capacity of the stream and in dealing with the complexities of stream self-purification developed in Chapters 4 through 8.

Organic Wastes

Organic wastes constitute by far the major stream pollution problem. Much of the urban and industrial sources comprise unstable organic matter subject to decay. In nature's economy the dead plant and animal organics are quickly oxidized by the chains of biological life to stable mineral forms to be used again by other living things. The stream itself provides the oxygen (DO) for this conversion. The direct measure of the organic waste load is therefore a function of the amount of oxygen required in the biochemical conversion, namely, the *biochemical oxygen demand.* The agricultural and natural sources of pollution likewise contribute organic fractions and in aggregate are large but widespread, whereas the urban industrial organic loads are concentrated as localized sources. Our concern with organic wastes is the rates at which the oxygen resources of the stream are demanded and replenished.

Microbial Wastes

Microbial wastes of primary concern in stream sanitation are bacteria, viruses, and other forms pathogenic to man. The major source is urban pollution from the community sewerage system, the storm water drainage, and the urban wash. Some industries, depending on the source and kind of raw product processed, may contain microbial contaminants in their wastewaters. Agricultural sources, particularly from livestock production are a further source. Small rural communities and seepage from individual household sewage disposal systems and in highly concentrated recreational areas are sources of localized contamination. Of increasing concern is contamination from large vessels and pleasure boats (e.g., at marinas during weekends) discharging high bacterial slugs directly into the watercourse. When it is recognized that the coliform-bacteria contribution per capita is on the order of 200 billion per day, it is easy to understand how a few sources of raw sewage can constitute a significant contamination.

Radioactive Wastes

Radioactive wastes are generally rigidly controlled at the source. However, the increasing use of radioactive tracers in industry and research, and the danger of accidental spill increase the potential for stream pollution. As atomic fuel increasingly replaces fossil fuel, radioactive wastes from steam turbine electric power plants and associated fuel processing will assume new proportions. Because of its great danger to man, radioactive waste will remain a special category, but as the industry grows its control and management will become of increasing concern to the water pollution control agency.

Inorganic Wastes

Inorganic wastes arise from all sources—urban, industrial, agricultural, and natural—as dissolved, colloidal, and suspended matter. Unlike other types, these wastes are relatively stable and do not decompose, decay, or dissipate. However, some fractions deposit or are utilized, but usually they are cumulative from source to source in progression down the river. Inert suspended matter may settle in pools and quiescent reaches; colloidal matter may be coagulated and likewise deposited.

Treatment removes practically all settleable matter subject to deposit from the urban sources connected to the community sewerage system, but a large proportion of the nutrients resulting from stabilization of the organics remain in the effluent. The urban wash, storm drains, and the combined sewers usually discharge large quantities of inert suspended inorganics, which form deposits in the streams. All urban sources contain a fraction of dissolved inorganic salts, and in cold climates the use of salts on streets and highways is a significant contribution.

The largest portion of inorganics arises from industrial sources, particularly chemical industries that use brines as raw material. The chlorides are persistent, cumulative, and resistant to treatment. In the petrochemical industry brines are usually returned to the earth through deep wells. Acid mine drainage is a serious pollution problem from unsealed abandoned coal mines. Some industrial wastes involve large quantities of inorganic suspended matter, which, if not removed by sedimentation, cause blanketing of the streambed.

The agricultural and natural sources, although widespread, contribute in aggregate large quantities of inorganics, silt from land erosion, and substantial quantities of residual agricultural chemicals. Silting from improper agricultural practices clogs stream channels and fills reservoirs, doing permanent damage.

Thermal Wastes

Heat wastes are almost entirely associated with the industrial source, primarily from the electric utility industry. An urban-industrial society depends on availability of large blocks of cheap electrical energy. To meet these demands the utility industry has been doubling its generating capacity at about 10-year intervals. The primary energy will depend on thermal rather than hydro generation. Both fossil and nuclear fuel steam electric power plants require large quantities of condenser water, resulting in enormous waste heat loads. The trend is toward concentration of generating capacity at increasingly larger stations. The result is an enormous discharge of hot water into rivers. For example, a 4000-MW generating station operating at full capacity would produce a waste heat load of 17.6 billion Btu/hr and would require over 2 million gal/min of circulating condenser water.

In addition to the steam electric power plants, other industries require cooling water in the various processes of manufacture, and this is rejected as waste heat to streams.

LOCATION AND MEASUREMENT OF WASTE LOADS

The Onshore Survey

Since the objective is to relate a specific waste loading to the water quality conditions of the receiving stream, the determination of waste loads should be obtained from an *onshore survey* and not derived from river sampling. After waste effluents are discharged into the stream, the identity of the source is obscured, and the difficulty of obtaining representative samples due to streaking and variation in streamflow leads to large errors. River samplings taken downstream at locations where loads are well mixed with the streamflow are reserved as an integrated check against the sum total of the onshore measurements.

Depending on the degree of accuracy desired, determination of the prevailing waste effluent loads may be considered in two categories, direct measurement and indirect estimates. For purposes of a preliminary evaluation indirect estimates may be adequate; where more refined evaluation and assessment to individual sources are required, a comprehensive onshore survey involving direct measurement is essential.

Outlets and Service Areas

In well organized communities with storm water drains and separate sewerage systems, good updated maps show locations of the outlets of storm drains and the industrial and municipal wastewater effluents. In addition, the locations of the major industries connected to the municipal system are shown on the sewer maps; data as to volume, characteristics of the wastes, production schedules, and flow sheets are available. In such communities the onshore survey is minimal, and spot checks of outfalls and analyses of operating records of the treatment systems provide the required data on loads and their variations.

Such ideal situations are of course seldom encountered, and usually the older and the larger urban-industrial complexes require substantial onshore field investigation, particularly where peripheral sections of the community and many industries are not connected to the municipal system. Where sewer and storm drain maps are not complete, recent aerial maps are invaluable in locating industries and serve as a base map on which to spot river outfalls. Shoreline investigations along the entire course of the river may be necessary to locate all significant sewer and storm drain discharges. Aerial reconnaissance by helicopter or river traverse by boat undertaken during low runoff stage aid in locating outfalls.

With outlets to the stream located, there remains the task of tracing to the source, definition of boundaries of the tributary collection system, and measurement of volume and characteristics of the discharge. If sewer maps are available, spot checks of boundaries define the tributary area, and zoning maps and field reconnaissance establish the nature of the service area occupancy. Three broad distinctions in service area are recognized: predominantly residential, predominantly commercial, and predominantly industrial. Where treatment works serve an area, the task of measuring the volume and characteristics of the effluent discharge is greatly simplified; where wastewaters are discharged without treatment, measurement of the volume and characteristics of wastes entails detailed effort.

Volume and Characteristics

Volume and characteristics differ substantially, depending on the composition of the service area. Municipal wastewaters follow reasonably uniform patterns, whereas independent industrial wastes are not only quite variable within an industry but are markedly different among different types of industries.

In evaluating the municipal service areas it may not be essential to measure all outlets. Results of carefully selected *representative sample* areas may be applied to similar groups and thus develop the discharge loads from all municipal system outlets. However, where service areas include a heterogeneous mixture of residential, commercial, and industrial sources, the representative sample approach may not be feasible and most outlets may require separate measurement.

The large industrial sources require intensive individual investigation. In establishments that employ full time water and waste management staffs, continuous records of volumes and characteristics of inplant sources are available and only spot checks are necessary; otherwise measurement of both volume and characteristics must be undertaken.

Measurement of Flow and Sampling

Measurement of volume per unit of time (rate of flow) requires knowledge of hydraulic principles. The selection of appropriate flow measuring devices depends on local conditions—whether in an open ditch, sewer pipe, free-falling discharge, or large conduit—and may involve employment of a velocity meter, dye tracer, various types of weirs, or a large Parshall flume. If measuring devices have not been previously installed, it is necessary to erect or construct an appropriate device. Most commonly weirs are employed with continuous recording of the head, which permits conversion to the rate of flow.

Sampling of wastewater must be coordinated with the rate of flow in order to convert concentration of constituents of the sample to absolute quantity discharged per unit time, such as per day. Since the flow and concentration of waste constituents vary through a time unit, samples are taken at regular intervals and in volume proportional to the rate of wastewater flow, and are accumulated as a *composite*. The analysis of such a 24-hr *weighted composite* provides a representative daily average. Some constituents, such as dissolved oxygen or other gases, pH, and chlorine residual, are altered by compositing over time, and hence analyses must be made immediately on the basis of individual *grab* samples.

Sampling equipment may range from a simple *manual* wide-mouth dipper to elaborate *automatic mechanical* devices. Where the concentration of constituents varies vertically or horizontally in the conduit, manual sampling permits flexibility to cover the cross section. Where constituents are well mixed throughout the cross section, automatic mechanical samplers equipped with devices to select portions in proportion to the rate of flow provide reliable, representative, composite samples. The sampling equipment employed also depends on the type of laboratory examination to be performed.

Types of Laboratory Examinations

Quantitative characterization of the constituents for which laboratory examinations are to be performed depends on the objectives of the survey, such as treatment plant design, quality criteria (physical, chemical, radioactive, bacterial, biological), specific stream analysis computations, or other purposes. *Standard Methods for the Examination of Water and Wastewater* [9] presents details, including precautions concerning sampling.

Since organic wastes constitute a major factor in stream analysis, special emphasis is given to the determination of BOD, particularly with respect to immediate demand, differentiation between the colloidal and dissolved fraction and the settleable solids fraction, and the development of the BOD rate curve. As discussed later, it is essential to make these distinctions in the BOD characterization of wastes for quantitative computations of the effect of organic waste loads on the dissolved oxygen profile of the stream. These are usually recognized, but no standard method has been defined for differentiation between the colloidal and dissolved fraction and the settleable solids fraction. An approximation that is satisfactory for stream analysis studies is the determination of the standard 5-day BOD of one portion of the total waste sample and of the supernatant of a second portion after 1 hr of quiescent settling; the difference in BOD of the two portions represents the settleable solids fraction, which, if discharged to the stream without treatment, constitutes the organic load subject to sludge deposit.

Since BOD is the dynamic chemical manifestation of a chain of biological life acting on the unstable organic matter serving as a food source, it is to be expected that considerable variation is inherent in test results among samples. This does not invalidate the BOD test; however, it does emphasize the necessity of examining a sufficient number of samples to derive a statistically reliable representative mean value.

The variation inherent in BOD is likewise inherent in bacteriological examination, such as determination of the coliform group density. Statistical methods for dealing with these are set forth in Appendices A and B.

Industrial Waste Survey

Although each industrial establishment constitutes a special case, there are a number of common elements in the conduct of an industrial waste survey in addition to the discussion concerning volume and characteristics. Efficient planning of the industrial waste survey requires familiarity with the production processes involved, the operating schedules and shifts, the raw materials, and the finished products. The period selected for the measurement of volume and sampling of wastes should be representative of normal production. A chronological record of past production relative to plant capacity serves as a guide to seasonal and day-to-day variation. Production beyond rated plant capacity usually results in a heavier waste load per unit of product. The record of production per shift or per day during the waste survey period affords a basis for relating volume and strength of waste load to unit product.

The selection of locations to measure volume and sample waste flow depends on the wastewater collection system layout. If an up-to-date map of the system is not available, a flow survey is necessary, tracing by dye the direction, interconnections, and outlets from the various unit processes of the plant. Measurement of volume and characteristics of the separate unit processes, in addition to the combined wastewater outlet, affords a better basis for evaluating the waste potential of the various units of the plant and serves as a guide to possible in-plant corrective measures.

The duration of the waste survey depends on the complexity of the industrial plant and its size. The objective is a representative measure of volume and characteristics. In small plants this may be achieved in 2 or 3 days; in large plants 5 days or more may be required. Measurement and sampling should reflect complete operating schedules through 8-hr shifts or through the entire 24-hr period. The extent of sampling is also limited by the capacity of the laboratory facilities and personnel available to carry out the examinations. Where several laboratories are utilized, care should be exercised to coordinate procedures and cross-check laboratory results.

Indirect Estimates

For some purposes the time and expense of direct measurement of sources of wastes are not warranted and indirect estimates are employed. In communities served by wastewater treatment systems the load of the major sewered area is available from the operating records. The unconnected fringe areas are estimated by various means. Where treatment is not provided but separate or combined sewer maps are available, a good estimate can be made by superimposing the latest population-census tracts and determining the population residing in each subarea tributary to outlets to the watercourse. Normally the census tract population may be taken to represent an ultimate BOD load of 0.24 lb per capita per day. In mixed industrial, residential, and commercial areas the census tract population may be adjusted upward, depending on the nature of the tributary industries.

In many cities good industrial directories are available and show the type of products; if production figures are not available, an indication of the size of the establishment is obtained from the number of employees, usually reported. If type and amount of product can be obtained, waste loading can be estimated from generalized knowledge of the BOD waste per unit of product, although it is recognized that such generalized unit loadings vary radically among establishments of an industry.

Mail questionnaire surveys are usually not very productive of reliable results; a well designed sample interview is superior. The difficulty in sample design is the lack of adequate information on the various waste contributing units of the community and the large sample that is necessary.

When estimates of various types are completed, if the problem warrants, a river-sampling survey may be undertaken as a confirmation of the estimated loads discharged. However, unless the survey is rather intensive under stable hydrology, results are not conclusive. An overall loading without differentiation among sources may be obtained by sampling at a section of the river sufficiently below the city to ensure good mixing of the wastes with the riverflow. If the data of previous river sampling surveys are available, they may be of supplementary assistance.

There are so many variables, within and among communities and industries, that affect the volume and characteristics of the waste loading that indirect estimates are fraught with uncertainties. Evaluations based on stream analysis procedures are only as reliable as the raw data on which they are based; consequently whenever possible direct measurement of waste loading is preferable.

Variation in Waste Load

Since wastes are a reflection of man's activity, it is to be expected that variation in loading occurs during the day, from day to day and from season to season. At best, therefore, the intensive onshore measurement is but a reflection of the load at a point in time or for a short interval of a few days. It has been said that an accurate reflection of man's working and living habits is inscribed on the flow chart and laboratory record of the wastewater treatment works. There was a time when Monday was clearly identified as "washday"; now, with the advent of the automatic home and coin laundries, washday is less clearly defined. Man's habits, like so many natural and biologic phenomena, although apparently highly variable, actually are quite orderly and follow the

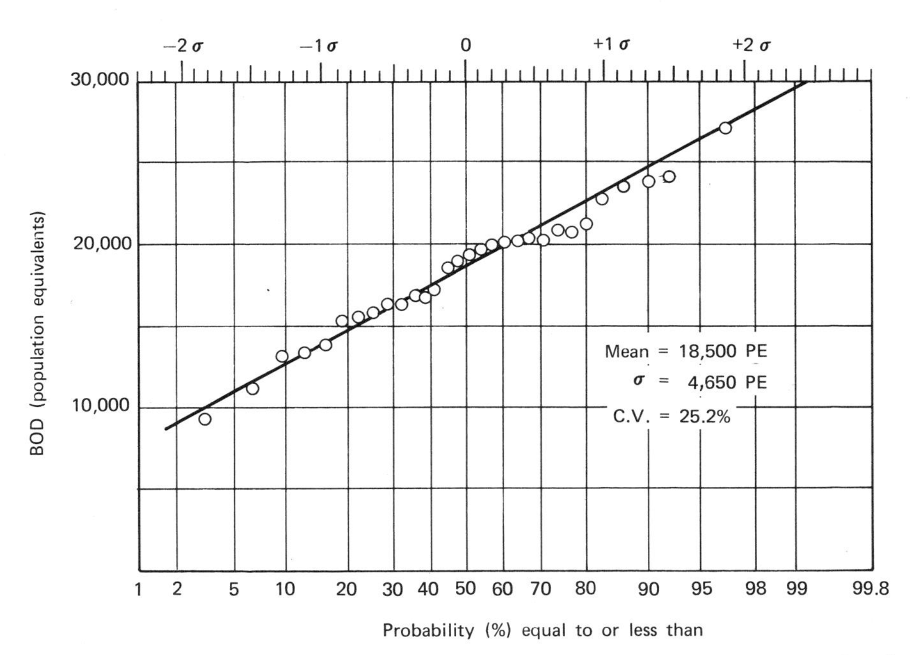

Figure 2–2 Normal distribution of daily BOD of influent for the month of September 1967 (residential community).

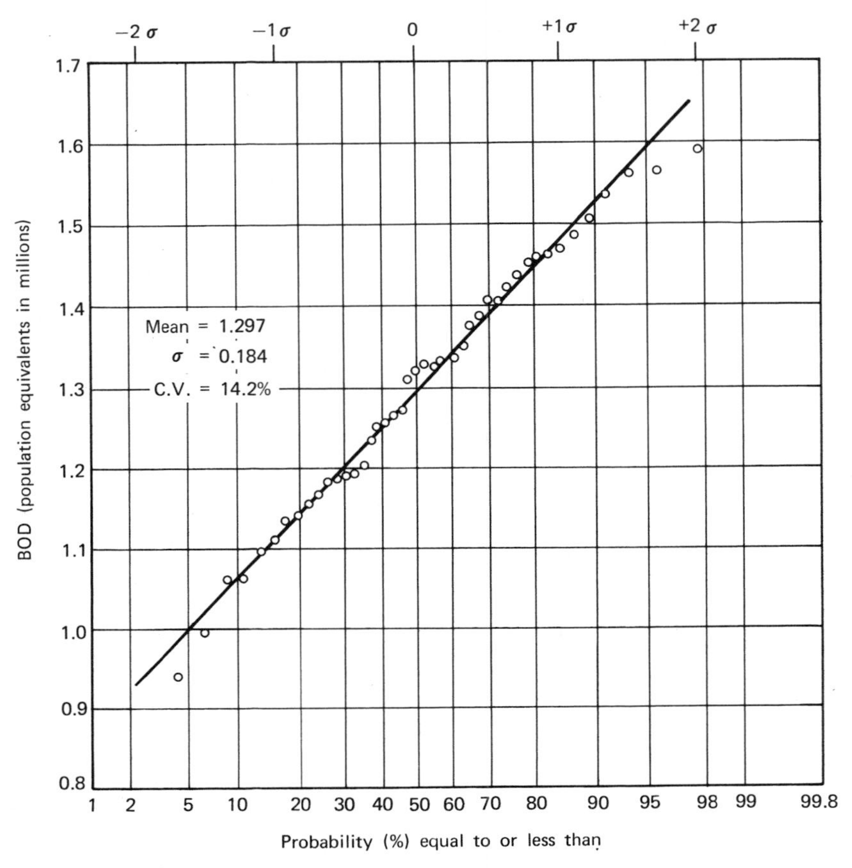

Figure 2–3 Normal distribution of daily BOD (large wet-process industry).

laws of chance. If adequate records are available (e.g., the daily operating reports of wastewater treatment works), these variations can be defined in statistical terms (see Appendix A).

In small communities the day-to-day variation in BOD is rather large, with a coefficient of variation (standard deviation as a percentage of the mean) of 30 to 40 percent; for moderate size communities the coefficient of variation is less, 20 to 30 percent; and for large cities the coefficient is usually least, 10 to 20 percent. Figure 2–2 shows the probability distribution of the daily BOD values of influent to the treatment works of a moderate size, predominantly residential community. The most probable BOD is 18,500 population equivalents, with a standard deviation of 4650 population equivalents, or a coefficient of variation of 25.2

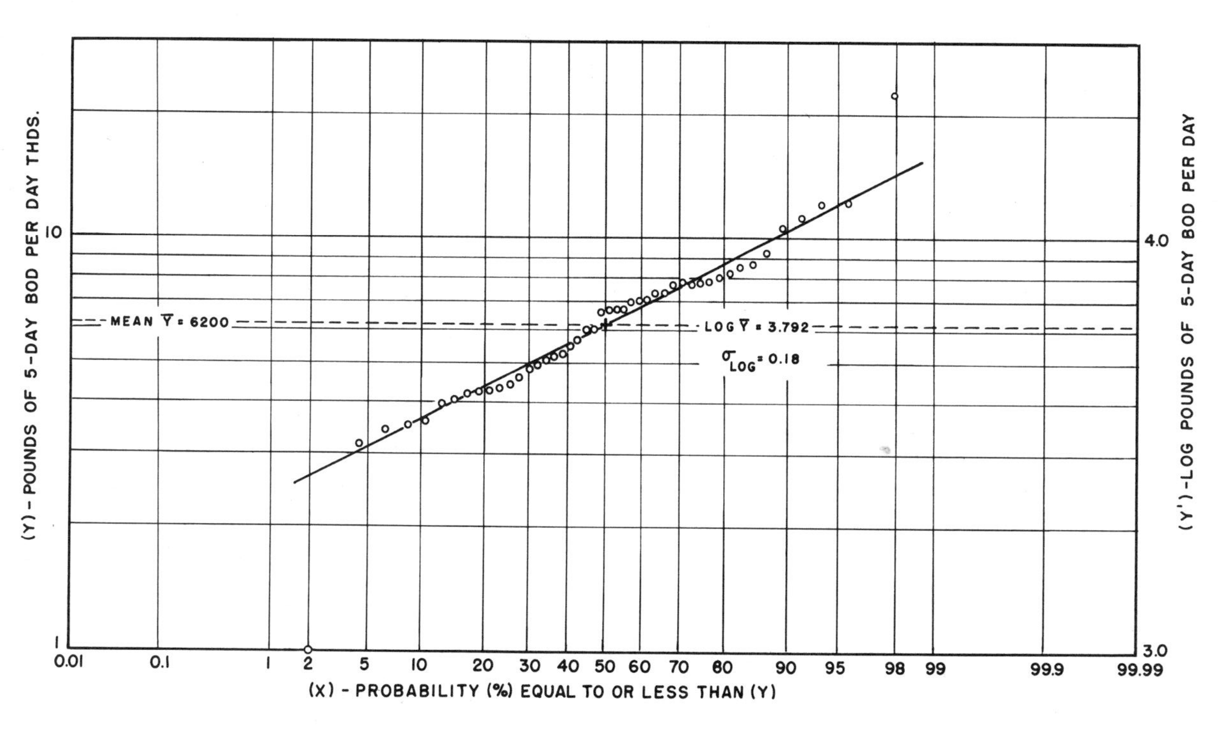

Figure 2–4 Variation in daily BOD load. Biological treatment of works effluent (large chemical plant).

percent. In this instance the most probable value is approximately equal to the connected population. Although the variation in daily BOD is rather large, part is due to the inherent variation in laboratory methods and part is associated with changes in the organic load of the community.

In an industrial plant, as in a community, many factors induce variation in the waste loading, both in volume and characteristics. Generally the variation in the daily waste production under a consistent production schedule follows a normal probability distribution. Figure 2–3 is a typical distribution of BOD of the combined untreated wastes of a large wet process industry, expressed in population equivalents based on daily values during a period of normal 24-hr, 7-day-per-week operation. The distribution forms a reasonably good straight line, indicating a normal variation with a mean of 1,297,000 population equivalents and a standard deviation of 184,000, or a coefficient of variation of 14.2 percent.

Figure 2–4 represents the distribution of a set of routine daily BOD effluent values of a large chemical manufacturing complex whose wastes receive a high degree of biological treatment. In this instance the distribution is skewed, approaching a logarithmically normal form for the bulk of the values, with a mean log of 3.792 (mean 6200) and a standard deviation ($\sigma_{\log}$) of 0.18. (Two values—the smallest, 990, and the largest, 22,400—are distinctly out of line for the otherwise log-normal distribution.) The log-normal distribution form is frequently associated with variation in effluents of biological treatment systems and may be taken as representative of good operating practice.

PROJECTING WASTE LOADINGS

If emphasis is to be given to *prevention* of excessive stream pollution, consideration must be given not only to the current situation but also to expected future conditions 20 to 50 years hence. This entails projecting the trend in waste loadings. Such projections are primarily a function of population and industrial and economic growth.

It is beyond the scope of this text to deal with the complexities of these projection methods except to emphasize the importance of long-range planning in defining water resource potentials and to describe briefly economic base studies and population forecasting tools.

The urban waste load, organic and microbial, is closely related to population, and hence population projection provides a good basis for estimating future municipal load. Industrial waste loads are more difficult to project because of the great variation among different industries and the changing technology of production. Radical changes in the types of industries can take place with shift from wet process to dry process (e.g., textile to electronic). In a free market such shifts are difficult to anticipate over extended periods.

Two approaches to projecting future waste loads are usually employed: the one based on *historical trends* of such parameters as population and economic growth; the other based on *systems analysis* of water resource potentials. In the former demographic projections and economic base studies are made without specific analysis of the basic water resource available, assuming water is not a controlling or limiting factor. The systems analysis of water resource potential is the reverse process; it defines water resources available under various plans of development, use, and management, and then fits community and industrial growth within this frame.

For purposes of quantitative evaluation of stream conditions expected from future community and industrial waste effluent loadings it is essential that projections be specific as to river basin, location of effluent discharges, types of waste, and amounts. This specificity is usually not available in the conventional regional growth projections of demographers and economists.

Each process of projection has its advocates, the demographers and economists generally favoring the former, planners and engineers usually favoring the latter; each approach has advantages and limitations, and, when both are employed, they complement each other and better projections are feasible.

Systems Analysis of Water Resource Potentials

The systems analysis approach to water resource development for optimal economic and social growth is a relatively new science and has not been extensively applied. The method involves extensive analyses, now possible with the aid of high speed computers, of numerous approaches to the development and use of land and water resources. It implies continuous appraisal of proposals and projections to guide and direct growth. Though current action and growth proceed more or less on a laissez-faire basis along lines of past trends, increasingly systems analysis studies are serving as rational guides in long-range policy decisions.

It is beyond the scope of this text to develop the general methods of systems analysis; however, the primary purpose of the text is presentation of the *rational method of stream analysis* (an essential tool in systems analysis) as a basis for guiding community and industrial growth to ensure satisfactory waste disposal and efficient and effective pollution control.

Projection of Historical Trends

Economic Base Studies

Economic base studies are made to arrive at a quantitative measure of economic opportunity and as a basis for projecting future growth potentials.

In its simplest form the economic base is described as those activities that provide the basic employment and income on which the rest of the community depends. The specific definition of the economic base depends on the variables selected for measurement. Unlike the more exact sciences, economic base studies permit disagreement as to the relevant variables—employment, population, personal income, gross regional product, flow of cash funds, number of new manufacturing establishments, sales of particular services, electric power production. As a further complication, certain of these factors are not precise by virtue of changes in method of assembly of statistical data over time.

There is generally reasonable agreement as to the present economic base, but as projections are extended into the future wide disagreements may emerge. In addition to analysis of trends of the historical data, the future can be changed radically by how physical resources (including water resources) are used, developed, and managed. Here social and political forces that come into play can invalidate trends reflected in the past historical economic base parameters. In highly developed regions drastic deviation from the past trends is less likely than in the partially or undeveloped regions. Thus economic base studies and projections are not to be taken as absolute; they provide guidelines and serve as a reference frame within which population and industrial projection are made.

Three general techniques of regional economic base analyses in common use are the export-base theory, the interindustry (or intersectional) flow method, and the theory of proportionate share by economic sector.

The export-base theory is founded on the concept that total employment is the best indicator of economic growth and that the level of total employment depends only on the level of employment in local exporting industries. The method does not consider the relative contribution of other industries and does not provide for impact of outside influence on growth.

The interindustry flow method is concerned with analysis of flows of products or inputs–outputs (either monetary or physical) between industries within the area studied and may be extended to consider interregional flows. The analyses are aided by development of input–output tables from which an estimate of contribution to gross national product can be determined.

The theory of proportionate share by economic sector focuses on the regional structure and direction of growth in terms of the region's historic proportionate share of a larger region's growth components, the large region commonly taken as the nation. Comparative analysis of regional growth is based on measurement of local growth performance against national standards, considering employment as the basic parameter. The premise is that exogenous national forces are the dominant determinants of regional employment and growth.

The major advantage of the proportionate share method over the other two methods is the emphasis on the outside factors affecting growth of a region and the relative ease of obtaining the basic data. The proportionate share method, however, does not explicitly define the composition and interdependence of various economic sectors of the region as does the interindustry method.

Combinations of these methods are possible and can lead to sophisticated regional economic mathematical models, but these do not necessarily provide more reliable forecasts, particularly where social and political decisions concerning the development of resources may change the past historical trends.

A sensitive reflector of economic change and industrial activity is the trend in electric power consumption in utility service areas. The annual increase of composite commercial and industrial energy, treated in a manner similar to decade census data, provides a basis of estimating saturation levels. Annual consumption expressed as percentage of saturation plotted on the logistic grid (Appendix A) approximates an overall logistic form. Recessions as well as booms are reflected more sensitively than they are by population or other economic parameters.

Community Population Projection

Basic Factors That Influence Community Growth

The primary elements that influence population growth are (a) birth and death rates, (b) migration (immigration and emigration), (c) economic opportunity, and (d) quality of the environment. Birth rates depend on the population structure, the proportion of women in the child-bearing age, and the so-called fertility rate. Life expectancy at birth in the United States has been dramatically increased by reduction in death rates, increasing from 35 years in 1850 to nearly 70 years at present; further improvement will be small and slow.

Whereas immigration from other countries is a minor factor today, migration among regions and communities within the country is a dominant factor in a fluid society such as the United States. The forces of attraction and repulsion are difficult to assess and are influenced by the quality of the community environment and basically by economic opportunity.

Economic opportunity is the underlying force that shapes the patterns of growth and reflects in birth and death rates and migration. Quite aside from biological factors, the will to have children is markedly influenced by economic conditions, as evidenced by the sharp decline in birth rates during the depression of 1929–1934 and the sudden rise associated with the subsequent boom years. The influence on death rate is not so easily detectable, but severe economic privation over years is certain to increase morbidity and mortality rates. Migration is extremely sensitive to economic opportunity. Communities are in a constant state of flux, some persons moving in as others move out, the net change depending largely on economic opportunity.

As the standard of living increases and economic opportunity is widespread, a strong force in community population growth is the quality of the physical and social environment. The attractive community, good schools, cultural and educational opportunities, good public services and utilities, good climate and recreational opportunities are factors that add to the quality of both working and living surroundings and are strong forces in growth. The bare necessities are not adequate to attract people. The quality of the social environment is likewise a repelling or attracting force. Social cleavages and unrest, racial conflict, poor government, and lack of integrity in public office are impediments to orderly community growth.

Since economic opportunity is a major factor in growth, population projections are best made on the basis of areas defined by the sphere of economic activity that supports the contiguous community population. In terms of the larger unit this may include the whole or parts of several states. Within the region are subdivisions of various corridors of economic opportunity differentiated as to type, such as urban-industrial or agricultural; in turn these units are further subdivided into specialized centers. With modern transportation, the distance between place of work and residence is ever widening. Although there is a marked trend from rural to urban living, there is simultaneously a strong movement of population from the central city toward the suburban periphery. Each community as an economic unit therefore overflows the central city and includes all or parts of satellite political units about it.

Since the economic sphere of influence changes with time and has imprecise geographical boundaries, it is difficult to composite the political units comprising community population from decade to decade. The smaller centers of population usually enclose the major local economic sphere, but annexations of fringe areas generally lag in time. The older definition of a metropolitan district as a contiguous area to a density of 150 persons per square mile took account of change in area and provided a reasonable census composite, but unfortunately this has been abandoned in favor of composites based on counties as standard metropolitan areas (SMAs). However, this difficulty is overcome to some extent by commencing with the large region and dealing successively with subdivisions as fractions of the larger whole. Differences among results of various methods of population projection stem in part from differences in definition of community composition over time.

Methods of Population Projection

The various methods of population projection involve as parameters one or more of the four basic factors influencing growth. The methods broadly divide into two classes, the *direct* and the *indirect*. The former employs the population census data directly, the latter relates population indirectly to economic projections. The direct methods range from simple graphical projections to mathematical models that include apportionment, rates of increase, ratios of smaller units to larger units, density per unit area, cohort survival or net migration and net births and deaths, and the logistic theory of growth.

Each of these methods has its advantages and disadvantages. The Bureau of the Census employs the cohort-survival approach, which, since it commences with the present population structure and specific birth and death rates and migration, is generally recognized as the most reliable method for short-range projections. This advantage soon diminishes when applied to long-range projections. Like any other method, it depends on ensuing changes in birth rates and migration, which are difficult to anticipate. The economists emphasize economic base studies, and, when population itself is not the primary parameter, the projections of other economic variables are useful guides in conversion to population growth.

The logistic method has the advantage of an underlying theory of growth that can be dealt with on a mathematical basis and is particularly applicable to long-range projections. In addition to projection in terms of numbers of persons it provides a means of defining characteristics of growth that are useful in comparisons among communities and in defining significant changes in patterns of growth. Application of the logistic theory of population growth is facilitated by the use of logistic population paper as developed and illustrated in Appendix A.

River Basin Projection

A frame of reference for projecting population growth in a river basin is provided by commencing with a larger region composed of several states in whole or part and successively subdividing the larger area into smaller units, working toward counties and cities of primary concern in the basin. Independent estimates are made for each unit, with the sum of subdivisions equating to the inclusive larger unit, ensuring consistency of the smaller areas with the patterns of the whole of which they are a part. This process requires successive trials to bring about a balance in saturation values and projections; some units provide well-defined logistic forms, and adjustments are then made in the less clearly defined units. Usually if a balance is achieved in saturation values, the projections for future decades beyond the last census approximate balances also. The subdivisions employed are made as far as possible on the basis of economic similarity, distinguishing between rural areas and logical urban-industrial centers or corridors. Economic base studies are an aid in the definition of units.

The coefficient of growth derived from the logistic plots affords a valuable tool in comparing similarities and differences in the characteristics of growth among units. There is a marked difference in characteristics between rural and urban-industrial areas, with higher coefficients disclosed for the latter. Some of the highly industrialized units disclose two-stage logistic forms, reflecting the change from an early agricultural base to the technological economy. Some rural areas show declines, reflecting the shift to the urban industrial areas.

Reliability of Projections

Events that are shaped largely by human decisions cannot be projected with any great degree of reliability, particularly in a free society. It is not surprising, therefore, to find considerable variation in the estimates of future economic and population growth. Since the parameters that govern growth are as yet not well defined, inevitably a degree of "bullish" or "bearish" subjectivity influences the forecaster; also the current event or recent trend overweights the future projection. The dampening effect of the depression reflected in the 1940 census led most demographers to forecast a sharp leveling to a "plateau" for the United States of 180 million between 1970 and 1990; the postwar boom reflected in the 1950 and 1960 census and concern over the world "population explosion" currently swing toward the upward trend, with projections exceeding 400 million by the year 2000. Economic and industrial growth projections are even more variable, with wide divergence among estimates prepared by different groups.

Projections for wastewater treatment plant design usually extend 10 to 20 years into the future; such forecasts are reasonably reliable if prepared by a rational method. More comprehensive water-resource projections may involve forecasts of 50 years or more and, as expected, are less reliable. In part, uncertainty and lack of precision in projections, if properly recognized, are compensated for by flexible design and construction programming to accommodate unforeseen changes as growth proceeds.

If society is to meet its future needs in a somewhat orderly manner, projections of growth are essential, and deviations and variations should not discredit or discourage effort. Considering the difficulties of the problem, the many forces at play, and the inadequacies of basic data, it is well to recognize these variations in projections and not accept or expect any one projection as absolute. Reliability is improved by considering projection as a continuing process, extending and refining studies as data become available.

Conversion of Projections to Specific Waste Loadings

In addition to detailed population and economic projections specific to service-area outlets to the stream, a current onshore survey of existing waste loads with direct measurement of volume and characteristics, and classification of the tributary service areas is usually required in the conversion of the projections to anticipated future waste loads. Interest generally centers on conversion to two types of waste, organic and microbial. The conversions are first made in terms of raw wastes conveniently expressed on a comparable basis in population equivalents (PE).

Microbial Population Equivalent

If interest centers on the bacteriological quality of the receiving stream as a source of water supply or as a recreation area, the waste load is *directly proportional to the tributary population.* As will be shown in Chapter 5, the average annual coliform bacteria contribution per capita is taken as 200 billion per day, with seasonal variations ranging from a warm-weather monthly average peak of 182 percent of the annual average to a cold-weather low of 45 percent. In addition to the population connected to the sewerage system, the bacterial load is increased by the uncontrollable source of urban wash discharged in the storm drains and small drainage channels. This source is highly variable, depending on the type of sewerage system and storm drains, the general quality of the community environment, and the frequency of rainfall. In the absence of more specific data it is reasonable to increase the projected population by 5 to 10 percent in communities served by separate systems to allow for urban wash. Where the community is served by *combined* sewers, it is well to keep in mind that even though dry-weather flow may be intercepted for treatment, the entire sewered raw load is bypassed to the stream during storms, without treatment.

Organic Population Equivalent

The organic load conversions of projections are based on 0.24 lb of ultimate biochemical oxygen demand (BOD_U) per capita per day as the basic population equivalent. This is taken as representative of a normal community composed of residential and commercial areas, with a moderate mix of industries, without domination of large wet-process industries with heavy organic waste loads.

The specific organic waste loads comprise three subdivisions: the normal community waste, the large wet-process industries connected to the community system and private systems, and the uncontrollable urban wash. If the communities are normal without special organic waste industries, the BOD conversions in population equivalents are *directly proportional to the projected populations;* to this is added the urban wash, which may be taken at 5 to 10 percent of the projected population.

Where large wet-process industries are connected to the community system, adjustment is made in the population equivalent. Three approaches lead to reasonable conversions, based on (a) a higher per capita unit oxygen demand, (b) population equivalents per unit product, and (c) population equivalent based on economic projection of employment by the industrial sector.

The onshore survey provides the current total BOD and service-area population from which a unit population equivalent is obtained. This unit is adjusted upward or downward according to the industrial mix projection and may range from 0.24 to over 0.5 lb per capita per day. The municipal and industrial complex tributary to various service areas in the Miami River basin, as determined by Driver [10], ranges from 0.24 to 0.42. The projected population is converted to population equivalents by multiplying by the ratio of the projected per capita unit to 0.24. For example, if the projected population to a tributary outlet is 10,000 for the year 1985 and the unit per capita oxygen demand is projected as 0.40, the 1985 organic load is increased to 16,670 PE $[10{,}000 \times (0.40/0.24)]$.

Where large industries are involved, an analysis of population equivalent (or BOD) per unit of product projected to future production also affords a basis for conversion. This procedure should be specific to a particular industrial establishment, as generalized values vary substantially among establishments of a given industry. The current onshore survey provides a good reference point in defining equivalent organic waste load per unit of finished product; analysis of past records and projection for the specific establishment in consultation with the technical and management staff provides a reasonable future estimate of organic load. If the large industry is connected to the community system with the service area composed of residential, commercial, and industrial sectors, the conversion then involves the sum of the three elements: the normal community service-area population; the large industry adjusted to normal population equivalents; and urban wash taken as 5 to 10 percent of the service-area population.

In service areas composed of a mix of wet-process industries economic projections are made by industrial sector in terms of *employment*. In turn projected employment is converted to community population on the basis of the expected number of persons supported per employee (currently 3.5 per employee).

The summation of the various industrial-sector conversions then constitutes the normal population equivalent of the industrial fraction. Such an approach involves extensive analyses, but where the problem warrants, it provides the most reliable conversion.

REFERENCES

[1] U.S. Department of Health, Education, and Welfare, Public Health Service, Division of Water Supply and Pollution Control, *Report on Pollution of the Detroit River, Michigan Waters of Lake Erie, and their Tributaries,* 1965.

[2] Michigan Department of Health, Michigan Water Resources Commission, and U.S. Department of Health, Education, and Welfare, Public Health Service, Detroit-Lake Erie Project, May 1963–August 1964, *Tabulations of Data Compiled from a Survey of Discharges from Combined Sewers and Separate Sewers at Detroit and Ann Arbor, Mich.*

[3] Weibel, S. R., Anderson, R. J., and Woodward, R. L. *J. Water Poll. Control Fed.,* **36,** 914 (1964).

[4] Sawyer, C. N., *J. New England Water Works Assn.* **61,** 109 (1947).

[5] Johnston, W. R., et al., *Proc. Soil Sci. Am.,* **29,** 3 (May–June 1965).

[6] *Proceedings of Conference on Research Needs and Approaches to the Use of Agricultural Chemicals from a Public Health Viewpoint,* University of California at Davis, 1964.

[7] *U.S. Department of Agriculture Year Book,* U.S. Government Printing Office, Washington, D.C., 1955.

[8] *Report of Proceedings at Public Hearing Regarding Proposed Classification for Waters of the Chowan Basin, N.C.,* Stream Sanitation Committee, 1955.

[9] *Standard Methods for the Examination of Water and Wastewater,* American Public Health Association, Inc., New York.

[10] Driver, B. L. *Some Relationships between Economic Growth and a Stated Water Quality Objective: The Miami Basin, Ohio, as a Case,* Ph.D. Thesis, University of Michigan, 1967.

3

HYDROLOGIC AND CLIMATOLOGIC FACTORS IN STREAM SANITATION

In dealing with the extensive subject of hydrology and climatology in this chapter we shall limit consideration to the aspects most relevant to stream sanitation, emphasizing quantitative definition of streamflow, characterization of the stream channel, and analysis of pertinent climatologic factors. As an aid to understanding the underlying forces, the concept of the interrelation of the entire hydrologic cycle, the balance of water, land, and vegetative cover, and physiography are supporting considerations.

THE HYDROLOGIC CYCLE

Nature's Balance

Although the character of water resources is dynamic, at a given time and place in the natural setting an equilibrium stability in the hydrologic environment is established through a complex interrelation of water, land, and vegetative cover. Stability of environment is relative, depending on the time period. In terms of geologic time changes have occurred, but whether a change is a downward pulsation in an otherwise upward trend or part of a well-defined cycle is difficult to establish. However, the history of the world in the past 2000 years does not lead us to expect sudden changes in nature's balance. Hence in the shorter historical view hydrologic stability may be accepted, and practical definitions of the water resources available can be determined from statistical analyses of current records; however, the pulsations reflected in geologic time should temper overconfidence in the parameters developed from short records as absolute values.

Within equilibrium stability, nature allows a wide range of variation in the interrelation of water, land, and vegetative cover and in the elements of the hydrologic cycle of vaporization, condensation, and precipitation. The elements of the hydrologic cycle are schematically represented in Figure 3–1. In tracing the paths of water in this complex chain it is easily appreciated how variation in time and place can occur. The sun's energy vaporizes water through the process of *evaporation* and *transpiration*. A portion of rainfall is evaporated in falling before it reaches the earth; a portion is intercepted by vegetation and evaporates. Evaporation takes place from exposed water surfaces of ponds, lakes, streams, and oceans, as well as from moist ground surfaces. The relative amount of evaporation depends to a great extent on *evaporation opportunity*. Exposed water surfaces offer continuous opportunity, but the area of ponds, lakes, and streams is small in relation to the area of land; in turn land area is large but evaporation opportunity is variable—high immediately after rainfall but diminishing as soil and ground-

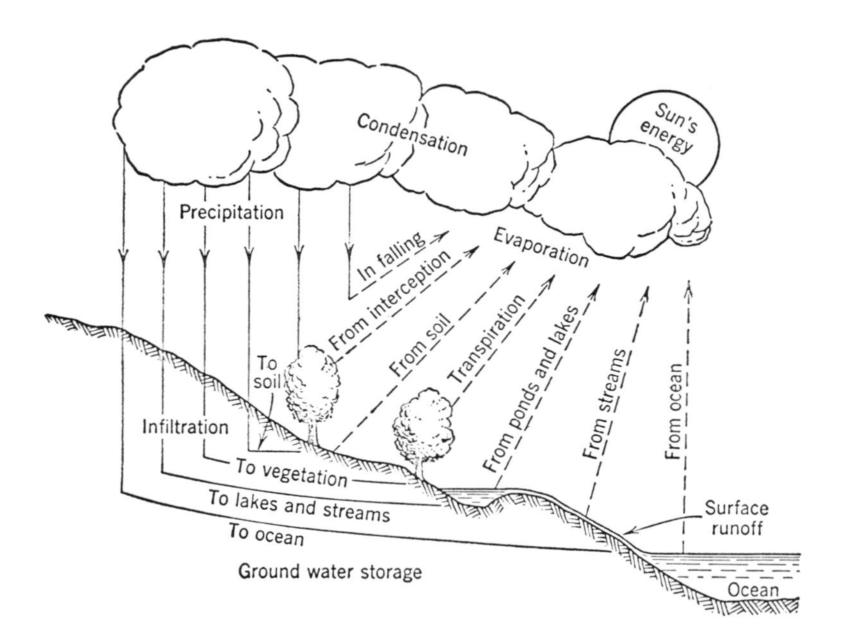

Figure 3–1 The hydrologic cycle.

water decline. Temperature, relative humidity, and wind velocity are also major elements involved in evaporation.

A substantial loss of water occurs through the phenomenon of transpiration by trees, shrubs, grasses, and cultivated crops. Transpiration opportunity varies with the availability of soil moisture, seasonal temperature, light, and type of vegetative cover. Average water losses for the growing season range from 4 to 6 in. for coniferous trees, 6 to 8 in. for small brush, and 8 to 12 in. for deciduous trees and crops. Of the residual rainfall reaching the earth some drains to streams as surface runoff; some infiltrates, temporarily retained in the soil, is lost by evaporation and transpiration, or percolates to the subsoil and replenishes the groundwater. Seepage from groundwater ultimately outcrops to ponds, lakes, or streams and is the major feeder to maintain flow during the dry periods.

Water vapor from evapotranspiration is lighter than air. In rising from the denser air at the earth's surface the water vapor expands, resulting in cooling and condensation. Thus water that is lost to the drainage area on earth cannot rise indefinitely into the atmosphere. Approximately 50 percent is condensed before ascending 1 mile; less than 10 percent is above 10 miles; all water vapor is confined below a ceiling of approximately 12 miles. Thus water in nature's economy is neither lost nor gained, and the hydrologic cycle of evaporation, condensation, and precipitation is perpetually repeated.

The sun's energy also plays a secondary role in the cycle. Unequal heating of the earth's surface induces convection currents and mass air movements that transport water vapor and condensed clouds over wide geographical areas in varying patterns. It is not surprising, therefore, that precipitation as rain and snow occurs unequally over different drainage basins at irregular intervals. Distribution in turn is influenced by physiographic land forms and location of the drainage basin relative to the usual patterns of air mass movement. For example, high mountains in the Pacific Northwest, in forcing moisture-laden air from the ocean upward, induce heavy *orographic* precipitation on the windward side, and in the descent on the leeward side little or no precipitation occurs. Each drainage basin has its unique pattern.

In considering the almost infinite variety of possible combinations of factors in the hydrologic cycle it is readily comprehensible that water is a variable, dynamic natural resource. What is less comprehensible is the fact that nature, within these variations, is quite orderly, following with remarkable consistency the laws of chance, establishing for each drainage basin an equilibrium stability in water resources. However, we do not always comprehend the consequences of man's impact on nature's balance of water, land, and vegetative cover, nor the potentialities for abuse, nor responsibility for intelligent use and management.

Physiography

Since physiography to a large extent determines the unique water resource characteristics among drainage basins as well as within a given basin, its definition contributes to practical consideration in use and development. Relevant features to be considered are size and shape of the drainage area, the river system and tributary streams, physiographic regions and land forms, vegetative cover, and geology and groundwater formations. These are illustrated in the following selections.

The Drainage Basin

The shape and size of the drainage basin and river systems are defined by topography. Essential are good topographic maps with contours, such as the U.S. Geological Survey quadrangle sheets at a scale of 1:62,500 (approximately 1 in. to 1 mile) or preferably 1:24,000. Tracing the divides by contours provides the watershed boundaries of the drainage area; planimetering these boundaries and the boundaries of tributary streams provide the increments of drainage area of the river system from headwaters to mouth. Accurate drainage figures for areas tributary to stream gaging stations established by the Geological Survey are available. Drainage area data are also available from survey reports of other agencies, such as the Corps of Engineers.

In dealing with practical problems it is necessary to subdivide major systems into appropriate sub-basins. These vary in size, shape, and arrangement of tributary streams. Some are long and narrow, with many minor tributaries along the course of the main stream; an example is the Trinity River (Figure 3–2). Others are long and narrow but comprise two major tributaries joining to form a common outlet in the lower reach—for example, the Apalachicola–Chattahoochee–Flint system (Figure 3–3). Still others are fanlike, with two or more tributaries joining to form the main stream, such as the Alabama–Tombigbee–Warrior system, with the major tributary area in the upper reaches (Figure 9–2); another example is the Saginaw system in Michigan, which comprises a series of tributaries that join near the mouth.

Physiographic Regions, Land Forms, and Vegetative Cover

The nature of a river system is radically influenced by the physiographic regions that cut the basin and the related land forms and vegeta-

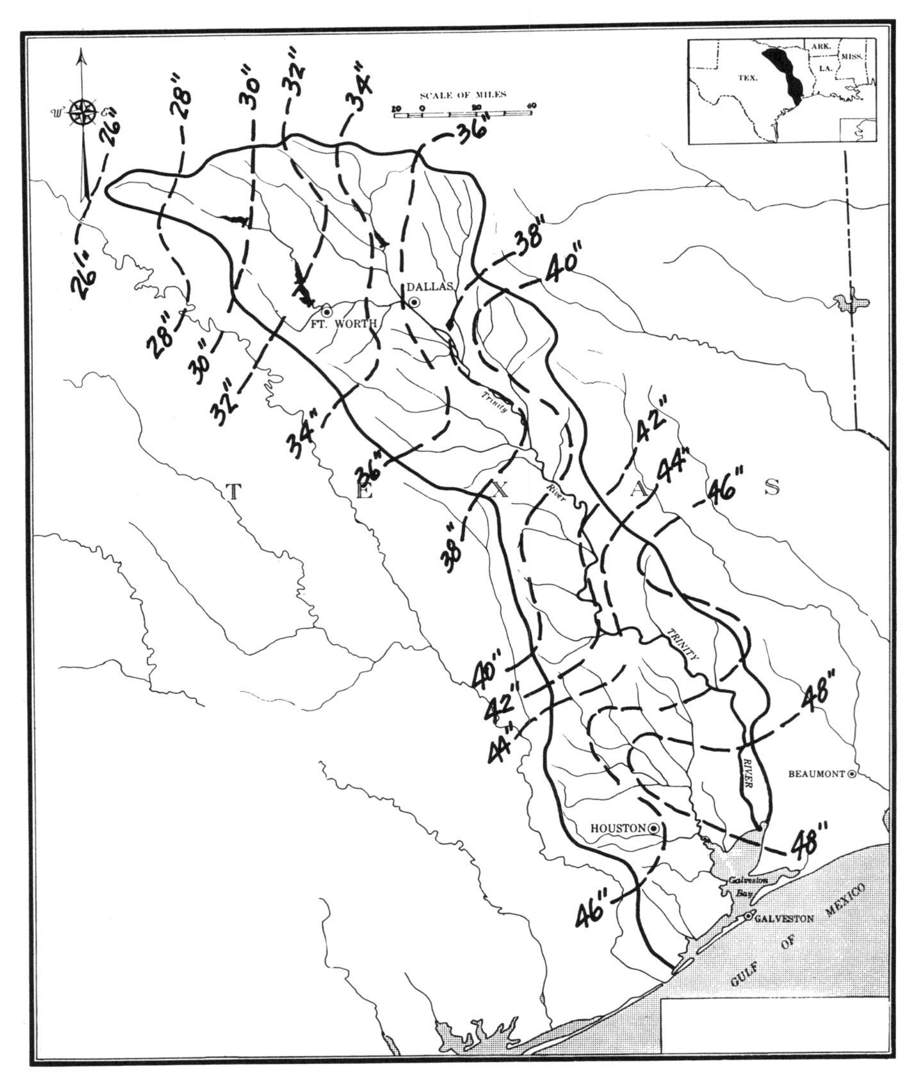

Figure 3–2 Trinity River basin.

tive cover. Some basins are homogeneous; others, such as the Apalachicola–Chattahoochee–Flint system, are crossed by several distinct regions (Figure 3–3). The Coastal Flatwoods are composed of swamp, low, poorly drained marshland, dense swamp forest, and little cleared land; they are generally nonagricultural. The Lower Coastal Plains are composed

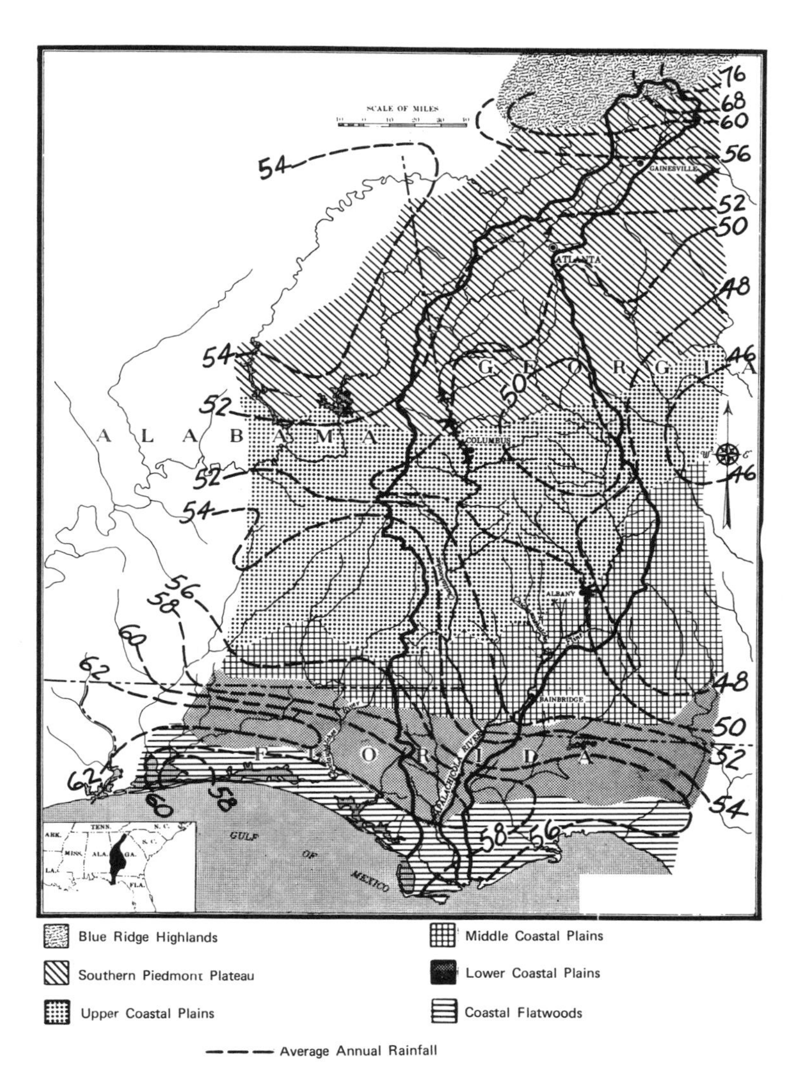

Figure 3–3 Apalachicola River basin.

of undulating to gently rolling hills of gray sandy loam, generally well drained, some swampy areas along the Chattahoochee and Flint rivers, and sink holes and outcropping springs. The Upper Coastal Plains are composed of gently to moderately rolling terrain, gray or red-brown sandy loams with friable red clay subsoils, occasional sands, some narrow belts of heavy soils or fine sandy loam, underlain at shallow depths by heavy plastic clay. The Piedmont Plateau is composed of rolling, steep hills, of Pine Mountain district (northeast of Columbus, Ga.) rising several hundred feet above the general surface of the plateau, of bedrock overlain with a thick blanket of residual soil of sandy or red clay loam leached and subject to erosion, and of cutover agricultural land. The southern margin of the Piedmont Plateau is marked by the *fall line,* the geologic boundary between recent poorly consolidated sediments of the Coastal Plains and the hard crystalline rocks of the older and higher Piedmont Plateau. The Blue Ridge Highland is composed of irregular steep mountains rising 2500 to 3000 ft above the general surface of the Piedmont, with rugged, wooded knolls rising 4000 to 4500 ft above sea level.

A similar physiographic complex of five major regions traverses the Alabama–Tombigbee–Warrior basin: Gulf Coast Flatwoods, Gulf Coastal Plains, Mississippi–Alabama Black Prairie and Clay Hills, Appalachian Border Plains, Cumberland Plateau, and Appalachian Valley and Piedmont Plateau. The southern boundary of the latter marks the fall line.

Geology and Groundwater Formations

Groundwater is a primary source of water supply for small and moderate-sized communities and industries, and in many instances is a major element in streamflow. For example, three geologic and groundwater formations underlie the Apalachicola basin (Figure 3–3), the hard crystalline rock above the fall line, the unconsolidated and semiconsolidated formations, and the consolidated rock aquifers below the fall line. The hard crystalline rock formation is a poor water-bearing source; wells yield generally less than 50 gpm and add little yield to streamflow. Below the fall line in the Coastal Plains groundwater yields are abundant and readily available. These characteristics have a marked effect on the severity of drought streamflow along the course of the streams, with rather sharp distinction between the Chattahoochee and the Flint rivers. Approximately one-third of the drainage area of the Flint lies in the Piedmont, whereas over one-half of the Chattahoochee basin is in this region of poor yield. Also, below the fall line the Flint is more predominantly spring fed.

CLIMATOLOGIC FACTORS

Although physiography in a given basin is reasonably stable, the climatologic factors—precipitation, temperature, wind velocity, vapor pressure, and solar radiation—are highly variable in time and place, and induce variation in stream runoff and in waste assimilation capacity.

Precipitation

Basin Distribution

Distribution of mean annual precipitation based on records at weather stations throughout a drainage basin develops a well-defined normal pattern. The patterns differ radically, however, from basin to basin. For example, the distribution pattern for the Trinity basin (Figure 3–2) shows average annual rainfall of 48 in. near the mouth, which progressively decreases inland to 26 in. in the headwaters.

In contrast, the distribution pattern for the Apalachicola basin (Figure 3–3) shows 56 in. near the mouth, decreasing inland to a low of 48 in. midway in the basin, and then increasing with elevation to a high of 76 in. in the Blue Ridge Highlands.

A radically different pattern, associated with orographic precipitation, occurs in the Deschutes basin and adjacent Middle Columbia River drainage areas of the Pacific Northwest. On the leeward side of the Cascade divide, average annual rainfall is 10 in. over wide areas, increasing moderately toward the adjacent eastern basins. Along the Cascade divide forming the western boundary of the Deschutes precipitation ranges between 80 and 90 in., and increases to 140 in. in the high elevations of Mount Hood.

Such patterns of basin distribution of precipitation aid in evaluating the runoff characteristics from tributaries along the course of a river.

Annual and Seasonal Variation

Although each drainage basin portrays a normal annual distribution pattern of precipitation over the basin, there are variations from year to year and through the seasons, variations that are highly important for waste disposal and stream pollution control. These variations are illustrated in the analysis of a 51-year record in the Roanoke basin. The average annual geographical distribution pattern over the basin is 50 in. in the Coastal Plains near the mouth, 40 in. in the Upper Piedmont Plateau, and increases up the eastern slope of the Blue Ridge Mountains to 46 in. at the rim of the headwaters.

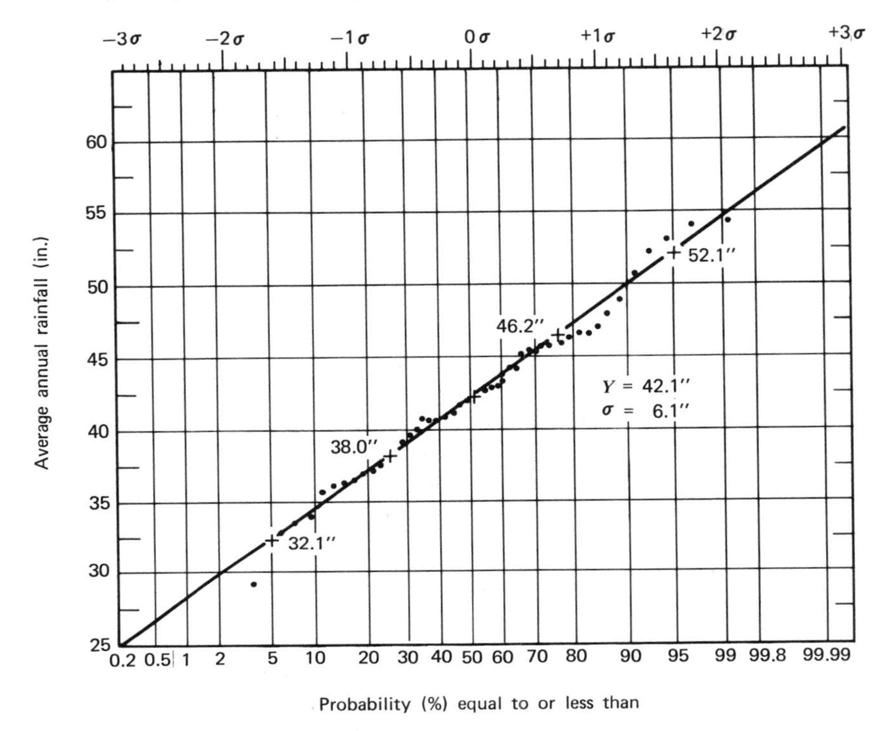

Figure 3–4 Probability distribution of annual rainfall as average depth in the Roanoke basin, 1891–1942.

The Army Corps of Engineers, in studies of the Roanoke, developed basin distribution patterns for each year for 51 years of record, and from these isohyets compiled annual values as an average depth over the entire drainage area. Statistical analyses of these annual basinwide depths afford an excellent definition of the nature of annual variation in precipitation. The annual values follow a normal distribution,* as shown in Figure 3–4, with a mean annual average basin depth of 42.1 in. and a standard deviation of 6.1 in. Thus on the average in the long run for 50 percent of the years the annual average basin depth can be expected to be within the range 38.0 to 46.2 in., and for 90 percent of the years within the range 32.1 to 52.1 in.

The seasonal pattern of rainfall through the months of the year for the Roanoke basin shown in Figure 3–5 is determined from a statistical analysis of the 51-year record of depths by months, as an average on

* For graphical statistical methods see Appendix A.

the drainage area based on isohyets for each month, as compiled by the Corps of Engineers. Considering each month of the year separately, variation from year to year for specific months is skewed, with logarithms of depth of rainfall generally forming normal distributions. Curve *A* of Figure 3–5 shows the most probable monthly rainfall expected on the drainage area for each month of the year, and from this data a normal seasonal pattern emerges. Normally, high rainfall is expected during the summer months May through August, with a most probable seasonal peak of 4.3 in. during July; a most probable seasonal low of 2.1 in. may be expected for October and November.

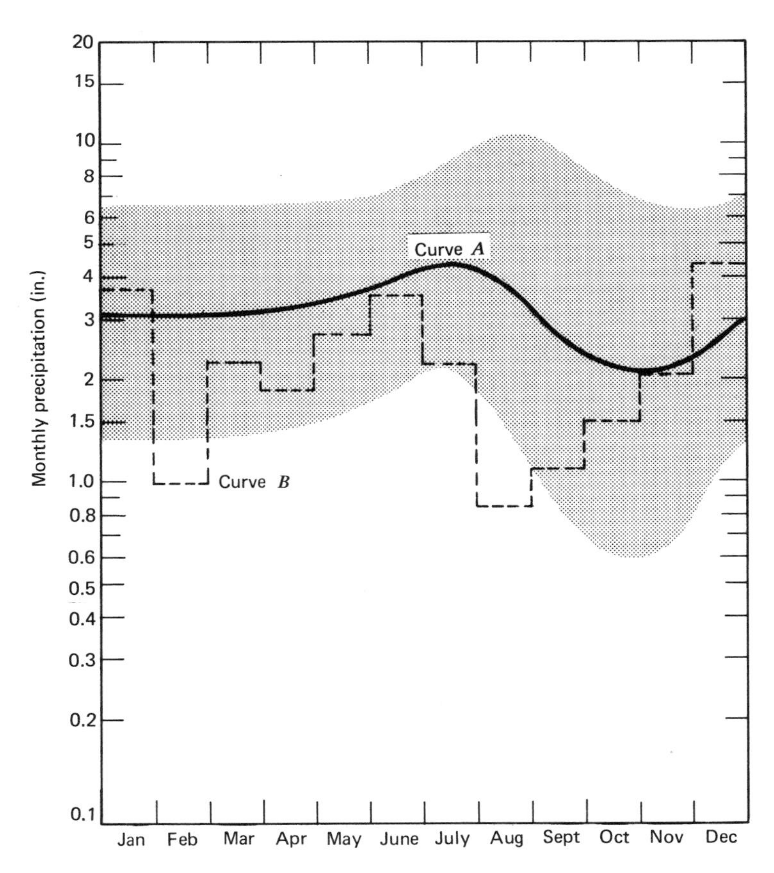

Figure 3–5 Roanoke basin. Seasonal pattern of precipitation. Monthly rainfall as average depth on drainage area. Curve *A*: most probable monthly rainfall; curve *B*: monthly rainfall during severe drought year of 1930; shaded area: range within which rainfall can be expected in 90 percent of years.

The shaded area shows the range within which monthly rainfall can be expected in 90 percent of the years as developed from the log-normal distributions for each specific month. From this it is noted that great variation can be expected in monthly precipitation, particularly during the critical summer–fall warm weather season.

The seasonal pattern for any specific year may deviate from this general pattern, since not all months through a year will experience rainfall of the same probability. Curve *B* shows the monthly rainfall for the specific year 1930, generally recognized as a year of severe drought.

Although these illustrations pertain to a specific basin, they are representative of the form of a general pattern and indicate a method of analysis useful in defining relevant characteristics for any particular basin with adequate precipitation data.

Air and Water Temperature

Temperature relates to stream sanitation in several ways. It is a primary element in the hydrologic cycle, principally in its influence on evaporation and transpiration and their interlacing with stream runoff. The temperature of the water of receiving streams has a direct bearing on organic waste assimilation capacity (see Chapter 4); the temperature of the air and other climatologic parameters govern the dissipation of waste heat (see Chapter 7); the temperature of water is important to aquatic life, and it influences the microbial self-purification of the stream (see Chapter 5).

Long-term records of water temperature are usually not available. However, the surface water temperature of many streams parallels reasonably the monthly mean air temperature; where observed water temperature records are for short periods or unavailable, the patterns developed from statistical analyses of long records of monthly mean air temperature serve as a reasonable approximation for many practical applications. This generalization, however, must be qualified in the light of specific hydrologic settings of particular drainage basins.

For problems involving the dissipation of waste heat in streams, cooling ponds, and reservoirs it is essential to approach water temperature on a more fundamental basis derived from the meteorologic variables involved. This is dealt with more fully in Chapter 7.

Seasonal Patterns, Air and Water

Figure 3–6 shows a comparison of air and water temperature records for the lower Arkansas River. Based on statistical analyses of the 66-year record of air temperature at Pine Bluff, Arkansas, considering each month of the year separately, the monthly average air temperature

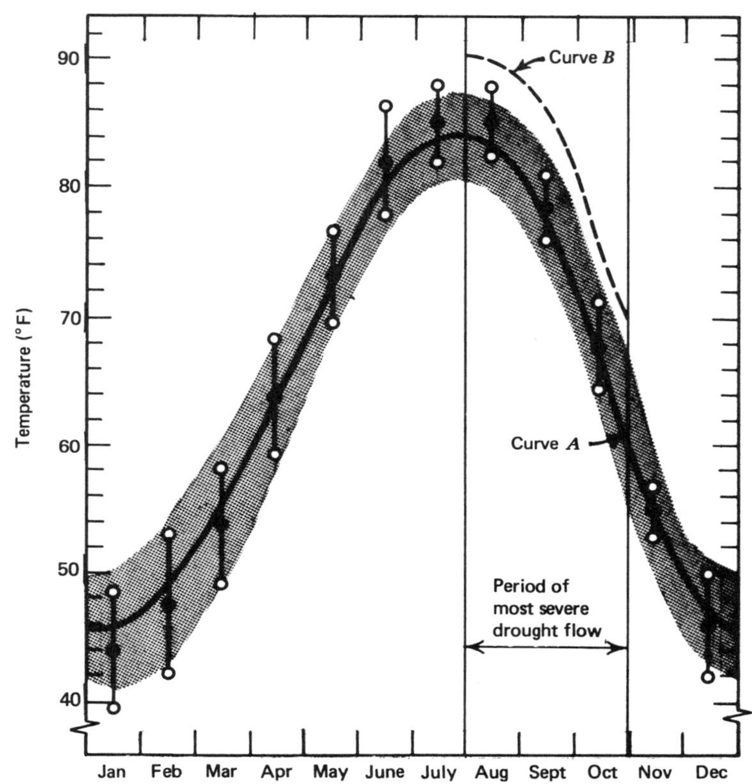

Figure 3–6 Arkansas River. Seasonal pattern of air and water temperature. Curve *A*: most probable monthly average air temperature, Pine Bluff, Arkansas; Curve *B*: critical weekly average during severe droughts; shaded area: range within which monthly average can be expected for 80 percent of the years. Monthly average river water temperature at Little Rock, Arkansas, based on records for 1945–1959; ●: most probable, ○: 80-percent confidence range.

values describe remarkably normal distributions from which the most probable values and any desired confidence range can be readily determined. With these values plotted at midmonth, smoothed seasonal patterns are drawn. Curve *A* shows the most probable monthly average temperature expected, and the shaded band about the most probable indicates the range within which monthly average air temperature can be expected for 80 percent of the years.

The observed river-water temperatures as monthly averages recorded at Little Rock, Arkansas, upstream but at about the same latitude as Pine Bluff, for the 14-year period October 1945 to September 1959, develop for each month of the year reasonably normal probability distributions. The most probable monthly average river water temperatures are shown in Figure 3–6 as solid circles, and the upper and lower limits of the 80 percent confidence range as open circles. It will be noted that there is reasonable agreement between the seasonal pattern of monthly average air and river water temperatures.

A more direct definition of the nature of variation in river water temperature is reflected in the statistical analyses of the continuous record of the Olentangy River at Delaware, Ohio, as shown in Figure 3–7;

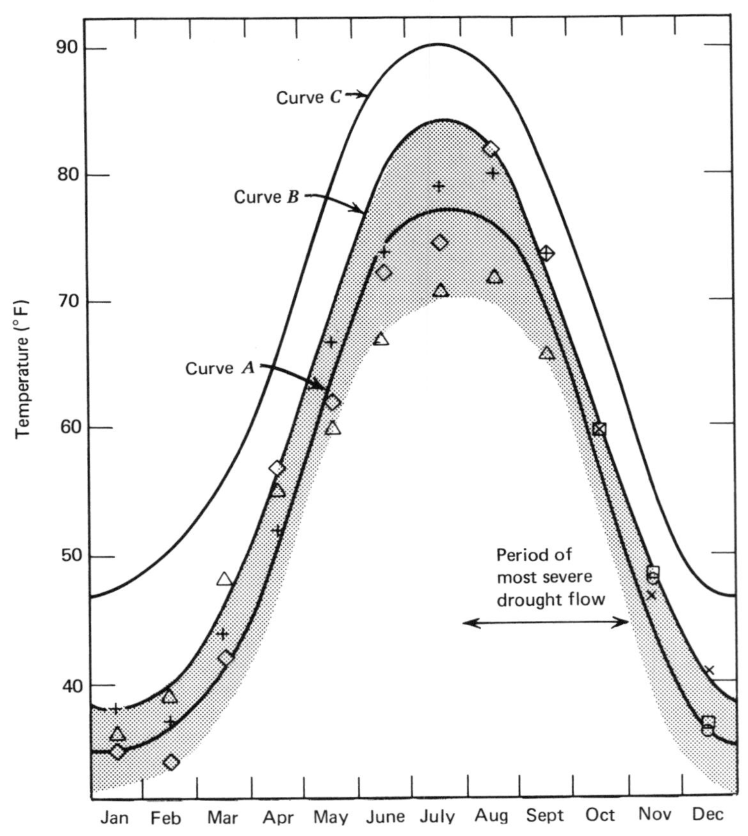

Figure 3–7 Olentangy River. Seasoial pattern of surface water temperature (based on record of Olentangy River at Delaware, Ohio). Curve *A*: most probable monthly mean value; curve *B*: once-in-10-year monthly mean value; curve *C*: once-in-10-year maximum 3-day average; shaded area: range within which monthly mean can be expected in 80 percent of years. × Mad River near Dayton, observed monthly mean, 1947 water temperature; △ Mad River near Dayton, observed monthly mean, 1948 water temperature; ⊙ Miami River at Hamilton, observed monthly mean, 1950 water temperature; + Miami River at Hamilton, observed monthly mean, 1951 water temperature; □ Miami River at Dayton, observed monthly mean, 1946 water temperature; ◇ Miami River at Dayton, observed monthly mean, 1947 water temperature.

curve A is the most probable monthly average water temperature, and the shaded band is the range within which the monthly mean can be expected in 80 percent of the years. On the average, once in 10 years the monthly mean water temperature at or above curve B can be expected. Of special interest is the analysis of the maximum 3-day average temperature series recorded for each month from which the once-in-10-years maximum 3-day average water temperature pattern is developed, as shown by curve C. The peak in water temperature occurs in July, with an expected most probable monthly mean of 77°F, a monthly mean expected once in 10 years of 84°F, and a maximum 3-day average peak expected once in 10 years of 90°F. As the period over which the temperature is averaged is shortened, the temperature increases. In general river basins in the same region and climatologic setting at approximately the same latitude have similar seasonal water temperature patterns. This is illustrated by the superposition in Figure 3–7 (Olentangy) of the short-term monthly mean observed river water temperatures reported for the adjacent Miami River.

From analysis of air and water temperature records in drainage basins throughout the United States a well-defined seasonal pattern prevails; in northern latitudes the seasonal peak occurs in June–July, and in the more southern latitudes the peak rises and shifts toward July–August. The variation from year to year in monthly means is consistently normal, from which a most probable value and any desired confidence range can be defined.

Wind Velocity, Vapor Pressure, Solar Radiation—Seasonal Variation

In addition to air temperature, of special significance in dealing with the dissipation of waste heat from industries and thermal power plants are wind velocity, relative humidity or related vapor pressure, and solar radiation. With the exception of solar radiation, extensive records of these climatologic factors are available at U.S. Weather Bureau stations throughout the country, and from these seasonal patterns and expected variations can be defined. As for air temperature, these climatologic data consistently describe normal probability distributions.

Selections from studies of the Tittabawassee basin based on records of the nearby Weather Bureau station at East Lansing, Michigan, provide typical illustrations.

Wind Velocity

The velocity of the wind is extremely variable in time, place, and location of recording instruments. Seasonal patterns are best reflected

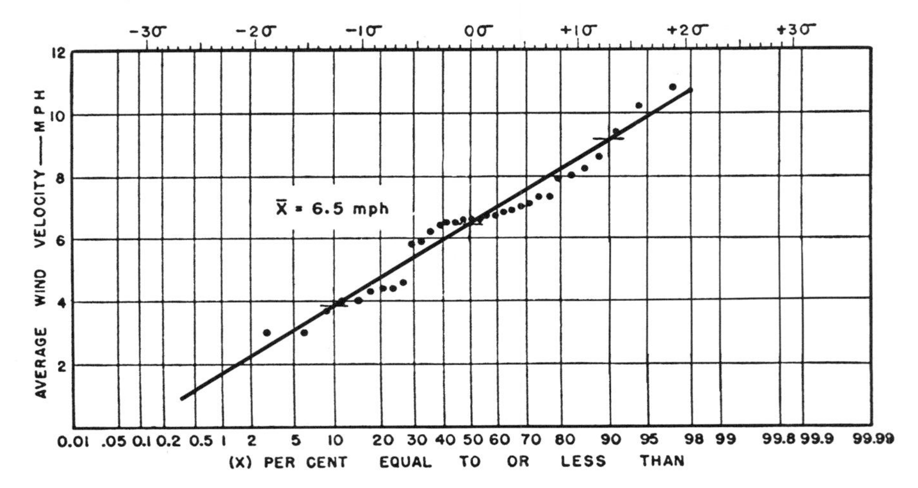

Figure 3–8 Monthly average wind velocity for July at East Lansing, Michigan, based on records from 1922 to 1954.

by statistical analyses of the monthly average records. Figure 3–8 represents the probability distribution for the month of July for the 33-year record at East Lansing. From such analyses of successive months of the year the seasonal pattern of variation emerges, as shown in Figure 3–9. The most probable monthly mean velocity is indicated by curve *A*, and the 80-percent confidence range is indicated by the shaded belt. High wind velocity in this locality is normally expected in the spring, peaking in March; low velocities consistently coincide with the hot summer season, and in this locality the minimum occurs in August.

Atmospheric Water Vapor Pressure

Absolute water vapor pressure of the atmosphere in inches of mercury, determined from the monthly mean relative humidity and monthly mean temperature, is employed in developing the seasonal pattern. Again, statistical analyses of the monthly averages, considering each month of the year separately, develop normal distributions, from which the most probable and the standard deviation are determined.

A quite orderly seasonal pattern of water vapor pressure emerges, as shown in Figure 3–10, with the most probable monthly mean designated as curve *A* and the 80-percent range indicated by the shaded belt. The seasonal peak of water vapor pressure consistently coincides

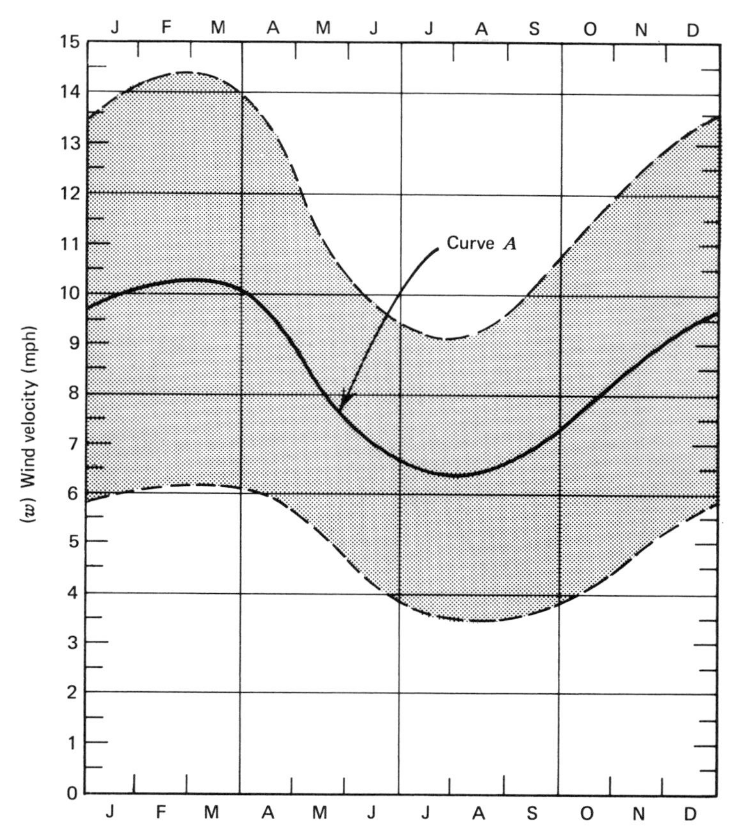

Figure 3–9 Expected seasonal pattern of monthly average wind velocity, based on East Lansing, Michigan, records from 1922 to 1954. Curve A: most probable monthly average; shaded area: range within which monthly average can be expected for 80 percent of years.

with the summer warm-weather period and the low coincides with the winter season.

Solar Radiation

Solar radiation as reported is the direct and diffuse, or scattered, sky radiation received on a horizontal surface. It is expressed in *langleys*, gram calories per square centimeter, usually reported as the total received per day. (Langleys in gram calories per square centimeter are converted to Btu per square foot by multiplying by the factor 3.69; langleys per day are converted to Btu per square foot per hour by multiplying by the factor 0.1535.)

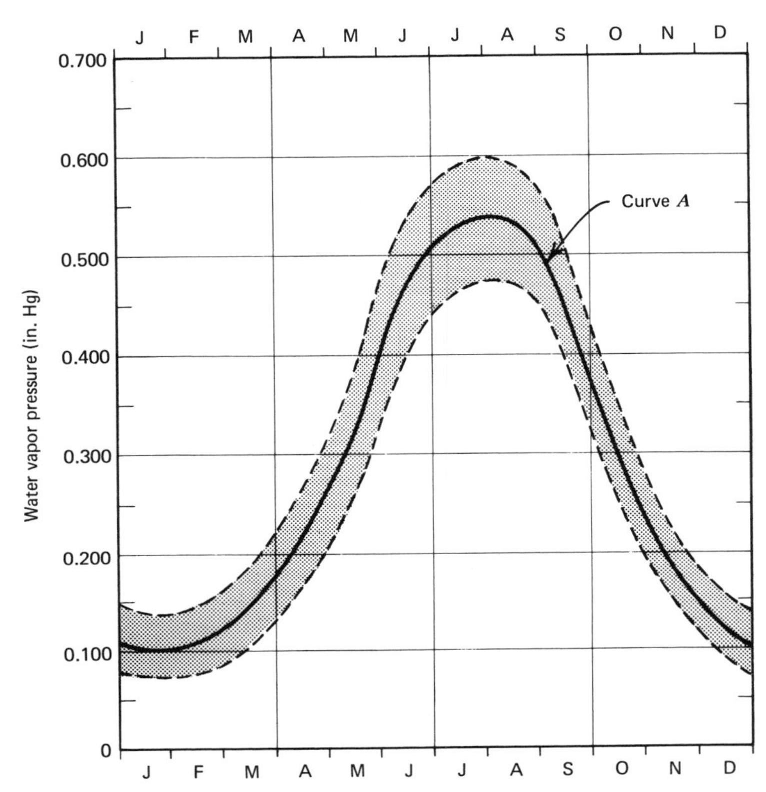

Figure 3–10 Expected seasonal pattern of monthly average water vapor pressure, based on East Lansing, Michigan, records from 1922 to 1954. Curve *A*: most probable monthly mean; shaded area: range within which monthly average can be expected for 80 percent of years.

Only a limited number of Weather Bureau stations are equipped for routine measurement of solar radiation, and records usually cover short periods. Where length of records permits statistical analysis, a well-defined seasonal pattern is determined from the monthly averages, considering each month of the year separately. As with other meteorologic factors, the monthly distributions are normal from which the most probable and standard deviation are obtained. Figure 3–11 is a typical probability distribution for the month of August, based on the relatively long record (1933–1957) at Madison, Wisconsin. Reasonably normal distributions develop even from short records, from which the local seasonal pattern can be constructed, as illustrated by Figure 3–12 for the 15-year East Lansing, Michigan, record.

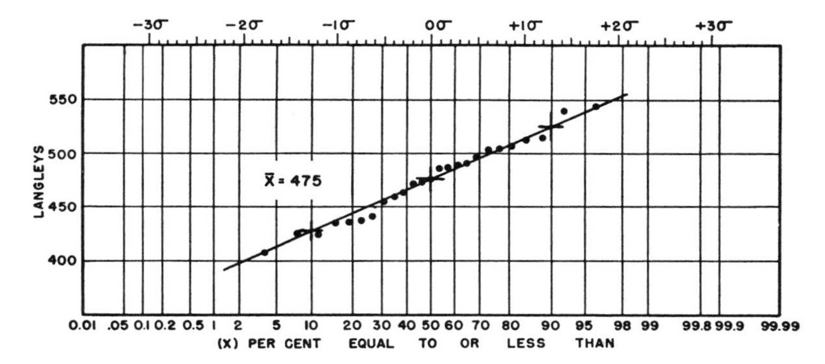

Figure 3–11 Monthly average solar radiation for August at Madison, Wisconsin, based on records from 1933 to 1957.

In view of the limited number of stations with continuous records, the results of statistical analyses of a number of key stations are summarized in Table 3–1, giving by months of the year the most probable monthly average, the standard deviation, and the coefficient of variation.

Since the angle at which the sun's rays are incident on the earth determines the amount of solar radiation received, the seasonal pattern varies with latitude, the maximum being at the equator and decreasing as latitude increases. The day-to-day variations are radically influenced by sky cover, which in turn is influenced by variation in the pattern of clouds, by elevation above sea level, and by induced air pollution.

Where exposure is similar, locations at approximately the same latitude have similar seasonal patterns of solar radiation, as illustrated in the correlation of the most probable monthly average values based on the relatively long records at the Blue Hill Observatory of Harvard University at Milton, Massachusetts (latitude 42°13′N, elevation 672 ft), with Madison, Wisconsin (latitude 43°05′N, elevation 974 ft), as shown in Figure 3–13. The most probable values approach the line of identical correlation as represented by curve *A* of Figure 3–13*a*. The standard deviation from the most probable does not show the same degree of high correlation, but the tendency is toward similarity as shown by curve *A* of Figure 3–13*b*. For many purposes these relations can be employed for locations of reasonably similar exposure and within a degree or so of these latitudes.

The influence of local air pollution is illustrated in the correlation of the most probable monthly average values of Blue Hill, Massachusetts, with those of the Boston station (latitude 42°21′N, elevation

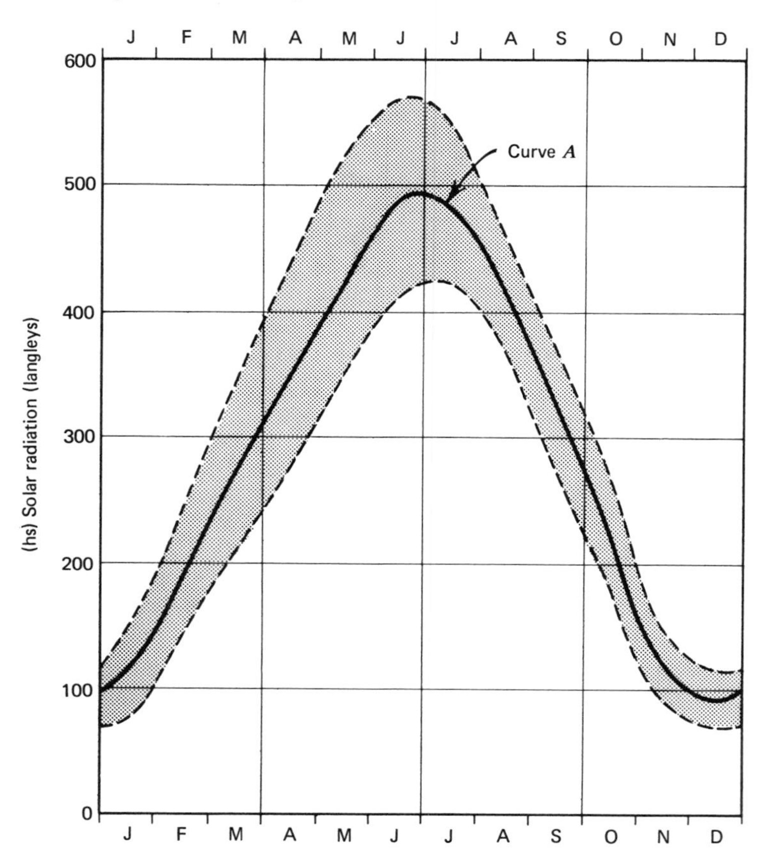

Figure 3–12 Expected seasonal pattern of monthly average solar radiation, based on East Lansing, Michigan, records from 1943 to 1957. Curve *A*: most probable monthly average; shaded area: range within which monthly average can be expected for 80 percent of years.

300 ft) which is located at an exposure subject to considerable induced sky cover from urban-industrial air pollution. Curve *A* of Figure 3–14 represents the line of identical correlation, which it is noted does not fit the data; curve *B* drawn through the data indicates that the Boston most probable monthly average values are consistently 94 percent of the Blue Hill values; that is, solar radiation is reduced by about 6 percent by local air pollution. Some of this difference may be attributed to the lower elevation of the Boston station, but the major factor is air pollution. This is further confirmed by major differences recorded on individual days of heavy smog.

Table 3–1 Seasonal Patterns and Variations in Monthly Average Solar Radiation

Month	Most Probable Monthly Mean (langleys per day)	Standard Deviation(σ)	Coefficient of Variation
Madison, Wisconsin (1933–1957); 43°05′N; elevation 974 ft			
January	147	19	.129
February	218	34	.156
March	317	40	.126
April	402	50	.124
May	478	58	.120
June	541	47	.087
July	559	37	.066
August	475	38	.081
September	366	45	.123
October	267	31	.115
November	153	24	.155
December	125	28	.220
Blue Hill, Massachusetts (1934–1964); 42°13′N; elevation 672 ft			
January	159	28	.178
February	230	24	.104
March	322	30	.094
April	398	40	.100
May	482	59	.122
June	524	62	.118
July	510	51	.100
August	455	30	.066
September	361	29	.080
October	266	37	.139
November	164	23	.140
December	138	19	.138
Boston, Massachusetts (1944–1963); 42°21′N; elevation 300 ft			
January	133	36	.271
February	202	17	.084
March	299	43	.144
April	374	47	.126
May	457	74	.162
June	502	72	.143
July	498	52	.104
August	425	30	.071
September	342	35	.102
October	246	38	.155
November	144	20	.139
December	121	21	.174

Table 3–1 (Continued)

Month	Most Probable Monthly Mean (langleys per day)	Standard Deviation(σ)	Coefficient of Variation
East Lansing, Michigan (1943–1957); 42°43′N; elevation 868 ft			
January	112	33	.295
February	188	38	.202
March	276	53	.192
April	346	72	.208
May	417	67	.160
June	490	60	.122
July	487	50	.103
August	415	35	.084
September	322	43	.133
October	230	36	.157
November	117	20	.171
December	91	17	.187
Indianapolis, Indiana, 39°44′N			
January	151	28	.185
February	225	52	.231
March	317	50	.158
April	406	45	.111
May	507	53	.104
June	565	47	.083
July	546	47	.086
August	494	23	.047
September	410	42	.102
October	291	49	.168
November	177	16	.090
December	131	36	.275
Nashville, Tennessee (1942–1965); 36°07′N; elevation 602 ft			
January	157	32	.204
February	233	46	.197
March	322	51	.158
April	445	41	.092
May	517	59	.114
June	552	58	.105
July	533	53	.100
August	484	45	.093
September	407	62	.152
October	322	40	.124
November	206	34	.165
December	148	34	.230

Table 3–1 (Continued)

Month	Most Probable Monthly Mean (langleys per day)	Standard Deviation(σ)	Coefficient of Variation
Atlanta, Georgia (1952–1965); 33°39′N; elevation 1010 ft			
January	228	25	.110
February	288	50	.174
March	386	50	.129
April	488	48	.098
May	558	52	.093
June	552	62	.112
July	535	51	.095
August	506	59	.117
September	414	54	.130
October	360	58	.161
November	266	21	.079
December	215	19	.088
Charleston, South Carolina (1951–1963); 32°47′N; elevation 9 ft			
January	250	20	.080
February	314	26	.080
March	401	25	.062
April	525	41	.078
May	560	46	.082
June	557	60	.108
July	533	46	.086
August	502	50	.100
September	420	40	.095
October	355	49	.138
November	292	26	.089
December	230	36	.156

STREAMFLOW

Runoff along the Course of the Stream

The size of a river is a dominant factor in stream sanitation. Size at a specific time and location in the river basin is comprised of two elements—tributary drainage area and the amount of water flowing past this location in a unit time interval at the specified period in chronological time, more commonly referred to as streamflow or runoff. The former is a fixed element, the latter a dynamic variable. Quantitative determination of runoff along the course of the river system is dependent on the availability of contour maps to determine the tributary drainage area

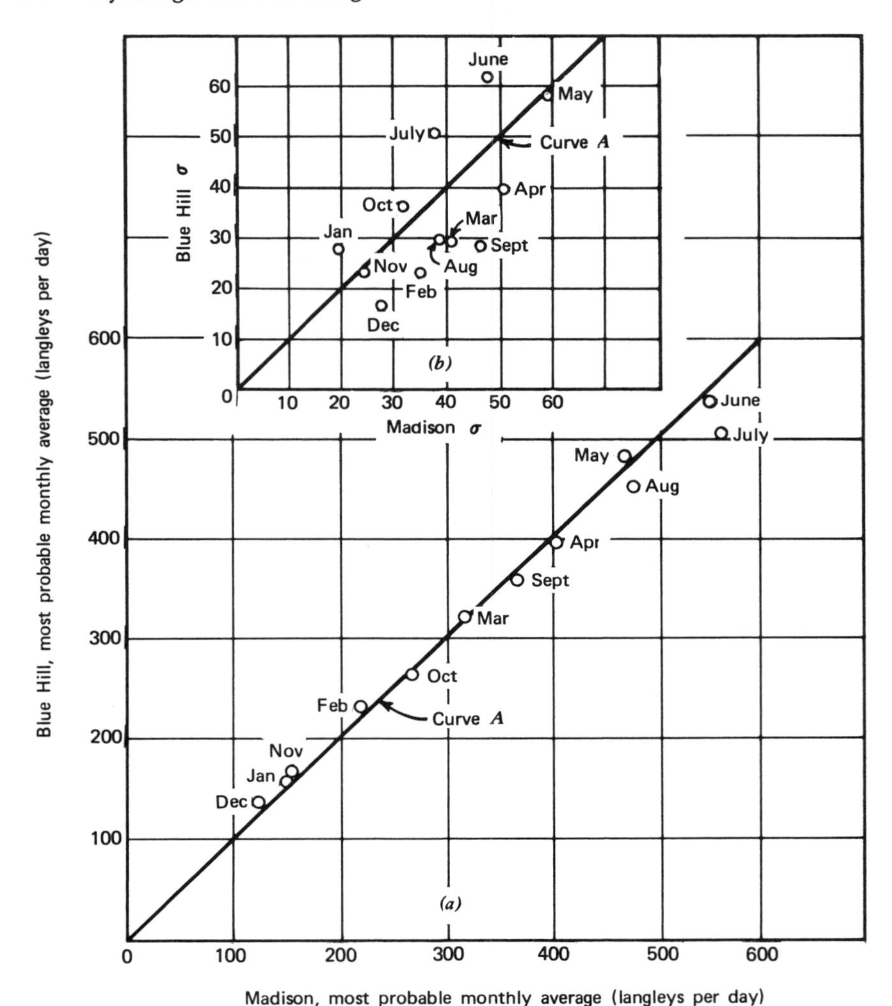

Figure 3–13 Solar radiation correlation: (*a*) most probable monthly average for Madison, Wisconsin, and Blue Hill, Massachusetts; (*b*) standard deviation.

and on records (over a period of years) of streamflow at a number of key locations in the basin. Some drainage basins are homogeneous in character, and the runoff along the course is directly proportional to the tributary drainage area. Runoff then is readily obtained as drainage area in square miles multiplied by the yield per square mile as measured at the key stream gaging station or as weighted yields at more than one station. Other drainage basins are composed of heterogeneous tributary areas of widely divergent yields, and estimating runoff is more complex. No phase of applied stream sanitation is more important, and more difficult, than determining runoff from location to location along the course of the stream.

Tributary Drainage Area

A convenient method of obtaining the drainage area tributary to any location along the course of the stream is a graph of cumulative area plotted against river miles, commencing in the headwaters and proceeding to the mouth. Tributary areas to official stream-gaging stations and other locations accurately determined by the U.S. Geological Survey, the Army Corps of Engineers, or other public and private agencies constitute control points. These are supplemented by planimetering delineated subareas from contour maps or aerial photos. Figure 3–15 illustrates such a cumulative drainage area graph developed for the upper French Broad River.

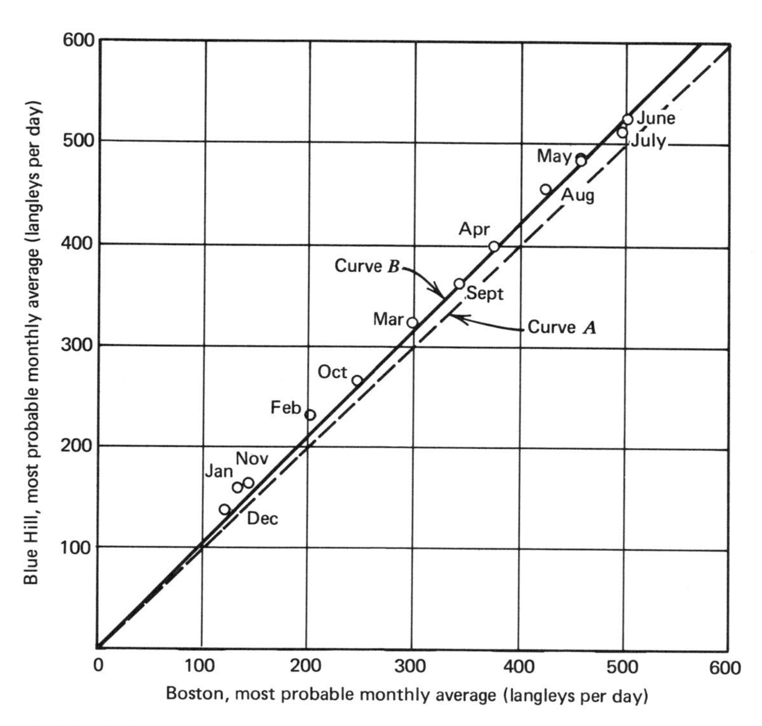

Figure 3–14 Solar radiation correlation for Blue Hill, Massachusetts, and Boston, showing the influence of air pollution.

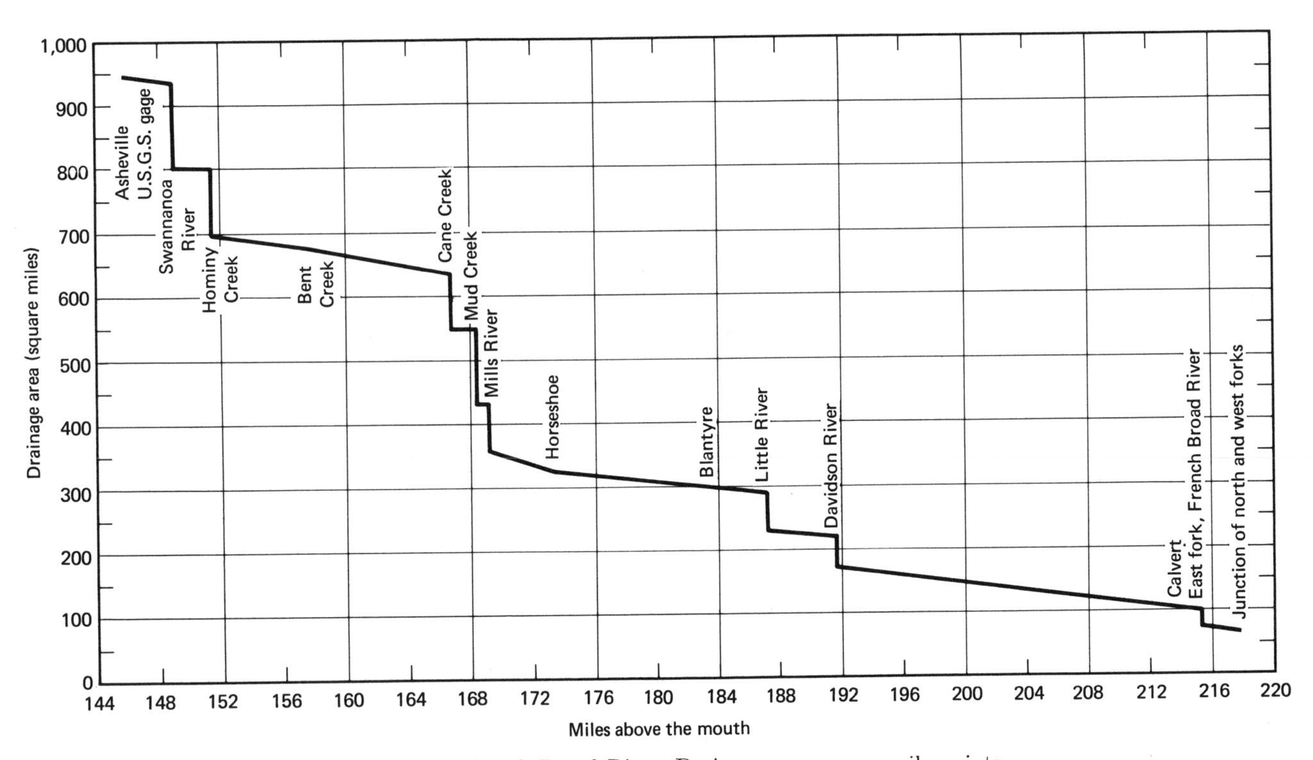

Figure 3–15 French Broad River. Drainage area versus mile points.

Stream Gaging and Runoff Records

The U.S. Geological Survey, in cooperation with federal, state, and private agencies, has primary responsibility for stream discharge measurements and the design of a national network. Coverage is extensive and in areas intensive, but in many basins inadequate. Length of records varies from a few years with gaps to excellent continuous long-term series. For statistical analysis and probability determination unbroken continuous long-term records provide the most reliable sample of hydrologic influences from which the nature of variation in streamflow can be defined.

Methods of stream gaging and instrumentation are presented in detail in publications of the U.S. Geological Survey and in hydrology texts. It suffices to indicate here the general method of measurement and the distinction in two types of gage. Automatic recorders provide a complete record of instantaneous variations in flow; manual gages are read on a staff or chain once or twice daily. In publishing the discharge records streamflow from these gage records is reduced to a daily average through the 24-hr period. Where natural runoff occurs uninfluenced by man-made river developments, variation through the hours of a day is slight and not much significant information is lost by averaging, except where the peak flood discharge is vital. However, where such river developments as navigation locks and particularly hydropower plants exist, hour-to-hour variations are often extremely great and the daily average as reported masks these important fluctuations. The evaluation of such diurnal fluctuations requires access to the continuous automatically recorded charts, obtainable on request from the U.S. Geological Survey.

The actual rate of stream discharge at a particular location is a function of the channel cross-sectional area and the mean velocity throughout the section ($Q = AV$) usually expressed in cubic feet per second (cfs). The stream gage indirectly measures Q by recording the elevation of the water surface behind some relatively stable control in the channel. In turn the gage height is related to actual measurements of mean velocity and section area taken at a location of stable channel below the control over a range of riverflows. From this a *rating curve* is developed for converting gage height (G.H.) to streamflow (Q). Figure 3–16 is a typical rating curve for the Tittabawassee River at Midland, Michigan.

As routine practice the U.S. Geological Survey verifies and adjusts the rating curve, particularly after floods, which might disturb the control and thus affect the gage height corresponding to a specific stream discharge or which might alter the channel cross section where velocity

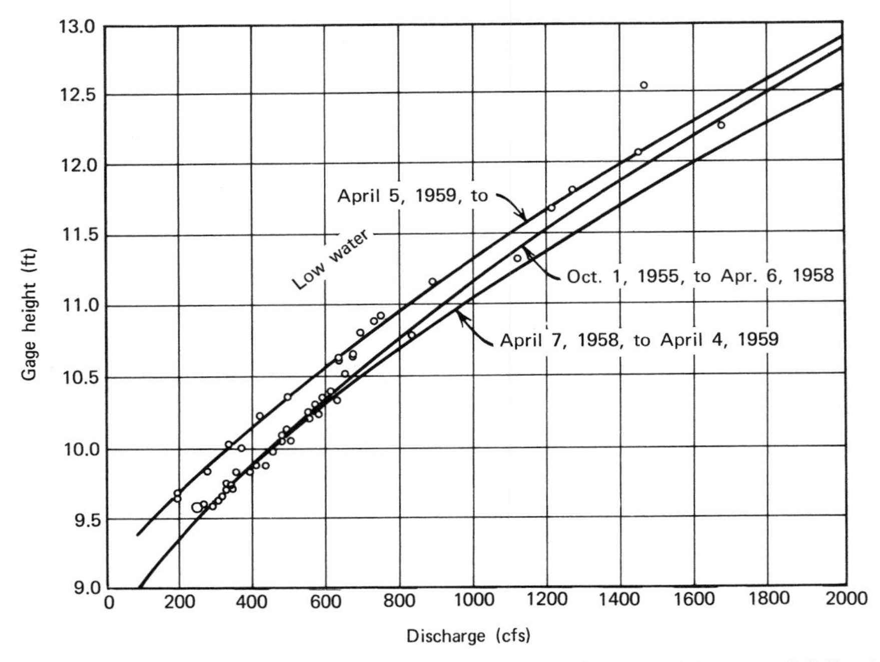

Figure 3–16 Discharge rating curve for the Tittabwassee River at Midland, Michigan.

measurements are made. Such adjustments in rating curve are usually indicated in publication of the stream-discharge records.

The records of stream gaging are published by the U.S. Geological Survey in *Water Supply Papers* by year. Early records are reported on the basis of the calendar year; since 1914 runoff is reported on the basis of the *water year* ending September 30. From time to time compilations of monthly average discharge have been published, including any revisions. Commencing with the water year ending September 30, 1961, the annual reporting of daily discharge records ceased; since 1961, a 5-year series has been used. To meet interim requirements local offices will issue reports annually, state by state. However, selected records can be obtained as punched cards or magnetic tapes for use with computers.

With carefully determined tributary drainage areas and a well-located key stream gage (or gages) representative of sub-basin runoff yield characteristics, it is possible to estimate the runoff at any desired location on tributaries or on the main stream. As an aid in selection of the appropriate runoff yield to apply at any river location, a transparent print of stream-gage locations superimposed on basin maps of physiographic regions, land forms, vegetative cover, geologic and groundwater formation, rainfall distribution, and so on, affords ready visual interpretation of interrelation of hydrologic elements. Estimating runoff along the course of a stream is as much an art as it is a science and depends on intimate knowledge of the drainage basin, tempered by experience and professional judgment.

Quantitative Characterization of Runoff

In a casual examination of the records of several years of daily average stream discharge at a gaging station the first impression is that day-to-day or season-to-season variation is too erratic to permit orderly characterization. It is difficult to convince the public, policy makers, and professional groups that, although highly dynamic, stream variation is actually quite orderly and can be defined in useful statistical terms. Skepticism in acceptance of statistical definition often leads to arbitrary policy decisions without benefit of a frame of reference of expected probability of occurrence or risks involved. Graphical statistical methods (Appendix A) are usually more readily accepted than conventional analytical methods. In this section we deal with a number of useful graphical characterizations of runoff, each of which reflects different elements of the hydrologic setting.

Monthly Hydrograph and Seasonal Pattern

Plotting the monthly average discharges through the year provides a monthly hydrograph. All elements of the hydrologic setting influence the monthly hydrograph, but principally it reflects climate, the geographic and seasonal distribution of precipitation, temperature, evaporation, and transpiration. From statistical analyses of long-term monthly average runoff records a seasonal pattern emerges, such as illustrated in Figure 3–17 for the Tittabawassee River at Midland, Michigan. In developing the seasonal pattern each month of the year is analyzed independently; the variation in monthly average quite generally is logarithmically normal and is readily defined by plotting on log-normal probability paper (Appendix A) from which the most probable monthly average and any confidence range, such as that expected for 80 percent of the years can be obtained.

Since the probability distributions are logarithmically normal, the seasonal pattern is plotted as a histogram on semilog paper, as shown on Figure 3–17. Curve *A* is the most probable or normal monthly average expected, and the shaded belt about the most probable is the range within which monthly average discharge for specific months of the year can be expected for 80 percent of the years; for 10 percent of the years the monthly average can be expected to be above the upper limit of the shaded belt, and for 10 percent of the years it can be expected to be below the lower limit. The seasonal pattern for any specific year may deviate from the normal pattern, since not all months through a year will experience runoff with the same probability of occurrence; it can be expected, however, that 80 percent of the yearly patterns will fall within the frame of the shaded belt. Curve *B* shows the monthly hydrograph for the specific year 1941, generally recognized as a year of severe drought in this region.

Referring to curve *A* of Figure 3–17, high runoff normally occurs in the spring with melting snow and rainfall; low runoff occurs in the

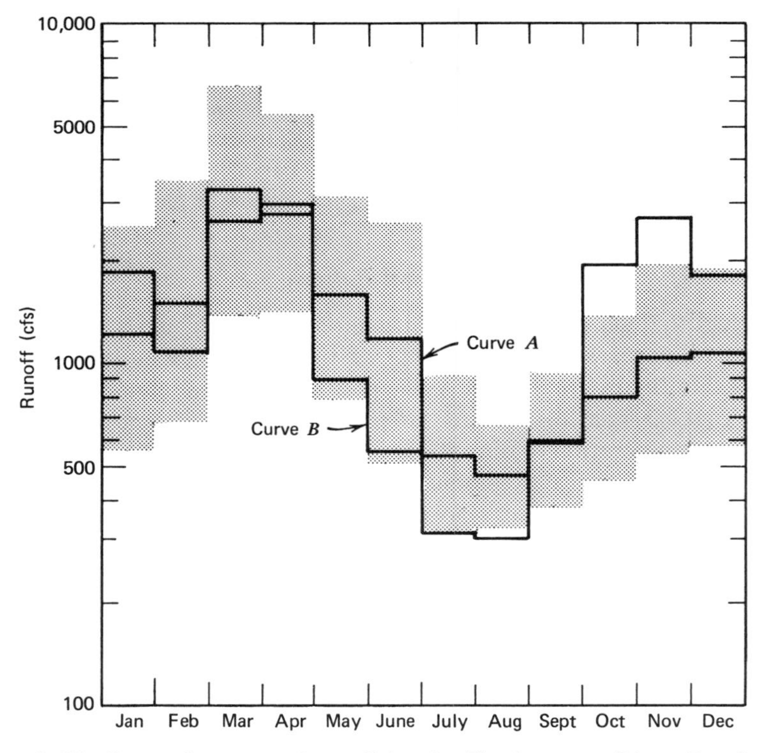

Figure 3–17 Seasonal pattern of runoff for the Tittabawassee River. Based on the U.S. Geological Survey gage Midland, Michigan, adjusted for diversion by the Dow Chemical Company. Curve *A*: most probable monthly average; curve *B*: recorded monthly averages (adjusted) for the severe drought year 1941; shaded area: range within which monthly average can be expected for 80 percent of years.

summer–fall season when precipitation losses are high, associated with heavy evaporation and transpiration; as temperature declines, precipitation losses decline; as soil storage is replenished, runoff increases.

The seasonal pattern is quantitatively defined by the most probable, and the expected variation about the normal constitutes a significant characterization of streamflow for comparisons among different areas within a basin and particularly among different basins. With a reasonable length of record a rather well-defined seasonal pattern can always be extrapolated, but, as is expected from the wide differences in hydrologic setting among basins, each has its unique seasonal runoff characterization. A few examples illustrate this individuality.

Figure 3–18 is the seasonal pattern of runoff developed for the Chowan River in the Atlantic Coastal Plain, in contrast with the pattern for the Tittabawassee River in Michigan (Figure 3–17). The tributary drainage areas are nearly the same size, 2465 square miles for the Chowan and 2400 square miles for the Tittabawassee; the long-term average runoff for the periods of record does not differ radically—2150 and 1540 cfs, respectively. Likewise, the most probable seasonal lows are practically the same (472 cfs for the Tittabawassee and 470 cfs for the Chowan), except that the low for the Chowan occurs later in the season. If seasonal patterns were limited to these characterizations, both rivers might be considered (erroneously) reasonably similar. However, a very important element of characterization so far has not been considered—namely, the expected pattern of *deviation* from normal.

The shaded belts about the most probable in each case represent the range within which a monthly average can be expected for 80 percent of the years. In comparing these ranges similarity between the two basins vanishes and a dramatic difference is evident. For the Tittabawassee the seasonal low is August, with a most probable monthly average of 472 cfs; the 80-percent confidence range is 320 to 680 cfs. For the Chowan the seasonal low is October, with a most probable monthly average of 470 cfs, but the 80-percent confidence range extends from 70 to 3100 cfs. The critical warm-season low in the Tittabawassee is much more stable, with a narrow range in deviation; on the average it can be expected that the August monthly average will not decline below 320 cfs more often than once in 10 years. The instability of runoff in the warm season for the Chowan is marked by occasionally expected extreme low and extreme high monthly average. On the average once in 10 years it can be expected that for the Chowan the monthly average of October will fall below 70 cfs and that of September below 60 cfs. However, although subject to risks of such extreme lows, it is also evident that in many years the warm-weather runoff of the Chowan will be

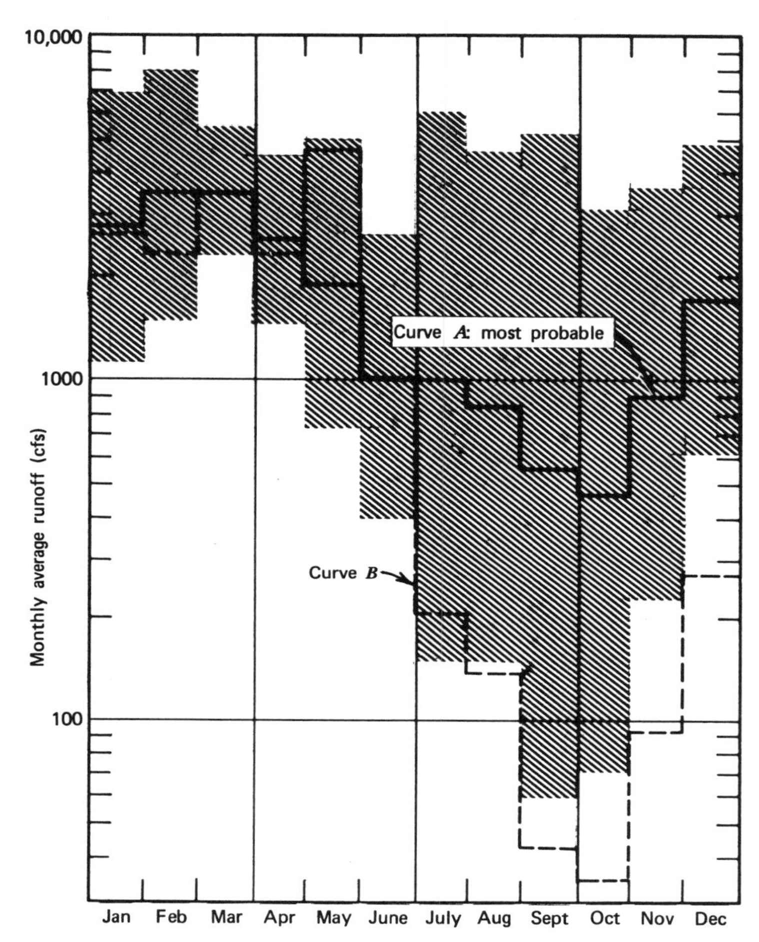

Figure 3–18 Chowan River—seasonal pattern of runoff (junction of Blackwater and Nottoway). Curve *A*: most probable monthly mean; curve *B*: monthly means for 1954; shaded area: range within which monthly mean can be expected for 80 percent of years.

very high, well above that even rarely expected in the Tittabawassee, and at a frequency of once in 10 years runoff will exceed 3100 cfs for October and 5200 cfs for September.

A river such as the Chowan with its severe droughts entails great risks in relying on natural streamflow. The value of regulating such a river to decrease the risks and ensure a much higher sustained runoff is obvious. The point to emphasize here is the importance of defining the degree of variation from normal in the development of a seasonal pattern.

In comparing seasonal patterns a useful transformation is to convert runoff in cubic feet per second to yield cubic feet per second per square mile of tributary drainage area, thus providing direct comparability for records of runoff from drainage areas of different size. There is a limit to this transformation because the size of the drainage area does affect yield. Within limits of reasonable size converting to yield permits superimposing one seasonal pattern on another for direct comparison.

Figure 3–19 represents the seasonal pattern for the Big Manistee River, based upon the gaging record at Sherman, Michigan. The logarithmic scale at the left ordinate expresses the monthly average runoff in yield; the scale at the right is the corresponding runoff in cubic feet per second. Of interest is the remarkable stability of the seasonal pattern with re-

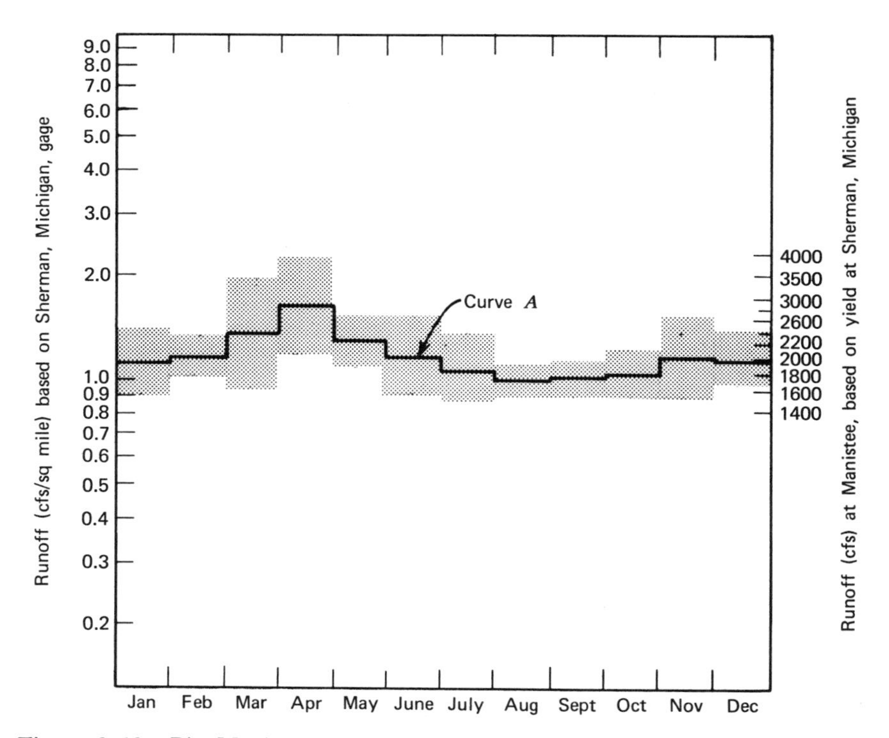

Figure 3–19 Big Manistee River. Seasonal pattern of runoff, based on record at Sherman, Michigan, 1933–1954. Curve *A*: most probable monthly mean for specific months of the year; shaded area: range within which monthly means can be expected for 90 percent of years.

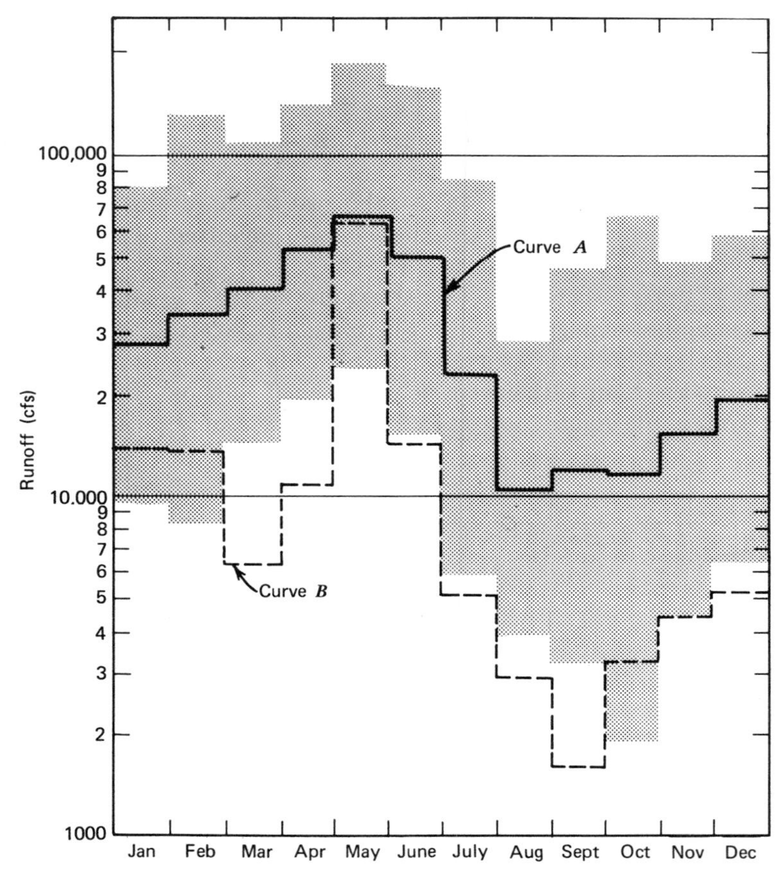

Figure 3–20 Seasonal pattern of runoff for the Arkansas River at Little Rock (based on record for 1934–1955). Curve A: most probable monthly mean for specific months; curve B: recorded monthly means for 1954; shaded area: range within which monthly means can be expected for 80 percent of years.

spect not only to the narrow range in deviation from the most probable but also to the unusually high sustained yield throughout the seasons of the year. In this instance physiographic characteristics of sandy soil, vegetative cover, and natural storage overshadow climatologic factors.

Figure 3–20 represents a seasonal pattern for a large nonhomogeneous drainage basin, the Arkansas River at Little Rock. Curve A is the most probable monthly average discharge, considering each month of the year independently; the shaded belt is the range within which the monthly average can be expected to fall for 80 percent of the years. At this location in the basin wide variation in monthly average can be expected throughout all seasons of the year.

The Daily Hydrograph

Plotting the daily average discharges through the year establishes a *daily hydrograph.* Again, the daily hydrograph is affected by all elements of the hydrologic setting, but it reflects primarily the flashy or stable character of streamflow, principally the physiographic elements—shape of the drainage area, rugged or flat topography, geology, soil, and natural storage in lakes, ponds, and swamps. Steep slopes of impervious soil formation with lack of natural storage give rise to a flashy stream with a daily hydrograph punctuated by sudden rises and rapid declines with each occurrence of rainfall. Intervening periods between rains are usually periods of extremely poor yield. Such a flashy stream is illustrated by the daily hydrograph of discharge of the James River at the Lick Run gage and of a tributary, the Jackson River, at the Falling Springs gage (Figure 3–21).

The converse is reflected in the daily hydrographs of discharge of the Big Manistee River (Figure 3–22), at the upstream (Sherman) and downstream (Manistee) gages. The stability of the Big Manistee is also reflected in the monthly hydrograph as shown in Figure 3–19. A stream with a stable hydrograph is an asset not only because its flow is stable but also because its yield is unusually high.

In addition to physiography, the daily hydrograph also reflects to some extent man-made influences, such as operation of hydroelectric plants. This is evident in the daily hydrograph of the Big Manistee at the Manistee gage (Figure 3–22) located below two hydroelectric plants. A weekly cycle is evident, particularly through the summer–fall season, with high daily average flow during weekdays and low daily average flow from 1 to 3 days during the weekend offpeak power demand. The weekly cycle at the Manistee gage contrasts with the steady hydrograph of the Sherman gage, located above the hydroelectric plants. The daily average discharge at Manistee on weekends is below the natural drought yield. The impact of such river developments is dealt with more fully in Chapter 9.

The Instantaneous Hydrograph

Whereas the daily hydrograph reflects to some extent man-made river developments, the *instantaneous hydrograph* presents the full impact on streamflow. The 24-hr average discharge, as reported in *Water Supply Papers,* does not reflect moment-to-moment variations in runoff, and, as already mentioned, these variations are detected only at gages with

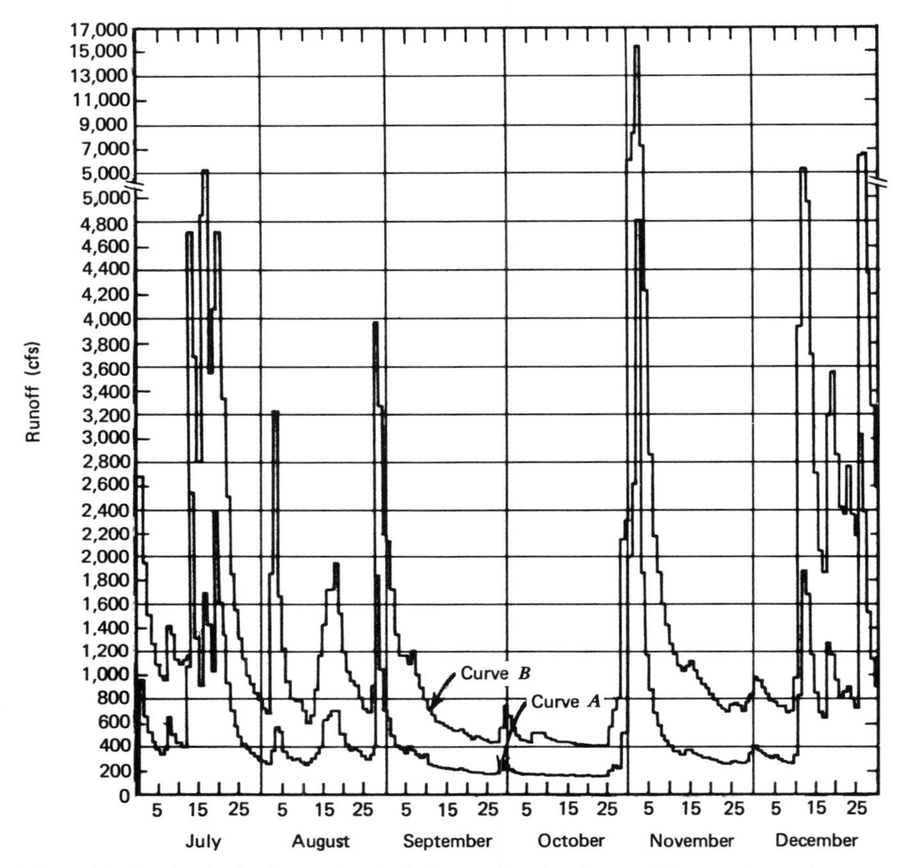

Figure 3–21 Daily hydrograph of discharge for the James River and a tributary, the Jackson River, for 1949. Curve *A* (shaded): Jackson River, Falling Springs gage (drainage area 409 square miles); curve *B*: James River, Lick Run gage (drainage area 1369 square miles).

continuous automatic recorders. Figure 3–23 represents the instantaneous hydrograph for the period July 15–27, 1953, for the Roanoke River at the Roanoke Rapids gage, which is located below hydroelectric power installations. The diurnal cycle during weekdays and the weekend shutdown are typical hydroelectric power operating practice in meeting the peak demands for electrical energy. The amplitude of the pulsation and extent of the weekend shutdown varies with the peak energy demand. On the Roanoke, in a matter of 3 hr on July 16, the flow increased from 2500 to 10,500 cfs and then gradually declined to the diurnal low

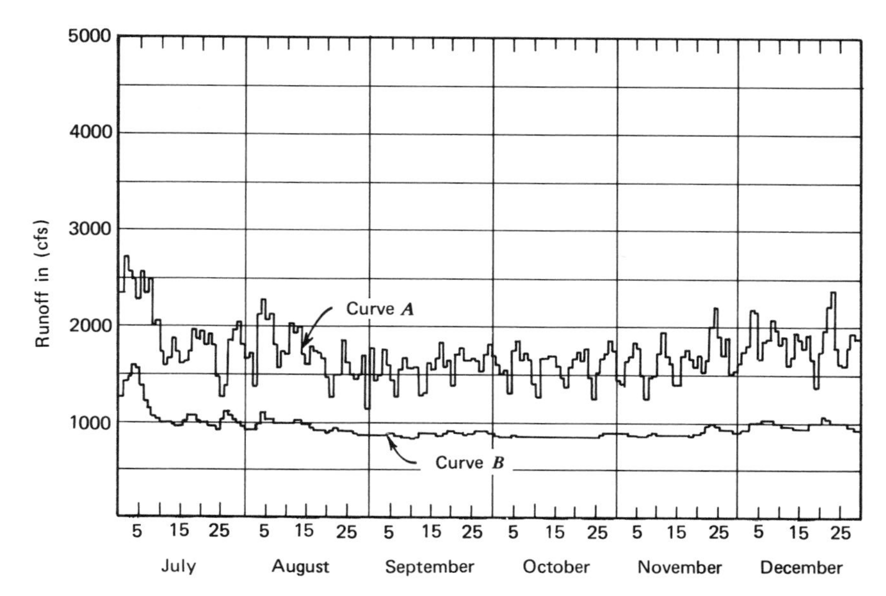

Figure 3–22 Big Manistee River—daily hydrograph for 1953. Curve *A*: daily average discharge at the Manistee gage (drainage area 1780 square miles); curve *B*: daily average discharge at the Sherman gage (drainage area 900 square miles).

again; the weekend peaking was reduced on Saturday, followed by 2 full days of low discharge at rates between 1000 and 2000 cfs.

Such diurnal cycles and weekend shutdowns persist as pulsation in stream discharge for miles downstream. Based on reports by the Army Corps of Engineers, Figure 3–24 represents typical fluctuations in river stage induced by hydropower practice on the Pee Dee River. The impulses induced by the Blewett Falls hydroelectric plant located at mile 195 are traced at three locations along the lower Pee Dee: curve *A* at Cheraw, South Carolina (mile 172.5); curve *B* at Pee Dee, South Carolina (mile 107.2); and curve *C* at Godfrey's Ferry (mile 74.5). The weekend shutdown at Blewett Falls induced a 6-ft decline in river stage at Cheraw 22.5 miles downstream, a 5-ft drop at Pee Dee 87.8 miles downstream, and a 3.5-ft drop at Godfrey's Ferry 120.5 miles downstream. From curve *A* it is noted that the weekend shutdown at Blewett Falls is reflected in a decline in stage at Cheraw commencing at 5 A.M. on Saturday, with the low stage at 7 P.M. on Monday; a sharp rise then ensued, reaching a peak at 5 A.M. on Tuesday; river stage then remained substantially above the weekend low with diurnal cycles until the next weekend shutdown. At Pee Dee, 87.8 miles below

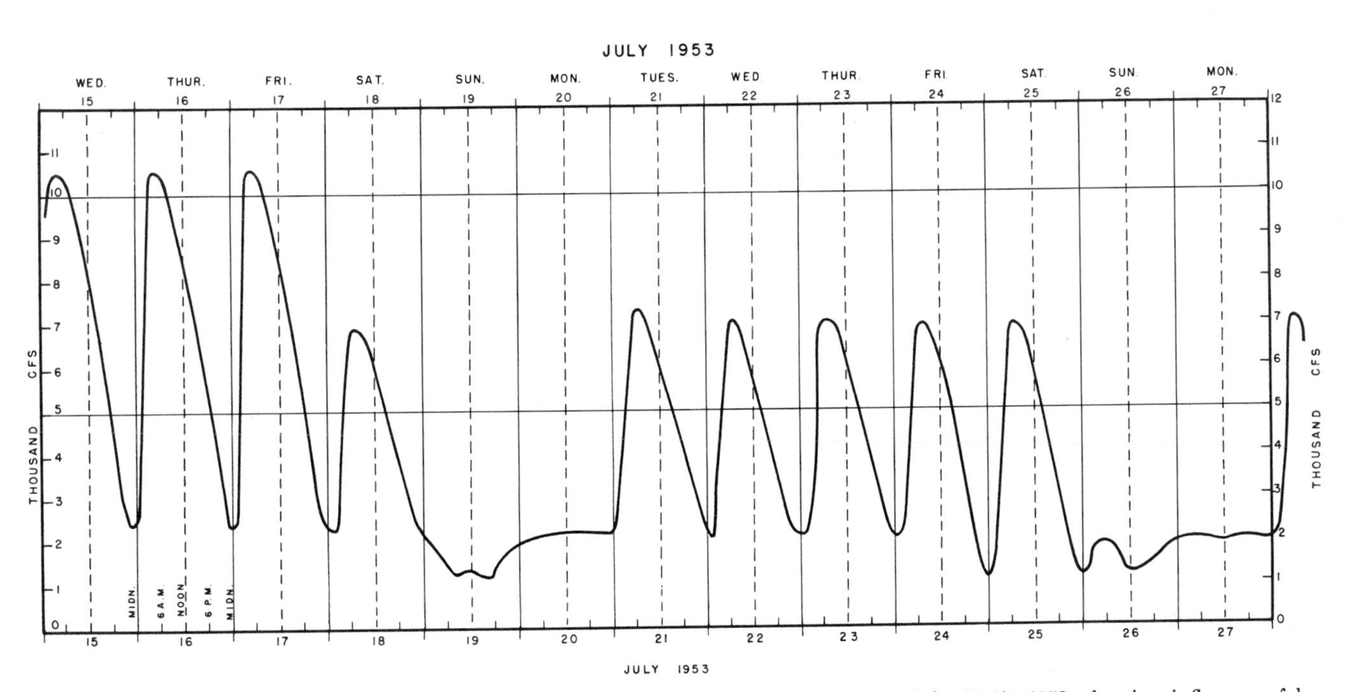

Figure 3–23 Roanoke River. Instantaneous hydrograph at the Roanoke Rapids gage, July 15–27, 1953, showing influence of hydroelectric power operation at John H. Kerr and Virginia Electric Power plants.

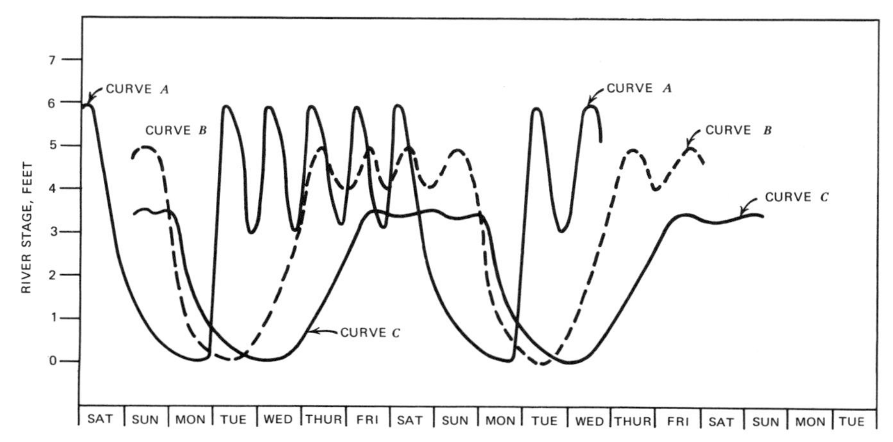

Figure 3–24 Pee Dee River. Typical flunctation in river stage induced by hydroelectric power plants during ordinary low runoff. Curve *A*: river stage at Cheraw, South Carolina (mile 172.5); curve *B*: river stage at Pee Dee, South Carolina (mile 107.2); curve *C*: river stage at Godfrey's Ferry (mile 74.5).

the hydroelectric plant, the decline in stage commences about noon on Sunday, reaches a low at about noon on Tuesday, and then rises to a high at about 4 A.M. on Thursday. Diurnal fluctuation during peaking on weekdays is foreshortened but is evident, although not pronounced. At Godfrey's Ferry, 120.5 miles downstream, the lag in weekend shutdown increases with the decline in stage commencing about midnight on Sunday, reaching a low at about 8 A.M. on Wednesday, then rises to a high at about noon on Friday; diurnal pulsation is practically damped out. Based on the rating curve, the 5-ft drop in stage at the Pee Dee gage associated with the weekend shutdown at Blewett Falls represents a decline in stream discharge of the magnitude of 3500 cfs.

It is evident that diurnal pulsation and weekend shutdown at hydropower plants is reflected along the course of a river for many miles downstream. The extent and degree of influence depends on channel characteristics. Weekend shutdowns in some locations have been observed to be drastic, with little or no flow released for as long as 3 days. The impact of such practices on stream sanitation is discussed more fully in Chapter 9.

Extremes in Streamflow

In addition to the quantitative characterizations of runoff reflected in the monthly, daily, and instantaneous hydrographs, the two extremes, floods and droughts, require further definition. The literature on floods is extensive; interest here is limited to problems of stage that relate to community and industrial plant location and to magnitude of flow as it affects channel scour. Relation of frequency of occurrence and severity of floodflow is best defined in statistical terms by the theory of extreme values. In terms of river stage, in defining the probabilities of flooding low-lying areas, long-term automatic stage records are required; from these the instantaneous peak stage for each year can be selected. In terms of streamflow, the maximum daily average discharge or the instantaneous peak discharge is selected from each year of record. Figure A–9 (Appendix A) shows the distribution of a 19-year record of annual instantaneous peak flood discharges of the Clark Fork River below Missoula, Montana. Since not all gaging stations are equipped with automatic recorders, the floodflow series is most commonly based on the maximum daily average discharge, which usually ranges from 10 to 15 percent below the instantaneous peak.

In stream sanitation interest in extremes centers principally on the minimum discharge, or the drought flow. We shall therefore deal with this in some detail.

Drought Flow Characteristics

Where urban and industrial development relies on the natural unregulated stream runoff, the character of the drought flow is the restricting element in growth potential. Any level of development carries a risk of unsatisfactory stream condition related to the probability of drought severity. Decisions as to the level of community and industrial development, the required degree of wastewater treatment, and downstream water quality objectives rest on the definition of drought flow expectancy.

In order to portray the full implication of drought flow it is desirable at this point to introduce some statistical concepts, which are more fully developed in Appendix A.

Adaptation of Gumbel Theory of Extreme Values

A fundamental approach to the statistical evaluation of the drought flow characteristics of streams stems from the theory of extreme values. It is well known that extremes of hydrologic phenomena, such as floods and droughts, do not follow a normal symmetrical distribution but are skewed. Gumbel [1], in analyzing floods, developed a *standard skewed distribution* based on the theory of largest values, which, when reduced to a straight-line form on extreme probability paper, as suggested by Powell [2], simplifies and facilitates the solution of many statistical problems dealing with series of extreme values.

The extreme probability grid initially developed for floodflows is adapted to droughts by the simple technique of dealing with values in their order of *severity* rather than in order of magnitude. For floodflows arranging in order of severity involves going from a low flow to a high flow, whereas for droughts it involves going from a high to a low flow. This places the most severe flood and the most severe drought in the same position in terms of the probability of severity. Thus the extreme-probability grid is equally adaptable to minima as well as maxima series arranged in order of severity.

In adapting the Gumbel theory of extreme values to drought flows certain modifying factors are taken into account. Three time elements are involved in the definition of drought flow:

1. The base unit of time from which a low flow is selected from the record.
2. The length of time over which a low flow is averaged.
3. The season in which the selection is made.

The theory is predicated on a base unit of time such that any extreme value selected is one of a large number of observations and such that extreme values selected from consecutive time units are independent of each other. It is common practice in dealing with streamflow data to employ the year as the base time unit. Since runoff records are reported as daily average discharge, this unit provides an extreme value as one most severe among 365 measurements, thereby meeting one part of the basic criteria.

In dealing with floods the one largest value selected in each year is practically independent of those selected from each succeeding year, and therefore both criteria are satisfied in that case. However, in dealing with droughts the base unit of one year does not always form a series of extreme values that in successive years are completely independent of each other. The time-lag factor in hydrologic cycle influencing drought flows may extend in some drainage basins over 2 or more years. Consequently, to obtain a true series of completely independent minima the base time unit should perhaps be greater than 1 year. However, considering the relatively short records available, practical considerations dictate the adoption of the base time unit as 1 year, recognizing that on this basis some of the less severe flows selected in the series in a true sense are not droughts. As will be shown later, this characteristic is reflected in the plottings of distributions, and a method is proposed for dealing with it.

The second time element, the period over which a low flow is averaged, may be taken on a calendar basis of a day, week, or month, or preferably on the basis of a consecutive period without regard to the calendar. For refined analysis it is desirable to select four separate drought flows—the minimum daily average and the minimum consecutive 7-, 15-, and 30-day averages. For many practical problems it is adequate to employ the minimum daily and minimum monthly averages and the minimum consecutive 7-day average; the former two are directly available in *Water Supply Papers,* and the latter can be readily located in the record. These provide a means for interpolating intermediate averages, as illustrated later.

The third time element, the seasonal factor, distinguishes between warm and cold weather stream discharge. Low flow associated with severe winter freezing and ice cover is a distinctly different extreme value from the low flow associated with the warm-weather drought season. For water resource uses, such as waste assimilation, irrigation, and recreation, the combination of high temperature and low streamflow is critical. While retaining the base time unit of the year, the minimum flows are selected from the warm weather *summer–fall* season. For northern latitudes this may be taken as June through October; for southern latitudes, May through November.

The theory of extreme values, in the development of Gumbel's standard distribution, is also based on the concept of unlimited distribution. In dealing with floods in ascending magnitude this criterion is met, but in dealing with droughts some streams actually go dry or approach zero discharge as a limit. To meet such situations Gumbel has proposed *logarithmic* extremal probability paper (Appendix A) whereon the data are plotted as the logarithms of the drought flows and are thus unlimited at the low end of the distribution. Experience has shown that better fits are obtained with this modification, and hence logarithmic extremal probability paper is usually adopted.

Some drought streamflow records are compounded, a combination of a variable factor and a constant due to natural or man-made influences, such as streams fed by abundant and stable groundwater, large natural lake and swamp areas, or artificial reservoir storage. Drought flow data from such streams produce distorted distributions on log-extremal probability paper, with curvilinear plots approaching a limit or base flow above that expected from the variable increment alone. In some instances where the natural flow is curtailed or withheld, such as may be associated with some hydroelectric power practices, the distribution of droughts may fall below what is normally expected in accordance with extremal theory. Both of these situations are dealt with by plotting the drought flows as recorded discharges and as adjusted drought flows, subtracting or adding a constant increment representing a base flow or an artificial retention. The constant is determined graphically by successive approximations until a straight line is obtained from the adjusted record. Drought flow severities for desired probabilities of occurrence are then obtained from the straight-line fit and reconverted to terms of the original data by adding (or subtracting) the constant base flow.

In the application of the theory of extreme values, preference has been given to graphical methods rather than rigid analytical procedures. In addition to a complete visual presentation of the data, graphical methods afford the opportunity to exercise professional judgement. Values in the series that are distinctly out of line are readily identified, and in fitting the distribution weights can be given to different segments of the series. For example, it is inherent in the adoption of a base time unit of a year that the less severe values selected are likely not to be droughts in the sense of independent extreme values. In the graphical fit of distributions greater weight is given to the more severe values to the right of the mode, which are more likely to reflect the true drought characteristics of the drainage basin.

Examples of Log-Extremal Probability and Distorted Distributions. A number of examples illustrate these characteristics encountered in drought flow analysis. Figure 3–27*a* shows the plot of the minimum daily average drought series for the 21-year stream discharge record of Battle Creek at the Battle Creek gage. Arranging the series in order of severity of drought and assigning plotting positions in accordance with Gumbel's refinement (see Appendix A) provides the coordinates for plotting on logarithmic extremal probability paper. A reasonably good straight line develops through the majority of the data points without recourse to adjusting the flows for a base-flow constant. The two least severe values, 95 and 88 cfs, are somewhat out of line, above the distribution. The less severe values to the left of the mode are likely to portray this type of deviation because selection of drought flows are made on a base time unit of only 1 year. Such deviations to the left of the mode are included in deriving plotting positions for the series but are ignored or given less weight in fitting the distribution line graphically. The minimum daily average drought severities expected for any probability are readily obtained from the distribution line. Since the return period is in units of the base-time interval of 1 year, the drought severities expected once in 20, 10, and 5 years are at the intersection of the 20-, 10-, and 5-year return period as 23, 28.5, and 35.5 cfs, respectively. The most probable minimum daily average drought expected (56 cfs) is at the intersection of the mode.

Where drought flow records are a combination of a variable factor and a constant due to natural or man-made influences, curvilinear probability plots develop. Where monthly change in reservoir contents are reported or rules of release are prescribed, the recorded streamflow can be adjusted to natural yields. However, unless evaporation is also taken into account, the adjusted drought flows may deviate substantially from true natural yields. A rational method of dealing with these situations is to employ a constant increment to produce adjusted drought flows that by successive approximations produce straight-line plots.

Figure 3–25*a* for Pine River illustrates a curtailment of minimum daily average flow by a hydroelectric power plant weekend shutdown; Figure 3–25*b* reflects moderate augmentation for the consecutive 7-day average drought flows. The unadjusted minimum daily average values tend to curve downward, indicating a curtailment of flow; by successive approximations of adding an increment to each recorded value a straight-line fit is achieved with the increment of 10 cfs. Under continuation of the pattern of operation reflected in the past recorded flow, an approximation of expected minimum daily average drought flow for any probability is then obtained from the straight-line fit from which is subtracted the 10-cfs adjustment increment. For example, the once-in-10-years value is 11 minus 10, or 1.0 cfs.

For the consecutive 7-day average drought flows weekend shutdowns are averaged out, and the weekday peaking practice to meet the higher power need results in recorded flows that are higher than expected. This is evidenced by the plotting of the recorded unadjusted values curving upward, as shown in Figure 3–25*b*. In successive approximations of subtracting an increment from each recorded value a straight-line fit is obtained for the 7-day averages by subtracting 25 cfs. The expected drought flows for any probability, assuming the operating pattern continues as reflected in the past, recorded flow, is approximated from the straight-line fit to which in this instance the adjustment increment is additive. For example, the expected 10-year 7-day average drought value is 57 cfs $(32 + 25)$.

As man-made river developments increase, more drought flow records are likely to be compounded. If the records cover a substantial period before and after operation of river developments, the record is divided into two parts; the portion prior to development is analyzed as a natural drought-flow sequence without adjustment; the portion compounded by river development is dealt with separately, as illustrated above. Comparisons are made between the two results and also with analyses of other gaging records of adjacent basins.

Figures 3–26*a* and *b* represent the drought flows of the Pike River

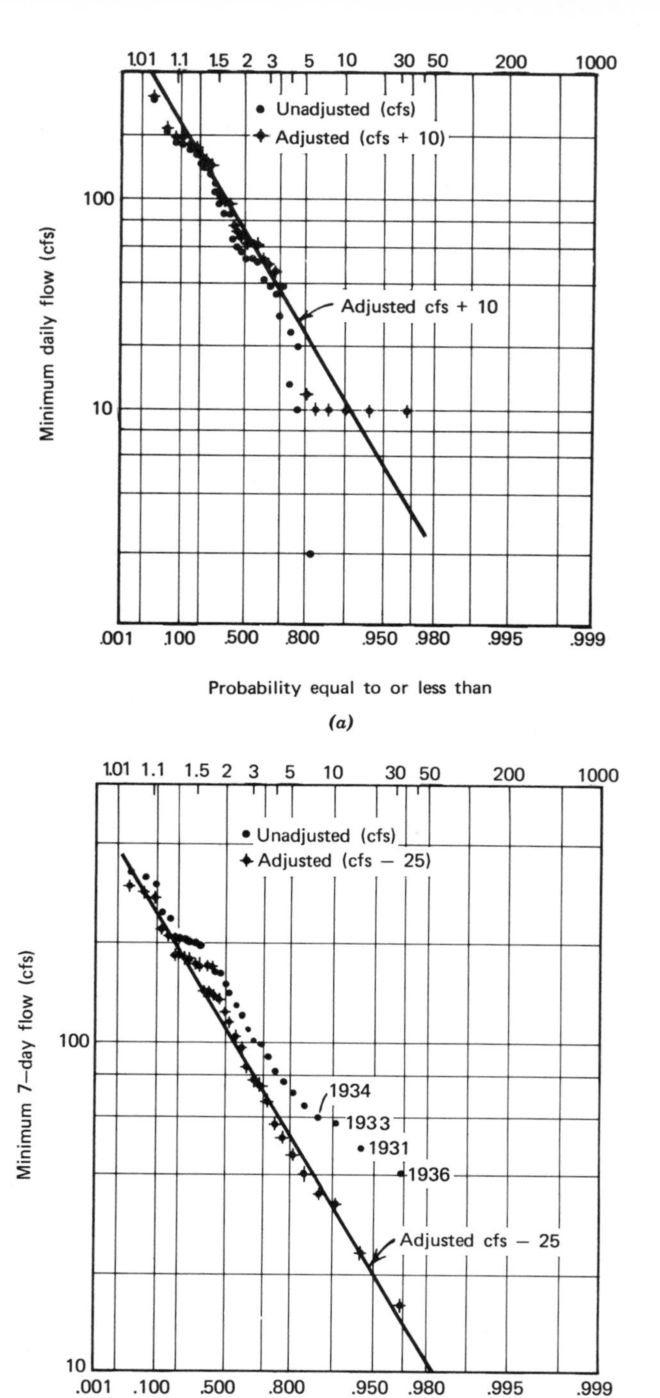

Figure 3–25 Pine River at the Pine River Powerplant near Florence, Wisconsin: (*a*) minimum daily flow (May through October) (5 unadjusted values 0.0 cfs); (*b*) minimum 7-day flow (May through October).

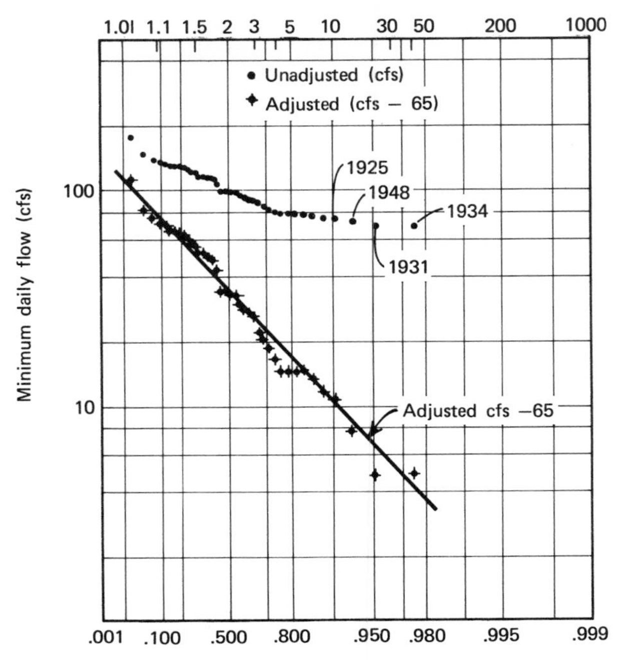

(a)

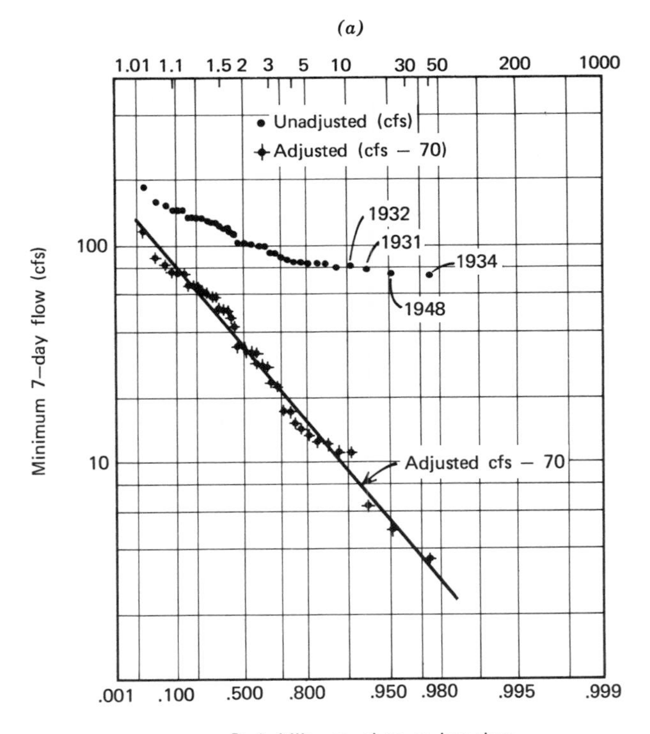

(b)

Figure 3–26 Pike River at Amberg, Wisconsin: (*a*) minimum daily flow (May through October); (*b*) minimum 7-day flow (May through October).

at the Amberg, Wisconsin, gage influenced by a natural base flow that augments the variable factor. The plots of the recorded values bend upward, reflecting the base flow. By successive approximations of subtracting increments from the recorded values, straight-line fits develop, with a deduction for base flow of 65 cfs for the daily average droughts and 70 cfs for the 7-day averages. The variable factor for any probability is obtained from the straight-line distributions to which is added the base flow. Thus the expected once-in-10-years minimum daily average flow is 76 cfs $(11 + 65)$ and the minimum 7-day average is 79 cfs $(9 + 70)$.

Representative Sample, Length of Record

A question that is certain to arise is, how long should a continuous record extend in years to warrant analysis by extremal theory? This is difficult to answer. The longer the record, the more reliable the sample of drought experience and the more reliable the estimates of expected severity at various probabilities of occurrence. Unlike floods, drought flow probabilities can tolerate a larger error with less serious consequence. Also decisions with respect to drought flow usually center around a severity expected on the average once in 10 years. In addition to the number of years of record is the element of degree of variability of drought flow from year to year. A short record of, say, 15 years for a relatively stable stream may be adequate; a record of 25 years may be necessary to ensure equal reliability where drought flows are known to be highly variable. However, where one is obliged to deal with short-term records, it is increasingly important that a rational method, such as the adaptation of Gumbel's extremal value theory, be employed in the analysis. (Other methods of analysis, devoted primarily to statistical curve fitting, do not afford an adequate basis for defining drought-flow characteristics.) Any record is better than no record; there are other approaches by which a short record can be related to other adjacent long records or synthetically generated.

As a further answer to the question of record length and reliability, an analysis is presented of the minimum daily average droughts of the long-term record of the French Broad River of Asheville, North Carolina, considered in four ways: the period 1903–1964, with the base unit of time as 1 year; the period 1903–1964, with the base unit of time as 2 years; the early 44-year period 1903–1946, with the base unit of time as 1 year; the late 18-year period 1947–1964, with the base unit of time as 1 year. The four sequences plotted on log-extremal probability paper indicate close agreement with the extremal value theory, and expected drought severities are tabulated for comparison in Table 3–2.

Table 3–2 Comparison of Drought Severities Expected from Long-, Intermediate-, and Short-Term Records—French Broad River, Asheville, North Carolina

	Minimum Daily Average Drought Severity Expected (cfs)			
Period of Record	Most Probable	Once in 5 Years	Once in 10 Years	Once in 20 Years
1903–1964 (1-yr base)	725	485	400	325
1903–1964 (2-yr base)		510	405	330
1903–1946 (1-yr base)	780	505	405	325
1947–1964 (1-yr base)	640	450	375	315

It may be seen from the results that applying the base time unit of 1 and 2 years to the long-term (62-year) record produces drought severities in close agreement. Any disagreement is likely to be at the left of the mode; hence, if distributions are fit to the data to the right of the mode, the time unit of 1 year is a reliable base. The results of the analysis of the 44-year record of intermediate length are also in close agreement with the 62-year record results. The late 18-year short-term record (1947–1964) includes 2 extremely low drought years, 1954 and 1956, with minimum daily averages of 252 and 350 cfs, respectively; these are the second and fourth most severe droughts experienced in the 62-year period 1903–1964. These values distort the distribution; hence this short record produces results somewhat more severe than that given by analyses of the longer record.

With short records it must be recognized that an undue proportion of less severe droughts may be enclosed in such a limited sample; on the other hand, the small sample may contain an undue proportion of more severe droughts. Short records should therefore be used with caution, and it is good practice also to investigate the longer records of nearby streams with similar hydrologic characteristics.

The Drought Flow Reference Frame

From the extreme probability distribution plots of drought flows a *reference frame* of expected drought flow severity is constructed; this constitutes a useful summary. As an example, reference is made to the analyses of the record of Battle Creek at the Battle Creek gage. Figure 3–27*a,b,c,d* shows the log-extremal probability plots of the minimum

1-, 7-, 15-, and 30-day averages from which the most probable, the severities expected once in 5, 10, and 20 years are derived. These are shown in Figure 3–28, in which the ordinate is drought severity and the abscissa is consecutive days over which the drought flow is averaged. Curve *A* represents the most probable drought expected as an average over any number of consecutive days from 1 to 30; similarly curves *B, C,* and *D* represent the drought severities expected on the average once in 5, 10, and 20 years, respectively.

As one shortens the period over which the discharge is averaged, drought severity increases. Conversely, extremely low flows do not prevail for extended periods. From the standpoint of pollution control, consecutive low flow over a period equal to the time of passage through the critical reaches of the stream is the significant interval. This will vary from stream to stream and may range from 3 days to 2 weeks or more, depending on the physical characteristics of the channel and the distribution of pollution loads along the course. Drought flow reference frames afford a convenient means for interpolating severities as an average for any number of consecutive days appropriate to the situation.

In analysis of many records throughout the United States the drought flow reference frames for natural undeveloped streams usually approach straight-line relations between 7- and 30-day averages, and in many instances the straight-line relation extends to the 1-day average. Where man-made river developments, such as hydropower plants, are located above the gaging station, generally the 1-day average is decidedly below the projections of the linear relation, as illustrated in Figure 3–29 for the Fox River at Rapide Croche Dam, Wisconsin. This is due to hydroelectric power plant weekend shutdown practices reflected in the minimum daily average flows; the degree of the deviation below the projection is an approximate measure of the extent to which such practice adversely affects the natural expected drought.

Each basin and sub-basin runoff produces its unique drought flow reference frame. For comparative study it is convenient to convert the drought flow in absolute terms to yield per square mile of drainage area.

In reaching decisions about pollution control requirements, water quality objectives, and levels of community and industrial development, it is obviously essential to take a position at some point within the drought flow reference frame. This position must be defined in terms of both coordinates, horizontally by the consecutive number of days over which flow will be averaged and vertically by the drought flow severity within the range from the most probable to, let us say, that

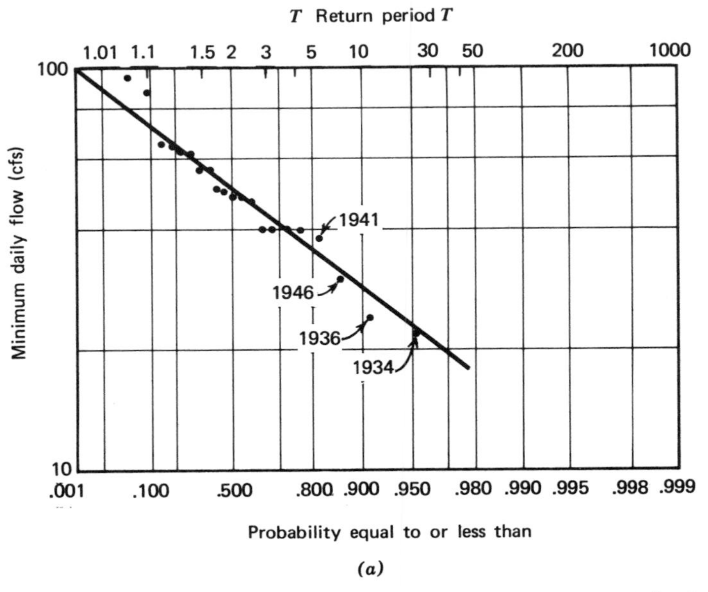

(a)

Figure 3–27*a* Battle Creek at Battle Creek, Michigan. Minimum daily flow (May through October).

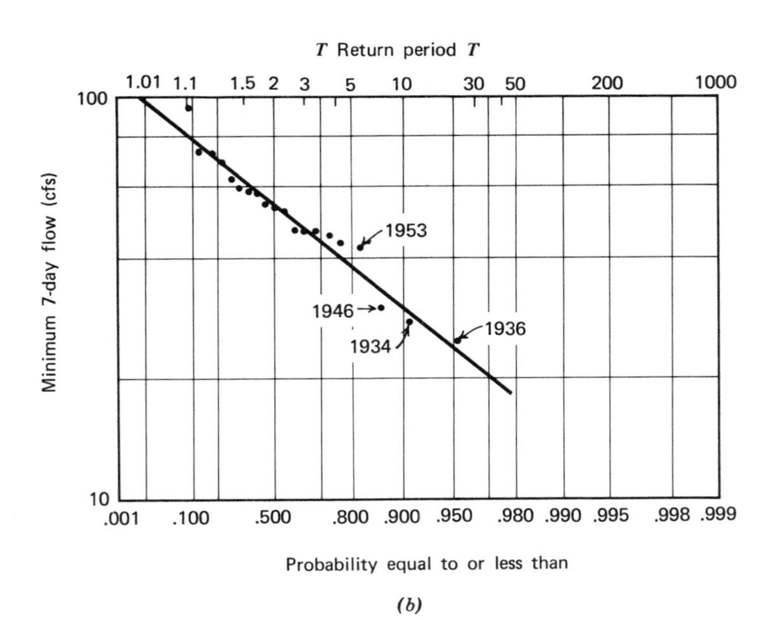

(b)

Figure 3–27*b* Battle Creek at Battle Creek, Michigan. Minimum 7-day flow (May through October).

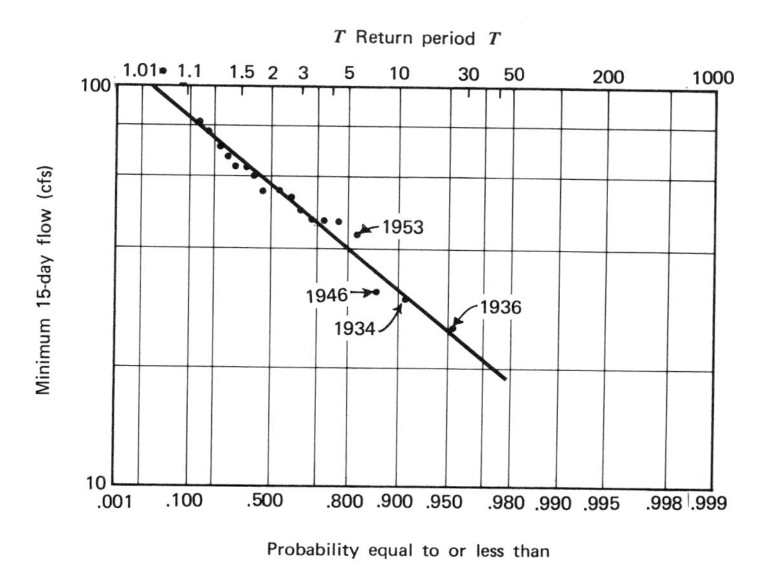

(c)

Figure 3–27*c* Battle Creek at Battle Creek, Michigan. Minimum 15-day flow (May through October).

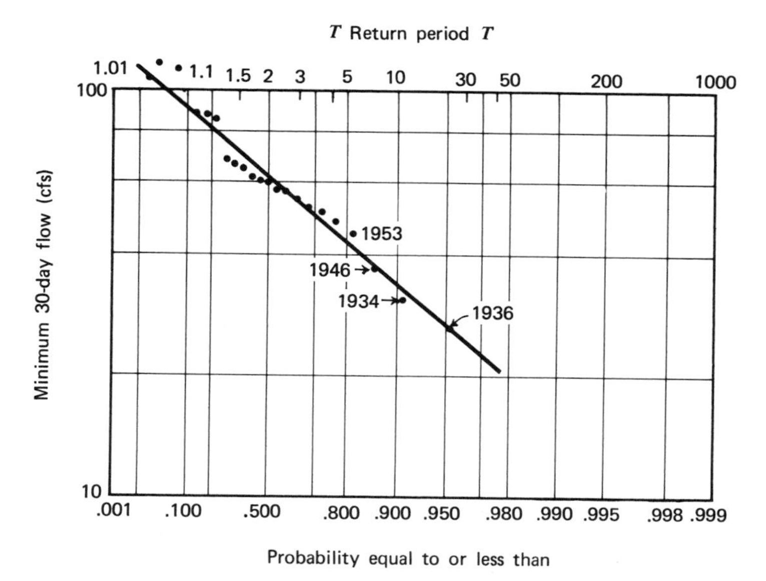

(d)

Figure 3–27*d* Battle Creek at Battle Creek, Michigan. Minimum 30-day flow (May through October).

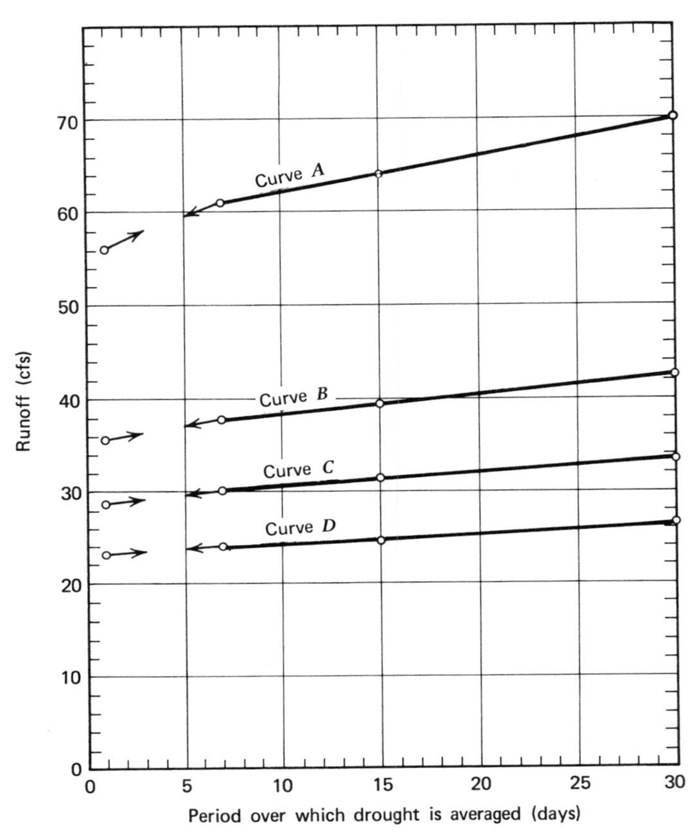

Figure 3–28 Battle Creek at Battle Creek, Michigan. Drought duration versus severity. Curve *A*: most probable drought; curve *B*: once in 5 years; curve *C*: once in 10 years; curve *D*: once in 20 years.

expected once in 20 years. If the decision is made on a consecutive 30-day average, it must be recognized that more severe droughts will surely be experienced for the shorter period averages. The 7- or 15-day average may be considered more appropriate. If the decision is made at, let us say, once-in-3-years severity, then it must be recognized that on the average in the long run once in 3 years drought flow will be more severe and streamflow will not be adequate. If such a frequency of falling short of desired streamflow is unacceptable, it is necessary to restrict activity to a lower flow, such as a severity that is expected once-in-10-years. On the other hand, to be so conservative as to take a position at, let us say, the minimal 1-day average expected once in 20 years, development may be drastically restricted. One must take a reasonable position within the natural drought flow reference frame and evaluate the consequences of that position or one must consider alternatives of relieving the restriction by *drought control.* This is discussed in more detail in Chapter 11.

Drought-Flow Indices

Two indices, the *yield* and the *variability ratio,* afford a basis for comparing drought flow characteristics within and among drainage basins.

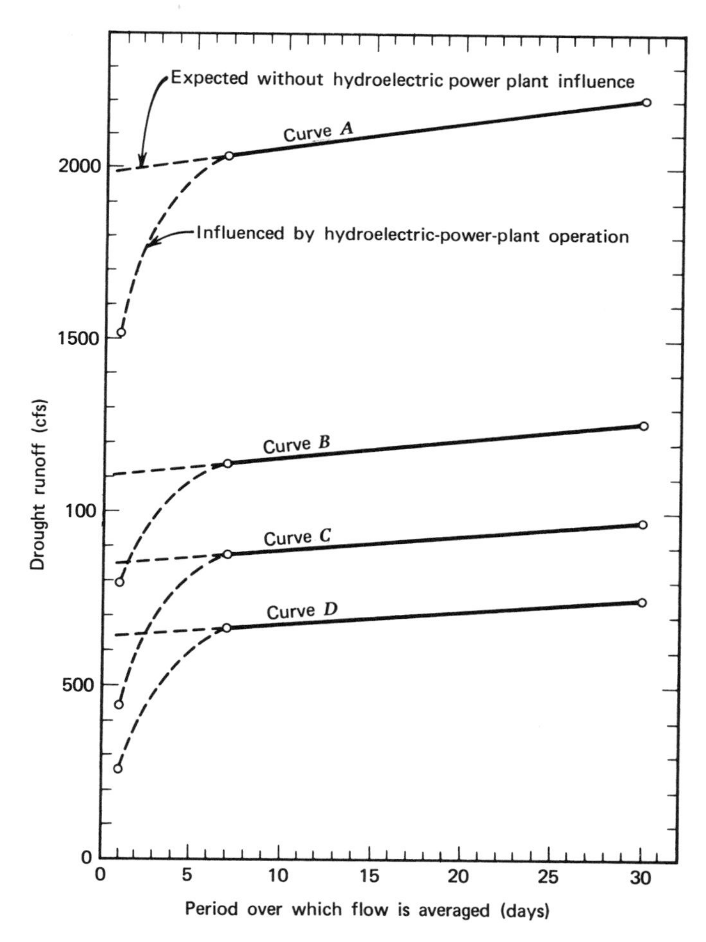

Figure 3–29 Lower Fox River Drought flow characteristics at Rapide Croche Dam (drainage area 6150 square miles). Curve A: most probable; curve B: once in 5 years; curve C: once in 10 years; curve D: once in 20 years.

The Yield. For rivers of moderate difference in size of drainage area tributary to gaging stations it is common practice to convert the absolute flow to a yield per square mile. Comparison on the basis of yield affords a useful index of differences in drought severity characteristics. For example, the Manistee River near Sherman, Michigan (drainage area 900 square miles) shows a once-in-10-years drought as a 7-day average of 730 cfs, a very high yield of 0.811 cfs per square mile. In contrast, the value for the River Raisin at Monroe, Michigan (drainage area 1034 square miles) for the same probability is 28 cfs, a very poor yield of 0.027 cfs per square mile, or one-thirtieth that of the Manistee.

The Variability Ratio. Some streams show wide variation from year to year in the severity of the drought flow, whereas others are relatively stable. The conventional measure of degree of variation in statistical terms is the standard deviation. With normal symmetrical distributions this is a meaningful index and would appear as the slope of the distribution line on normal probability paper. However, in dealing with asymmetrical distributions, such as drought flows, the use of the standard deviation becomes complicated and less meaningful. The steepness of the slope of the line of the distribution on log-extremal probability plots obviously remains a good and rational measure of variation, but the expression of the slope no longer is represented simply by the standard deviation. Also, as has been already pointed out, in order to reduce the distribution to a straight-line fit in some instances it may be necessary to adjust the data. Under these conditions any conventional statistical expression of variation becomes mathematically complex and of little practical value.

In order to present an understandable and useful index that is free of these complications the concept of a simple ratio is employed; it answers directly the question, *what proportion of the normal drought flow is the rare drought flow?* The variability ratio in a standardized form is defined as the ratio of the once-in-10-years drought flow to the most probable drought flow. Thus a variability ratio of unity would indicate perfect stability with no variation; a small ratio would indicate a high degree of variation in drought flow from year to year, with great risk of occasional extremely severe low flow.

The usefulness of the variability ratio as an index of relative variation in drought flow characteristics between two streams is illustrated in further comparison of the Mainistee and Raisin Rivers. The variability ratio for the 7-day average drought flow of the Manistee River is about 0.9, which is to say that the once-in-10-years drought flow is 90 percent of the most probable, or a flow almost as large as the normal expected—indication of an unusually stable stream. In contrast, the River Raisin record develops a variability ratio of about 0.3, which is to say that the once-in-10-years 7-day average drought is only 30 percent of that normally expected, indicating a river of high variability from year to year and subject to occasional drought flows of considerable severity. Similarly the variability ratio may be applied to the 1-day average, the 30-day average, or any other intervening drought flow as an average for a period of days.

Comparison of Drought-Flow Characteristics of Streams

In comparing the drought potential of streams both indices—the yield and the variability ratio—contribute to quantitative characterization. The yield defines the level or magnitude of flow; the variability ratio defines the degree of variation, or stability, and the degree of risk of severe deviation below normal.

Extensive studies of the drought flow characteristics of streams throughout the United States disclose that the theory of extreme values is consistently applicable and that drought flow indices are usable devices. These are briefly summarized under three geographic bases: state-wide, regional, and river basin analyses.

A State-Wide Analysis. The increasing competition for the use of streams in the State of Michigan and the pressing need for establishing a floor of low flow available as a guide to rational development and multiple use prompted a state-wide analysis of drought flow characteristics of all stream gaging stations whose continuous discharge records cover 10 or more years [3]. Excerpts from this study delineate significant divergence in drought severity among the streams. The minimal 1-day average and the consecutive 7-, 15-, and 30-day average summer–fall droughts were analyzed as log-extremal probability distributions in accordance with the methods outlined above. The consecutive 7-day average drought flows are considered to be representative. Table 3–3 summarizes the drainage area, the most probable yield, the once-in-10-years yield, and the variability ratio for 81 stream-gaging stations.

It is apparent that there is wide divergence in drought yield and variability ratio among the streams of the state. In the northwestern portion of the Lower Peninsula streams show unusually high yields, with normal or most probable yields of 0.5 to 1.0 cfs per square mile. Also, as reflected by the variability ratio, these streams show unusual stability, with great dependability from year to year. Similarly, good yields are shown for streams that drain the northeastern sandy region of the Lower Peninsula. In contrast, extremely poor drought-flow yields are shown for the streams that drain the central and southeastern por-

Table 3–3 Drought Indices of Michigan Streams

Name	Drainage area (sq miles)	Consecutive 7-Day Average: Most Probable Yield (cfs)	Consecutive 7-Day Average: Once-in-10-Years Yield (cfs)	Variability Ratio
Streams tributary to Lake Superior:				
Montreal River near Saxon, Wisconsin	281	0.427	0.066	0.155
Presque Isle River:				
At Marenisco	175	0.342	0.087	0.256
Near Tula	260	0.224	0.092	0.410
Middle Branch Ontonagon River:				
Near Paulding	175	0.594	0.408	0.687
Near Trout Creek	225	0.204	0.156	0.769
Near Rockland	670	0.349	0.255	0.731
East Branch Ontonagon River:				
Near Mass	265	0.517	0.294	0.569
West Branch Ontonagon River:				
Near Bergland	160	0.389	0.099	0.254
Cisco Branch Ontonagon River:				
At Cisco Lake Outlet	50	0.144	0.011	0.078
South Branch Ontonagon River:				
At Ewen	320	0.672	0.275	0.409
Ontonagon River near Rockland	1290	0.496	0.341	0.688
Otter River near Elo	175	0.497	0.420	0.845
Sturgeon River near Arnheim	680	0.454	0.341	0.751
Streams tributary to Lake Michigan:				
Manistique River:				
At Germfask	341	0.677	0.478	0.706
Near Blaney	704	0.472	0.334	0.708
Near Manistique	1100	0.480	0.305	0.636
Duck Creek near Blaney:	92	0.121	0.057	0.476
West Branch Manistique River:				
Near Manistique	322	0.422	0.301	0.713
Indian River near Manistique	302	0.728	0.457	0.627
Brule River near Florence, Wisconsin	380	0.663	0.400	0.603
Paint River at Crystal Falls	616	0.411	0.235	0.573
Michigamme River near Crystal Falls	670	0.457	0.154	0.338
Menominee River:				
At Twin Falls near Iron Mountain	1790	0.528	0.302	0.572
Below Koss	3790	0.409	0.174	0.426
Near McAllister, Wisconsin	4020	0.435	0.248	0.568

Table 3-3 (Continued)

Name	Drainage area (sq miles)	Consecutive 7-Day Average: Most Probable Yield (cfs)	Once-in-10-Years Yield (cfs)	Variability Ratio
Pine River near Florence, Wisconsin	528	0.314	0.109	0.346
Pike River at Amberg, Wisconsin	253	0.454	0.313	0.689
East Branch Coldwater River at Coldwater	60	0.078	0.004	0.051
St. Joseph River:				
At Mottville	1860	0.352	0.189	0.537
At Niles	3620	0.420	0.238	0.566
Elkhard River at Goshen, Indiana	580	0.272	0.122	0.449
Battle Creek at Battle Creek	241	0.253	0.124	0.492
Kalamazoo River:				
Near Battle Creek	849	0.369	0.225	0.610
At Comstock	1010	0.426	0.230	0.540
Near Allegan–Fennville	1540 1600	0.394	0.152	0.386
Grand River:				
At Jackson	174	0.216	0.105	0.488
At Lansing	1230	0.133	0.052	0.390
At Grand Rapids	4900	0.227	0.119	0.524
Portage River below Little Portage Lake near Munith	55	0.064	0.012	0.180
Orchard Creek at Munith	49	0.099	0.039	0.393
Cedar River at East Lansing	55	0.077	0.030	0.390
Lookingglass River near Eagle	281	0.117	0.073	0.628
Maple River at Maple Rapids	434	0.048	0.025	0.532
Thornapple River near Hastings	385	0.227	0.122	0.537
Muskegon River:				
At Evart	1450	0.277	0.200	0.723
At Newaygo	2350	0.366	0.256	0.700
Pere Marquette River at Scottville	709	0.550	0.472	0.859
Big Sable River near Free Soil	127	0.760	0.661	0.870
Manistee River:				
Near Grayling	159	1.038	0.953	0.918
Near Sherman	900	0.906	0.811	0.896
Streams tributary to Lake Huron:				
Sturgeon River near Wolverine	164	0.896	0.759	0.847
Indian River at Indian River	583	0.674	0.511	0.758
Pigeon River at Afton	159	0.484	0.412	0.851
Cheboygan River near Cheboygan	865	0.464	0.245	0.529

Table 3–3 (Continued)

Name	Drainage area (sq miles)	Consecutive 7-Day Average		
		Most Probable Yield (cfs)	Once-in-10-Years Yield (cfs)	Variability Ratio
Black River:				
Near Tower	313	0.399	0.283	0.708
Near Cheboygan	597	0.199	0.094	0.471
Rainy River near Onaway	79	0.0042	0.0005	0.112
Thunder Bay River:				
Near Hilman	232	0.565	0.478	0.847
Near Bolton	588	0.422	0.347	0.823
North Branch Thunder Bay River near Bolton	184	0.060	0.010	0.166
Middle Branch Au Sable River at Grayling	110	0.514	0.432	0.841
Rifle River at Michigan Highway 70 near Sterling	320	0.430	0.372	0.865
Shiawassee River:				
At Owosso	538	0.129	0.050	0.388
Near Fergus	637	0.130	0.068	0.524
Farmers Creek near Lapeer	57	0.051	0.015	0.300
Flint River:				
At Genesee	593	0.098	0.037	0.383
Near Flint	927	0.082	0.035	0.426
Near Fosters	1120	0.094	0.039	0.419
Cass River near Frankenmuth	848	0.049	0.019	0.390
Salt River near North Bradley	138	0.041	0.017	0.410
Chippewa River near Mount Pleasant	416	0.252	0.147	0.581
Pine River at Alma	288	0.184	0.080	0.436
Tittabawassee River:				
At Midland	2400	0.104	0.060	0.578
At Freeland	2530	0.194	0.077	0.400
Tributary to the St. Clair River:				
Black River:				
Near Fargo	475	0.026	0.006	0.223
Near Port Huron	634	0.024	0.008	0.322
Tributary to Lake St. Clair:				
Clinton River at Mount Clemens	733	0.134	0.052	0.389
Tributary to the Detroit River:				
River Rouge at Detroit	193	0.074	0.033	0.444
Streams Tributary to Lake Erie:				
Portage Creek near Pinckney	79	0.087	0.032	0.371
Huron River at Barton–Ann Arbor	723	0.141	0.044	0.309
River Raisin at Monroe	1034	0.091	0.027	0.297

tions of the Lower Peninsula, with normal or most probable drought yields in the range 0.2 to 0.1 cfs per square mile and many less than 0.1 cfs per square mile. In addition to normally poor yields, the variability ratios indicate highly unstable streams, with great risk of occasional extreme drought flows decidedly below normal. The streams that drain the southwestern portion disclose moderate yields and moderate stability. The streams that drain the Upper Peninsula, with some exceptions, show moderately good normal drought yields, but the variability ratios disclose considerable variation from year to year and risk of occasional severe drought.

Since there are no radical geographic differences in the normal patterns of precipitation and temperature in the State of Michigan, the differences in drought flow characteristics reflect primarily physiographic differences among the drainage basins (land forms, soil mantle, surface and subsurface geology, and vegetative cover) rather than climatologic differences.

A Regional Analysis. Intensive drought flow analyses developed from log-extreme probability distributions of 42 gaging records of streams tributary to the Ohio River, principally in the Wabash, the Miami, and the Scioto basins, disclose significant regional characteristics. Although this north central region of the United States is in a humid belt, the drought yields are very low, with the most probable 7-day average being less than 0.1 cfs per square mile and many of the once-in-10-years droughts approaching 0.02 cfs per square mile. The degree of variation is also great, as reflected by low variability ratios. An exception to this generally poor drought yield is the Mad River, a tributary of the Miami River, whose most probable yields approach 0.4 cfs per square mile, whose once-in-10-years drought level is about 0.25 cfs per square mile, and whose relatively high variability ratio of .6 indicates stable as well as high yields. This is due principally to the unusual geologic formation of underlying gravel and a sustained streamflow during the drought season from large groundwater seepage.

Of special interest is the close agreement disclosed in these studies between the results of early analyses made from shorter records and that of the longer record at the same gaging station, extended in general by 9 years. The yields are not changed significantly, and the variability ratios are changed but slightly.

River-Basin Analyses. Studies made in a number of river basins throughout the United States also disclose wide differences in yield and variability ratio, not only among basins but in some cases within a basin. However, where climatologic and physiographic factors are similar, yields and variability ratios are also reasonably similar.

These applications confirm the value of the theory of extreme values as a significant tool in identifying the drought flow characteristics of streams and emphasize the necessity of intensively analyzing all available records in any given river basin.

Estimating Drought Flow along the Stream Course

Estimating drought flow along the course of a stream is as much an art, tempered by experience and professional judgment, as an application of statistical methods in the science of hydrology. Unlike ordinary runoff or flood discharge, the hydrologic influences that underlie severe-drought conditions are widespread and encompass an entire river basin or region. Thus it can be assumed that all parts of the basin are subject to the same probability of occurrence, the main stem as well as the tributaries. The specific yields and variability ratios at various locations in the basin may differ, but, if at all locations they are determined for a specific probability of occurrence (such as that expected on the average in the long run once in 10 years), then the accumulation of flow along the course of the river will also represent a once-in-10-years drought flow pattern.

Application of Yield and Variability Ratio. As already defined, the two elements essential to estimating runoff at a location other than a stream gaging station are the drainage area tributary to the location and the runoff characteristics at one or more key gaging stations. The uncertain element is entailed in the choice of the key gaging station, or stations. There are no firm rules for this choice; intimate knowledge of the river basin and professional judgment lead to rational decisions.

If there are no gaging stations on a tributary, data for an adjacent stream of similar character for which a yield and variability ratio are available can be applied. However, this requires careful investigation of the degree of similarity if serious error is to be avoided. For example, if the once-in-10-years 7-day average drought yield of the Mad River at Dayton (0.195 cfs) were applied to a location on the adjacent Stillwater River at Englewood, 646 square miles, the drought flow of the Stillwater would be 126 cfs, whereas actually it is known that the poor yield of the Stillwater produces a once-in-10-years drought flow of only 13.5 cfs. Both tributaries are in the Miami basin, but the Mad River has excellent groundwater feed whereas the Stillwater has singularly little to maintain drought flow.

Usually variations in yield and variability ratio are greatest for tributary streams and more uniform for the main stream. If two gages are available on the main stream, drought flow at intermediate locations can be based on a graduated yield for the intervening area commencing at the upstream gage and balancing at the downstream gage. If key gages are located on two major tributaries joining to form a main stream, a combined weighted yield in proportion to the drainage area of each tributary provides a basis for estimating drought flow along the main stream. If a key gage is located in the upper reaches of the main stream and a number of tributary streams enter along the course, the yields of the tributaries may differ from that of the main stream gage and incremental drought flows are therefore developed from the gaged tributaries.

In many instances it is desired to know the drought severity in absolute terms for various probabilities of occurrence at a specific location; the complete extremal probability distribution can be constructed from the appropriate most probable yield and variability ratio. Assuming the distribution to be log-extremal probability, the most probable drought in cubic feet per second is located at the mode; the once-in-10-years drought, the product of the most probable and the variability ratio, is located at the 10-year return period. The straight line connecting these two points on the log-extremal probability paper defines the entire distribution from which drought flow at any other desired probability or return period can be obtained.

Drainage Area–Drought Flow Relation. In homogeneous basins a reasonably well-defined relationship develops between the drought flow in absolute terms (cfs) and the size of the tributary drainage area along the stem of the main stream. This provides a convenient means of estimating drought flow at any location along the rivercourse. Figure 3–30 is an illustration of the relation for the Wisconsin River, based on log-extremal probability distributions of minimum monthly average drought flows at six stream gaging stations adjusted to natural flow by change in reservoir contents. A linear relation is shown between the log of drought flow in cubic feet per second and the log of the drainage area in square miles; curve *A* represents the most probable drought flow, and curve *B* is the drought flow expected on the average in the long run once in 10 years. If the drainage area tributary to any location on the main stream is known, the drought flow at such location is readily obtained from the graph. For example, at a location with a tributary drainage area of 2000 square miles the most probable monthly average drought flow from curve *A* is 880 cfs and the once-in-10-years drought from curve *B* is 470 cfs, or yields of 0.440 and 0.235 cfs per square mile, respectively. From these the variability ratio is 470:880, or .534. These indices are consistent with those developed at the stream gages above and below the location. Drought flows at other probabilities of occurrence are readily obtained by reconstructing the distribution on log-extremal probability paper. From such a reconstruction of the distribution, the once-in-5-years and once-in-20-years monthly average droughts are 580 and 380 cfs, respectively.

The Nonhomogeneous Basin. Some drainage basins have radically different drought indices for subdrainage areas within the basin. Each tributary, if gaged, is analyzed separately, and the cumulative flow along the main river is synthesized from the successive increments of the tributaries. The Willamette basin illustrates such nonhomogeneity; the eastern portion of the basin drains the snow-capped Cascade Mountain Range, and the western portion drains the relatively dry Coastal Range. Tributaries fed by the melting snows and orographic precipitation of the Cascades produce high yields, whereas those from the Coastal Range are poor in yield. Figure 3–31 is a diagrammatic presentation of the estimated streamflow along the Willamette from Salem to Portland, Oregon, for the once-in-5-years minimal consecutive 7-day average drought severity. The last downstream gage on the main river is at Salem, reflecting runoff from the headwaters drainage area of 7280 square miles. The once-in-5-years 7-day average drought yield at Salem is 0.391 cfs per square mile, or a flow of 2850 cfs. The yields of the Yamhill and the Tualatin, which enter from the Coastal Range, are poor at

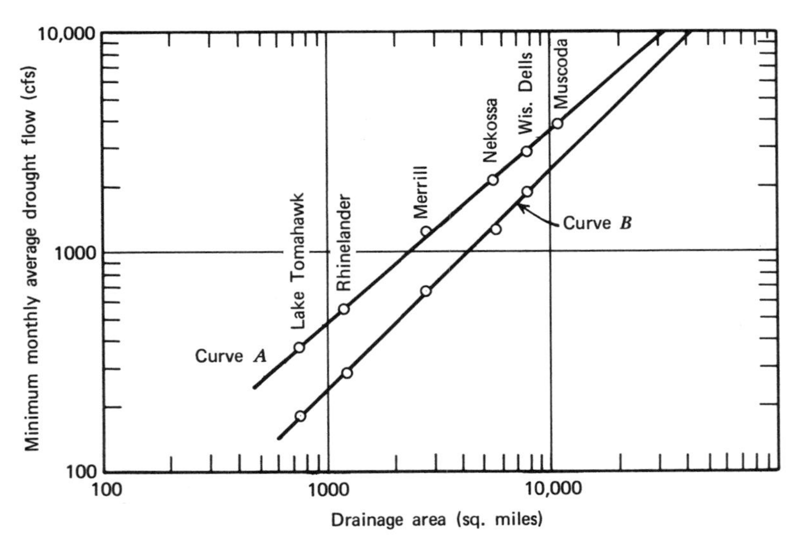

Figure 3–30 Wisconsin River. Relation between drainage area and minimum monthly average drought flow. Curve *A*: most probable; curve *B*: once in 10 years.

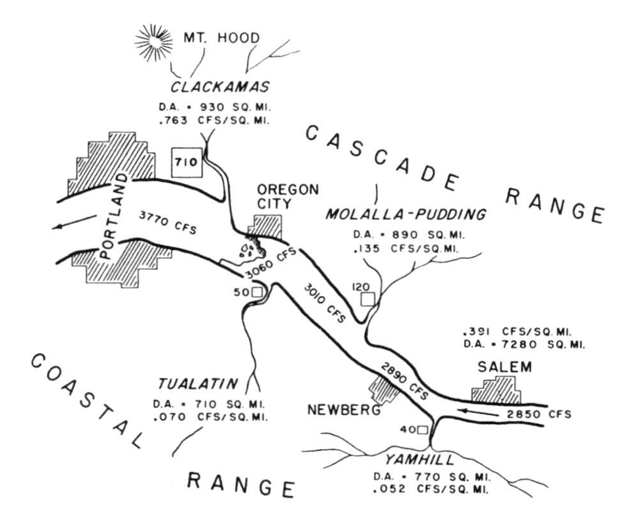

Figure 3–31 Streamflow available along the Willamette River at once-in-5-years minimum weekly average drought severity.

0.052 and 0.070 cfs per square mile, respectively; in contrast the yields of the Molalla-Pudding and the Clackamas, which enter from the Cascade Range, are 0.135 and 0.763 cfs per square mile, respectively. The drought yield of the Clackamas, whose headwaters are at Mount Hood, is about 15 times that of the Yamhill for the same probability of occurrence. The cumulative once-in-5-years 7-day average drought flow along the course is estimated as 2850 cfs at Salem, increasing to 2890 cfs below the junction with the Yamhill, to 3010 cfs below the Mollala-Pudding, to 3060 cfs below the Tualatin, and, with the significant increment from the Clackamas, to 3770 cfs. These flows represent the natural drought condition before augmentation from upstream reservoirs, which currently ensure flow on the order of 7000 cfs through Portland. Obviously, if the differences in yields of the tributaries are neglected and drought flow along the course of the river is based only on the yield of the Willamette at the Salem gage, a substantial error is introduced, particularly in the reach to Oregon City.

STREAM-CHANNEL CHARACTERISTICS

Physical and Hydraulic Factors

A phase of hydrology that is essential to quantitative computation of waste-assimilation capacity is the detailed characterization of the stream channel. Usually these data are not available. Early river survey reports may contain profiles of the main channel bed and low-water gradients, but usually they are not related to a specific runoff. Although these are valuable, especially when later good river maps or aerial photos provide configuration and river widths, they are not adequate for refined computation. Occasionally river surveys made by the Corps of Engineers, particularly for navigation improvements, include detailed channel cross-section soundings at frequent intervals along the course, located on good river strip maps and referred to a specific low water runoff regime. Such surveys provide the basic channel characterization required.

Natural Channel Types

Rivers vary radically in configuration, cross-sectional shape, depth, channel bed, and hydraulic gradient. For purposes of applied stream sanitation two broad classifications are encountered in the natural hydrologic setting—namely, the *pool and riffle type,* and the *regular gradient type.* The smaller tributaries and headwater streams, traversing the more rugged terrain, descend in irregular gradients consisting of a series of deep, quiet pools interspersed with steep, shallow reaches of riffles, rapids, or sharp falls. In contrast, the larger main channel rivers that traverse the lower portion of the drainage area descend gradually in regular, flatter gradients without exposed riffles or rapids. In the pool and riffle type the radical differences in channel cross-sectional depth and velocity are visually evident. In the regular gradient type the changes in channel are submerged and are not visually evident, except to the experienced river man. Actually the channel cross-sectional shapes and depths are anything but regular, and sounding at frequent sections discloses deep holes and shallow reaches along the course. One of the greatest sources of error in the quantitative computation of waste assimilation capacity is failure to take these changes in channel characteristics into account reach by reach along the course in both types of channel.

As river development increases, these classifications of natural channels are masked, but not lost, by the construction of dams and reservoirs, and particularly by the construction of navigation locks and dams, which create series of deep slackwater pools extending many miles upstream. Such man-made changes in natural channel characteristics must also be defined, including the old submerged streambed configuration.

Channel Cross-Section Sounding

Where good channel cross-section soundings are not available, they are best obtained by echo-sounding equipment or by rod or sounding lead. Cross sections taken at approximately 500-ft intervals along the course, and more frequently at irregular configuration, are desirable. This is not an insurmountable task with modern echo-sounding instrumentation. It is not necessary to locate the sections precisely by survey traverse since it is usually possible to obtain maps or aerial photos on which locations can be marked by the field survey party. If maps or photos are to a reasonable scale, river widths can be scaled; if not, widths are obtained by stadia. Cross-section soundings should be undertaken at reasonably low water conditions and a steady hydrograph. Runoff is recorded at least daily during the survey at the nearest key gaging station.

In addition to the discharge rating curves at key gaging stations, it may be necessary to establish and rate a number of auxiliary river stage gages along the course. The purpose of the stage rating curves is to permit adjustment of the river cross-section soundings as taken in the field to any runoff regime by adding or subtracting a change in stage associated with a change in runoff regime.

Where hydroelectric power installations are involved, radical pulsations in stage may occur. It is extremely difficult to correct soundings for these pulsations in stage; arrangements should be made with the operating utility to stabilize streamflow during the period of cross-section sounding.

For small streams of the pool and riffle type echo sounding cross sectioning or rod sounding is employed in the pool reaches, but the riffles and rapids are too irregular and shallow. Dye tracer techniques with timing of passage at frequent locations provide indirectly the desired channel measurements.

When echo-sounding equipment is employed, the channel cross-section is recorded as a continuous chart and can be planimetered directly and with the scaled or stadia river widths is converted to cross-sectional areas. Figure 3–32 shows a set of typical echo-sounding cross sections taken on the Ouachita River by boat traverse at uniform speed from shore to shore. If cross-section soundings are tabulated as depths, the data are plotted for planimetering or are punched on cards for computer adaptation.

Channel Parameters and Time of Passage

In the quantitative evaluation of stream self-purification three relevant channel parameters are the occupied channel volume, surface area, and the effective depth, reach by reach. From these are derived two additional vital factors—time of passage and mean velocity, reach by reach. All five factors are related to stream runoff and at any specific runoff regime are determined from the channel cross-section soundings.

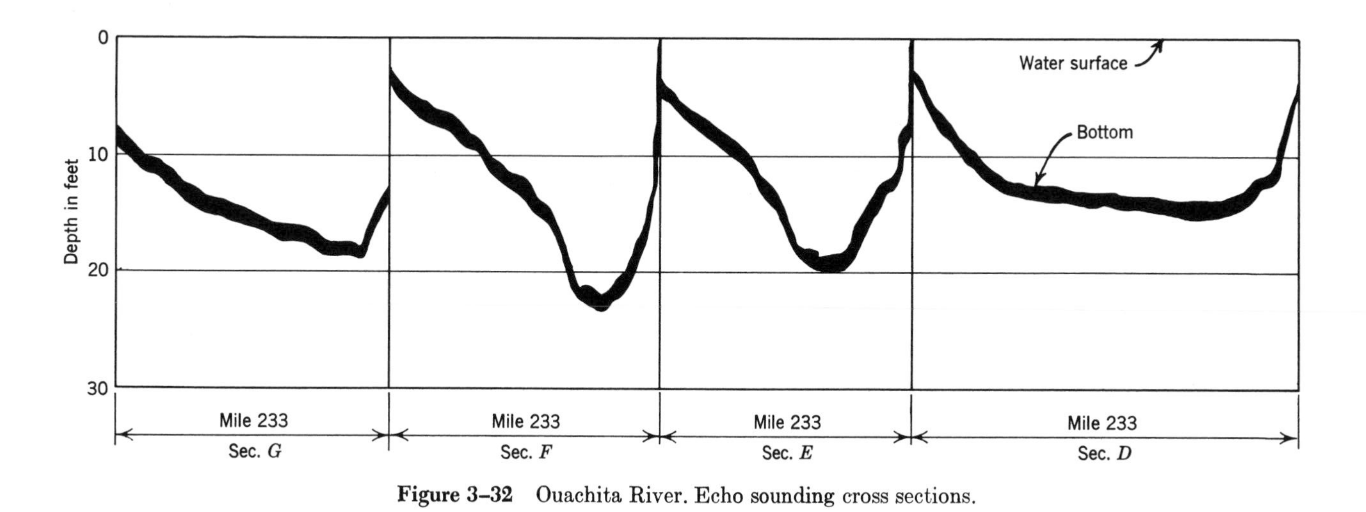

Figure 3–32 Ouachita River. Echo sounding cross sections.

Channel Parameters

Since the cross-section soundings are obtained over a period during which runoff has possibly varied somewhat from day to day, the first step is to convert each day's soundings to a specific runoff regime by subtracting or adding an increment of depth based on the appropriate stage rating curve. With soundings adjusted to the reference runoff regime, occupied channel volume in cubic feet is computed as the average cross-sectional area in square feet of adjacent cross-sections multiplied by the distance in feet between section stations. A mean depth is readily derived at each cross-section as the section area divided by the channel width at the water surface. However, where good maps or aerial photos are available, it is preferable to determine *effective depth* for the reach as the occupied channel volume divided by the water-surface area, where surface area is determined as the average of a number of scaled widths multiplied by the distance between the sections.

Time of Passage

In considering the time of passage along the course of a stream a distinction is made between the traverse of the *hydraulic crest,* the time at which changes in the rate of discharge are noted at locations down-river, and the *mass movement* of the body of water. It is the latter that is the measure of the time of passage of wastes from reach to reach along the course. The hydraulic crest, more in the nature of translation of a surface wave, outstrips movement of waste contained in the body of the water, particularly where large volumetric displacements are involved through deep reaches or pools behind dams.

This distinction is strikingly evidenced in analysis of reported data on the Androscoggin River, as shown in Figure 3–33. Curve A is the traverse time of the hydraulic crest, as reported by operators of hydro-power plants along the river; curve B is the time of passage of the mass movement of the body of water, as reflected by chemical tests associated with the shutdown and resumption of industrial operations. It is noted that the hydraulic crest traverses the run-of-the-river channel to the large Gulf Island pool 1.5 days sooner than the time of arrival of wastes; and the hydraulic crest reaches the dam, skimming through the pool, 10 days ahead of arrival of the wastes. These two times of passage have a significant bearing on stream analysis—the hydraulic crest in relation to interpretation of river sampling along the course, and the time of passage of waste products as the time element involved in satisfaction by self-purification processes. It is the latter that is the

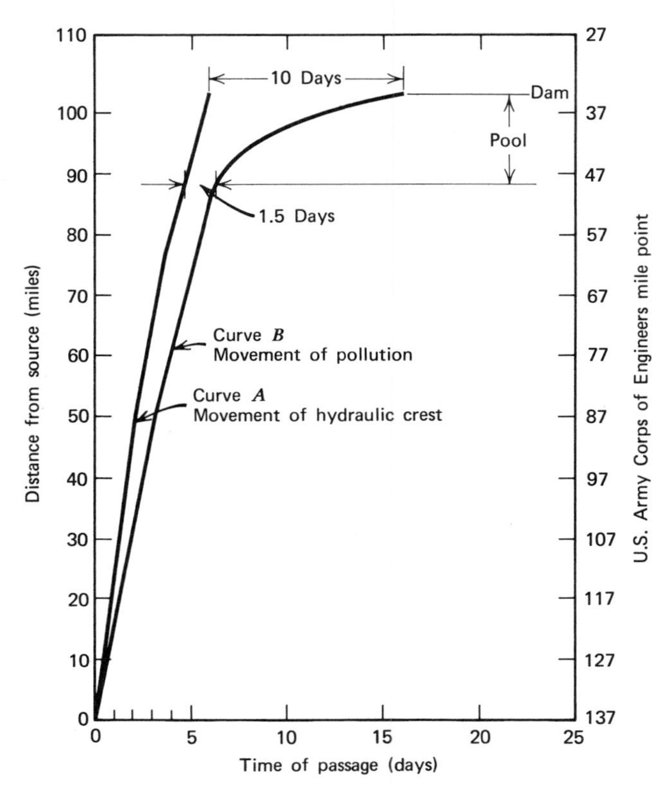

Figure 3–33 Androscoggin River. Movement of pollution and of the hydraulic crest.

more important, and in this text the term "time of passage" pertains to this mass movement of the body of water.

The time of passage is determined at any specific runoff regime as the occupied channel volume reach by reach divided by the tributary runoff to each successive reach. Occupied channel volume is obtained from the channel cross-section soundings, as outlined above; runoff along the ourse of the channel at any location is determined by the tributary drainage area and the appropriate yield, as outlined heretofore. With occupied channel volume expressed in cubic feet and runoff in cubic feet per second, time of passage is in seconds; it is customary, however, to convert the time of passage to days. The summation of the increments of time from cross-section to cross-section provides the cumulative time along the course, whose plot against river mile locations provide the *time of passage curve.*

Each runoff regime describes its own specific time of passage curve; if the runoff regime is identified as to probability of occurrence, so also is its time of passage curve. Accordingly the pattern of time of passage is represented by a nest of curves for the range of runoff under study. In some instances this is only the range of runoff identified by the drought flow reference frame; in others it may cover the entire expected seasonal variation in runoff.

Figure 3–34 shows time of passage curves for the nontidal reach of the Willamette River from Salem to Oregon City Falls based on detailed channel cross-sections. In this instance data for each cross-section sounding were punched on cards and a program was written for computer adaptation to adjust sounding for each runoff and to execute the extensive computations.

Figure 3–35 shows time of passage curves for a reach of the Ouachita River based on recorded echo-sounding cross sectioning of the channel as illustrated in Figure 3–32. The planimetered cross-sectional areas as recorded are adjusted to square feet by a conversion factor that depends on the scale of the recording chart and the measured or scaled river width of the section. A program is written on the basis of cross-sectional areas, with adjustment for change in area by gage height differential for changes in runoff regime; the computations are completed by computers. If computers are not available, the solutions can be executed without difficulty in reasonable time with a slide rule or desk calculator. Computer adaptation has the advantages of speed and less chance for error in the many calculations involved.

In addition to the cumulative time of passage curve it is desirable to develop corresponding cumulative occupied channel volume and surface area curves progressively from section to section along the course; these volume and surface area curves permit subdividing the river into any succession of reaches that may be required in computing the waste assimilation capacity.

The impact of man-made river developments on time of passage and channel parameters is shown by the data in Table 3–4 for a reach of the Coosa River. Based on excellent river strip maps with cross-section soundings taken at frequent intervals of 200 to 500 ft along the course, the time of passage along the natural channel at the drought yield of 0.26 cfs per square mile in 30 miles is 2.5 days. A proposed dam at about mile 60 would increase the low water profile associated with the drought yield of 0.26 cfs by 47, 26.5, and 17.5 ft at miles 60, 31, and 15, respectively. With these increases in occupied channel volumes the time of passage for the runoff regime of 0.26 cfs per square mile, assuming no stratification, is greatly increased from 2.5 to 64 days in the

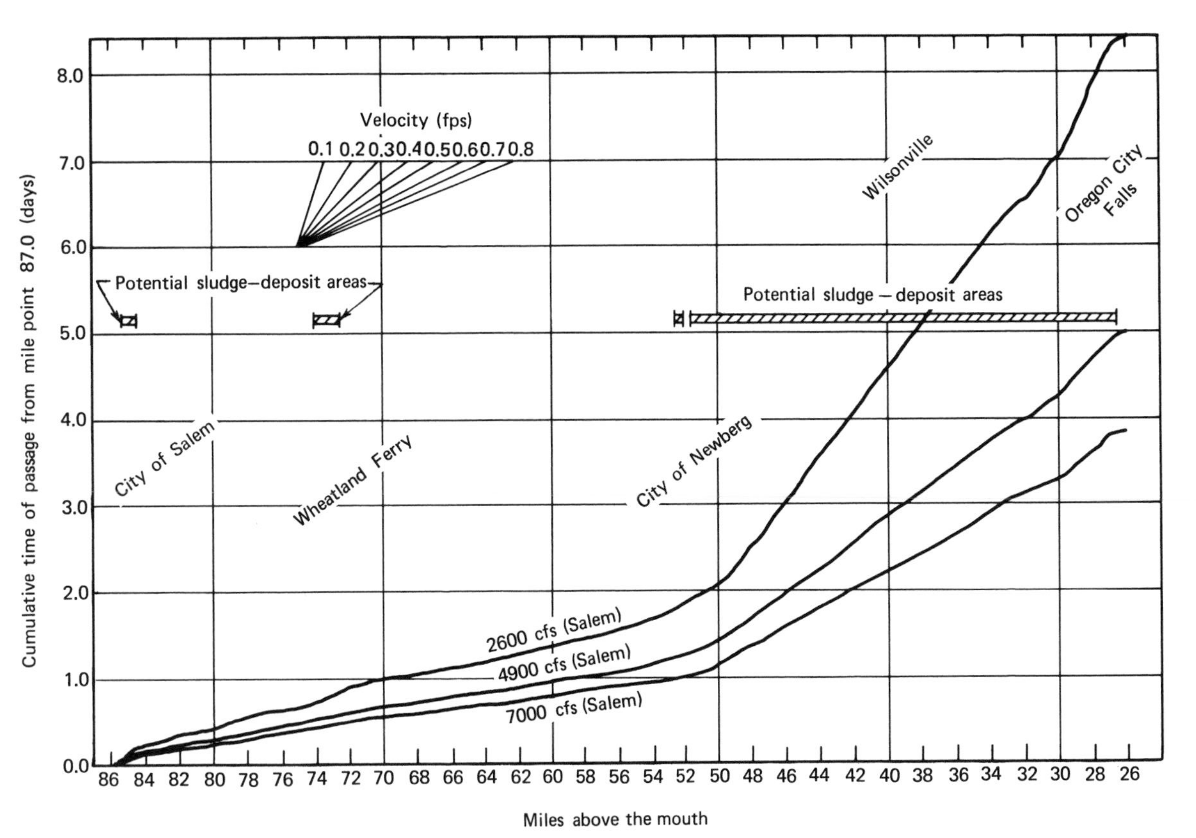

Figure 3–34 Willamette River. Time of passage from Salem to Oregon City at various runoffs.

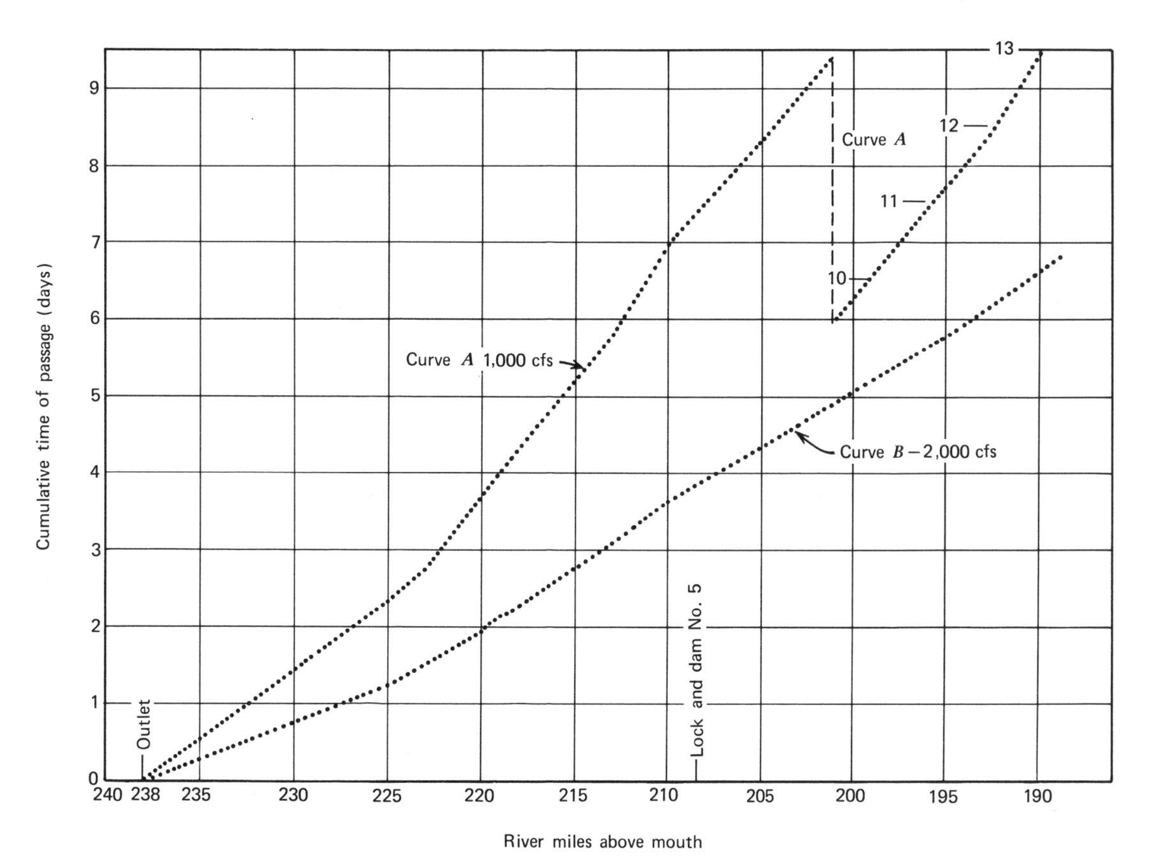

Figure 3–35 Ouachita River. Cumulative time of passage from mile 238.

30-mile reach. Such drastic changes in channel have a profound effect on self-purification characteristics and waste assimilation capacity, as will be shown later.

An illustration of another type of river development effect on the time of passage is shown in Figure 3–36 for the Miami River in its flow through the urban-industrial complex extending from Dayton, Ohio, to the Ohio River. Curve *A* is the computed time of passage for the once-in-10-years 7-day average drought flow of 316 cfs referred to the Hamilton gage, based on detailed channel cross-section soundings taken at about 500-ft intervals along the course. Curve *B* is the time of passage based on the once-in-10-years 7-day average yield augmented

Table 3–4 Coosa River—Influence of Proposed Dam on Time of Passage at Runoff Yield of 0.26 cfs per Square Mile

	Time of Passage (days)	
Mile Point	Natural Channel	After Construction of Proposed Dam
15	0	0
20	0.3	3
30	1.2	10
35	1.6	20
40	2.1	41
45	2.5	64

by upland storage to provide an ensured flow at the Hamilton gage of 800 cfs. In comparing curves *A* and *B* a marked decrease in the time of passage from Dayton to the mouth is expected (under similar drought probability) with the low flow augmentation—from 13.5 to 6.5 days. The benefits of drought control by flow regulation are substantial, as discussed more fully in Chapters 4 and 11.

A comparison of the time of passage along the course of the Miami River below Dayton, measured by the dye tracer technique and by channel volume displacement, is shown in Figure 3–37. Curve *A* is the time of passage based on fluorimetric measurement of the apparent maximum concentration of dye reported in a test carried out jointly by the U.S. Geological Survey and the Miami Conservancy District during July 1965 at a runoff of 542 cfs (Miamisburg gage). Curve *B* is the

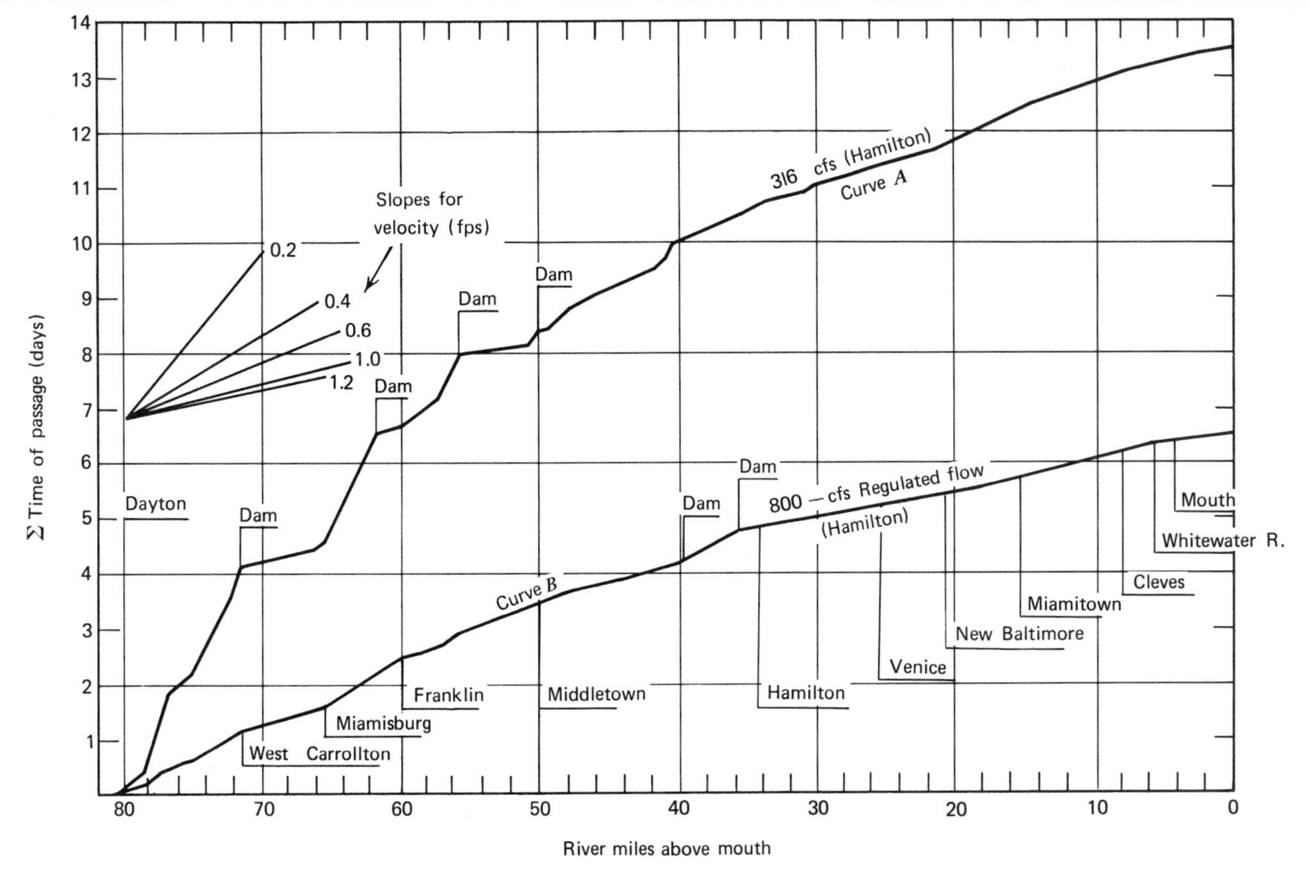

Figure 3–36 Miami River. Time of passage from Dayton, Ohio, to the mouth (volume displacement by runoff). Curve A: computed time of passage for the once-in-10-years 7-day average drought flow; curve B: time of passage based on the once-in-10-years 7-day average yield augmented by upland storage to provide an ensured flow at the Hamilton gage of 800 cfs.

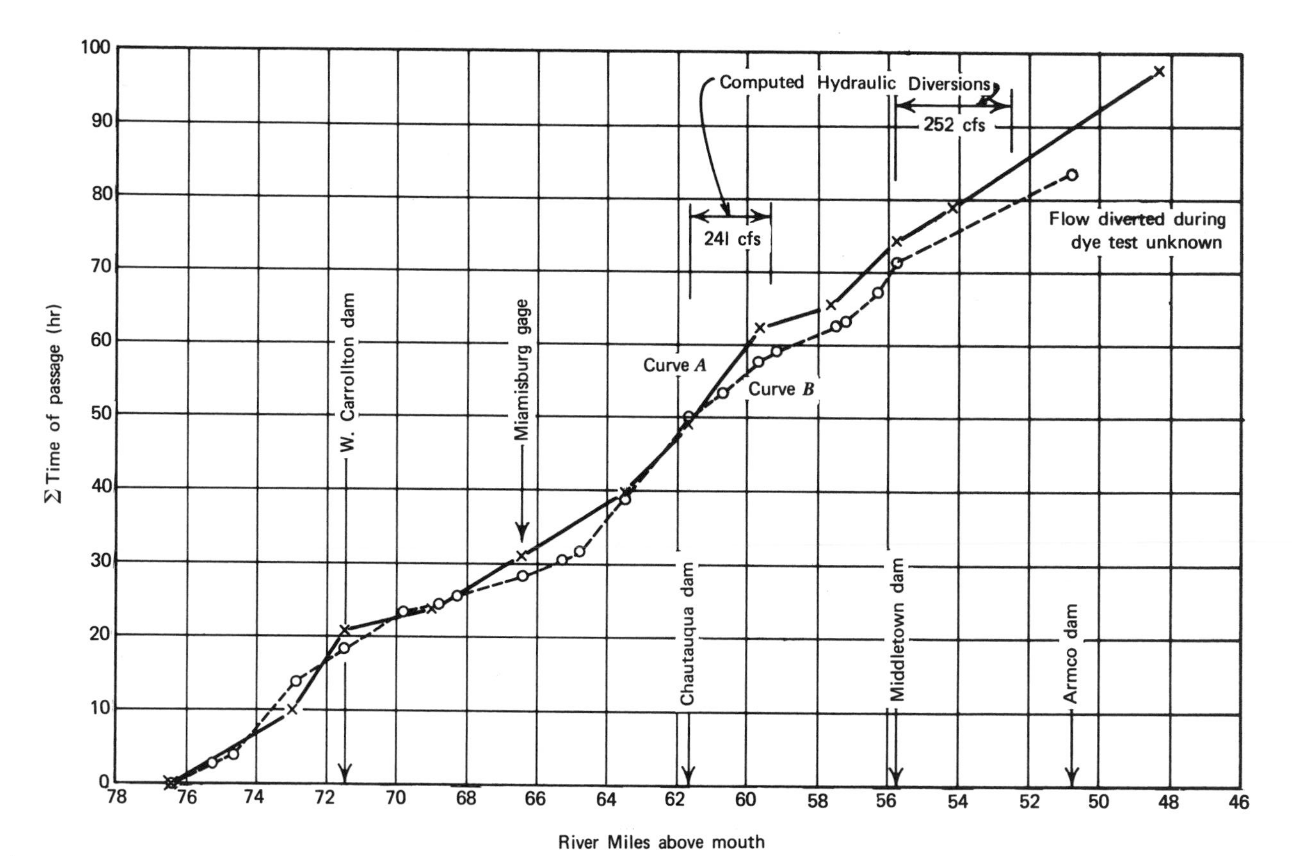

Figure 3–37 Miami River. Comparison of the time of passage by the dye tracer technique (curve A), and computed by channel volume displacement (curve B); runoff 542 cfs (Miamisburg gage).

computed time of passage for the same runoff based on channel-volume displacement by runoff. Considering the complexities of the Miami channel and unknown diversions through hydraulic channels, there is good agreement between curves *A* and *B*, confirming the validity of the volume displacement method.

In dealing with the time of passage so far it has been assumed that the entire occupied channel volume is displaced in blocklike manner by runoff. This assumption, it is recognized, is not strictly valid, since a certain degree of short-circuiting and longitudinal dispersion take place. However, where channel configuration is reasonably regular, occupied channel volume displaced by runoff is a valid measure of the mean time of waste passage along the course. This is verified by close agreement with dye tracer results and confirmed by analyses of resultant stream conditions. Two situations that require modification in the occupied channel volume method deserve special consideration—namely, channels with wide flats or isolated bay areas and channels of great depth.

Some streams have a well-defined main channel bounded by shallow, weedy flats, either as a natural configuration or induced by backwater from a downstream dam. The actively flowing stream predominates in the main channel section, and the adjacent wide, shallow flats are practically stagnant. The extent to which such dead areas should be deleted from the channel cross section depends on judgment and close field examination. Similarly some streams affected by backwater from run-of-the-river dams are composed of a reasonably well-defined channel of moderate depth, but along their courses there are small tributaries that act as isolated bays. In such instances it is customary to eliminate the back bays and not include these reaches in the cross section of the main active flowthrough channel.

The second modification in channel cross section pertains to deep natural channels or deep channels formed by backwater from dams that are subject to vertical stratification. During the critical warm weather drought season it is not uncommon that the actively flowing section is only the upper 15 to 20 ft. The volume below this depth from the standpoint of the time of passage is virtually dead area. It is difficult to determine what portion of the cross section should be eliminated. Magnitude of the stream runoff relative to the channel configuration is a factor. Some rivers with high runoff, even though composed of deep reaches, do not stratify, and the entire cross section remains active. Temperature and dissolved oxygen are sensitive indicators that aid in identifying the boundaries of the actively flowing channel area to be considered in computing the time of passage. Determination of the time of passage by dye or radioactive tracer techniques also assists in defining the limits of active channel volume displacement.

Large deep reservoirs and lakes cannot be generalized; the patterns of flow are varied and difficult to define, being governed principally by seasonal density gradients (see Chapter 7).

Stream Velocity, Deposit, and Scour

The time of passage curve affords a convenient method of determining the mean velocity of flow at any location or reach along the course of the stream. Since the cumulative time of passage is plotted against distance along the course, the tangential slope at a specific mile point location on the time of passage curve or the slope through a reach is the measure of velocity. Construction of a nest of slopes at specific velocity intervals on the time of passage graph provides a ready means for investigating the velocity of streamflow at any location for any runoff regime to which the time of passage curves are constructed. From this the potentialities of sludge deposit and scour in any reach along the river are readily identified.

For example, the time of passage curves in Figure 3–34 for the range in drought flow on the Willamette River show the velocity slopes in feet per second. A mean channel velocity of 0.6 fps is the critical velocity at which deposits of organic settleable solids occur and accumulate. By translating the 0.6-fps slope along the time of passage curves, four potential sludge deposit areas are located, as shown in Figure 3–34. There is a good theoretical basis for this method of identifying the potential zones of sludge deposit, and field observations confirm these locations.

In a natural stream the limiting factor that determines the occurrence and accumulation of sludge deposits is the velocity of flow at which scour commences. Shields [4] has defined this critical velocity as

$$V_c = \left[\frac{8Bg(s - 1)d}{f}\right]^{\frac{1}{2}},$$

where V_c is the mean channel velocity at which scour commences, B is a constant, g is the gravity constant, f is the Weisbach–Darcy friction factor, s is the specific gravity of the particle, and d is its diameter.

From what is known of the performance of grit chambers, B for fresh deposition of organic matter is 0.06, and a critical velocity of about 0.6 fps is required to scour organic matter granular in shape and up to 1 mm in diameter. Under natural stream conditions greater turbulence due to a higher value of f would allow an increase in B up to 0.22, with comparable results. However, in the event channel velocities fall below 0.6 fps over extended periods, organic settleable solids will deposit and in the process of accumulating will cohere and compact, resulting in an increase in B. If B is assumed in such instances to increase to 0.8, the velocity required to induce scour of the accumulation would increase to 1.15 fps, or practically double that where accumulation and compaction did not occur. Also, during accumulation deposited particles no longer remain discrete, since they coalesce to form larger aggregations; hence the size to be scoured increases and the velocity required to induce scour will be further increased. To offset these tendencies, however, gases produced during decomposition of organic deposits will agitate the accumulation and lift some particles into the flowing stream. Thus fresh organic sludge deposits undergoing digestion may be readily scoured, probably at velocities between 0.6 and 1.0 fps.

For resettled partially or wholly digested material for which active gasification has subsided, the above mentioned adverse to scour tendencies hold, and higher velocities, such as occur during freshets or floods, are required. Taking an extreme situation where B is 1.0 and organic deposits are stabilized material, the velocity required to scour a 1-mm-diameter particle would be approximately 1.3 fps.

An opportunity to observe such critical balances between channel velocity, deposit, and scour was afforded in studies of the Kalamazoo River. From time of passage curves two deposit areas separated by a stretch of clean bed were identified. A general relation between mean channel velocity and runoff (referred to the reference gage) for each of the three reaches is shown in Figure 3–38. for the reach in the river in which no deposits occurred (curve A) it will be noted that a channel velocity of 0.6 fps would not occur until the runoff declined to about 130 cfs. From probability studies it was determined that a runoff as low as this could be expected on the average once in 20 years; hence for practical purposes sludge deposit and accumulation, according to theory, were rarely permitted. The cross-section soundings through this reach revealed a complete absence of organic matter and a clean sand and gravel bottom.

The upper deposit area below the city of Kalamazoo is indicated by curve B. Here, it will be noted, a channel velocity of 0.6 fps or less is attained at a runoff below 700 cfs, which occurs frequently each year; consequently it can be expected that at times sludge of organic character will accumulate. It will be further noted, however, that with runoff between 700 and 1500 cfs, which also occurs frequently each year, velocities would range between 0.6 and 1.0 fps, sufficient to scour fresh, actively decomposing sludge. An examination of cross-sections taken during a period when the runoff was 835 cfs indicated the presence of organic sludge but no large accumulations.

Further downstream in a pool area a large accumulation of relatively stabilized material was observed. For this area (curve C) channel veloci-

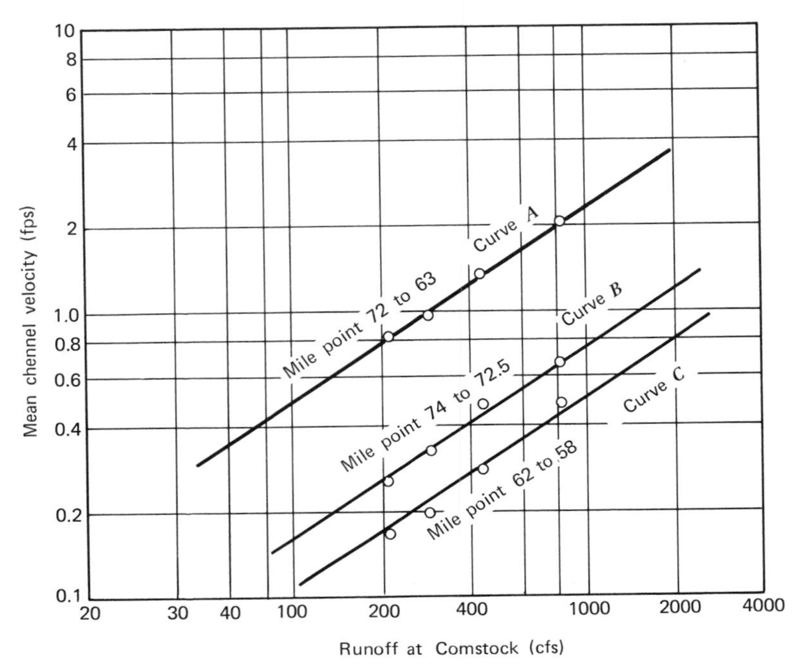

Figure 3–38 Kalamazoo River. Relation between mean channel velocity and runoff.

ties of 0.6 fps or less occur when the discharge is below 1500 cfs. Flows below this level occur for protracted periods, with opportunity for extended accumulation. Thus this zone receives the deposit that bypasses or is scoured from the upper area. Scour velocities in the pool area occur only during freshets and the flood season.

Similar verification of the relation of channel velocity, deposit, and scour has been obtained for a number of other rivers. Quite apart from the element of time, the time of passage curves therefore serve as a valuable guide in estimating potentialities for sludge deposit and scour.

REFERENCES

[1] Gumbel, E. J., *Ann. Math. Statistics,* **12,** 163 (1941).

[2] Powell, R. W., *Civil Engineering,* **13,** 105 (1943).

[3] Velz, C. J., and Gannon, J. J., *Drought Flow Characteristics of Michigan Streams,* Michigan Water Resources Commission, June 1960.

[4] Shields, in *An Analysis of Sediment Transportation in the Light of Fluid Turbulence,* by H. Rouse, U.S. Department of Agriculture, Sedimentation Division, 1939.

4

ORGANIC SELF-PURIFICATION

The practical waste assimilation capacity of streams as a water resource has its basis in the complex phenomenon termed stream self-purification. We shall identify the primary factors and their interrelation, leading to the quantitative definition of the range of the available waste assimilation capacity. Since this is a dynamic phenomenon that reflects hydrologic and biologic variations, the interrelations are not yet fully understood in precise terms. However, sufficient knowledge is available to permit quantitative definition of resultant stream conditions under expected ranges of variation to serve as a practical guide in decisions dealing with water resource use, development, and management. Several useful theories are available, but here emphasis is placed on practical solutions that give consistent results, verified against observed stream quality under a wide variety of conditions.

As discussed in Chapter 2, for purposes of defining specific stream self-purification, the heterogeneous combination of waste products that reaches the stream is classified into four broad types: organic, microbial, radioactive and inorganic, and thermal. These are dealt with in Chapters 4 through 7, respectively, in relation to freshwater bodies; Chapter 8 explains modifications induced by tidal action in estuaries.

Although the self-purification characteristics of each classification are presented separately, it should be recognized that the stream must handle the heterogeneous complex simultaneously. Depending on relative magnitudes, one classification may have a marked influence on another, for instance, heat on organics.

In magnitude and complexity the organic constituents of the end products of community and industrial activities comprise the most significant problem in satisfactory waste disposal. Usually if the waste assimilation capacity of the stream is adequate to handle the organic effluent residual, the self-purification capacity is adequate for the other classifications of waste as well.

THE NATURE OF ORGANIC SELF-PURIFICATION

There are so many facets to organic self-purification that it may be well first briefly to sketch the nature of the phenomenon in the familiar financial terms so aptly employed by Phelps. The best single measure of organic waste is the quantity of oxygen required for stabilization by natural biological processes of decay, or biochemical oxygen demand (BOD). The oxygen required to satisfy the demand of unstable organic matter discharged into a watercourse is obtained from the river water, which normally contains a quantity of dissolved oxygen. If the demand for oxygen is in excess of that available, the dissolved oxygen is exhausted and the stream goes into bankruptcy. Thus BOD may be regarded as liabilities, or outstanding debts, and the dissolved oxygen of the river water as assets. To evaluate the solvency or condition of the stream along its course we must know the amount of debt outstanding and its amortization schedule, and we must also know the amount of assets on hand and the rate of income from the asset sources. Stream solvency then depends on the balance between outgo and income.

Assets

The asset side of the ledger is governed primarily by physical and chemical processes. There are but two dependable sources of oxygen income to the stream—the dissolved oxygen contained in the streamflow and any increments added from tributaries along the course, and reaeration from the atmosphere. The quantity of dissolved oxygen that is contributed by runoff depends on the size of the river, the rate of streamflow. When the stream discharge is high, the contribution is high; during periods of low flow the contribution diminishes. However, the contribution from runoff also depends on the condition of the river. If upstream pollution exercises a demand, the river may not carry its full quota of dissolved oxygen. The amount of dissolved oxygen that is contributed by runoff depends also on temperature, which controls the saturation level of water exposed to the atmosphere. During the heat of summer the stream is capable of holding at saturation only about one-half the dissolved oxygen it can carry during the cold winter season. Cash income from runoff, then, is least during the warm weather drought period.

The rate of oxygen income from reaeration is a highly variable and complex asset. The rate of income is dependent on the amount of cash on hand; if the bank is full (water saturated), there is of course no income; if the oxygen is completely exhausted, the rate of income reaches its maximum; between these extremes the rate is directly proportional to the oxygen deficit below saturation. Although oxygen deficit is the throttle that regulates the rate of income, the actual amount received from reaeration depends on physical and hydrologic characteristics, water temperature, depth, occupied channel volume, and stream turnover. Knowing these basic factors, reaeration for any particular stretch of river can be computed.

Liabilities

The liability side of the ledger is governed primarily by biological processes. Nature supplies the labor force to carry out oxidation of unstable organics to stable mineralized forms. The process is a living biochemical reaction in which the organic matter serves as the food supply for a complex chain of biological life, each group of organisms performing its allotted share of work in oxidation. As long as some dissolved oxygen remains in the stream this process is controlled predominantly by the aerobes, which carry out the work without nuisance. If, however, the organic load is so great that dissolved oxygen is exhausted, the anaerobes take over. The organics are oxidized, but the anaerobes take the necessary oxygen from previously stabilized matter, such as sulfate, and in the process reduce it to hydrogen sulfide. The result is a foul, putrid ruin. The crux is to keep the work in the hands of the aerobes with always some cash on hand.

Amortization of the organic debt (the rate of BOD satisfaction) is really then a reflection of the rate at which the biological organisms eat. As in so many natural phenomena, this chain of life eats at a *constant rate* rather than a constant amount. This is probably part of nature's wise scheme of survival. Eating at a constant rate always preserves some fraction of food for the lean days. It is also an economical way of dealing with an unusually large organic load and quickly bringing it to a stabilized form. The specific rate of eating depends on the nature of the food supply and increases or decreases with rise or fall in temperature. Payment of debt at a constant rate is a very simple rule, but it establishes an unbalanced amortization schedule with heavy payments in the early periods. Normally at 20°C about 20 percent is amortized per day of the outstanding debt remaining from the previous day; at 30°C the rate of amortization increases to about 30 percent per day.

The liability side of the ledger is therefore essentially a biological system, and amortization is a time–temperature function.

Abnormalities That Affect the Amortization Schedule

Under natural stream conditions there may be a number of abnormalities that affect nature's otherwise quite normal straightforward amortization schedule. The mixed organics of municipal waste usually are amortized in accordance with a normal schedule. Certain industrial wastes may demand very high rates of amortization or in other instances highly treated biological effluents may exact a subnormal rate.

If sewage or waste contains chemical compounds or products of decomposition that are immediately capable of chemical reaction with the oxygen dissolved in river water, an *immediate demand* may be exercised on the stream assets without the time element in the biological oxidation process. A portion of the debt accordingly is amortized immediately on discharge into the stream, thereby further increasing the drain on assets in the early stages of amortization.

If the BOD of sewage or waste contains a fraction in the form of settleable solids and if the stream velocities are sufficiently low (as in pool areas behind dams), settleable solids will deposit on the bottom of the river and, if not scoured by freshets, will accumulate. Food supply continues to accumulate until the amount eaten from the pile per day just balances the daily increment of fresh organics deposited. When this equilibrium is reached, the combined daily payment of debts from the accumulation is equivalent to paying in 1 day the total debt of the deposit of that day. This places an excessive demand on the assets of the stream in the area of the deposits and greatly adds to the burden of amortization in the early stages.

In shallow streams, where the mass of water in relation to the surface area of the streambed is small, the biological growth attached to the bed has the facility of extracting and storing in its mass an organic food supply. This natural mechanism resembles a trickling filter within the stream channel. In streams of moderate depth biological floc is formed and dispersed throughout the channel by currents, increasing in size in transit, with excess floc depositing along the shore and in eddy areas. This resembles an activated sludge system within the river channel. As with sludge deposits, this extraction and accumulation will build to equilibrium storage so that the amount amortized per day from the accumulated storage just balances the daily amount extracted. This likewise places a very high burden on the oxygen resources of the stream in such a region. These abnormalities are dealt with in more detail later.

The Dissolved Oxygen Balance

To summarize, for any river solvency in transit along its course is reflected by the balance of outgo and income dictated by four primary self-purification factors:

1. Stream runoff, including condition as to residual unpaid debt and dissolved oxygen assets arriving at the head of the reach and from tributaries along the channel course.
2. Time of passage, the time element in the amortization of debts from section to section along the course of the stream from sources of waste effluent discharge.
3. Temperature, the natural or induced temperature of the river water from reach to reach.
4. Reaeration, the amount of dissolved oxygen contributed from the atmosphere to the occupied channel volume from section to section along the river.

The key factor for a particular stream is runoff, since to a high degree the other parameters are related to it, influencing both the asset and the liability sides of the ledger. A change in runoff, in addition to its direct effect, exerts an indirect effect on the river stage, altering the channel elements involved in the time of passage and reaeration.

Therefore self-purification, or waste assimilation capacity, is not a fixed quantity but a dynamic variable that changes from day to day and season to season, paralleling the patterns of hydrologic variation peculiar to each river system.

The dissolved oxygen profile resulting from specific organic waste loading therefore has stability only to the extent that the waste loading remains steady and a stable river regime is established and prevails during the time of passage downriver. Normally stable river hydrographs do not prevail for long intervals; hence dissolved oxygen profiles and the associated waste assimilation capacity are dynamic variables that must be defined in statistical terms. This reality of hydrologic and biologic variability is fundamental to rational analysis and is basic in reaching practical solutions to waste disposal and pollution control.

QUANTITATIVE DEFINITION

The Rational Method

From the briefly sketched nature of organic self-purification it is evident that the liability side of the ledger, deoxygenation, is primarily a biochemical process, whereas the asset side, reoxygenation, is essentially a physical and chemical process. Although the resultant dissolved-oxygen balance is the combined effect of both systems, dealing with deoxygenation and reoxygenation separately permits refined quantitative determination of each, employing the basic parameters that are specific to each system. The separately determined assets and liabilities, reach by reach, are then combined to give the dissolved oxygen profile along the course of the river. The methods of quantitative computation for each system are therefore developed separately for practical application. This approach is referred to as the *rational method of stream analysis.*

Phelps' Deoxygenation Law

We are indebted primarily to Phelps [1] for a generalized expression that serves as a basis for the quantitative determination of the liability fraction in organic self-purification: "*the rate of biochemical oxidation of organic matter is proportional to the remaining concentration of unoxidized substance, measured in terms of oxidizability.*" Phelps' law then is expressed in differential form as

$$\frac{-dL}{dt} = KL,$$

which may be integrated to

$$\log_e \frac{L_t}{L} = -Kt$$

or

$$\log \frac{L_t}{L} = -0.434Kt = -kt,$$

$$\frac{L_t}{L} = 10^{-kt}, \qquad (4\text{–}1)$$

where L is the initial oxidizability at time zero (the initial total outstanding organic debt); L_t represents the corresponding value at time t. The term L_t/L represents the fraction of the oxidizable organic debt remaining, and $1 - (L_t/L)$ is the fraction that has been oxidized, or amortized. The rate of reaction, the constant k, represents the rate at which the amortization proceeds. If time t is taken in days, k is found experimentally to have a value of 0.1 at 20°C.

The Normal Rate of Biochemical Oxidation

Phelps' monomolecular reaction formulation is generally referred to as the *normal rate* of satisfaction of BOD. The normal amortization schedule expressed in percent of the total initial outstanding BOD debt is listed in Table 4–1.

Table 4–1 Normal Amortization Schedule of Organic Debts at 20°C, $k = 0.1$

	Percentage of Total Initial Debt		
Time (days) (1)	Remaining at Beginning of the Day (2)	Paid During the Day (3)	Cumulative Payment (4)
0	100		0
		20.6	
1	79.4		20.6
		16.4	
2	63.0		37.0
		13.0	
3	50.0		50.0
		10.2	
4	39.8		60.2
		8.2	
5	31.6		68.4
		6.6	
6	25.0		75.0
		5.0	
7	20.0		80.0
		4.2	
8	15.8		84.2
		3.3	
9	12.5		87.5
		2.5	
10	10.0		90.0
		2.1	
11	7.9		92.1
		1.6	
12	6.3		93.7
		1.3	
13	5.0		95.0
		1.0	
14	4.0		96.0
		0.8	
15	3.2		96.8
		0.7	
16	2.5		97.5
		0.5	
17	2.0		98.0
		0.4	
18	1.6		98.4
		0.3	
19	1.3		98.7
		0.3	
20	1.0		99.0

With k of 0.1 the normal biochemical oxidation of organic matter proceeds at such a rate that it is 20.6 percent completed $(1 - 10^{-0.1})$ at the end of the first day; and during each succeeding day, in accordance with Phelps' law, the remaining unoxidized organic matter is further reduced by 20.6 percent of its value. Although the rate of amortization is constant, the amount paid on each succeeding day diminishes, as may be seen from column 3 of Table 4–1, which shows the decidedly unbalanced payment schedule, with the heavy payments in the early periods.

At this point it is well to emphasize the distinction between instantaneous rate of amortization and the amount amortized over the 24-hr period, namely, 20.6 percent. The instantaneous rate in differential form expressed as K in natural logarithms is 2.3×0.1, or 0.23, which implies that, if this instantaneous rate persists over 24 hr, 23 percent of the debt will be amortized. The instantaneous rate persists, but the residual declines from instant to instant through the 24-hr period; hence the amount amortized, taking into account this declining residual over 24 hr, is 20.6 percent of the residual at the beginning of the day.

Plotting as coordinates column 1 and 2 of Table 4–1 on semilog paper (Figure 4–1) provides a linear relation between time t in days and log of the percentage of debts remaining (Y), represented by the straight-line equation

$$\log Y = 2.0 - kt, \tag{4-2}$$

in which the reaction rate k is the slope of the line $\Delta Y'/\Delta X$. This is a convenient form for graphical integration of multiple sources of organic load along the course of a stream, as will be illustrated later.

Half-Life and the Standard BOD

In the normal amortization schedule, of special interest are two periods—3 days and 5 days. At the end of 3 days 50 percent of the debt is paid and 50 percent remains outstanding; thus for the normal decay rate the *half-life* of the organic debt is 3 days. Again in accordance with Phelps' law, during each succeeding 3-day interval the remaining debt is reduced by 50 percent; at the end of 6 days 25 percent remains, and at the end of 9 days 12.5 percent remains, and so forth.

The debt paid at the end of 5 days is referred to as the standard biochemical oxygen demand. This derives from laboratory procedure prescribed in *Standard Methods* [2] wherein the amount of oxygen utilized in a 5-day period of incubation at 20°C is determined. Referring to the amortization (Table 4–1), it will be noted that at the normal rate in 5 days about 32 percent of the debt remains unpaid and 68 percent of the total debt has been amortized. Hence the normal standard

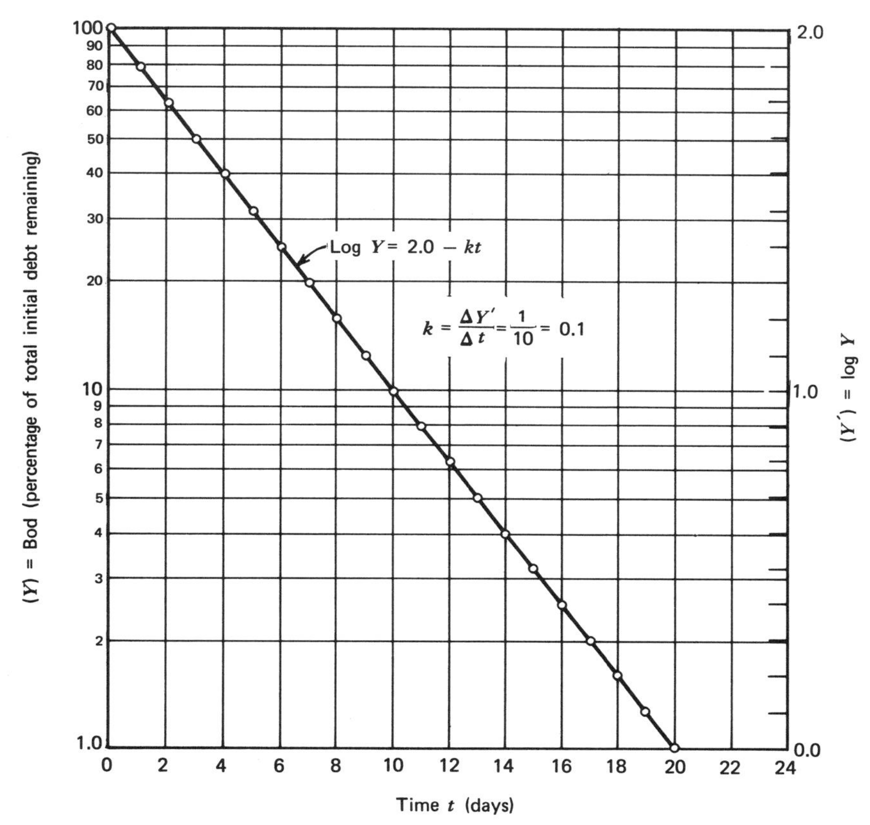

Figure 4–1 Normal amortization of organic BOD debt at 20°C, $k = 0.1$.

BOD represents only 68 percent of the total. Thus, if the reaction rate is normal ($k = 0.1$), the standard 5-day BOD value is divided by 0.68 to convert to the total initial BOD debt that ultimately must be satisfied.

It is unfortunate that in much of the literature BOD values are stated without reference to the period of incubation (1, 3, 5, or 10 days) or indication whether a value has been converted to total ultimate by dividing by the proportion utilized in the period of incubation. Confusion can be avoided by indicating the time of incubaton by a subscript, BOD_5 representing the standard 5-day 20°C value, BOD_3 representing the half-life 20°C value, and so forth, and BOD_U representing the total initial debt, or the *ultimate demand*. If the temperature of incubation deviates from the standard 20°C, the specific temperature should also be indicated.

In addition to the period of incubation, the specific reaction rate must be known in order to determine the proportion of the total BOD that is exercised in the period of incubation. Usually mixed organic wastes of the urban complex and many industrial wastes approximate the normal reaction rate, $k = 0.1$. Effluents of biological treatment systems and some industrial effluents may deviate significantly from normal, necessitating determination of the specific reaction rate before converting the standard 5-day (BOD_5) value to the total ultimate debt (BOD_U).

The Effect of Temperature

A general expression for the effect of temperature on the reaction rate k is

$$\frac{k_1}{k_2} = \theta^{(T_1 - T_2)},$$

where k_1 and k_2 represent the reaction constants at temperature T_1 and T_2, respectively, and θ is the temperature coefficient. The accepted coefficient based on experimental results over the usual range of river temperatures is taken as 1.047. Thus the BOD reaction rate k at any working temperature that deviates from 20°C is

$$k_T = k_{20°C} \times 1.047^{(T-20)}. \tag{4-3}$$

A graphical illustration of the effect of temperature on the amortization schedule of organic debt is shown in Figure 4–2. Curve A is the normal rate ($k = 0.1$) at 20°C, with a half-life of 3 days and a BOD_5 of 68 percent (100 — 32). Increasing the temperature to 29°C increases k to 0.15 (curve B), speeding up the amortization, reducing the half-life to 2 days, and increasing BOD_5 to 82 percent (100 — 18). Conversely, decreasing the temperature to 14°C reduces k to 0.075 (curve C), slowing the amortization, increasing the half-life to 4 days, and decreasing BOD_5 to 58 percent (100 — 42).

In addition to its effect on the BOD reaction rate, temperature may also have a slight effect on the ultimate oxidizability of the organics, thereby increasing the ultimate total demand. Early laboratory studies suggest 1 to 2 percent increase per degree of temperatures above 20°C. However, although field studies confirm the increase in reaction rate, they do not indicate a detectable increase in ultimate demand with increasing temperatures. In practical application such refinement at this stage of knowledge is not justified; any error thus introduced is well within the limits of day-to-day variation in waste loads and the ability to determine precisely the time of passage along the stream channel.

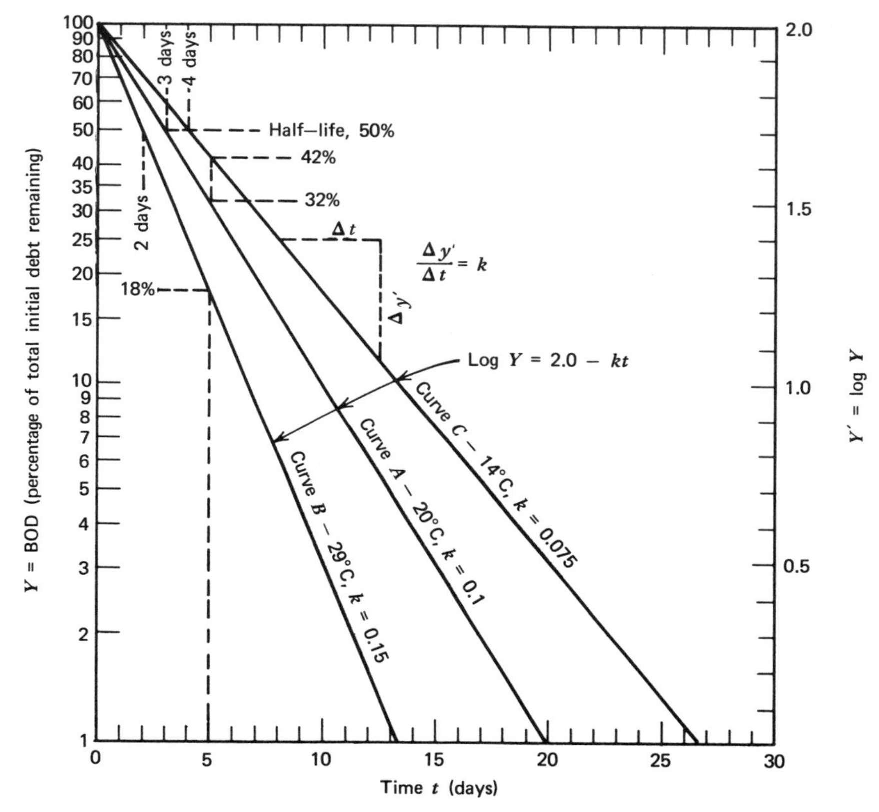

Figure 4–2 Normal BOD amortization and the effect of temperature [$k_T = k_{20°C} \times 1.047^{(T-20)}$].

Deviations from Normal

Since the rate of satisfaction of organic debt is a complex biological phenomenon, it is not unexpected that variations are observed. These variations do not preclude the useful application of the basic theory to stream analysis. Practical experience and professional judgment are important elements in interpreting these variations, in identifying significant abnormalities, and in deciding on an applicable rate or rates to be adopted. At times these abnormalities in reaction rate are lost when the waste is intermixed with the river water or with other organic wastes; in other instances they persist and must be taken into account.

Two general methods—the laboratory investigation and the stream survey—are available for identifying abnormalities and determining the specific BOD reaction rate. In laboratory investigation time and the temperature of incubation can be closely controlled, but it is difficult to simulate the dynamics of the river environment, including the biological chains. In the stream survey the river environment is that actually encountered, but temperature and particularly the time of passage along the course are highly variable from section to section because of instability in river hydrology. Again, practical experience and professional judgment dictate which method to employ; frequently both the laboratory and the stream survey are essential to distinguish and quantify significant deviations from normal.

Five significant abnormalities to deoxygenation are dealt with in some detail: abnormal specific BOD reaction rate, immediate demand, nitrification, sludge deposits, and biological extraction and accumulation.

Determination of the Specific Reaction Rate

Where the problem warrants and foreknowledge indicates a BOD reaction rate above or below normal, it is desirable to determine the specific reaction rates of major waste sources and of river water after intermixture of the waste effluents. Generally the laboratory method is employed to determine the specific BOD reaction rate k for raw waste, effluents, and river waters. This involves a series of BOD determinations made over time and applying analytical or graphical methods of evaluation of the laboratory results. For strong raw waste the laboratory procedure involves the usual dilution method, extending incubation from 1 to 10 or 20 days at equal or unequal time intervals. If wastes or river samples are low in total BOD, the jug technique is preferred since dilution is not involved. Since the amortization schedule is heavily unbalanced toward the early period, it is highly desirable to start BOD readings at 6- or 12-hr intervals for 2 or 3 days, followed by 1-day intervals to 5 to 7 days, and then at 2- or 3-day intervals up to 10 or 20 days. More recently continuous BOD recording has become possible by using dissolved oxygen electrodes. This provides a complete consecutive record from which appropriate increments can be selected.

The literature is replete with analytical methods for computing the BOD reaction rate, and there is considerable variation in the resultant k value among these methods applied to a given set of data. The rigidity of analytical methods may introduce further variations because of inability to take into account such factors as immediate demand, initial lag, or individual BOD values that are decidedly erratic or out of line. Such factors in many cases distort an otherwise quite normal rate. If the reaction is quite regular, the Reed–Theriault least-squares analytical procedure provides the most consistent results but, it is tedious and time consuming. Gannon and Downs [3] have provided a program for digital computer application for rapid solution. Other analytical methods are approximations introduced to facilitate computation and to reduce the labor involved. The shortcomings of analytical methods prompt adoption of graphical procedures.

Lee's Graphical Method. A unique graphical method is that proposed by Lee [4]. A series of graphs covering a range of k values is constructed on the basis of Phelps' law, so that a time series of BOD values that follow a specific reaction rate plots as a straight line on the grid constructed for that rate. Linear relation between the BOD on the Y-axis and time t on the Lee grid as an X-axis is obtained by positioning t on any arbitrary linear distance (such as 10 in.) in proportion to $(1 - 10^{-kt})$. Table 4–2 illustrates the solution of Lee's grid for $k = 0.1$, column 2 giving the values of $(1 - 10^{-kt})$, the proportion of total BOD satisfied corresponding with the time in days listed in column 1. The position on a linear plotting distance of 10 in. (scale factor 10) is listed in column 3. For example, for time t of 0.25 day the value of $(1 - 10^{-kt})$ is 0.056 (representing 0.056 of the total ultimate BOD) and is positioned on the linear scale of 10 in. at 0.056×10, or 0.56 in. Similarly 0.50 day is positioned at 1.09 in., etc., to 30 days positioned at 9.99 in., with the position 10 in. representing the total, or ultimate, BOD_U. Any desired scale factor may be employed, and the distances on that scale are column 2 multiplied by the factor. A series of grids for a range of k values at intervals of 0.02 is readily developed in a similar manner and can be ruled for reproduction and ready use.

In using Lee's graphical method successive trial plots of the laboratory data are made on selected grids. If the plot bends downward, it indicates that the selected k grid is too low; if the plot bends upward, the grid is too high. Subsequent selection of plotting grids leads quickly to the best straight-line fit for the data, identifying the appropriate BOD reaction rate k. A persistent curvature may indicate a two-stage reaction, usually associated with the onset of nitrification.

Extrapolation of a single stage line to the right to ultimate time gives directly the magnitude of the ultimate first-stage demand. Extrapolation of the line to the left, if intersecting the Y-axis at zero time, indicates that the reaction has proceeded normally from initial incubation. An intercept on the Y-axis at zero time indicates the presence of an immediate demand, with the Y intercept giving directly the magnitude of the immediate demand; an intercept on the X-axis indicates a lag in the start of the reaction, with the X intercept giving directly the magnitude of the time lag.

No adjustment of the direct laboratory results is required. All data,

Tables 4–2 Lee's Grid, BOD Reaction Rate Determination (k = 0.1)

Time t (days) (1)	$(1 - 10^{-kt})$ (2)	Plotting Distance (in.) (Scale Factor, 10 in.) (3)
0.25	0.056	0.56
0.50	0.109	1.09
0.75	0.159	1.59
1.00	0.206	2.06
1.25	0.250	2.50
1.50	0.291	2.91
1.75	0.332	3.32
2.0	0.369	3.69
2.5	0.438	4.38
3.0	0.500	5.00
3.5	0.553	5.53
4.0	0.602	6.02
4.5	0.646	6.46
5.0	0.684	6.84
6.0	0.749	7.49
7.0	0.800	8.00
8.0	0.842	8.42
9.0	0.874	8.74
10.0	0.900	9.00
12.0	0.937	9.37
16.0	0.975	9.75
20.0	0.990	9.90
30.0	0.999	9.99
∞	1.0	10.0

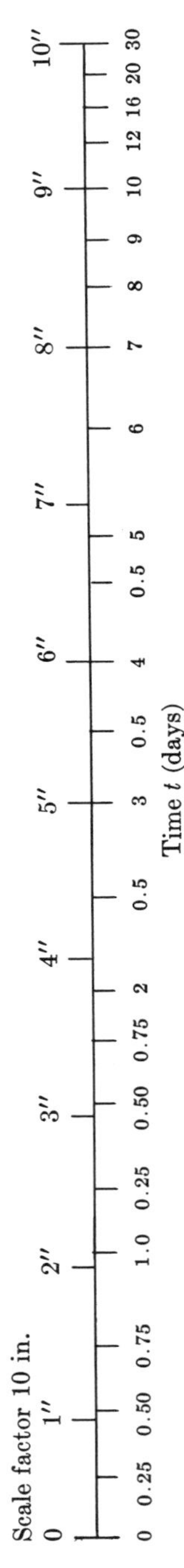

Lee's grid, k = 0.1, scale factor 10 in.

taken either at equal or unequal intervals, can be plotted. Also, the user has opportunity to exercise professional judgment in fitting the line, giving less weight to one or more values that are obviously out of line. Identification of such values also affords opportunity to reexamine the laboratory procedure for possible explanation of the deviation.

As a further verification of the reaction rate identified by Lee's grid, replotting of the data on semilog paper in terms of the BOD remaining unsatisfied can be readily developed by subtracting from the BOD_U the respective laboratory results. The slope of this line $(\Delta Y'/\Delta X)$ gives the value for k.

Table 4–3 BOD Rate Determinations*

Incubation Time (days)	BOD at 20°C (ppm)	BOD at 25°C (ppm)
0.71	3.0	3.0
1.00	6.5	7.5
1.83	9.0	12.0
2.84	13.0	20.5
4.0	19.0	26.5
5.0	22.0	29.0
6.0	25.5	31.0
8.0	30.5	35.0
10.0	34.5	37.0
11.0	35.0	38.0
12.0	36.0	39.0
14.0	38.0	41.5
18.0	41.5	43.5

* Sample March 4, 1958; initial NO_3, 0.01 ppm; final NO_3 (18 days), 0.01 ppm.

Experience indicates that Lee's graphical method can be used to determine the BOD reaction rate k to the nearest 0.02. This is adequate for application to stream analysis and well within sampling error and variations encountered in the pertinent biologic and hydrologic factors. The ability to identify visually and take into account the unusual characteristics in the BOD reaction series is a major asset in favor of the graphical method of analysis.

The following examples illustrate the application of Lee's graphical method to a time series of BOD determinations of samples of biologically treated industrial waste effluent.* Table 4–3 lists the laboratory BOD

* Data courtesy of the Michigan Water Resources Commission.

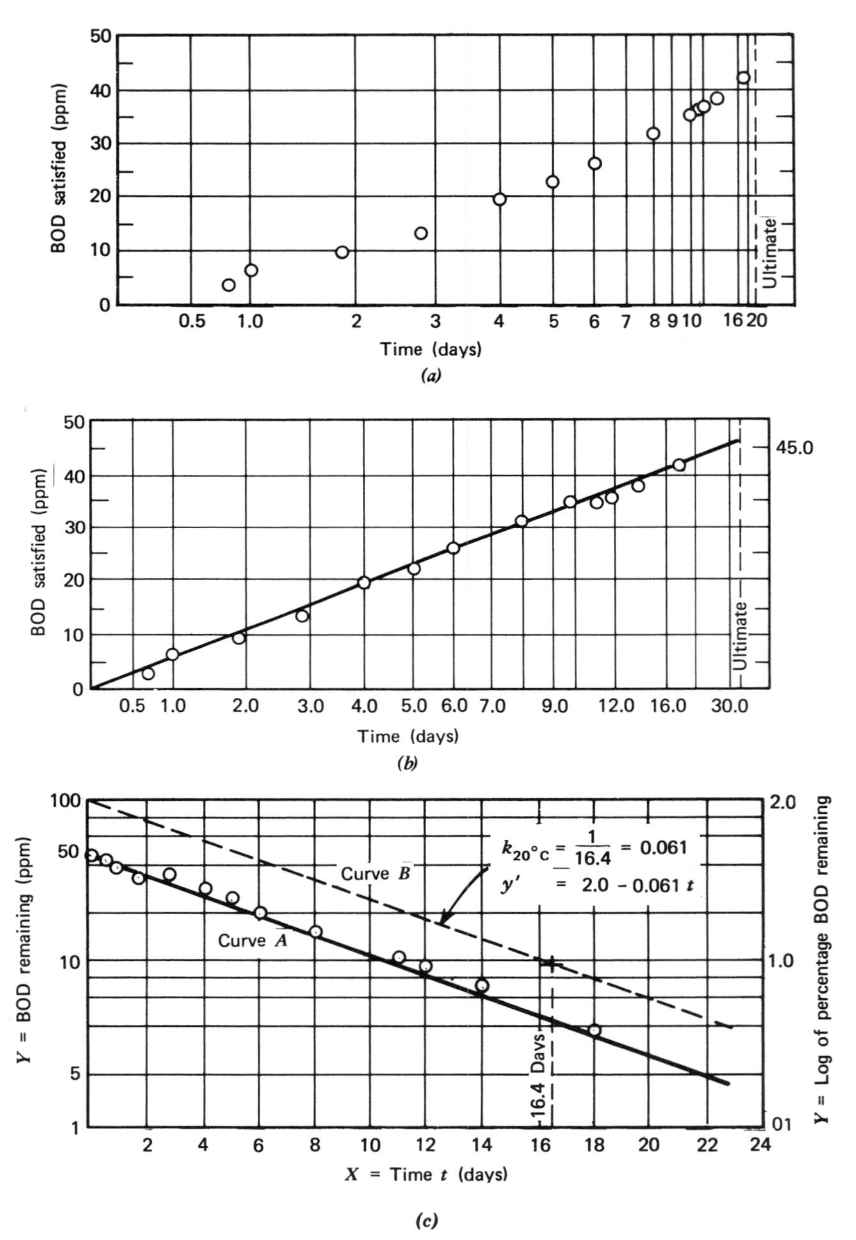

Figure 4–3 Rate of satisfaction of BOD at 20°C (sample March 4, 1958): (*a*) Lee's grid for $k = 0.10$; (*b*) Lee's grid for $k = 0.06$; (*c*) semilog plot.

time series of two portions of a sample, incubated at 20 and 25°C (columns 2 and 3, respectively). Figure 4–3*a* is a plot of 20°C incubation results on the Lee grid for $k = 0.10$ as a first assumption that the BOD reaction rate may be normal. It is observed that the data do not approach a straight line and show a definite bend upward, indicating that a k value of 0.10 is too high. Replotting the data on the k grid of 0.04, on the other hand, discloses a tendency to bend downward, indicating the selection of too low a k grid. Replotting on the k grid of 0.06 (Figure 4–3*b*) develops a good straight line, with no indication of two-stage association of nitrification, which was confirmed by laboratory determination run for nitrogen on samples. The line extrapolates to zero BOD at zero time and to a BOD_U of 45 ppm.

Subtracting the laboratory BOD time-series values from the ultimate 45 ppm and plotting on semilog paper (Figure 4–3*c*) provides a good straight-line fit in terms of the BOD remaining unsatisfied at the respective incubation periods. Transferring curve A in ppm BOD to percentage of total remaining (curve B) and obtaining the time intercept for a decline through a log cycle of 16.4 days gives a slope $\Delta Y'/\Delta X$ of 1/16.4, or $k = 0.061$. This is good verification of the selected k grid of 0.06.

The second portion of the effluent sample incubated at 25°C develops a reasonably good straight-line fit on a k grid of 0.08, as shown in Figure 4–4, extrapolating to zero and to a BOD_U of 45 ppm. Replotting of the BOD_U of 45 ppm in terms of the BOD remaining of the slope $\Delta Y'/\Delta X$ of 1/12.3, or $k = 0.081$ in verification of the selected k grid of 0.08.

Of special interest is the identical ultimate BOD developed for both 20 and 25°C incubation, indicating no increase in total BOD with increase in temperature. Temperature increase, however, increases the BOD

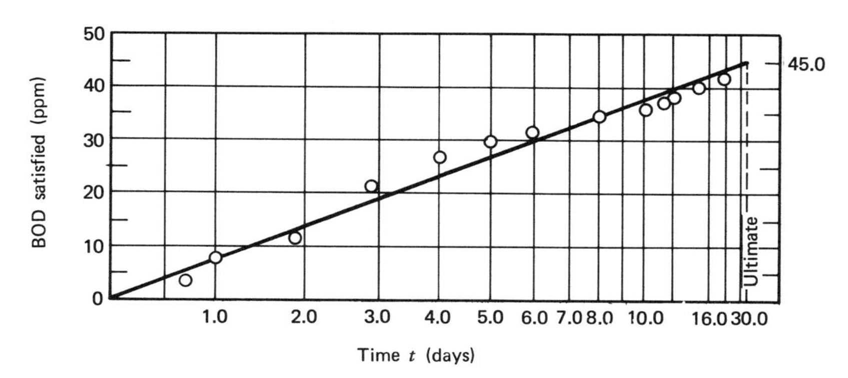

Figure 4–4 Rate of satisfaction of BOD at 25°C. Lee's grid for $k = 0.08$.

reaction rate from 0.06 (0.061) to 0.08 (0.081). Correcting $k_{25°C} = 0.081$ to a temperature of 20°C in accordance with equation 4–3 gives $k_{20°C} = 0.064$, which is in close agreement with $k_{20°C} = 0.061$ developed from the initial series incubated at 20°C.

Abnormality of Immediate Demand

Compounds Avid for Oxygen. Municipal or industrial organic wastes transported over long distances in sewers or in long outfalls undergo anaerobic decomposition in transit and may contain reduced products, such as hydrogen sulfide, which on discharge to the stream exercise an *immediate demand* (ID) on the oxygen resources of the stream. Other industrial wastes likewise may contain inorganic chemical products that are avid for oxygen.

Such demands may in instances comprise 5 to 15 percent of the ultimate demand. If discharge to the stream is direct, without treatment, this can cause a sharp initial drop in the dissolved oxygen at or within a short distance below the point of discharge, greatly adding to the burden of amorization in the early stages. Lee's method of BOD reaction-rate determination is a means of identifying such immediate demand.

Effluents with Abnormally Low Dissolved Oxygen. A sharp depression in dissolved oxygen is also associated with discharge of waste effluents with abnormally low dissolved oxygen when the flow of waste relative to the stream runoff is large. On small streams during the drought season it is not uncommon that practically the entire streamflow is diverted as community or industrial water supply, and the returned wastewater, even though receiving biological treatment, may contain as little as 1 to 2 ppm of dissolved oxygen.

Of increasing concern is the use of large volumes of condenser circulating water at steam-electric-power plants. If the plant is located on an unregulated river, the entire dry-weather flow may pass through the condensers more than once, with a rise in temperature of 10°C and at critical times up to 20°C. The temperature rise reduces the dissolved-oxygen saturation level and results in a loss of dissolved oxygen in the condenser discharge. For example, if the intake water during the critical warm-weather season is 25°C and dissolved oxygen is at saturation, a 10°C rise in water temperature results in a dissolved oxygen decline from 8.5 to 7.1 ppm, or a loss of 1.3 ppm; a 20°C rise results in a decline to 6.1 ppm, or a loss of 2.3 ppm. However, if the condenser-intake water is not at saturation, the loss due to increased temperature may not be pronounced. If some organic matter is contained in the condenser water, losses may be greater.

Abnormality of Nitrification

So far we have dealt with amortization of organic matter in terms of stabilization of the carbonaceous matter, which represents the true demand or permanent drain on the dissolved oxygen of the stream. Simultaneously or as a delayed reaction nitrogenous matter may be converted to oxidized nitrates. Although this will draw on the dissolved oxygen of the stream, it does not represent a permanent loss of oxygen, since the nitrate oxygen is available as *stored dissolved oxygen,* a reserve asset that is again available when the dissolved oxygen is depleted.

Nitrification represents a series of associated reactions by which ammonia and simple amine compounds are oxidized to nitrite and then to nitrate. Unlike the carbonaceous reaction, which is dominated by the rather rugged group of heterotrophic organisms, oxidation of the nitrogenous material is carried out by highly specialized groups of organisms, more restricted in numbers and much more sensitive to environmental conditions. Consequently, except under unusual conditions, the nitrification phase is not manifest in the natural stream environment. It is, however, a factor that is often encountered in the laboratory determination of the BOD reaction rate.

Two-Step Conversion. Although nitrification is a series conversion, it occurs in rather distinct steps carried out by two specialized autotrophic groups of organisms, as compared with the smooth operations of the multitude of organisms involved in stabilizing carbonaceous matter. The oxidation of ammonia to nitrite is carried out by bacteria of the genus *Nitrosomonas,* and the second phase, the conversion of nitrite to nitrate, is carried out by the genus *Nitrobacter.*

The stoichiometric relation of oxygen to ammonia nitrogen in conversion to nitrite is represented by

$$\begin{array}{lll} NH_3 + & 1\tfrac{1}{2}O_2 \rightarrow & HNO_2 + H_2O \\ 14 & 48 & \\ 1 & 3.43 & \end{array}$$

and conversion to nitrate by

$$\begin{array}{lll} HNO_2 + & \tfrac{1}{2}O_2 \rightarrow & HNO_3 \\ 14 & 16 & \\ 1 & 1.14 & \end{array}$$

or in combination

$$\begin{array}{lll} NH_3 + & 2O_2 \rightarrow & HNO_3 + H_2O \\ 14 & 64 & 62 \\ 1 & 4.57 & \end{array}$$

Thus 1 mg/l of ammonia nitrogen is equivalent to 4.57 mg/l of BOD; 1 mg/l of nitrate produced by the oxidation of ammonia nitrogen is equivalent to 1.03 mg/l of BOD.

It is to be expected that some variation from the strictly stoichiometric relations will be evidenced in the biological system due to carbon dioxide fixation, probably reducing the ratio of oxygen to ammonia nitrogen from 4.57 to 4.3 to 4.4.

The conversion is not always synchronized since each group has a different response at various levels of such environmental conditions as temperature and pH. Although autotrophic bacteria of the family Nitrobacteraceae are widely distributed in nature, the specialized requirements and long generation time of *Nitrosomonas* and *Nitrobacter* normally do not permit populations that are large enough to dominate in the stream environment. The initial attack is on the carbonaceous matter; then after a delay nitrification may take over under proper environmental conditions and adequate population, or it may proceed at a very slow rate by a small work force.

In biological systems for the treatment of wastes, particularly the activated sludge system, substantial nitrification can take place, with end product of nitrites and nitrates in the effluent. Where such effluents constitute a large proportion of the streamflow, an adequate population of the nitrifying bacteria is initially present, and at optimal temperature (25–30°C) nitrification quickly dominates. If stream temperature and pH are not optimal, nitrification may be insignificant or a substantial lag may occur, followed by a slow nitrification rate. Depending on the amount of ammonia in the effluent, the nitrification phase under optimal conditions can depress the dissolved oxygen substantially below the level accounted for by the residual carbonaceous BOD of the effluent.

Such a condition, described and analyzed by Courchaine [5] for a reach of the Grand River below the Lansing, Michigan, activated-sludge treatment plant, reveals the variable character of the nitrification reaction rate and illustrates the application of the rational method of analysis in defining the effect of the abnormality on the dissolved oxygen profile.

The Logistic Nitrification Rate. Where nitrification is known to be a potentiality, a differentiation is made between the carbonaceous and the nitrogenous factors in the laboratory determinations of BOD. If incubation is at 20°C, nitrification is slowed and a considerable lag usually occurs before this phase is apparent; hence the BOD rate determination should be extended to 20 days or more. Dual incubation with and without a nitrification inhibitor, such as methylene blue or ally-thiourea, provides the differentiation; the difference between the two gives the nitrogenous fraction.

The carbonaceous BOD fraction follows the law of constant rate and can be determined graphically by Lee's method. Since the nitrogenous reaction depends on establishment of an adequate population of the specialized bacteria, the 20°C BOD is best represented by the logistic curve; the *coefficient of growth* (CG) reflected by the slope of the straight line on logistic paper constitutes an overall measure of the nitrogenous BOD reaction rate.

Figure 4–5 reproduces Courchaine's [5] laboratory results for the 20°C BOD of a sample of the Grand River taken (February 6, 1961) about 1 mile below the Lansing treatment works outlet. Curve *A* represents the total uninhibited BOD; curve *B*, the nitrogenous; and curve *C*, the carbonaceous BOD. The logistic form of the nitrogenous BOD is evident, with an ultimate demand (or population saturation) of 12.5 mg/l. The respective nitrogenous BOD values, converted to percentage of the ultimate plotted on logistic paper (curve *A* of Figure 4–6), confirm the logistic form of the reaction. The coefficient of growth, the slope

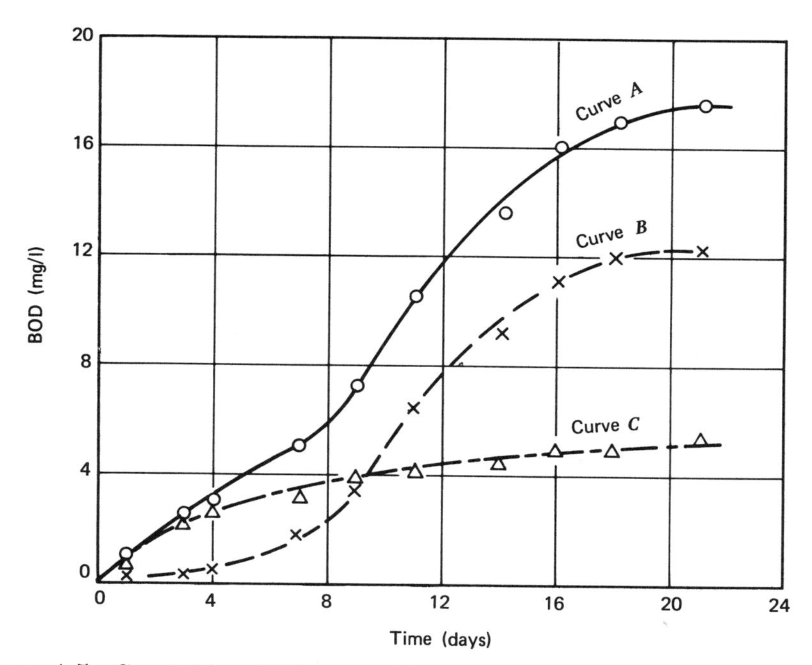

Figure 4–5 Grand River BOD curve (North Waverly Road sample, February 6, 1961). Curve *A*: total BOD; curve *B*: nitrogenous BOD; curve *C*: carbonaceous BOD.

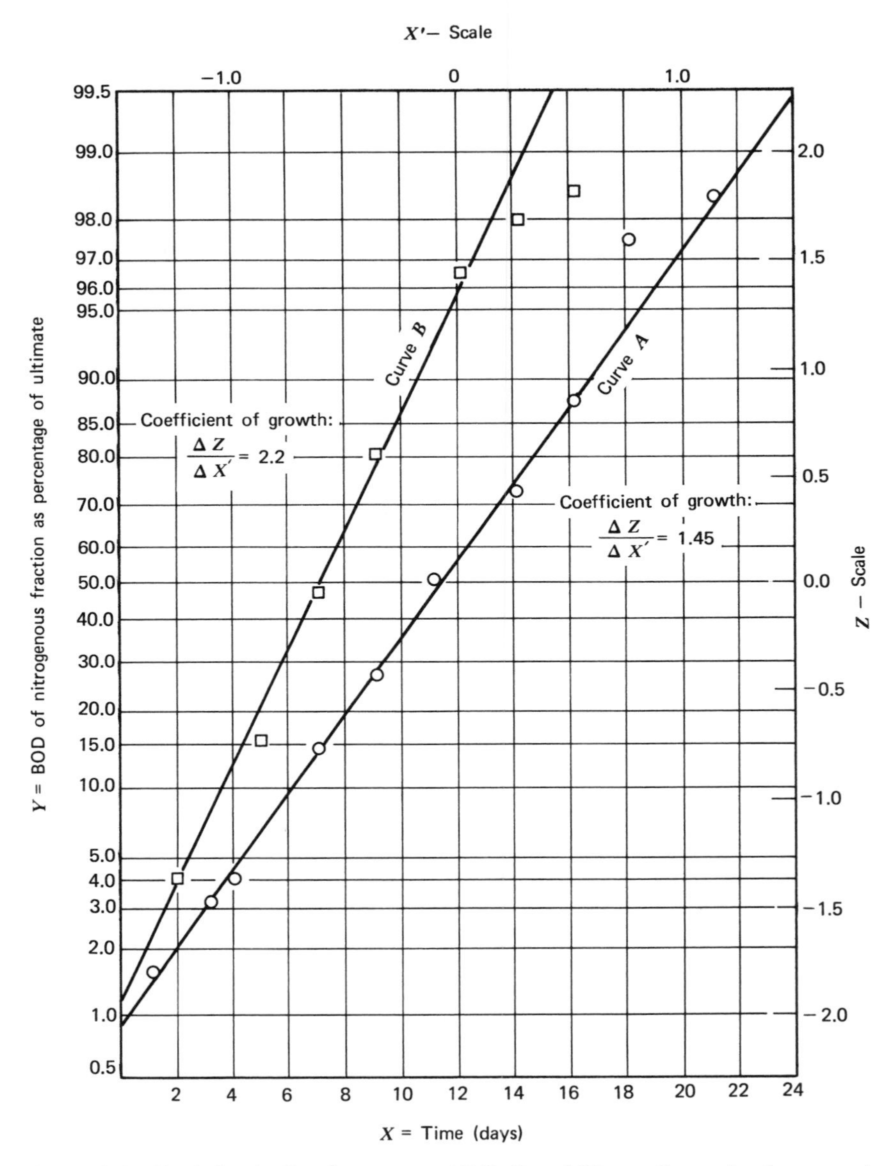

Figure 4–6 Logistic nitrification rate at 20°C, Grand River. Curve A: river sample taken on February 6, 1961; curve B: river sample taken on February 8, 1961.

$\Delta Z/\Delta X'$ as plotted with $Z = 8$ days, is 1.45 (with $Z = X' = 1$ day, $\Delta Z/\Delta X' = k = 0.181$). The midpoint of most active growth at 50 percent occurs 11 days after the beginning of incubation of 20°C. In contrast, curve B of Figure 4–6 is the logistic plot of BOD values for a sample taken at the same location on February 8, 1961. The form is logistic, but the coefficient of growth is much higher at 2.22, with the midpoint occurring earlier in the incubation period (at 7 days).

Similarly Figure 4–7 is the logistic plot of the nitrogenous BOD of a sample of the activated sludge treatment plant effluent, also taken on February 8, 1961. A two-stage nitrification is evident, with the growth coefficient of the first stage at 1.12 extending to about the 13th day, followed by a much higher rate with a growth coefficient of 2.52. Conceivably the two-stage reaction reflects the steplike division between the nitrite reaction carried out by the *Nitrosomonas* and synchronization after 13 days with the nitrate reaction carried out by the *Nitrobacter*. It is noted that the reaction rate of the second stage as represented by the growth coefficient of 2.52 is about the same as the rate reflected in the river-water sample taken on the same day, with a growth coefficient of 2.22. The laboratory results for samples of river water, although variable in terms of the growth coefficient, do not show two-stage growth, principally because in the river environment both groups of nitrifying organisms are established and have had opportunity to attain reasonable synchronization.

Nitrification Rate in the River Environment. The sensitivity of the nitrifying bacteria to environmental conditions results in a substantial difference between the laboratory 20°C reaction form and rate and that shown by the observed decline in BOD along the river course. Within the optimal river environment the populations of nitrifiers are established and synchronized, and consequently the oxidation of nitrogenous matter proceeds in accordance with the law of *constant rate* without the necessity of passing through the logistic growth phase. This is shown in the August 1960 Grand River survey results. The nitrogenous phase (based on nitrate production along the river course) proceeded at a very high rate, 90 percent oxidized in 0.62 day time passage, or $k_R = 1.6$. Courchaine [5] ascribes this high observed river rate to a combination of three factors: established populations of nitrifiers from effluents of three upstream activated-sludge plants; high nitrogenous BOD relative to the carbonaceous fraction and low streamflow; and optimal river temperatures, 28.5 to 26.4°C, induced by waste heat discharge from a steam electric-power plant.

Courchaine further verifies the abnormality of nitrification and the

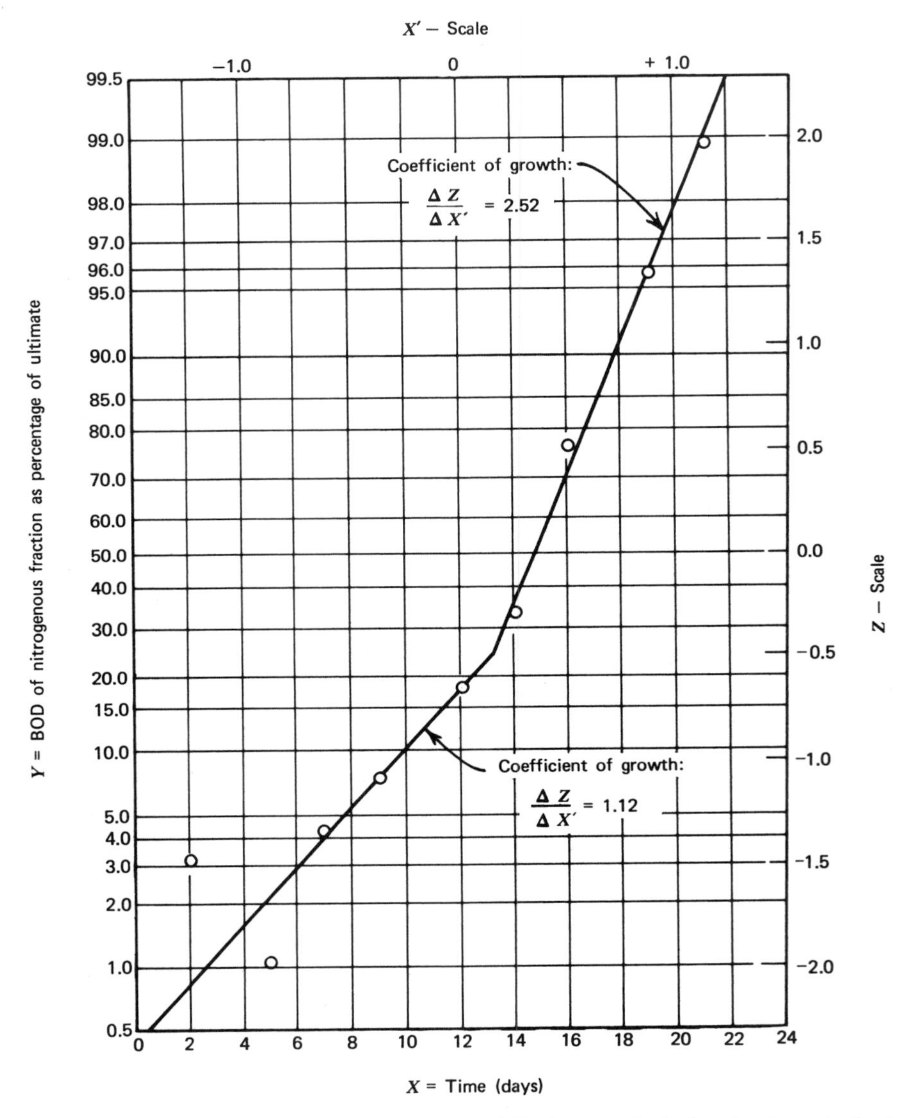

Figure 4–7 Logistic nitrification rate at 20°C. Activated sludge treatment plant effluent (sample taken on February 8, 1961).

high rate of BOD satisfaction by comparing computed dissolved oxygen profiles (employing the rational method) with that observed in intensive river sampling, as shown in Figure 4–8. The profile (curve A) that is based only on the carbonaceous BOD is distinctly above the observed dissolved oxygen; curve B, based on the combined carbanaceous and nitrogenous BOD, integrated at the river rate, is in reasonable agreement with the observed dissolved oxygen level; curve C, the profile based on the separate integration of the carbonaceous and the nitrogenous fractions determined from the observed increase in nitrates is also in reasonable agreement with the observed dissolved oxygen content.

In addition to nitrification, the abnormality of biological extraction and accumulation appears also to be involved in the Grand River, the net effect represented by the difference between curves B and C of Figure 4–8. This is further supported by the high river rate of the carbonaceous fraction at $k_R = 1.2$, compared with the laboratory BOD rate of the effluent of 0.08. Since the carbonaceous BOD fraction is small, the abnormality of nitrification dominates and is the primary factor in determining the shape of the dissolved oxygen profile.

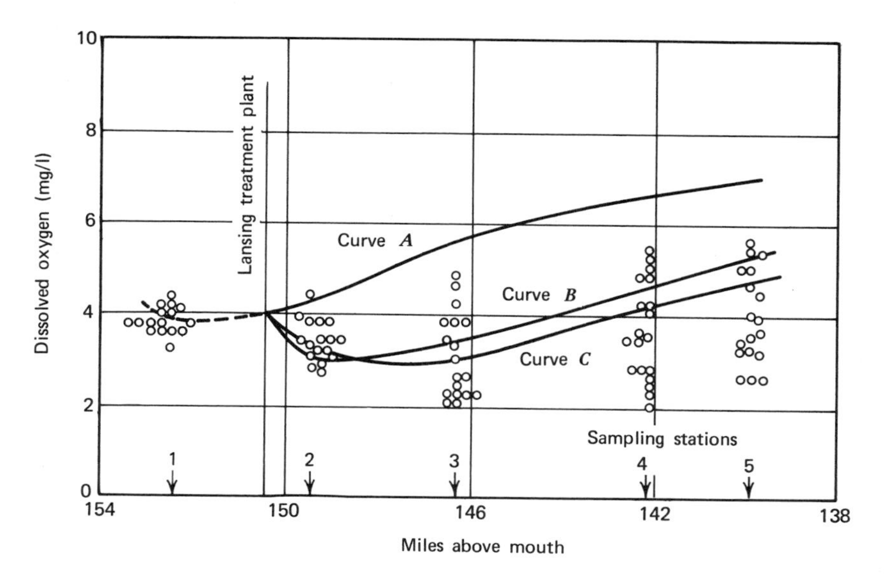

Figure 4–8 Grand River study. Observed dissolved oxygen (○) versus computed dissolved oxygen profiles (——). Curve A: carbonaceous BOD; curve B: combined carbonaceous and nitrogenous BOD; curve C: profile based on the separate integration of the carbonaceous and nitrogenous fractions, determined from the observed increase in nitrates. After Courchaine [5].

Where biologically treated effluents are discharged to small streams, the potentiality of nitrification abnormality exists and should be investigated. The determination of ammonia and organic nitrogen of the effluent affords a measure of the oxygen demand. The nitrogenous BOD rate determinations of the effluent and the river water, incubated at optimal temperature (28°C) employing the jug technique, provide an approximation of the expected river reaction rate.

Abnormality of Sludge Deposit

In Chapter 3 we dealt with the hydraulic conditions involved in channel deposit and scour. The slope of the time of passage curve defining velocity along the course of the channel identifies reaches of potential sludge deposit for any particular runoff regime. We now deal with the abnormality of sludge deposit as it influences the amortization of organic debts and alters the otherwise normal demand on the oxygen resources of the stream.

Types of Deposit. A distinction should be made between the fresh organic deposit arising primarily from current municipal and industrial waste discharges and the old compacted accumulations of partially stabilized organic residue and river muds from surface wash after flood periods. The latter we shall call *benthal deposit;* the former, *sludge deposit.* Actually where sludge deposits occur there usually also are benthal deposits underlying the sludge, and to some extent one classification blends into the other. However, for purposes of practical stream analysis we make the arbitrary separation, with the sludge as our major concern. The decomposition of these two classes of deposit is quite different and in turn exerts quite different effects on the oxygen resources of the stream.

Sludge deposits undergo active decomposition of a semianaerobic character, with end products readily leaching into the overflowing stream and utilizing dissolved oxygen. Gasification in the deposit results in sludge boils, and on occasion large floating masses rise to the surface, inducing a process of refloating and redepositing until active decomposition is practically completed. Where the sludge deposit is large in relation to the overflowing stream, dissolved oxygen may be completely exhausted and the gas bubbles may rise and break at the water surface, releasing hydrogen sulfide and other gases to the atmosphere. In such cases part of the oxygen debt is paid by the stream and part by the atmosphere.

In contrast, benthal deposits as here defined are relatively inactive; the residual organic matter remaining at the bottom is amortized at a very slow rate. Consequently, although benthal deposits over a long period of time may depress the level of oxygen dissolved in the stagnant water of stratified lakes and reservoirs, they exercise a negligible demand on the dissolved oxygen of flowing streams.

Accumulation and Demand. The effect of sludge deposit on the dissolved oxygen of the stream and methods for the quantitative measurement of that effect depend on the relation between accumulation of the deposits and the demand from these accumulations. If, following a scouring freshet, there is opportunity of only a day or two for deposit to accumulate, the effect on the dissolved oxygen is negligible. However, if opportunity to deposit and accumulate continues uninterrupted for 40 or 50 days, the demand is pronounced. The quantitative determination of deoxygenation involves the interrelation of five factors:

1. The BOD of the sources of waste and fraction subject to deposit.
2. The amortization of these BOD fractions in transit to the sludge deposit area.
3. The net daily addition of BOD to the sludge-deposit area.
4. The level of accumulation of BOD in the sludge pile.
5. The amortization of the BOD in the accumulation.

In practical application to stream analysis each of these factors is considered separately.

In dealing with sludge in quantitative terms it is the BOD content (not the solids bulk) of the deposit that requires evaluation. Three types of sludge deposit involve significant BOD fractions: settleable solids; flocculation or coagulation of colloids and nonsettleable solids; and biological extraction and accumulation. In most instances the primary source is the settleable solids of municipal and industrial wastes. In untreated wastes this usually constitutes about one-third of the total BOD, but with some industrial wastes that are high in suspended solids content may aggregate as high as 80 percent. A direct measure can be approximated by laboratory standard BOD of the total waste sample and of the supernatant after 1- or 2-hr settling, the difference constituting the fraction of BOD subject to deposit.

The second source, flocculation or coagulation of colloids and fines, is not as easily determined. In a freshwater stream this may constitute about 5 percent of the total BOD; in brackish or seawater, particularly in estuaries carrying inorganic turbidity, coagulation of organic fines and colloids may aggregate 25 to 35 percent of the total BOD of the waste sources.

The third source of organic sludge is the abnormality of biological extraction and accumulation that is usually associated with infestation of *Sphaerotilus*. The biological mass acting as a trickling filter or activated-sludge system has the capacity to extract suspended fines, colloids, and dissolved organics. Depending on channel characteristics, the excess sludge may be spread along the river course, depositing in eddy areas along the shore, or it may be transported to a downstream area of low velocity where it accumulates as a localized sludge deposit. Such a deposit reacts like other types of organic sludge, but the abnormality of biological extraction warrants further consideration and will be dealt with more fully later.

The second and third factors—namely, amortization in transit to the sludge deposit area and the amount of the daily deposit added to the sludge pile—involve the integration of the BOD sources down the river at the time of passage for the specific runoff and temperature, to the area where the potential deposit zones have been identified from the time of passage curve. This is a straightforward aerobic amortization in accordance with Phelps' law; no distinction is made in specific BOD reaction rate between the settleable BOD fraction and the nonsettleable portion. However, each fraction is integrated separately. The residual BOD of the settleable fraction from the integration to the deposit area provides directly the magnitude of the BOD load added daily to the sludge deposit; the difference between the upstream initial and the residual at the deposit area is the amount of debt amortized in transit in the flowing stream.

A simplified example that illustrates the procedure is shown in Figure 4–9. The city at station A contributes a total BOD load of 100,000 population equivalents (PE) of which the supernatant as the colloidal and dissolved fraction after 2-hr settling was determined as 60,000 PE. The difference of 40,000 PE is the fraction of BOD subject to deposit. Downriver at station B the industry discharges a total BOD load of 50,000 PE, of which 40,000 PE is the colloidal and dissolved fraction and 10,000 PE is the settleable fraction subject to deposit. The runoff is 500 cfs, with an increment of 50 cfs contributed by the tributary at station D. The time of passage from A to B is 0.5 day; from B to C, 1 day; and from C to D, 2 days; the river-water temperature is 25°C. A small dam located at C, providing a power intake, reduces velocity below 0.6 fps, allowing deposits to occur behind the dam.

The colloidal and dissolved fraction and the settleable fraction of the BOD loads are integrated separately on semilog paper at the time of passage for the normal amortization (adjusted to 25°C) as curves *A* and *B*, respectively. The settleable fraction commences at zero time at A at 40,000 PE, arrives at B after a time of passage of 0.5 day with a residual of 34,600 PE. At B, 10,000 PE is added from the industry,

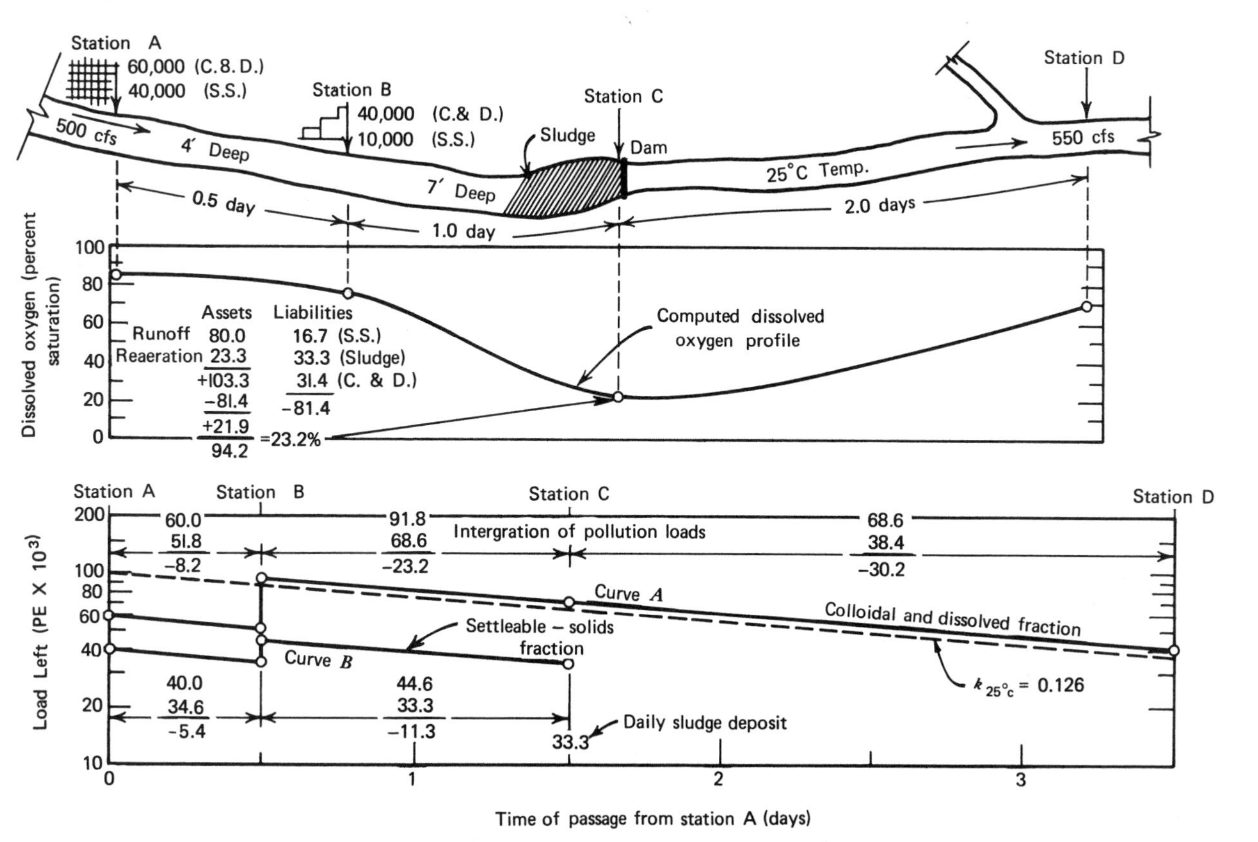

Figure 4–9 Example of sludge deposit abnormality procedure. Abbreviations: S.S.—settleable solid; C. & D.—colloidal and dissolved.

increasing the settleable fraction to 44,600. This combined debt then is integrated (at the same slope $k_{25°C} = 0.126$) an additional 1 day to station C, with the residual settleable BOD debt of 33,300 PE arriving over the sludge deposit area behind the dam. This then constitutes the residual BOD of the settleable portion of the debts that is daily added to the sludge deposit area. The portion of the settleable BOD debt amortized in transit in the flowing stream is the difference between the initial settleable debt and the residual at the deposit area, or 16,700 PE (50,000 — 33,300), 5400 PE between A and B, and 11,300 PE between C and D.

The separate integration of the colloidal and dissolved fraction of the debts (not involved in determination of the amount of the daily load deposited behind the dam) exercises a demand on the dissolved oxygen in transit from A to B to C, as shown by curve *A* of Figure 4–9. Integrating at the same slope the 60,000 PE at A is amortized to 51,800 PE at B where an additional 40,000 PE is added, giving a total at B of 91,800 PE. This in turn arrives at C with a residual of 68,600 PE. The amount of the colloidal and dissolved fraction amortized in transit, then, is the difference between the initial and the residual, or 31,400 PE (100,000 — 68,600), distributed as 8200 PE between A and B, and 23,200 PE between B and C. Thus the total debt, settleable plus colloidal and dissolved fraction, amortized in transit from A to B to C is 48,100 PE (16,700 + 31,400).

If prevailing hydrology remains constant for a protracted period and depositing continues at the rate of 33,300 PE per day, an equilibrium condition is approached so that the demand induced by the chain of biological life working on the sludge accumulation will equal the total BOD of the daily addition to the pile. Under the special condition of equilibrium accumulation the combined demand on the dissolved oxygen resources of the stream down to station C is that exercised in transit plus that from the sludge accumulation—namely, 81,400 PE (48,100 + 33,300). However, for any condition other than that of equilibrium the demand from sludge depends on the level of accumulation and the rate of amortization within the sludge pile.

The Sludge-Deposit Reaction Rate. Although conditions and biological organisms are different, and the reaction rate is not normal, the same general rule laid down in Phelps' law governs the amortization of BOD contained in sludge undergoing active decomposition. Expressed in differential form,

$$\frac{-dS}{dt} = K'S. \qquad (4\text{–}4)$$

Integrating to

$$\log_e \frac{S_t}{S} = -K't \tag{4-5}$$

or

$$\log \frac{S_t}{S} = -0.434K't = k't, \tag{4-6}$$

whence

$$\frac{S_t}{S} = 10^{-k't}. \tag{4-7}$$

Here S is the initial oxidizability of the sludge at time zero, S_t represents the corresponding value at time t, and S_t/S represents the fraction of the oxidizable debt remaining; $1 - (S_t/S)$ is the fraction that has been amortized.

According to Streeter [6], the reaction rate k' at 25°C, taking the day as the time unit, may range downward from 0.05 or 0.03 to less than 0.01, according to the relative age of the deposits. For active decomposition of current deposits, the author has found that at 25°C a mean rate of 0.03 is applicable to most river situations. Of more significance than precise knowledge of the reaction rate k' is the interval of stable hydrologic conditions permitting depositing and accumulation between occurrence of scouring freshets.

As with aerobic organics in the flowing stream, the amortization of BOD in sludge deposits can be represented graphically as a straight line on semilog paper with the slope of the line $\Delta Y'/\Delta X$ as k'.

The instantaneous rate at 25°C, expressed as K', is 2.3×0.03, or 0.069, which implies that, if this rate persists over 24 hr, 6.9 percent of the debt will be amortized. The instantaneous rate persists, but the residual declines from instant to instant through 24 hr; hence the amount amortized taking into account the declining residual over a 24-hr period is 6.67 percent of the residual at the beginning of the day $(1 - 10^{-0.03})$. With $k' = 0.03$, the biochemical oxidation of organic matter in active sludge decomposition proceeds at such a rate that it is 6.67 percent complete at the end of the first day; during each succeeding day the remaining unoxidized organic matter is further reduced by 6.67 percent of its current value.

Phelps reasoned that, notwithstanding the dominating influence of anaerobic reactions within the sludge accumulation, diffusion of the end products to the overflowing stream for supply of dissolved oxygen, the temperature coefficient is more like that for aerobic conditions. Thus the sludge reaction rate k' for the usual temperature ranges encountered in streams may be taken as

$$k'_T = k'_{25°C} \times 1.047^{(T-25)}. \tag{4–8}$$

Accumulation and Storage. There remains the determination of the accumulation and storage for conditions other than the special case of equilibrium. The condition of continuous oxidation taking place simultaneously with continuous depositing is expressed in differential form as

$$\frac{-dS}{dt'} = K'S_{t'} - D, \tag{4–9}$$

where D is the rate of deposition expressed in the same units as K' and $S_{t'}$ is the accumulated storage.

After integrating and transposing, the general condition representing cumulative storage of BOD in the sludge pile for any finite period of continuous oxidation and depositing is given by

$$S_{t'} = \frac{D}{2.3k'}(1 - 10^{-k't'}). \tag{4-10}$$

Time (t') in this case is not the time of passage downstream; it is the interval over which continuous depositing can take place and accumulate without scour or interruption. Taking the usual time unit of the day for the deposition and oxidation rates, and $k' = 0.03$ at 25°C, the period required to attain various stages of accumulation and the corresponding demand exercised by the accumulation based on this equation are expressed in Table 4–4 as percentage of the BOD of the daily deposit.

Since theoretically an infinite time is required to attain equilibrium, the general equation reduces to the specific equilibrium form

$$S_{t'} = \frac{D}{2.3k'} = S_e, \tag{4–11}$$

which for a unit rate of addition of BOD per day is

$$S_e = \frac{1}{2.3 \times 0.03} = 14.5. \tag{4–12}$$

Thus at equilibrium at 25°C the accumulated BOD in storage in the sludge is 14.5 times that of the daily addition of settleable solids BOD depositing to the sludge pile. It will also be noted that, if k' increases with increasing temperature, accumulation S_e declines; conversely, as

Table 4–4 Sludge BOD Accumulation and Demand at 25°C, $k' = 0.03$

Time (days)	Accumulation as Percentage of the BOD of the Daily Deposit	Daily Demand from the Accumulation as a Percentage of the BOD of the Daily Deposit
2	187	12.9
3	271	18.7
4	350	24.1
4	423	29.2
10	724	49.9
20	1086	74.9
30	1267	87.4
40	1359	93.7
50	1404	96.8
60	1427	98.4
70	1438	99.2
80	1444	99.6
90	1447	99.8
Ultimate	1450	100.0

temperature declines, accumulation in storage increases toward a higher equilibrium level.

Although theoretically an infinite time is required to attain equilibrium, it is seen from Table 4–4 that in 40 to 50 days equilibrium is practically attained, exercising a demand of 93.7 to 96.8 percent of that expected at full equilibrium. For short periods of opportunity to deposit and accumulate the demand exercised is drastically below that exercised at equilibrium; at 5 days the accumulation is only 4.23 times the daily load added and the daily demand exercised on this accumulation is only 29.2 percent of the daily addition to the pile; at 10 days the accumulation is about one-half that at equilibrium and the daily demand exercised is therefore only half that at equilibrium.

A convenient graphical method of representing accumulated BOD in storage at any time and the demand exercised from the storage is afforded by plotting the values of Table 4–4 in terms of their deficits below equilibrium conditions. Both deficits are represented by a single linear relation if expressed as percentage of their equilibrium values. This method of representation necessitates knowing the equilibrium storage value, which is readily available from the ratio of the daily deposit rate to the daily amortization rate D/K'. Referring to Table 4–4, values of accumulated storage of column 2 are expressed as $100[(S_e - S_t')/S_e]$, and values of demand as $100[(D_e - D_{t'})/D_e]$. Plotting these deficits on semilog paper provides the single straight line for both deficits. The slope of this line, $\Delta Y'/\Delta t'$, is fixed by k'. A separate line for each value of k' for a range of temperatures from 10 to 35°C at 5°C intervals provides the nest of curves shown in Figure 4–10.

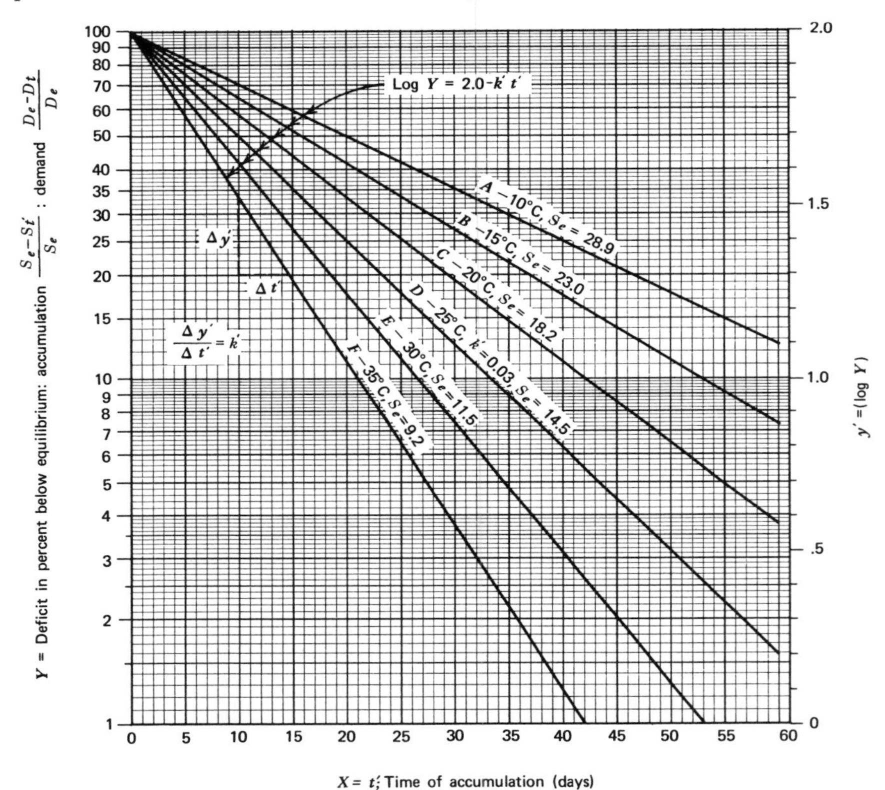

Figure 4–10 Sludge BOD accumulation and demand, and influence of temperature $[k'_T = k'_{25°C}\ 1.047^{(T-25)}]$.

Where continuous depositing occurs without interruption, these curves afford a ready means for evaluating sludge accumulation and demand. For example, if we wish to know the amount of accumulated BOD in storage and the demand exercised from it after 20 days of initial depositing at a temperature of 25°C, from curve D at 20 days we read the deficit on the Y-axis as 25.1 percent. The equilibrium storage by equation 4–12 is 14.5 times the daily deposit; the accumulated BOD in storage after 20 days of continuous depositing then is $[(100 - 25.1)/100]$

$\times$ 14.5, or 10.86 times the BOD of the daily deposit. In turn the demand exercised from accumulated storage of this magnitude as a percentage of the BOD of the daily deposit is 100 — 25.1, or 74.9 percent. If the magnitude of the daily deposit rate is known in terms of pounds of BOD or population equivalents, as determined in the previous illustration with 33,300 PE, the daily deposit rate and the storage and demand are obtained in absolute terms as 10.86 $\times$ 33,300 and 0.749 $\times$ 33,300 PE, respectively. Thus accumulated storage and demand are readily obtained graphically for any period of continuous accumulation under

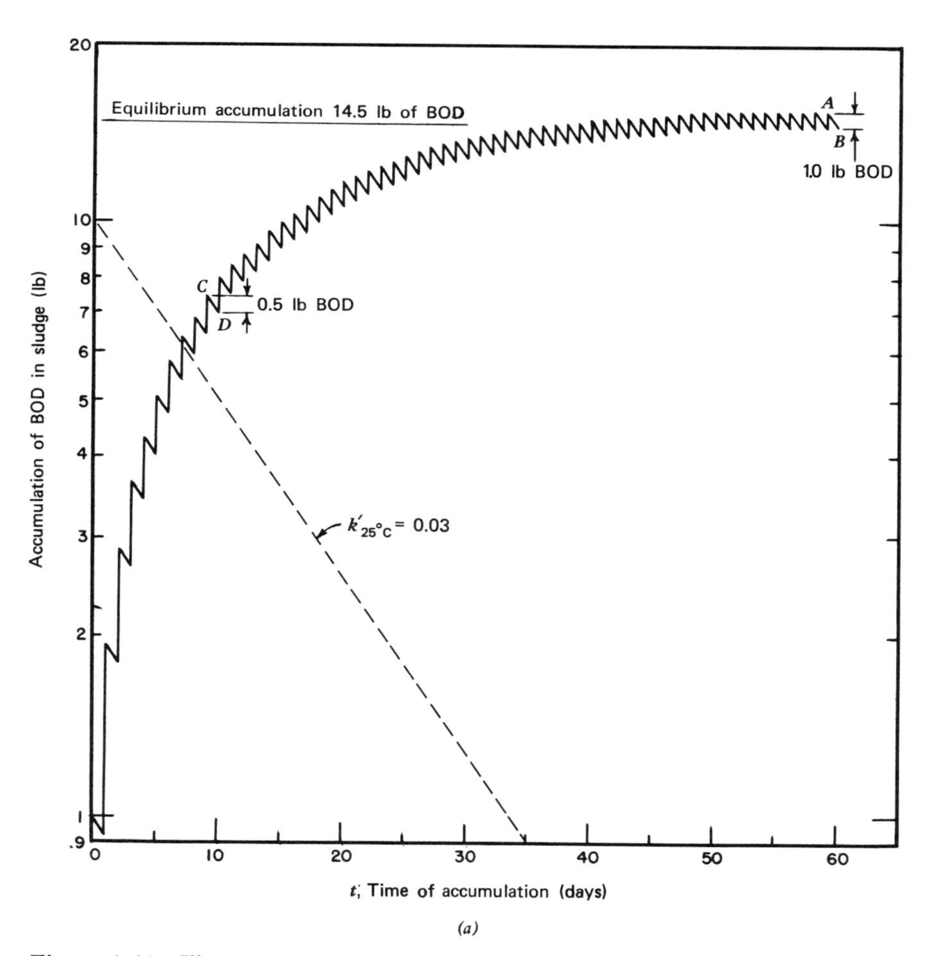

Figure 4–11 Illustrations of sludge accumulation and demand: (*a*) cases 1 and 2 (k = 0.03 at 25°C); (*b*) case 3; (*c*) case 4; (*d*) case 5; (*e*) case 6; (*f*) case 7.

any temperature from these curves or intermediate curves laid off for other specific k' rates.

Graphical Integration of Sludge Accumulation. In practical stream application the nature of the past daily hydrograph is of primary importance because it usually controls the opportunity to deposit for varying periods and thus establishes the level of accumulation. If the hydrograph remains low (velocities through the deposit areas remaining below 0.6 fps for 40 or 50 days), the sludge pile approaches equilibrium. If scour intervenes from a freshet, such accumulation may be completely flushed. If it is desired to check an observed dissolved oxygen stream condition

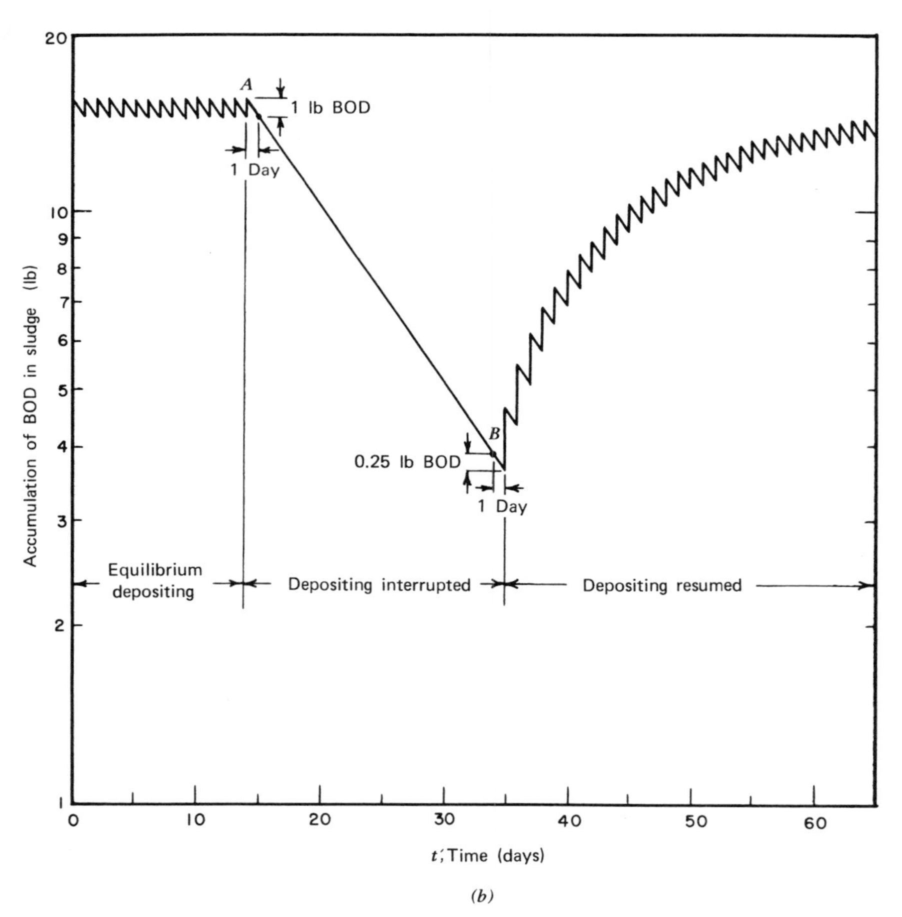

(b)

Figure 4–11 (Continued).

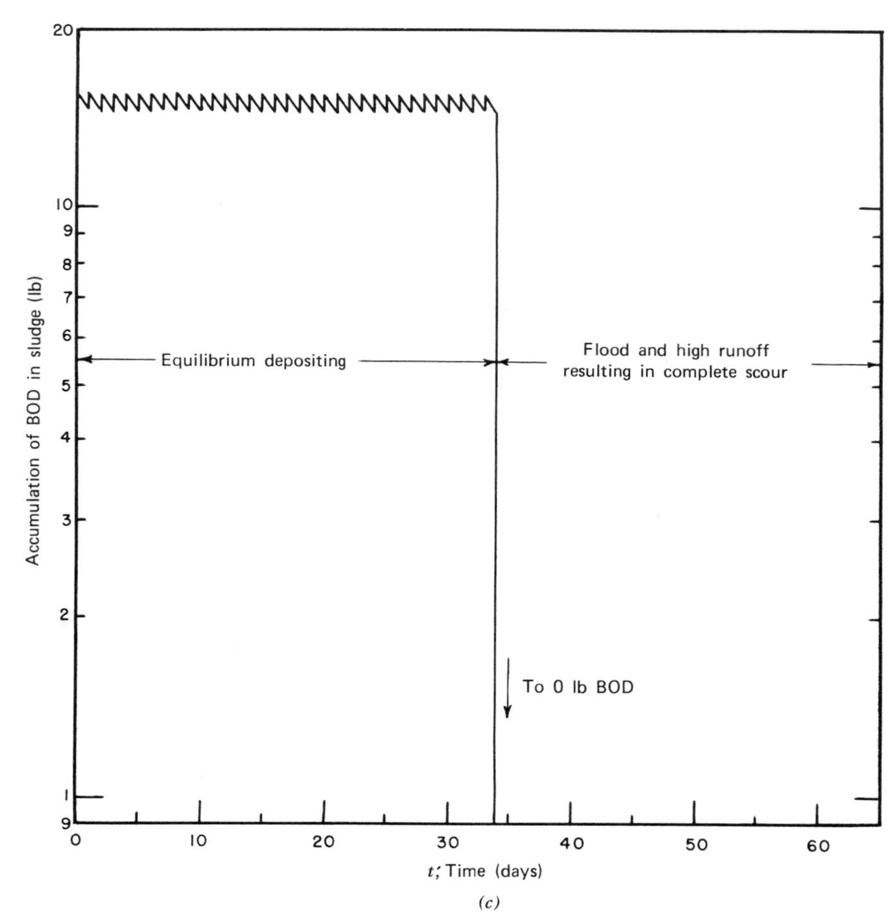

(c)

Figure 4–11 (**Continued**).

for which only a few days of prior accumulation has been possible, the accumulation must be integrated to the particular day on which the dissolved oxygen values were observed. A simplified graphical approximation permits ready evaluation of a variety of sludge deposit situations.

In the graphical method of integrating the accumulation, daily deposit is considered as a slug added at the beginning of each day rather than as a continuous contribution throughout the day. Each day's addition to the residual accumulation of prior days is then amortized through the day on semilog paper at the appropriate k' slope for the temperature involved. The level of accumulation is thus directly available at any period, and the demand exercised on any particular day is the difference between the accumulation at the beginning of the day and that at the end of the day. A variety of sludge deposit situations, taking into account river hydrology and changes in settleable waste load, can be readily investigated, as shown in the following illustrations (Figure 4–11):

Case 1. This is a case of uninterrupted daily additions and a constant rate of decomposition, approaching equilibrium. The daily increment of BOD is taken as unity ($D = 1$) and the rate of amortization within

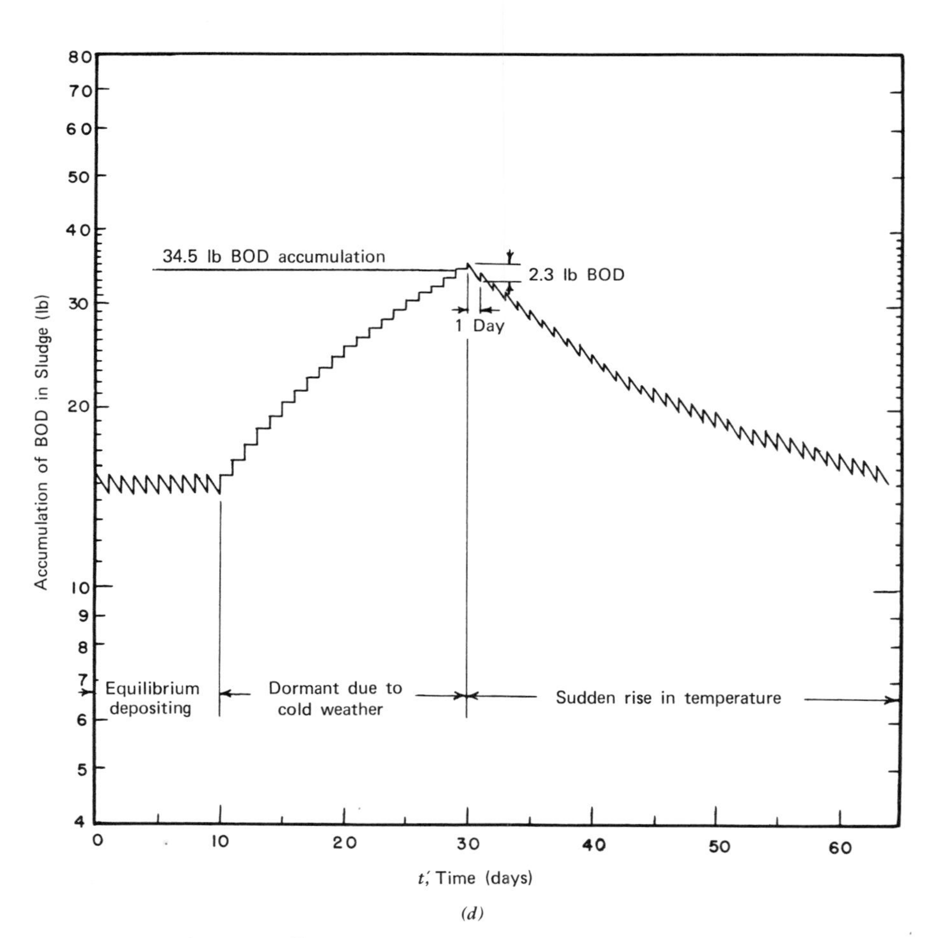

Figure 4–11 (**Continued**).

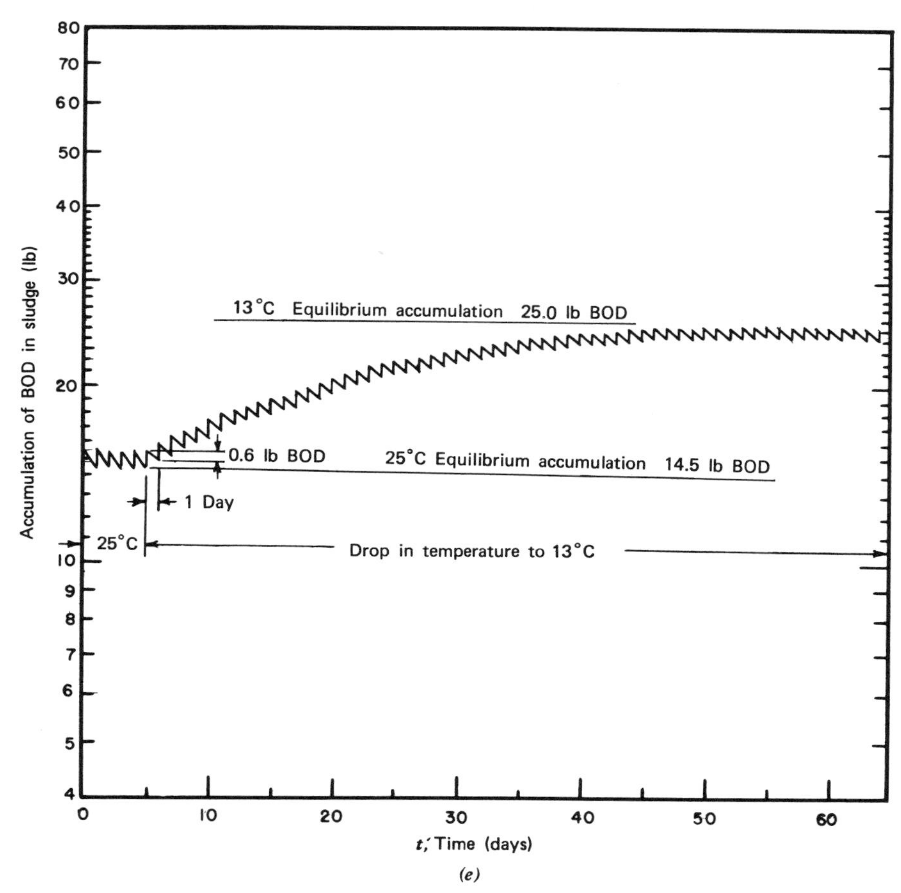

Figure 4–11 (**Continued**).

the sludge pile is laid off for $k' = 0.03$ at 25°C, as shown by curve *B*. The daily increments added to the preceding day's residual are successively integrated graphically through a 60-day accumulation interval. It will be observed that at the relatively slow rate of amortization the BOD residuals of the daily additions of deposit accumulate rapidly at the early stages and then gradually approach equilibrium level of 14.5 times the unit daily deposit. In 40 to 50 days the demand per day exercised in decomposition in the large accumulation is practically 1 unit of BOD which counterbalances the 1 unit of BOD added to the sludge pile per day.

Case 2. In many instances river hydrology will not permit 40 to

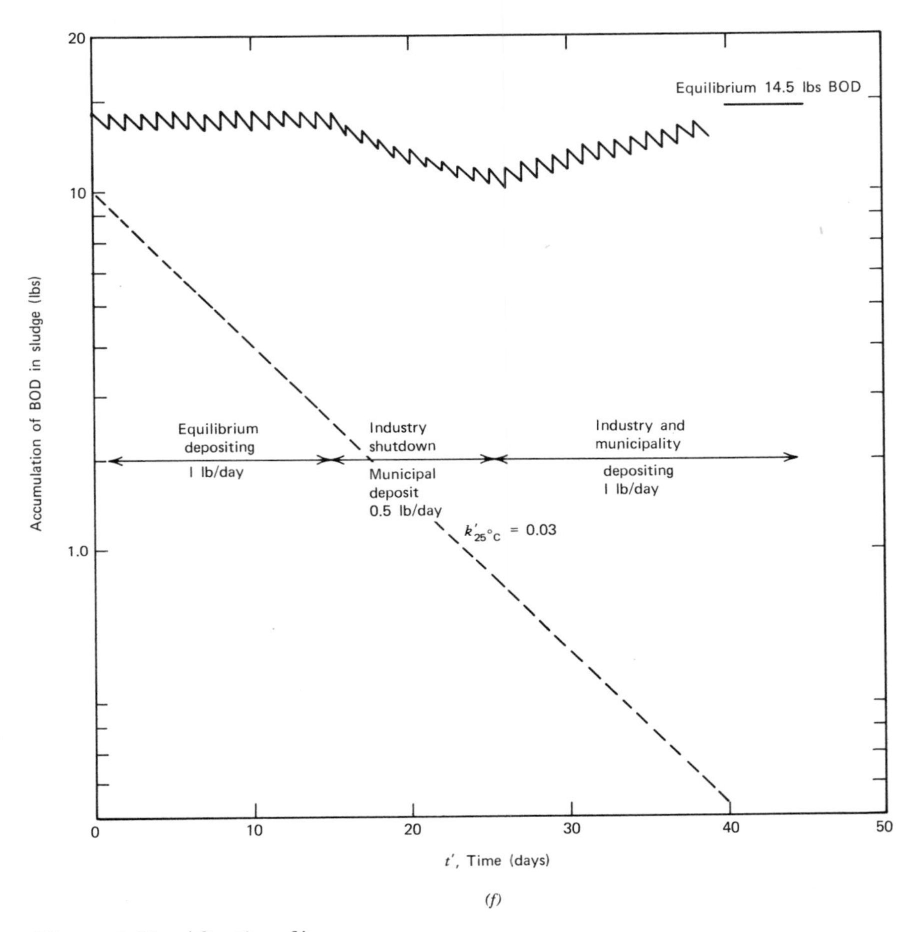

Figure 4–11 (**Continued**).

50 days of depositing uninterrupted by scour, and in such cases the demand will be less than the BOD of the daily deposit. An integration of the daily deposit for the stipulated period will provide the demand. For example, on the tenth day accumulated storage is about 7.2 units of BOD, and the demand exercised during the tenth day is about 0.5 unit, one-half that of the BOD added daily to the sludge pile.

Case 3. This may be taken as the common instance of irregular interrupted depositing, such as might occur when runoff is at times sufficiently high to bypass the area or when sources of pollution cease to discharge or are intermittent. At A, preceded by a prolonged period attaining equilibrium accumulation, deposit is interrupted and biological decomposition continues on the past accumulation, reducing the BOD in storage to B, at which point depositing is resumed. It will be noted that the demand at A at equilibrium is 1 unit per day, but at B, 20 days later, the storage has been reduced to less than 4 units and the demand has declined to about 0.25 unit per day.

Case 4. A fourth case might be considered in which complete scour associated with flood is followed by a period of runoff that is sufficiently high to keep velocity above 0.6 fps. During such a period there will be no sludge accumulation and therefore no unusual demand on the overlying water. All of the BOD discharged from sources will remain in suspension in the flowing stream and continue along the channel.

Case 5. A fifth case may occur in a situation where a sudden post-winter rise in temperature acts on a large accumulation of dormant sludge, inducing active decomposition. Storage accumulates to 34.5 lb, and the high temperature demand on the excess accumulation is greater than the BOD of the daily deposit. If it were conceivable that suddenly the temperature should rise to 25°C, the demand from this excessive accumulation would be 2.3 units per day. This demand would prevail only for a short period since the excess accumulation would decline to the equilibrium level of the new temperature. Such a situation occurs very seldom, as fortunately spring floods scour the winter's accumulation before warm weather sets in; moreover temperature usually increases gradually.

Case 6. A sixth case may occur as a reverse of case 5. Here a sudden decline in temperature causes a falling off in the rate of decomposition, and the demand is less than the BOD of the daily deposit. Assuming the accumulation is at equilibrium at 25°C, a sudden drop in temperature to 13°C would drop the demand to 0.6 unit. Accumulation, however, now increases and, as the new equilibrium level at the lower temperature is approached, the daily demand again approaches that of the BOD of the daily deposit.

Case 7. A frequent occurrence is the case of a temporary industrial shutdown. As an illustration, let us assume that one-half the deposit (0.5 lb) is from the industry and one-half (0.5 lb) from municipal sources. After a prolonged opportunity to accumulate from both sources, deposits attain equilibrium at 25°C, or 14.5 lb. The industry then shuts down for a 10-day period, leaving the river to respond only to the 0.5 lb of daily deposit from municipal sources. Under these conditions the accumulation in 10 days would decline to about 10.9 lb, and the demand from this reduced accumulation in a day would be approximately 0.7 lb. When the industry resumes operation, the daily deposit again becomes

1 lb per day and gradually the accumulation increases, again approaching the equilibrium of 14.5 lb.

On occasions river surveys are made before and after an industrial shutdown in an attempt to differentiate municipal pollution conditions from that of the industrial load. If sludge deposits are involved, such surveys do not in themselves show a true differentiation. Obviously the past accumulation of sludge continues to exercise a demand even though there is no load of any kind currently being discharged from industry.

Abnormality of Biological Extraction and Accumulation

Nature of Biological Extraction and Accumulation. The phenomenon of biological extraction and accumulation of organic matter, which is usually associated with infestation of filamentous growth, is exemplified in the conventional trickling filter and activated sludge treatment systems. In shallow streams the biological growth attached to the streambed extracts organic matter from the stream flowing above, performing in a manner analogous to the trickling filter. Other rivers, depending on channel characteristics, may perform as natural activated sludge systems because the growth may be dispersed as biological floc throughout the flowing stream. The biological floc agglomerating in transit builds in size and density, and deposits as excess activated sludge on the shores, behind obstructions, and in eddy areas. Enough sludge is returned to the stream by scour from these areas of accumulation to keep the process active along the course. In both instances the effect on the stream is removal of BOD far in excess of the normal satisfaction in the time of passage through the reach, and under equilibrium accumulation a drastic demand is exerted on dissolved oxygen.

The abnormal BOD removal from the stream actually comprises two simultaneous independent processes—normal biochemical satisfaction and biological extraction, with the latter dominating. The BOD removal is not oxidized at the instant of extraction but is merely transferred from the flowing stream by biophysical contact with the biological floc; part is stored in the biomass, and part by anabolic reactions (that do not require oxygen) is synthesized as additional biomass. Satisfaction of the extracted BOD then takes place slowly within the accumulated floc mass, usually as a sludge deposit. If extraction and accumulation of deposit along the course continue uninterrupted, an equilibrium is approached, and then the daily BOD satisfied from the accumulation equals the BOD extracted and added. At equilibrium an abnormal demand is exercised on the dissolved oxygen of the stream. This is akin to the usual sludge deposit discussed above, except that the amount of the deposit is a function of the rate of BOD extraction from the flowing stream, and usually the zone of deposit is spread along a substantial reach of river rather than concentrated in localized areas.

As demonstrated from trickling filter and activated sludge performance, not all of the BOD load in the streamflow is extractable [7]. Depending on the opportunity for contact and the ability to sustain floc or attached growth, biological extraction ends at some zone along the course, and further BOD removal reverts to normal satisfaction during transit in the flowing stream.

In a sense biological extraction and accumulation through the reach of its activity serves as a natural treatment works, relieving the burden on downstream reaches. If the upstream resources can be sacrificed to some extent in the interest of downstream users, then biological extraction, usually regarded as a liability, may assume the aspect of an asset.

Quite apart from abnormal influence on deoxygenation, the filamentous growth may interfere with commercial fishing; they readily attach to nets, causing them to sink. Furthermore, although dispersed floc in transit in the flowing stream does not appear to induce adverse physiological response in adult fish, in spawning reaches the filamentous growths may encase fish eggs and prevent hatching. The primary damage of course results from blanketing the streambed with sludge and from the concomitant deoxygenation.

Although the phenomenon of filamentous growth is known to occur in many areas of the world and is predominantly associated with *Sphaerotilus* (largely due to the pleomorphic characteristics of the organism), current knowledge of the environmental factors that cause its formation is insufficient to predict with reasonable assurance its occurrence or nonoccurrence for a specific river situation. However, observation of the effect on the stream and a knowledge of biological waste treatment afford the basis of a working theory to permit practical stream analysis evaluation.

Quantitative Relations. The working theory applicable to quantitative definition of the phenomenon may be stated as follows: *The rate of removal of organic matter in a unit time of passage through a biologically active reach of stream is proportional to the remaining concentration of organic matter measured in terms of its removability.* This is similar to the theory that holds for biological filters [7] and will be recognized as the familiar form of Phelps' law of biochemical oxidation, expressed differentially as

$$\frac{-dL}{dt} = BL; \qquad (4\text{–}13)$$

integrating to

$$\log_e \frac{L_t}{L} = -Bt; \qquad \frac{L_t}{L} = e^{-Bt}, \tag{4–14}$$

whence

$$\log \frac{L_t}{L} = -0.434Bt; \qquad \frac{L_t}{L} = 10^{-bt}, \tag{4–15}$$

where L represents the total removable fraction of BOD, t is the time of passage, and L_t represents the corresponding quantity of BOD remaining at the time of passage t. The ratio L_t/L is the proportion remaining in the flowing stream and $[1 - (L_t/L)]$ is the proportion removed.

Again, a straight-line graphical representation on semilog paper can be obtained by plotting BOD remaining against time in days:

$$\log Y = 2.0 - bt, \tag{4–16}$$

where Y is the percentage of BOD remaining at any time t and b is the specific BOD removal rate.

Since the overall rate of BOD removal from the flowing stream actually reflects two independent processes acting simultaneously,

$$B = B' + K,$$

where B' is the rate of biophysical extraction and K is the normal rate of BOD satisfaction. The proportion of BOD remaining in the flowing stream at time t is

$$\frac{L_t}{L} = e^{-(B'+K)t} = 10^{-(b'+k)t} \tag{4–17}$$

and the proportion removed is

$$1 - e^{-(B'+K)t} = 1 - 10^{-(b'+k)t}. \tag{4–18}$$

In the above equations B' appears to be a function of contact opportunity that is dependent on the mass of biological floc and on the physical and hydraulic characteristics of the stream. The term K, the normal rate of satisfaction of BOD, is a well-known temperature function, independent of stream characteristics. Since both processes are acting simultaneously, less BOD is satisfied in an interval of transit time than if the normal rate of satisfaction K were acting alone on the BOD.

The BOD removal assignable to normal satisfaction is therefore the ratio of the normal rate to the total removal rate applied to the total removal, or

$$\left(\frac{K}{B' + K}\right)(1 - e^{-(B'+K)t}). \tag{4–19}$$

Similarly, the BOD removal assignable to biophysical extraction is

$$\left(\frac{B'}{B' + K}\right)(1 - e^{-(B'+K)t}). \tag{4–20}$$

The value of B' is usually large relative to K; hence extraction dominates. For example, a specific BOD reaction rate run on a sample of river water develops a normal rate of amortization k of 0.1, whereas the river survey discloses a removal of 80 percent of the BOD in a 1-day time of passage through the reach of active biological extraction at a temperature of 20°C. The BOD removal rate B based on the river survey then is

$$\log (100 - 80) = 2.0 - bt,$$

or

$$b = 2.0 - 1.3 = 0.7; \qquad B = 2.3 \times 0.7 = 1.61.$$

Since K is 0.23 (2.3×0.1), B' is 1.38 $(1.61 - 0.23)$. Accordingly, the BOD removal of 80 percent is divided between normal amortization and biological extraction in the ratio of 0.23/1.61, or 0.143, and 1.38/1.61, or 0.857, respectively. If normal amortization were acting alone, in a 1-day time of passage the BOD satisfied would be 20.6 percent. However, since biological extraction is simultaneously at play, normal amortization is less, 0.143×80, or 11.4 percent. The amount extracted is dominant, 0.857×80, or 68.6 percent (about six times as large as normal satisfaction).

Since the BOD removed by biological extraction is not instantaneously oxidized, the demand on the stream's dissolved oxygen depends on the period over which the extracted BOD as excess biological floc has opportunity to deposit and accumulate. We can therefore utilize the general solution of sludge deposit previously developed, substituting the rate of extraction B' for the rate of depositing D. For protracted opportunity to deposit without interruption or scour, equilibrium accumulation is

$$S_e = \frac{B'}{K'}. \tag{4–21}$$

Cumulative storage of extracted BOD in sludge deposit for any finite period of continuous depositing below equilibrium level is given by

$$S_{t'} = \frac{B'}{K'}(1 - e^{-K't'})$$

or

$$S_{t'} = \frac{b'}{k'}(1 - 10^{-k't'}). \tag{4–22}$$

The biological mass depositing is considered to be amortized at the same rate as other organic sludge, namely, $k' = 0.03$ at 25°C.

For the special condition of equilibrium accumulation, since the amount that is extracted and deposited per day is equal to the BOD amortized in the sludge accumulation per day, it is not necessary to make the distinction between normal amortization and biological extraction. Solution is simplified by considering the total BOD removal rate B as a high specific amortization rate K. While doing so it is well to keep in mind that both phenomena are actually at play; this simplification is valid only for the condition approaching equilibrium (such as after protracted periods of low streamflow uninterrupted by scouring freshets). For short periods of opportunity to accumulate below equilibrium level it is essential to make the distinction between normal satisfaction and biological extraction.

To continue the previous example: where 80 percent of the BOD is observed to be removed in a 1-day time of passage, 11.4 percent was assigned to normal amortization and 68.6 percent to biological extraction. To express the problem in absolute terms it is assumed that the BOD_U at the upstream station is 200,000 PE. Then 11.4 percent, or 22,800 PE, is normal amortization in transit and 68.6 percent, or 137,200 PE is extracted and deposited; the total removal in the river reach is 160,000 PE (22,800 + 137,200), or 80 percent. Under equilibrium the amortization of BOD in the sludge accumulation just balances the 68.6 percent extracted and deposited per day. The critical reach of the river is therefore subject to deoxygenation of 80 percent (11.4 + 68.6), or in absolute terms 160,000 PE. This demand is equivalent to amortization in a 1-day time of passage at a high specific BOD rate of $k = 0.7$.

However, if a scouring freshet is followed by a rapid decline in hydrograph to the same low flow producing a 1-day time of passage between the two river sampling stations, and extraction and depositing occur for only 10 days, then the accumulation of sludge is below equilibrium level and the demand exercised is less than the 68.6 percent added to the sludge pile. This set of factors is conveniently solved by reference to Figure 4–10. The demand exercised from 10 days' accumulation is 42 percent (100 — 58) of that of the daily deposit of 137,200 PE, or 57,600 PE. Thus the deoxygenation under 10 days of opportunity to accumulate is normal amortization in transit, 22,800 PE, plus the lower demand from sludge deposit amortization of 57,600 PE, or a total of 80,400 PE, which is about one-half that under equilibrium conditions. The demand on oxygen assets below that at equilibrium produces an entirely different dissolved oxygen profile—depending on the level of sludge accumulation.

Where biological extraction occurs, the dissolved oxygen profile may be very sensitive and subject to wide variations, particularly in the early period of accumulation. In the later stages of accumulation (30 days or more), as equilibrium is approached the dissolved oxygen profile, although substantially depressed, becomes more stable.

Observed Biological Extraction Rates. The parameters that govern biological extraction rates are not sufficiently precise to permit predicting a specific rate of extraction for a particular river situation. Reliance must be placed on the stream survey to determine the time of passage and corresponding BOD residuals, from which b is obtained. Laboratory investigation of waste sources and mixed river water defines the specific amortization rate of organics in transit, k and from these the biological extraction rate b' is then obtained ($b' = b - k$).

The rate of biological extraction b' varies widely among rivers, ranging from 0.2 to 1.0 or more, depending on opportunity for contact between the biological mass and the water mass.* A number of complex factors influence contact opportunity; these may be conveniently considered in two general classes, those relating to the biological growth and those relating to the characteristics of the river channel. High rates of extraction are consistently associated with high concentrations of active, well-growing biomass. If the growths are attached, surface contact area is increased by undulating streamers; if the biomass is dispersed, turbulent distribution in the channel cross-section increases contact opportunity. As with coagulation phenomena, size of floc and agglomeration of gelatinous mass are involved in extraction and deposition.

Channel characteristics, as reflected in such hydraulic parameters as velocity, depth, hydraulic radius, and Reynolds number, affect contact opportunity. Deep, slow-moving streams afford little or no opportunity for extraction. A shallow stream with a high roughness coefficient from

*Higher rates for short reaches of very brief time of passage (e.g., in flumes) have been reported, but such rates do not persist throughout the entire active reach observed in rivers.

an irregular, rocky bed or submerged vegetation favors attached growth. Between these extremes is a wide variety of conditions of turbulence, channel shape, and cross-section that appear to have a significant role in establishing a specific rate of extraction for a particular reach of river.

Temperature does not appear to be a controlling factor in establishing a biological extraction rate. The abnormality occurs in cold and warm climates and persists through all seasons of the year. The total BOD removal rate b appears to be affected only insofar as k is altered by temperature; this indicates that b' is not appreciably influenced by temperature. Thus the value of b' that is determined for a particular river appears to hold through the seasons without change. In stream-analysis computations, therefore, the BOD removal rate b is obtained for any temperature as $b = b' + k_T$, where k_T is adjusted by the usual temperature relation.

Although each river appears to establish its unique biological extraction rate b', if the stream channel or the character of the waste source is substantially altered, b' can be expected to be influenced. However, although the extraction rate may be altered, the phenomenon is known to persist after drastic changes, particularly in small streams in which waste effluent flow is a major portion of the stream runoff.

REOXYGENATION

We now turn to the consideration of the factors that affect the asset or income side of the dissolved oxygen ledger. For practical application of stream analysis these factors may be classified in five groups: (a) dissolved oxygen from increments of stream runoff, (b) reaeration, (c) photosynthesis, (d) stored dissolved oxygen, and (e) negative dissolved oxygen. The first two of these are the most important.

Stream *runoff* is the medium that holds the oxygen asset. In a sense the size of the river as measured by its rate of discharge is the size of the oxygen "bank." A small runoff implies a limited banking institution; a large runoff is a large bank, capable of handling a large turnover in assets.

Reaeration from the atmosphere is the only dependable primary source of oxygen income, and it is a complex asset. The rate of income is dependent on the amount of cash on hand. If the bank is full (water saturated), there is no income; if oxygen is completely exhausted, the rate of income from reaeration reaches a maximum; between these extremes, the rate of income is directly proportional to the oxygen deficit. Although oxygen deficit is the control that regulates the rate of income, the actual amount received from reaeration depends on the physical and hydrologic characteristics of each particular stretch of river course, temperature, water depth, occupied channel volume, and stream turnover.

Photosynthesis, the process by which aquatic plants, principally algae, produce pure oxygen, is also a primary source, but it is not dependable, nor always available, nor particularly amenable to quantitative determination. When present in large numbers, algae may add substantially to temporary income, but the income is available only during daylight hours, because respiration at night institutes an outgo. Moreover, when dependable assets are most needed, during hot, dry weather, large masses of algae may die, and, instead of a living asset, the decomposing plants become a dead liability.

Oxygen stored in the form of nitrates is in the nature of reserve assets. This oxygen is available as a last resource to prevent complete bankruptcy when the dissolved oxygen supply is exhausted. In some special instances (e.g., the Androscoggin River in Maine) nitrates may actually be introduced into the stream as a means of avoiding more serious degradation. Nitrates, however, constitute a plant nutrient and may cause excessive weed or algal bloom. It is not good business practice to draw on reserve assets for current operations. Much better to consider these reserves as a factor of safety against complete disaster and to run the business on current income in the form of dissolved oxygen with a margin of cash on hand to meet anticipated current liabilities and moderate variations in day-to-day obligations.

Negative dissolved oxygen is operation in the red. It is a reflection of more current obligations to be met than cash and reserves available. The result is hydrogen sulfide, which, if it remains dissolved in the flowing stream, may place an extra burden on the resources downstream. On the other hand, if hydrogen sulfide escapes to the atmosphere, where the debt is paid by oxygen of the air rather than from the river water, such a loss of negative dissolved oxygen in a sense can be considered as bankruptcy proceedings, where at least the red is washed from the books and the river starts again from scratch. Although such mismanagement of the business of waste disposal actually on occasion has been known to occur, today there is no excuse for such degradation.

As with the liability side of the ledger, insofar as possible we shall deal separately with each of the factors influencing income rather than lump all of them into a single catchall reoxygenation constant.

Oxygen Resources of Stream Runoff

The size of the banking institution to carry on the business of self-purification is determined primarily by the magnitude of stream runoff. As we have seen in Chapter 3, measured in tributary drainage area the size of a river is fixed by quite definite boundaries, but measured as runoff from the drainage area it is highly dynamic, varying from season to season and from day to day. Thus the oxygen bank is not fixed in size but flexible, shrinking or expanding as variations occur in stream runoff. However, as we have also seen, although highly variable, runoff in the natural hydrologic setting is orderly, following the laws of chance occurrence. The size of the oxygen banking institution, in turn, is definable only in statistical terms. In dealing with dissolved oxygen assets it is therefore always essential to specify runoff probability.

In addition to variation in runoff expected at a given location, the size of the oxygen bank is enlarged as it proceeds from headwaters to mouth down the river channel, acquiring increments of tributary drainage area en route. Depending on topography, the increments of drainage area usually occur at junctions of tributary streams; intervening areas without defined drainage, however, can contribute as seepage more or less continuously along the river channel. The analyst must bear in mind that not all increments of drainage area yield streamflow at the same rate. Some tributaries have high yields per unit of area, whereas others, of different drainage characteristics, produce poor yield. Obviously basins of homogeneous regimen tend toward uniformity in yield; heterogeneous basins of varied topography, geology, vegetative cover, temperature, and rainfall distribution produce widely different yields among tributaries (for an example see the description of the Willamette River in Chapter 3).

Another factor is the influence of man-made river developments. Diversions for water supply or irrigation may substantially reduce the natural expected runoff during critical drought periods, whereas storage reservoirs holding floodwaters for release during the dry season can greatly augment critical flows. Hydropower operations oriented to meeting peak power demands induce marked pulsations in flow during the hours of the day, and weekend shutdowns may drastically curtail flow. Any man-made river development makes an impact, which, depending on how the development is planned and operated, may be either detrimental or beneficial.

The first phase in quantitative definition of oxygen resources is therefore the determination of the size of the bank as stream runoff from section to section along the river channel. This quantification entails an intimate knowledge of the drainage basin, good maps, and experienced professional judgment in interpreting meagre hydrologic data. Streamgaging records are usually relatively short, with limited coverage, and for some basins good topographic maps are not available. Estimating runoff at frequent locations along the river courses other than at gaging stations is essential and yet it is the task that is subject to greatest error. Useful methods for the quantitative definition of runoff are presented in Chapter 3, and applicable statistical tools are discussed in Appendix A.

Oxygen Resources of Reaeration

The primary source of dissolved oxygen in streams is the atmosphere. When dissolved oxygen declines below saturation level, the process of reaeration replenishes it. The earliest and still the most practical direct basis for calculating the amount of oxygen contributed by atmospheric reaeration is Phelps' mathematical formulation, which employs readily measurable physical characteristics of the stream [8].

Phelps' Reaeration Formulation

Reaeration is governed by two fundamental laws: the law of *solution* and the law of *diffusion.* The rate of solution is proportional to the dissolved oxygen deficit below saturation level; the rate of diffusion through water and between two points is proportional to the difference in concentration at these two points. On these two laws and the experimental determination of the coefficient of diffusion in quiescent water, Phelps developed the mathematical formulation for the quantitative determination of reaeration from the atmosphere:

$$D = 100 - \left[\left(1 - \frac{B}{100}\right) \times 81.06 \left(e^{-K} + \frac{e^{-9K}}{9} + \frac{e^{-25K}}{25} + \cdots\right)\right], \tag{4-23}$$

where D is the final dissolved oxygen content (average for the depth as percentage of saturation), B is the initial dissolved oxygen content, and K is a constant dependent on

$$\frac{\pi^2 at}{4L^2},$$

where t is the time of exposure in hours, L is the depth of water in centimeters, and a is the diffusion coefficient for a specific temperature. The expression $[1 - (B/100)]$ represents the law of solution, and the summation of the series represents the law of diffusion.

Phelps developed the equation for diffusion from Fick's law of hydrodiffusion. The rate of diffusion of a dissolved substance in terms of the concentration gradient is

$$\frac{ds}{dt} = -aqx\frac{dc}{dx},$$

where ds is the amount of substance passing any point x through a cross-sectional area q in time dt when the concentration gradient is dc/dx, c is the concentration at any point x, and a is a constant for a given substance at a given temperature (diffusion coefficient).

For quiescent water that is initially devoid of dissolved oxygen, integrating for the total oxygen absorbed between the limits of the surface and a depth L leads to Phelps' basic reaeration equation for the influence of diffusion. The solutions are involved, and expansion by means of Fourier series is adapted:

$$D = 100 - \frac{800}{\pi^2} \sum_{n=0}^{n=\infty} \frac{1}{(2n + 1)^2} e^{-ab^2 t}, \tag{4–24}$$

where D is the final dissolved oxygen content (average for the depth as percentage of saturation), t is the time of exposure in hours, n is an integer (1, 2, 3, etc.), a is the diffusion coefficient for a specific temperature, and b is a constant dependent on

$$\frac{(2n + 1)\pi}{2L},$$

where L is the depth of water in centimeters.

Substituting the expression for b in equation 4–24 gives

$$D = 100 - \frac{800}{\pi^2} \sum_{n=0}^{n=\infty} \frac{1}{(2n + 1)^2} \exp\left[\frac{-at\pi^2}{4L^2} (2n + 1)^2\right]. \tag{4–25}$$

Substituting K for $at\pi^2/4L^2$ gives

$$D = 100 - \frac{800}{\pi^2} \sum_{n=0}^{n=\infty} \frac{1}{(2n + 1)^2} \exp\left[-K(2n + 1)^2\right]. \tag{4–26}$$

Equation 4–26 leads to Phelps' expression

$$D = 100 - 81.06 \left(e^{-K} + \frac{e^{-9K}}{9} + \frac{e^{-25K}}{25} + \cdots\right) \tag{4–27}$$

or

$$D = 100 - 81.06 \left(0.779^c + \frac{0.105^c}{9} + \frac{0.0019^c}{25} + \cdots\right), \tag{4–28}$$

where

$$c = \frac{at\pi^2}{L^2}.$$

On the basis of extensive experimental results covering a temperature range of 6 to 30°C Phelps determined the diffusion coefficient a. Taking time in hours and depth in centimeters, the mean value of a was found to be 1.42 at 20°C. A reanalysis of the influence of temperature discloses an approximate linear relation of the form

$$\log a = 0.04125T - 0.672 \tag{4–29}$$

or

$$a_T = 1.42 \times 1.1^{(T-20)}. \tag{4–30}$$

Subsequent studies by others [9,10] indicate that Phelps' value for the diffusion coefficient a may be too large. Possibly his value reflects in addition to molecular diffusion some uncontrolled eddy diffusion due to convection currents, and absolute quiescence did not prevail in his experiments. However, this does not invalidate practical application, since any error in a is offset by the empirical derivation of mix interval.

Depth, it is noted, has a dominant influence in summation of the series, since c varies inversely as L^2. Fortunately this important physical characteristic of the river is readily measurable in the field, and its variation from reach to reach can be taken into account (see Chapter 3).

Adaptation to Stream Conditions. Phelps' original formulation postulates quiescence of the body of water during exposure to the atmosphere, with no mixing induced by turbulence or convection currents. This is a situation that is almost never encountered in natural bodies of water and flowing streams. Because of this conceptual difficulty and the cumbersome lengthy computations necessitated in expanding to a great number of terms the slowly converging series that define diffusion, the basic formulation was largely ignored for some 30 years in favor of the simplified oxygen sag equation (limitations of the sag equation will be discussed later). A re-examination of the mechanism of reaeration and a convenient ready solution by graphical methods reassert the value of the original Phelps formulation as a practical measure of field conditions encountered in natural waters.

Oxygen from the atmosphere enters a body of water in the form of dissolved oxygen at the air–water interface as an infinitely thin film that is instantaneously completely saturated. If the law of solution were the only one at play, the underlying water would be sealed against further oxygen solution. Hence the importance of the mechanism of subsurface diffusion in carrying oxygen to lower depths. The concept of mixing induced by stream turbulence does not alter the basic function of diffusion. When it is realized that the surface film of contact is infinitely thin, mixing is merely an admixture of very thin layers of saturated water and relatively thick layers of unsaturated water. It is well established that no amount of agitation or mixing per se can bring about molecular interchange involved in true solution. Subsurface solution in every case is achieved only by the process of diffusion.

Phelps' formulation for a quiescent state may be applied to turbulent conditions by either of two different methods: making a function of change in depth or taking pure time into consideration. Turbulence in mixing saturated surface layers into lower depths in effect decreases the depth through which diffusion must operate. The net overall change in depth is difficult to reduce to quantitative terms, but clearly it depends on the frequency with which surface layers are transported to the subsurface regions and is therefore a function of time.

The other more practical concept is a pure time effect, which the author applied in adaptation of the graphical solution of the reaeration formulation. Turbulent state is considered as consisting of successive periods of quiescence between which there is instantaneous and complete mixing. This interval between mixes, termed the *mix interval,* becomes the time of exposure in the diffusion formula, and an approximation of the total reoxygenation per day is the sum of the successive increments, or the amount per mix interval times the number of mixes per day.

The Standard Reaeration Curve

To facilitate the application of Phelps' formulation a *standard reaeration* curve has been developed [11,12]. The basic equation 4–28 was solved for six arbitrary values of the exponent c, expanding the series to as many as 500 terms, with the following results:

c	D
0.1	11.36
0.01	3.59
0.001	1.14
0.0001	0.367
0.00001	0.121
0.000001	0.0382

D = percentage of dissolved oxygen saturation absorbed by reaeration as function of c (water initially devoid of dissolved oxygen).

With these coordinates the standard reaeration curve (Figure 4–12, curve A) was constructed, and, employing the principle of the slide

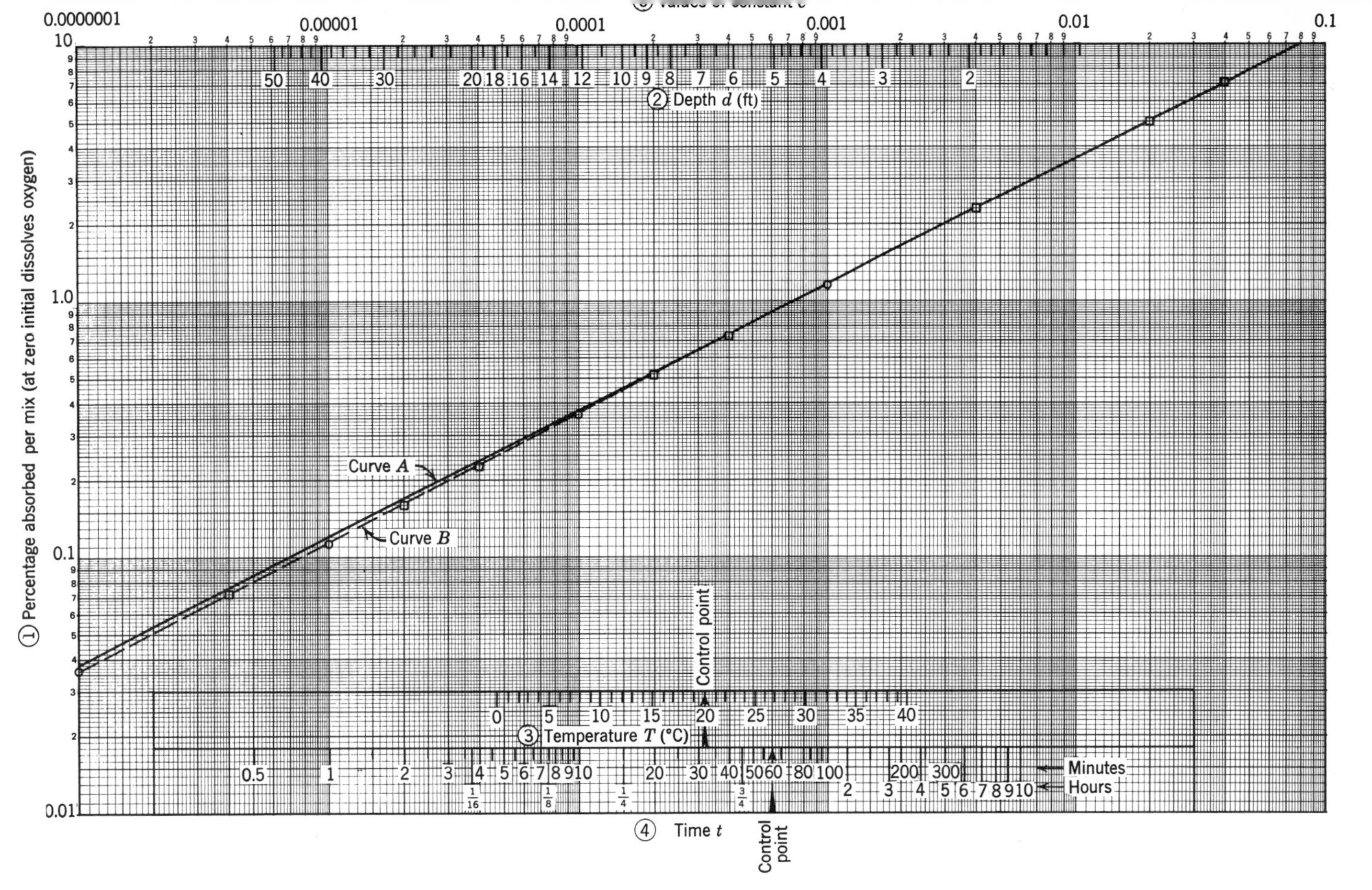

Figure 4–12 Standard reaeration curve (see text for instructions on using).

rule, c is readily derived graphically for any combination of stream depth d in feet, water temperature T in degrees Celsius, and mix interval t in minutes. For example, for a stream depth of 10 ft, water temperature 25°C, and mix interval 24.2 min, dissolved oxygen as percentage of saturation absorbed per mix interval at zero initial dissolved oxygen is quickly obtained as follows: slide the control of scale 3 to coincide with 10 ft on scale 2; opposite 25°C on scale 3 locate a position on scale 5; slide control of scale 4 to this position and opposite 24.2 min on scale 4 locate on scale 5 $c = 0.0001$; drop vertically from $c = 0.0001$ to the standard reaeration curve (curve A) and read on scale 1 dissolved oxygen of 0.37 percent saturation.

With the advent of high-speed computers it has become possible to extend the expansion of Phelps' diffusion series to a great number of terms. Downs [13] substituting for equation 4–28,

$$D = 100 - 81.057 - \sum_{j=0}^{\infty} \frac{\exp\left[-(2j+1)^2 k\right]}{(2j+1)^2},$$

performed the summation of the series by computer to end term of 10^{-8} for nine values of K (where c in Figure 4–12 is $4K$) with the following results:

K	c	D
0.05	0.20	16.062713
0.01	0.04	7.183437
0.005	0.02	5.079446
0.001	0.004	2.271584
0.0001	0.0004	0.718372
0.00005	0.0002	0.508010
0.00001	0.00004	0.227348
0.000005	0.00002	0.160885
0.000001	0.000004	0.072404

D = percentage of dissolved oxygen saturation absorbed by reaeration as a function of K (water initially devoid of dissolved oxygen).

Downs' values shown in Figure 4–12 describe the more refined curve B.

Somewhat later Deininger [14], using Laplace transforms to the basic formulation, successfully developed a direct solution that avoids the necessity of the laborious expansion of the Fourier series. Deininger's direct solution reduces to the form

$$D = 200\left(\frac{c}{\pi^3}\right)^{1/2}, \tag{4–31}$$

where D is the dissolved oxygen reaeration as percentage saturation absorbed per mix interval by water initially devoid of dissolved oxygen and c is the exponent employed in equation 4–28. The standard reaeration curve (Figure 4–12) by equation 4–31 reduces to straight-line form throughout, and the coordinates practically coincide with curve B.

There is close agreement between curves A and B through the region of practical application, with slight deviation for smaller values of c. Deininger's work, an outstanding achievement, strengthens the standard reaeration curve and facilitates computer adaptation. Thus equation 4–31 or curve B of Figure 4–12 is taken as the exact relation.

With ready graphical solution of the fundamental reaeration formulation, it is practicable to compute reaeration on a quantitative basis reach by reach, taking into consideration three measurable river channel characteristics—namely, the dominant factor stream depth, occupied channel volume, and water temperature. Stream depth for a reach of the stream, effective depth, is defined as the occupied channel volume divided by the surface area. Since occupied channel volume and surface area change with runoff, effective depth reach by reach is specific for each runoff regime. Occupied channel volume and surface area are obtained from the detailed channel cross-section sounding and adjusted to a specific runoff regime as described in Chapter 3. Water temperature, taken as the mean for each reach, is obtained from seasonal patterns or from field surveys, or is computed from prevailing meteorologic conditions. Where waste heat loads are discharged, as from steam electric-power plants, the river's temperature profile is computed and taken into account reach by reach.

The generally accepted tables of oxygen solubility in freshwater exposed to the atmosphere are those published in *Standard Methods* [2]. These tables also provide the basis for correction for barometric pressure and for salinity; the former is of slight consequence, except at very high altitudes; the latter is of significance in dealing with seawater and brackish waters of tidal estuaries. A useful tabulation of saturation values in freshwater at 0.1°C intervals prepared by Streeter [15] is presented in Table 4–5.

Recent work has been devoted to verification of the oxygen solubility saturation values, but deviations from earlier work have been slight. In practice we deal with waters polluted to varying degrees; the effect on the air–water interfacial resistance results in saturation values that

Table 4–5 Solubility of Oxygen in Freshwater (after Streeter [15])

	Dissolved Oxygen (mg/l) (at normal atmospheric pressure, 760 mm Hg)									
T (°C)	0.0	0.1	0.2	0.3	0.4	0.5	0.6	0.7	0.8	0.9
0	14.62	14.58	14.54	14.50	14.46	14.42	14.39	14.35	14.31	14.27
1	14.23	14.19	14.15	14.11	14.07	14.03	14.00	13.96	13.92	13.88
2	13.84	13.80	13.77	13.73	13.70	13.66	13.62	13.59	13.55	13.52
3	13.48	13.44	13.41	13.38	13.34	13.30	13.27	13.24	13.20	13.16
4	13.13	13.10	13.06	13.03	13.00	12.97	12.93	12.90	12.87	12.83
5	12.80	12.77	12.74	12.70	12.67	12.64	12.61	12.58	12.54	12.51
6	12.48	12.45	12.42	12.39	12.36	12.32	12.29	12.26	12.23	12.20
7	12.17	12.14	12.11	12.08	12.05	12.02	11.99	11.96	11.93	11.90
8	11.87	11.84	11.81	11.79	11.76	11.73	11.70	11.67	11.65	11.62
9	11.59	11.56	11.54	11.51	11.49	11.46	11.43	11.41	11.38	11.36
10	11.33	11.31	11.28	11.25	11.23	11.21	11.18	11.15	11.13	11.11
11	11.08	11.06	11.03	11.00	10.98	10.96	10.93	10.90	10.88	10.86
12	10.83	10.81	10.78	10.76	10.74	10.71	10.69	10.67	10.65	10.62
13	10.60	10.58	10.55	10.53	10.51	10.48	10.46	10.44	10.42	10.39
14	10.37	10.35	10.33	10.30	10.28	10.26	10.24	10.22	10.19	10.17
15	10.15	10.13	10.11	10.09	10.07	10.05	10.03	10.01	9.99	9.97
16	9.95	9.93	9.91	9.89	9.87	9.85	9.82	9.80	9.78	9.76
17	9.74	9.72	9.70	9.68	9.66	9.64	9.62	9.60	9.58	9.56
18	9.54	9.52	9.50	9.48	9.46	9.44	9.43	9.41	9.39	9.37
19	9.35	9.33	9.31	9.30	9.28	9.26	9.24	9.22	9.21	9.19
20	9.17	9.15	9.13	9.12	9.10	9.08	9.06	9.04	9.03	9.01
21	8.99	8.98	8.96	8.94	8.93	8.91	8.89	8.88	8.86	8.85
22	8.83	8.81	8.80	8.78	8.77	8.75	8.74	8.72	8.71	8.69
23	8.68	8.66	8.65	8.63	8.62	8.60	8.59	8.57	8.56	8.54
24	8.53	8.51	8.50	8.48	8.47	8.45	8.44	8.42	8.41	8.39
25	8.38	8.36	8.35	8.33	8.32	8.30	8.28	8.27	8.25	8.24
26	8.22	8.20	8.19	8.17	8.16	8.14	8.13	8.11	8.10	8.08
27	8.07	8.05	8.04	8.02	8.01	7.99	7.98	7.96	7.95	7.93
28	7.92	7.90	7.89	7.87	7.86	7.84	7.83	7.81	7.80	7.78
29	7.77	7.75	7.74	7.73	7.71	7.70	7.69	7.67	7.66	7.64
30	7.63	7.61	7.60	7.59	7.57	7.56	7.55	7.54		

T (°C)	Dissolved Oxygen (mg/l)	*T* (°C)	Dissolved Oxygen (mg/l)
31	7.5	41	6.5
32	7.4	42	6.4
33	7.3	43	6.3
34	7.2	44	6.2
35	7.1	45	6.1
36	7.0	46	6.0
37	6.9	47	5.9
38	6.8	48	5.8
39	6.7	49	5.7
40	6.6		

Note: correction for barometric pressure B (mm Hg) multiply by $B/760$; correction for salinity S (chloride, mg/l) multiply by $[1 - (S/100{,}000)]$.

deviate, but only slightly, from the standardized tables developed for pure water. Hence these minor refinements, although important in scientific research, are of little practical significance in stream analysis.

Where seasonal temperature variation is wide, the saturation capacity range is substantial. Saturation at summer temperatures is only about half that under winter conditions. Thus the critical period is normally the warm-weather drought season when the dissolved oxygen content endures a double jeopardy: the oxygen bank shrinks by virtue of low stream runoff, and at the same time the capacity of the runoff itself shrinks by virtue of lowered oxygen saturation level. In extreme northern climates, on the other hand, the critical season may occur in winter, when low runoff coincides with heavy ice cover, which prevents reaeration.

The one remaining parameter that is required in reaeration computation, the *mix interval,* is not directly measurable in the stream but has been empirically developed from comparison of computed and observed stream dissolved oxygen profiles on a number of streams for which the other basic data concerning BOD loading and channel characteristics were well defined. In accord with the basic concepts, mix interval is related to stream velocity and particularly to stream depth. For practical application streams are divided into three broad classes: (a) those of moderate depth and velocity, (b) those of shallow depth and relatively high velocity, and (c) tidal estuaries subject to ebb and flood translation. The mix interval has been related to effective depth, a measurable channel parameter. Curve *A* of Figure 4–13 may be taken as representative of the large majority of streams of moderate depth and velocity; curve *B* is applicable to tidal estuaries in which tidal currents and reversals of ebb and flood translation tend to shorten the mix interval for a given effective depth. Deep, sluggish streams may experience stratification, such as occurs in lakes and reservoirs. In these instances the depth is limited to the active flowthrough region, for determinations both of deoxygenation and of reaeration. Stratification is usually most pronounced during the critical warm-weather season, with active flowthrough restricted to the upper 10 to 20 ft. For shallow, relatively high velocity streams the mix interval is shortened and varies considerably, depending on the degree of turbulence. Curve *C* of Figure 4–13 may be taken as a first approximation, but verification should be made by comparing computed and observed dissolved oxygen profiles.

Flow over a dam (unless it is of the free fall weir type) actually does not contribute significantly to reaeration; similarly short reaches of riffles or rapids contribute less than is usually thought but are taken into account in determining the occupied channel volume, surface area, and the resultant effective depth. It is important, however, to subdivide

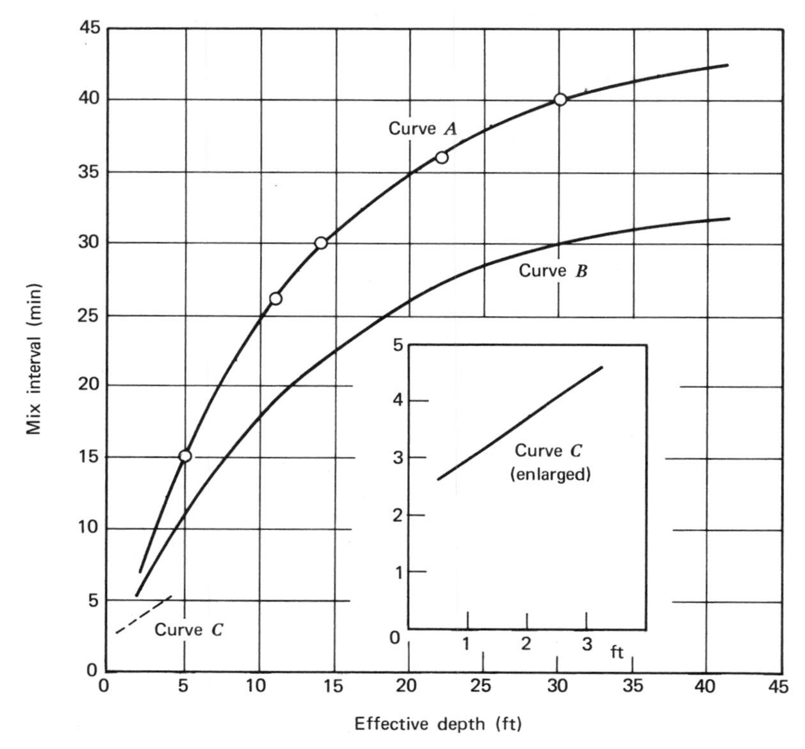

Figure 4–13 Relation between effective depth and mix interval. Curve A: usual freshwater streams; curve B: tidal estuaries; curve C: shallow, relatively-high-velocity streams.

the river channel into reaches of approximately similar depth and to avoid including in a reach both deep and shallow stretches.

The contribution to dissolved oxygen by reaeration is computed as the amount absorbed from the atmosphere per day. Since the dissolved oxygen profile is at equilibrium for the given conditions, the profile through any river reach remains constant through the day; consequently each successive mix interval through the day for a particular reach absorbs the same amount of oxygen. Reaeration per day then is the amount per mix interval times the number of mix intervals in a 24-hr period. It should be recognized, therefore, that, although the amount of oxygen absorbed per mix interval declines as the mix interval is shortened, the number of mix intervals in a 24-hr period increases and the total amount per day may actually increase; conversely, as the mix interval is lengthened, the oxygen absorbed per mix increases, but the number of mix intervals per day decreases.

Extending the previous example on the use of the standard reaeration curve, the amount of oxygen absorbed per day for a particular reach of river for the specified conditions is obtained as follows:

Occupied channel volume of the reach	175 MG
Effective depth (volume/surface area)	10 ft
Mean water temperature of the reach	25°C
Mix interval (Figure 4–13, curve *A*)	24.2 min
Mixes per day (1440/24.2)	59.5
Mean dissolved oxygen deficit of the reach based on the dissolved oxygen profile	28%
Reaeration per mix interval at 100% dissolved oxygen deficit (curve *A*, Figure 4–12)	0.37%
Reaeration per day at 28% deficit (0.37 × 59.5 × 0.28)	6.16%
Dissolved oxygen saturation at 25°C, 8.38 mg/l or	69.9 lb/MG
Dissolved oxygen reaeration per day for the reach (0.0616 × 69.9 × 175)	754 lb/day

Thus, as long as the dissolved oxygen profile and other conditions remain stable, reaeration contributes to the reach 754 lb of dissolved oxygen per day. Obviously to maintain a stable dissolved oxygen profile with this daily gain deoxygenation must be taking place simultaneously.

OXYGEN BALANCE AND THE STREAM DISSOLVED-OXYGEN PROFILE

The Rational Method Accounting System

With organic waste loads, hydrology, deoxygenation, and reoxygenation established, we are now in a position to set up an accounting system to strike an oxygen balance and define dissolved oxygen profiles for a variety of conditions likely to be encountered.

The first phase is an orderly accounting of the liability side of the ledger, deoxygenation. Form A is convenient system for summarizing these computations; page 1 identifies the specific conditions, and page 2 summarizes the BOD amortization. Runoff is referred to a key gage identifying drought probability and significant man-made alterations to natural flow, such as diversions, hydropulsation, or augmentation. Waste loading comprises two elements, residual from upstream sources and landwash, and the BOD loads from known sources along the course. In the absence of available stream-survey data from which to compute residual loading and, if reconnaissance has disclosed no significant man-

FORM A: DEOXYGENATION

Page 1
Computation No.
Computer
Date

STREAM
RUNOFF CFS AT REFERENCE GAGE
DROUGHT PROBABILITY
SPECIAL CONDITIONS

BOD LOADING: Residual at mile from upstream sources and landwash

WASTE LOADS

			BOD Population Equivalents (PE) Discharged to Stream			
River Mile (1)	Source (2)	Type of Treatment (3)	Immediate Demand (4)	Colloidal and Dissolved Fraction (5)	Settleable Solids Fraction (6)	Total (7)
	Upstream residual					

made sources, it is reasonable to assume 1.0 ppm of BOD_5. Natural stream runoff usually ranges from less than 1 to 2 ppm BOD_5, depending on the level of runoff and the nature of the drainage area. Tributaries are designated as sources if their residuals contribute a significant load. Man-made waste loads are determined as discussed in Chapter 2, and, depending on the purpose of the investigation, it may be necessary to subdivide the total BOD load into components of immediate demand, colloidal and dissolved fraction, and settleable solids fraction in order to deal with abnormalities. Where wastes receive suitable treatment before discharge, such abnormalities are generally not involved.

The integration of the BOD loads along the course of the stream is summarized on page 2 of Form A. The time of passage is specific for each runoff regime and is determined directly from the hydrologic analysis, as discussed in Chapter 3. In addition to each successive effluent outlet, a number of other significant items should be included: junctions of major tributaries, dams, hydro or thermal power plants, breaking points in channel configuration and effective depth, and critical stream

Form A: (Continued)

Page 2
Computation No.

Station (river mile) (1)	Mean Water Temperature (°C) (2)	k at Mean Temperature (3)	Time of Passage from Initial Station (days) (4)	BOD (Population Equivalents) Discharged at Station (5)	Residual from Station Upstream (6)	Total at Station (7)	Satisfied in Stream between Stations (8)	Total Satisfied above Station (9)

velocity. If in advance locations are selected where oxygen balance is to be computed, these are also included. Water temperature is the mean between stations, determined either from stream survey or computed as equilibrium temperatures from climatologic records. Where heat loading is a factor—for example, below a thermal power plant—the river temperature profile is computed and taken into account in short reaches below the waste heat discharge. The integration of BOD loading from station to station for the time of passage at the reaction rate k adjusted to the mean temperature may be graphical or analytical. Where precise results are required, a convenient logarithmic solution is afforded by equation 4–1, substituting for Y population equivalents remaining. However, it is always desirable to include a graphic integration as it provides a complete continuous BOD profile along the course of the stream and is a good check on analytical computations.

The second phase is the accounting of the asset side of the ledger and striking a net balance between assets and liabilities to provide the dissolved oxygen profile along the course of the stream. Form B is a convenient system for summarizing these computations. The stations employed need not include all locations involved in the liability calculations reported on Form A. The division of the river channel into a number of reaches for reaeration calculation and oxygen balance is contingent on the desired refinement in the dissolved oxygen profile and such factors as time of passage, channel changes, junction of tributaries, source of immediate demand, and location of concentration of sludge deposits. Since the dissolved oxygen profile in the river is curvilinear, reaches should be short enough for the profile to be adequately defined as straight lines through reaches. Particular consideration should be given

FORM B: REOXYGENATION AND OXYGEN BALANCE

Computation No.
Computer
Date

Station (river mile) (1)	Runoff at Station (cfs) (2)	Mean Temperature (°C) (3)	Effective Depth (ft) (4)	Frequency of Turnover (min/mix) (5)	River Volume (MG) (6)	Total Demand Satisfied above Station (PE) (Form A, Column 9) (7)	Dissolved Oxygen (Population Equivalents)							Percentage of Saturation at Station (Column 14 Divided by Column 10) (15)
							Added by Runoff between Stations (8)	Total Added by Runoff above Station (9)	Total Runoff at Station at Saturation (10)	Net at Station (Column 9 minus Column 7) (11)	Added by Reaeration between Stations (12)	Total added by Reaeration above Stations (13)	Net Oxygen Balance at Station (Column 11 plus Column 13) (14)	

to identifying the critical dissolved oxygen, or low point in the sag. A sensitive profile with a sharp decline and recovery requires shorter reaches than a profile of gradual change. On the other hand, to take excessively short reaches (e.g., every section at which the stream channel was cross sectioned), may lead to irrational irregularities in the profile. Natural stream changes are gradual, and adjacent channel changes, except at unusually sharp breaks, blend into a mean condition over a considerable reach. Where computer adaptation is available, the tendency may be toward overrefinement. A primary consideration in the selection of reaches is uniformity of effective depth—reaches that combine shallow and deep channel stretches should be avoided.

At the junction of a tributary double entry is provided, without and with allowance for the oxygen assets of the tributary runoff increment. The sharp rise in the dissolved oxygen profile assumes complete mixing of the two streams at the junction. Actually the distance below the junction at which the two streams become completely intermixed varies, depending on those channel configurations (particularly bends) that aid in the mixing. There is no exact formula for deciding on the subdivision of the channel into reaches for the reaeration and oxygen balance computations; experienced professional judgment leads to appropriate subdivisions for the situation in hand.

Runoff is obtained from the hydrologic study of yields along the course of the stream and the tributary drainage areas. Temperature is the mean for each individual reach. Effective depth need not correspond with the reaches employed in Form A; summation curves of occupied channel volume and of surface area provide a ready means of determining the effective depth for each reaeration reach. Frequency of turnover, or mix interval, is a function of effective depth and is obtained for each reach from the appropriate curve of Figure 4–13. River volume for each reach is obtained from the summation curve of occupied channel volume. Column 7, the total demand satisfied above the station, is obtained direct from Form A, column 9. If reaches do not coincide with the stations on Form A, the required values can be obtained from a graphical plot of column 9. Columns 8 through 14 are dissolved oxygen assets of both tributary runoff and reaeration expressed in population equivalents. Columns 8 and 9 are the oxygen assets of runoff, a function of saturation value for the prevailing temperature and the extent to which increments are below saturation as they enter the respective reaches. Seldom are natural unpolluted streams fully saturated with oxygen, and in the absence of field survey data, 85 percent of saturation is a reasonable value to assume. Where field information is available the appropriate percent saturation is applied for each successive runoff increment, with column 9 representing the accumulation. Column 10 represents the total oxygen capacity at saturation for the runoff and temperature. Column 11 is a partial balance between the assets of runoff and the total liability, neglecting for the moment reaeration. This balance may be either negative or positive. This is as far as we can proceed with the balance sheet until reaeration computations are made.

Reaeration Computation Applied to Stream Conditions

Reaeration in one of its basic parts is a function of dissolved oxygen deficit. Since the subdivision of the channel has been made to provide reaches through which the dissolved oxygen profile can be assumed as a straight line from the upper to the lower end of the reach, the dissolved oxygen deficit is obtained as the average of the dissolved oxygen at the beginning and the end of each reach. Before a computation of reaeration for the reach can be made it is necessary to know the dissolved oxygen at the end of the reach.

As in similar situations, we deal with this dilemma by successive approximations. Assume a dissolved oxygen value at the lower end, which with the previously determined value at the head of the reach provides a dissolved oxygen deficit necessary for the first reaeration calculation; make successive approximations until the assumed value is in close agreement with the computed one. We start at the head of the first reach usually with foreknowledge of the stream's condition as it arrives. With the knowledge of the net asset at the end of the reach (column 11 of Form B) a reasonable estimate of the dissolved oxygen at the downstream end can be made as the first approximation. The result of the first computation of reaeration discloses whether the approximation was too low or too high, and this constitutes the basis for a closer approximation in the second try. With experience, about three approximations usually lead to agreement within 0.1 percent of saturation, which is an adequate tolerance. The second and subsequent computations can be made quickly, since all other conditions remain constant and reaeration is adjusted in proportion to change in oxygen deficit.

After reaeration computations for the first reach are balanced, the running summary is tabulated in columns 12, 13, and 14 of Form B. The net dissolved oxygen balance at the end of the reach is converted to percentage of saturation in column 15 by dividing the net balance of column 14 by the saturation capacity of the runoff at the station as determined in column 10, expressed in percent.

We are now in a position to proceed with the computation of dissolved oxygen for the second reach. The computed value for the station at the end of the first reach becomes the beginning of the dissolved oxygen value for the second reach, and again, by assuming a value for the end of the reach and balancing by successive approximations, the dissolved oxygen is determined. Thus reach-by-reach computations proceed down the river, defining an independently computed dissolved oxygen profile, reserving observed results from river surveys as a verification of the self-purification factors adopted.

Each computed dissolved oxygen profile is specific for a particular hydrologic regime of runoff, temperature, and BOD loading.

Since the procedure for computing deoxygenation and reoxygenation takes into account fundamentally the primary factors involved, it is possible to compute reliable reflections of expected stream conditions for a great variety of different combinations, such as runoff over the whole range of drought probability or seasonal pattern, diverted or augmented flow, variation in seasonal temperature, man-made changes in the river channel (dredging, navigation locks, dams), increase or decrease in existing waste loads, interception of waste sources for various treatment schemes, proposed new locations of industries, hydropower influences, and concentrated heat loads for atomic or fossil fuel power plants.

Illustrated Example

The following abstract from a detailed study of River A illustrates the steps in determining the oxygen balance and the resulting dissolved oxygen profile. Page 1 of Form A lists the BOD loading, which includes a sludge deposit and an immediate demand; page 2 summarizes the deoxygenation integration determined graphically (Figure 4–14). Time of passage was determined from detailed channel cross sections (discussed in Chapter 3) for a runoff 4900 cfs at the reference gage. Form B summarizes the reoxygenation and oxygen balance, and detailed computations are shown in Table 4–6. The channel characteristics of each reach, effective depth, and river volume were determined for a runoff of 4900 cfs based on the channel cross-sections. The computed dissolved oxygen profile is shown in Figure 4–15.

THE SIMPLIFIED DISSOLVED-OXYGEN-SAG CURVE

The difficulties inherent in the mathematical solution of Phelps' original formulations for reaeration led to the simplification proposed by Streeter and Phelps [16], in which the relevant physical and dynamic stream characteristics are lumped into a single reaeration coefficient K_2. This simplification merely states that the rate at which the oxygen deficit is being reduced is always proportional to that deficit where K_2 is the

Page 1
Computation No.
Date

FORM A: DEOXYGENATION

STREAM: River A
RUNOFF: 4900 CFS AT REFERENCE GAGE
DROUGHT PROBABILITY
SPECIAL CONDITIONS Mean runoff during stream survey period

BOD LOADING Residual at mile 87 from upstream sources and Landwash
0.77 ppm BOD_5
$4900/1.547 \times 0.77/0.68 \times 8.34/0.24 = 125{,}000$ PE
WASTE LOADS:

			BOD Population Equivalent (PE) Discharged to Stream			
River Mile (1)	Source (2)	Type of Treat-ment (3)	Imme-diate Demand (4)	Colloidal and Dissolved Fraction (5)	Settle-able Solids Fraction (6)	Total (7)
87.0	Upstream residual			125,000		125,000
85.9	Municipal and Industrial			23,100		23,100
85.15	Industrial			133,200	41,200	174,400
83.28	Municipal			53,500		53,500
78.60						
72.57						
64.00						
54.86	Tributary B					
51.30						
50.05	Industrial		52,200	527,800		580,000
46.10						
38.68						
35.38	Tributary C					
29.99						
28.43	Tributary D					
26.59						

Page 2
Computation No.

River A

Station (river mile)	Mean Water Temperature (°C)	k at Mean Temperature	Time of Passage from Initial Station (days)	BOD (Population Equivalents)				
				Dis-charged at Station	Residual from Stations Upstream	Total at Station	Satisfied in Stream between Stations	Total Satisfied above Station
(1)	(2)	(3)	(4)	(5)	(6)	(7)	(8)	(9)
87.0			0		125,000	125,000		0
	21	0.125					500	
85.9			0.0226	23,100	124,500	147,600		500
	21	0.125					1,100	
85.15			0.0546	41,200(sludge at equilibrium)			41,200	
	21	0.125						
85.15			0.0546	133,200	146,500	279,700		42,800
	21	0.125					9,700	
83.28			0.1677	53,500	270,000	323,500		52,500
	21	0.125					14,000	
78.60			0.3387		309,500	309,500		66,500
	21	0.125					20,500	
72.57			0.5870		289,000	289,000		87,000
	21	0.125					19,000	
64.00			0.8207		270,000	270,000		106,000
	21	0.125					22,000	
54.86			1.112		248,000	248,000		128,000
	21	0.125					13,000	
51.30			1.314		235,000	235,000		141,000
	21	0.125					7,000	
50.05			1.423		228,000	228,000		148,000
	21	0.125		52,200(immediate demand)			52,200	
50.05			1.423	527,800	228,000	755,800		200,200
	21	0.125					115,800	
46.10			1.965		640,000	640,000		316,000
	21	0.125					170,000	
38.68			3.045		470,000	470,000		486,000
	21	0.125						
							200,000	
26.59			4.986		270,000	270,000		686,000

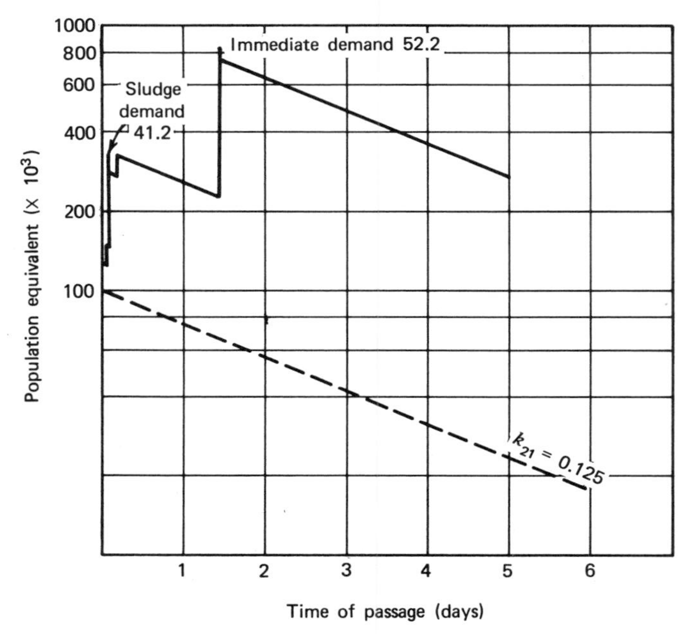

Figure 4–14 River A. Integration of BOD loading (reduced scale).

proportionality factor. The gain in simplicity, however, is a loss in generality.

This simplification led to the dissolved oxygen sag equation, which integrates deoxygenation and reaeration into a single expression. Its value in practical stream analysis is limited because of inability to adequately determine K_2 from measurable stream parameters. Lacking a rational basis for determining K_2, there has developed a tendency to misuse it in the sag equation as though it were constant over long reaches of river, whereas from the fundamental formulation it is known to be variable from reach to reach. Also there is a tendency to use the sag equation without determining the time of passage from reach to reach on the basis of measured physical characteristics of the channel. This produces a dissolved oxygen profile in relation to the time of passage but one that is undefined in relation to geographic location along the course.

More recent attempts at direct computation of reaeration center on relation to stream turbulence. Unfortunately indices of turbulence are no more susceptible to direct measurement in the field than is rate of reaeration itself. A promising possibility is the use of radioactive tracers, which are amenable to field measurement along the changing river course.

FORM B: REOXYGENATION AND OXYGEN BALANCE

Computation No.
Computer
Date

River A

Station (river mile) (1)	Runoff at Station (cfs) (2)	Mean Temperature (°C) (3)	Effective Depth (ft) (4)	Frequency of Turn-over (min/mix) (5)	River Volume (MG) (6)	Total Demand Satisfied above Station (PE) (Form A, Column 9) (7)	Dissolved Oxygen (Population Equivalents): Added by Runoff between Stations (8)	Total Added by Runoff above Station (9)	Total Runoff at Station at Saturation (10)	Net at Station (Column 9 minus Column 7) (11)	Added by Re-aeration between Stations (12)	Total Added by Re-aeration above Stations (13)	Net Oxygen Balance at Station (Column 11 plus Column 13) (14)	Percent-age of Saturation at Station Divided by Column 10 (15)
87.0	4900							840,980	989,380	840,980		0	840,980	85.0
		21	6.06	17.4	172.8		0				3,260			
85.15	4900					42,800		840,980	989,380	798,180		3,260	801,440	81.0
		21	8.75	22.0	358.1		0				4,690			
83.28	4900					52,500		840,980	989,380	788,480		7,950	796,430	80.5
		21	5.71	16.6	541.2		0				12,500			
78.60	4900					66,500		840,980	989,380	774,480		20,450	794,930	80.5
		21	5.98	17.1	785.9		0				17,700			
72.57	4900					87,000		840,980	989,380	753,980		38,150	792,130	80.0
		21	6.26	17.6	739.9		0				16,000			
64.00	4900					106,000		840,980	989,380	734,980		54,150	789,130	79.7
		21	6.51	18.2	921.2		0				19,400			
54.86	4900					128,000		840,980	989,380	712,980		73,550	786,530	79.5
		21					14,120							
54.86	4981.3					128,000		855,100	1,005,800	727,100		73,550	800,650	79.6
		21	7.90	21.0	651.1		0				10,300			
51.30						141,000		855,100	1,005,800	714,100		83,850	797,950	79.4
		21	11.08	26.0	351.7		0				3,660			
50.05						148,000		855,100	1,005,800	707,100		87,510	794,610	79.0
50.05						200,200		855,100	1,005,800	654,900		87,510	742,410	73.8

Table 4–6 Illustrated Oxygen Balance Computations—River A
(Solution by slide rule, standard reaeration curve *A* of Figure 4–12)

Mile point 87.0–85.15

Volume = 172.8 MG
Depth = 6.06 ft
Temperature = 21°C
Mix interval = 17.4 min (Figure 4–13)
Dissolved oxygen per mix at 100% deficit = 0.430%

17.5% deficit | First try | ←85% | ←80% (assumed) | 165 / 2 = 82.5% average

0.175 × 0.430 × (1440/17.4) = 6.23% per day at 17.5% deficit
(4900/1.547) × 8.99 × (8.34/0.24) = 989,380 PE at saturation
0.85 × 989,380 = 840,980 PE at 85% saturation at mile point 87.0
8.99 × (8.34/0.24) = 312 PE/MG at saturation
0.0623 × 312 × 172.8 = 3,360 PE reaeration in reach
798,180 PE (column 11 of Form B)
801,540 PE
801,540/989,380 = 81% (*too high*)

17.0% deficit | Second try | ←85 | ←81% | 166 / 2 = 83% average

(17/17.5) × 3360 = 3,260 reaeration
798,180
801,440
801,440/989,380 = 81% (*satisfactory*)

Mile point 85.15–83.28

Volume = 358.1 MG
Depth = 8.75 ft
Temperature = 21°C
Mix interval = 22 min
Dissolved oxygen per mix at 100% deficit = 0.335%

19.5% deficit | First try | ←81% | ←80% (assumed) | 161 / 2 = 80.5% average

0.195 × 0.335 × (1440/22) = 4.260% per day at 19.5% deficit
0.0426 × 312 × 358.1 = 4,750 reaeration in reach
3,260 previous reaeration
8,010
788,480
796,490
796,490/989,380 = 80.5% (*too high*)

19.25% deficit | Second try | ←81% | ←80.5% | 161.5 / 2 = 80.75% average

(19.25/19.5) × 4750 = 4,690
3,260
7,950
788,480
796,430
796,430/989,380 = 80.5% (*satisfactory*)

Table 4–6 (Continued)

Mile point 83.28–78.60

Volume = 541.2 MG
Depth = 5.71 ft
Temperature = 21°C
Mix interval = 16.6 min
Dissolved oxygen per mix at 100% deficit = 0.44%

19.75% deficit | First try | ←80.5%, ←80.0% — 160.5/2 = 80.25% average

0.1975 × 0.44 × (1440/16.6) = 7.54% per day at 19.75% deficit
0.0754 × 312 × 541.2 = 12,700 reaeration
7,950 previous reaeration
20,650
774,480
795,130

795,130/989,380 = 80.5% (*too high*)

19.5% deficit | Second try | ←80.5%, ←80.5% — 80.5% average

(19.5/19.75) × 12,700 = 12,500
7,950
20,450
774,480
794,930

794,930/989,380 = 80.5% (*satisfactory*)

Mile point 78.60–72.57

Volume = 785.9 MG
Depth = 5.98 ft
Temperature = 21°C
Mix interval = 17.1 min
Dissolved oxygen per mix at 100% deficit = 0.435%

19.75% deficit | First try | ←80.5%, ←80.0% — 160.5/2 = 80.25% average

0.1975 × 0.435 × (1440/17.1) = 7.23% per day at 19.75% deficit
0.0723 × 312 × 785.9 = 17,700 Reaeration
20,450 previous reaeration
38,150
753,980
792,130

792,130/989,380 = 80.0% (*satisfactory*)

Table 4–6 (Continued)

Mile point 72.57–64.00

		←80.0%	Volume = 739.9 MG
			Depth = 6.26 ft
20.0%	First	80.0% average	Temperature = 21°C
deficit	try		Mix interval = 17.6 min
		←80.0%	Dissolved oxygen per mix at 100% deficit = 0.42%

0.20 × 0.42 × (1440/17.6) = 6.88% per day at 20.0% deficit

0.0688 × 312 × 739.9 = 15,860 reaeration
38,150 previous reaeration
54,010
734,980
788,990

788,990/989,380 = 79.7% (*too low*)

		←80.0%	
20.15%	Second	79.85% average	(20.15/20.0) × 15,860 = 16,000
deficit	try	←79.7%	38,150
		159.7	54,150
		2	734,980
			789,130

789,130/989,380 = 79.7% (*satisfactory*)

Mile point 64.00–54.86 (above junction of tributary B)

		←79.7%	Volume = 921.2 MG
			Depth = 6.51 ft
20.4%	First	79.6% average	Temperature = 21°C
deficit	try	←79.5%	Mix interval = 18.2 min
		159.2	Dissolved oxygen per mix at 100% deficit = 0.418%
		2	

0.204 × 0.418 × (1440/18.2) = 6.75% per day at 20.4% deficit

0.0675 × 312 × 921.2 = 19,400 reaeration
54,150 previous reaeration
73,550
712,980
786,530

786,530/989,380 = 79.5% (*satisfactory*)

Mile point 54.86 (below junction of tributary B)

←79.5% above

727,100
73,550
←79.6% below 800,650

800,650/1,005800 = 79.6%

Table 4–6 (Continued)

Mile point 54.86–51.30

23.7% deficit	First try	←79.6% 76.3% average ←73.0% 152.6 2	Volume = 651.1 MG Depth = 7.90 ft Temperature = 21°C Mix interval = 21.0 min Dissolved oxygen per mix at 100% deficit = 0.36%

0.237 × 0.36 × (1440/21) = 5.86% per day at 23.7% deficit
0.0586 × 312 × 651.1 = 11,900 reaeration
73,550 previous reaeration
85,450
714,100
799,550
799,550/1,005,800 = 79.5% (*too high*)

20.5% deficit	Second try	←79.6% 79.5% average ←79.4% 159.0 2	(20.5/23.7) × 11,900 = 10,300 73,550 83,850 714,100 797,950

797,950/1,005,800 = 79.4% (*satisfactory*)

Mile point 51.30–50.05 (without immediate demand)

20.65% deficit	First try	←79.4% 79.35% average ←79.3% 158.7 2	Volume = 351.7 MG Depth = 11.08 ft Temperature = 21°C Mix interval = 26.0 min Dissolved oxygen per mix at 100% deficit = 0.29%

0.2065 × 0.29 × (1440/26.0) = 3.32% per day at 20.65% deficit
0.0332 × 312 × 351.7 = 3,640 reaeration
83,850 previous reaeration
87,490
707,100
794,590
794,590/1,005,800 = 79.0% (*too low*)

20.8% deficit	Second try	←79.4% 79.2% average ←79.0%	(20.8/20.65) × 3,640 = 3,660 83,850 87,510 707,100 794,610

794,610/1,005,800 = 79.0% (*satisfactory*)

Table 4–6 (Continued)

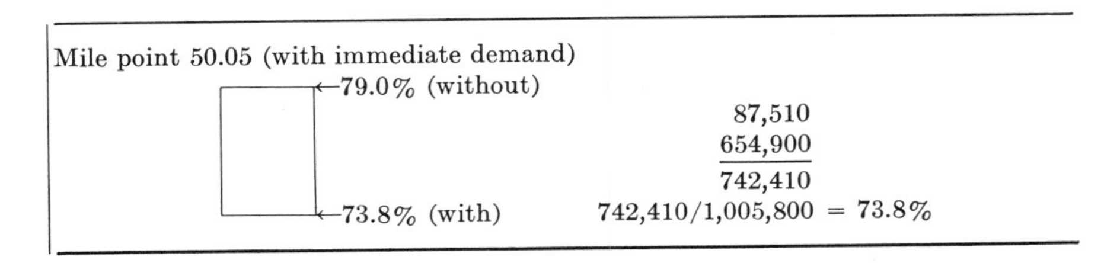

Mile point 50.05 (with immediate demand)

←79.0% (without)

←73.8% (with)

	87,510
	654,900
	742,410
742,410/1,005,800 = 73.8%	

Other recent approaches employ empirical formulations based on statistical analyses of observed stream conditions. Such methods are limited to the conditions of the initial observations, and it is hazardous to generalize by extrapolation. These approaches, however, reemphasize the important role of stream depth as initially included in Phelps' basic formulation.

Although the author does not recommend the use of the integrated sag equation, the method is summarized and comparisons with the rational method are presented.

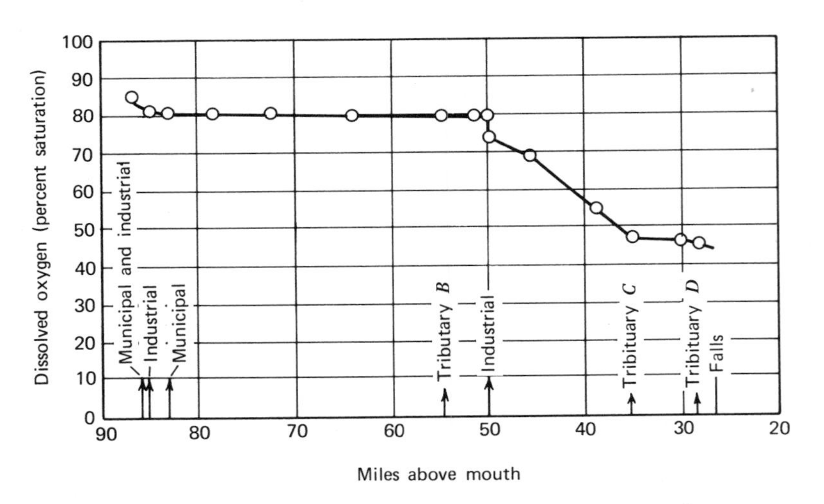

Figure 4–15 River A. Computed dissolved oxygen profile.

The Sag Equation

The reader is referred to *Stream Sanitation* [17] for a full development of the sag equation. For convenience we present the more commonly used forms:

$$D = \frac{k_1 La}{k_2 - k_1}(10^{-k_1 t} - 10^{-k_2 t}) + Da \times 10^{-k_2 t}.$$

In the customary units D is the oxygen saturation deficit in parts per million at time t and Da is the initial deficit; La is the initial BOD in parts per million in the streamflow; t, is the time of passage in days; k_1 and k_2 are deoxygenation and reaeration rates based on common logarithms and the day as the time unit. Solutions for a series of time of passage intervals from an initial BOD source provide the points for constructing the dissolved oxygen sag curve.

Of special interest is the *critical* oxygen deficit and its location below a BOD source. The critical time t_c is given by

$$t_c = \frac{1}{k_2 - k_1} \log \left\{ \frac{k_2}{k_1} \left[1 - \frac{Da(k_2 - k_1)}{Lak_1} \right] \right\}$$

and the critical deficit is given by

$$D_c = \frac{k_1}{k_2} La \times 10^{-k_1 t_c}.$$

The application of these equations is restricted to the sag below a single composite source of BOD and without increments of runoff from tributaries; k_2 and k_1 remain constant from reach to reach. Hence their practical usefulness is limited. For more complex situations the general sag equation is employed, subdividing the stream into reaches and taking into account changes in the time of passage and in the reaeration coefficient k_2 with changes in the river channel. This requires good channel cross-sections from which the time of passage can be determined and factors affecting k_2 identified reach by reach.

Reaeration Coefficient

A number of theoretical and empirical bases for estimating the appropriate reaeration coefficient k_2 have been proposed. Streeter and Phelps [16] related k_2 to the following variables: depth H, velocity V, and the two constants c and n, which depend in part on channel slope and irregularity. They suggested the form

$$k_2 = c\frac{V^n}{H^2},$$

with adjustment for temperature as

$$k_{2(T)} = k_{2(20°\mathrm{C})} \times 1.047^{(T-20)}.$$

Later work by Streeter, Wright, and Kehr [18] refined the temperature adjustment to

$$k_{2(T)} = k_{2(20°C)} \times 1.015^{(T-20)}.$$

O'Connor and Dobbins [19] developed two theoretical expressions for k_2—one for shallow, turbulent streams (nonisotropic) and another for deeper, low velocity streams (isotropic)—generalized into a common form as

$$k_2 = 127 \frac{(D_L U)^{1/2}}{H^{3/2}},$$

where D_L is the coefficient of molecular diffusion at 20°C expressed in square feet per day, U is mean velocity in feet per second, and H is the mean depth of flow in feet.

A temperature adjustment for D_L is

$$D_{L(T)} = 0.00192 \times 1.04^{(T-20)} \text{ (ft}^2\text{/day)}.$$

More recently Churchill, Elmore, and Buckingham [20] developed, on the basis of extensive field measurements in relatively shallow streams at high runoff and velocity, the empirical relation, employing multiple correlation with a number of hydraulic factors:

$$k_{2(20°C)} = 5.026 \frac{V^{0.969}}{H^{1.673}},$$

where V is the mean velocity in feet per second and H is the mean depth in feet. Based on experimental results adjustment of k_2 for temperature is given as

$$k_{2(T)} = k_{2(20°C)} \times 1.0241^{(T-20)}.$$

Each of these formulations has its limitations. The Streeter–Phelps equation and the O'Connor–Dobbins equation need verification against observed stream conditions; the equation of Churchill et al., although derived from observed stream conditions, cannot with assurance be applied to conditions other than those observed.

Comparison of the Rational and the Sag-Equation Methods

The lack of precision in the usual application of the sag equation in comparison with the recommended rational method is illustrated by the following examples of dissolved oxygen profile computation. The conditions for a 20-mile reach of River B of the pool-and-riffle type are as follows:

Runoff	1150 cfs, with no appreciable tributary increments
Temperature	25°C
Da	1 ppm
La	12.15 ppm (311,000 PE)
k_1	0.144, adjusted to 0.181 for 25°C

Computation of the dissolved oxygen profile by the rational method is based on measurements of stream depth, width, and volume (at 1-mile intervals based on river maps, with the channel low water profile adjusted to a 1150-cfs flow regime) from which the time of passage is developed. The channel is characterized by relatively shallow reaches and a number of deeper pool areas, with depths ranging from 2 to 16 ft. The profile computed by the rational method (curve *A* of Figure 4–16) shows a succession of sags associated with the changing channel characteristics, following closely the deep, low velocity pool areas identified in the time of passage curve. The critical dissolved oxygen level declines to 3.2 ppm at mile 13, 2.17 days' time of passage below the waste source.

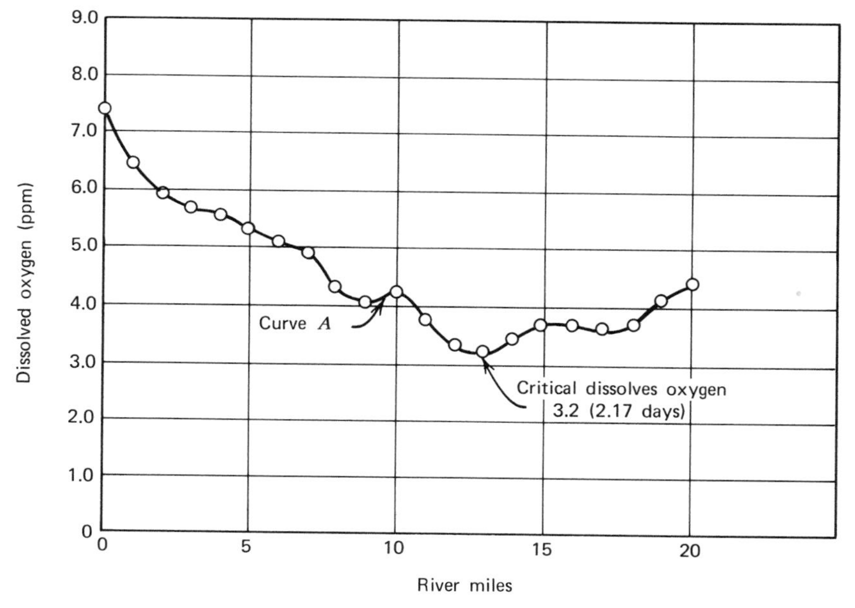

Figure 4–16 River B. Dissolved oxygen profile computed by the rational method. Temperature 25°C, BOD load 311,000 PE, $k_{20°C}$0.144.

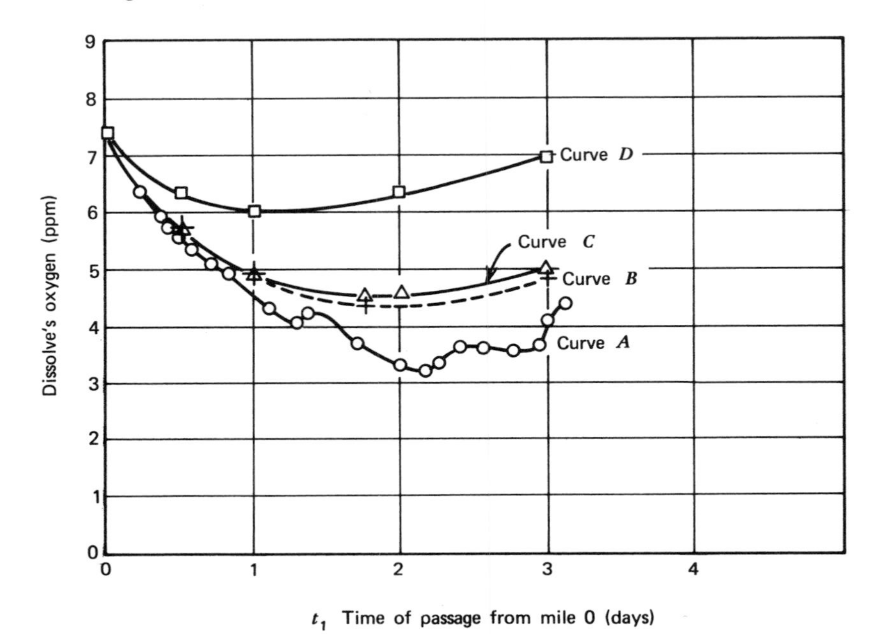

Figure 4–17 River B. Comparison of dissolved oxygen profiles computed by the rational method (curve *A*) and by the simplified sag equation. Curve *B*: Streeter–Phelps k_2; curve *C*: Churchill k_2: curve *D*: O'Connor–Dobbins k_2. Runoff 1150 cfs, temperature 25°C. BOD load 311,000, $k_{1(25°C)}$0.181.

Figure 4–17 shows the dissolved oxygen profiles computed by the simplified sag equation for the same conditions, except that the reaeration coefficient k_2 was developed without benefit of the detailed channel characteristics and was based on general hydraulic relationships of depth and velocity measured in the upper reach (miles 2–7). Adjusted to the runoff of 1150 cfs, a mean velocity of 0.6 fps and a mean depth of 4 ft were adopted to apply for the entire 20-mile reach of river. Based on this velocity and depth, the reaeration coefficient $k_{2(25°C)} = 0.6$, determined from the O'Connor–Dobbins equation, gives the dissolved oxygen profile, (curve *D*), with a critical value of 6.0 ppm at a point 1.02 days downstream. Similarly the reaeration coefficient $k_{2(25°C)} = 0.273$, determined from the Churchill equation, gives the dissolved oxygen profile (curve *C*), with a critical value of 4.5 ppm at a point 1.75 days downstream. The sag (curve *B*) is based on $k_{2(25°C)} = 0.26$, estimated from a modified Streeter–Phelps equation as

$$k_{2(25°C)} = \frac{6.5V}{H^2}.$$

The critical dissolved oxygen value is computed as 4.4 ppm at a point 1.78 days downstream.

For comparability with the curves derived from the simplified sag equation the dissolved oxygen profile computed by the rational method, now referred to time of passage below the waste source, is shown as curve *A* in Figure 4–17. There is a marked difference among the four profiles, each presumed to reflect expected stream conditions for the same 20-mile river reach resulting from the same waste load, runoff, temperature, and rate of BOD satisfaction. Of particular significance is the difference between profile *A*, derived by the rational method, taking into account changes in the channel along the course, and the profiles *B*, *C*, and *D*, derived from the sag equation and based on single k_2 values. This illustrates an all too common misuse of the sag equation, a k_2 value estimated from limited measurements of channel characteristics and applied as a single value over too long a reach of river.

To further emphasize the importance of taking into account changes in channel, profiles *C* and *D* of Figure 4–17 have been recomputed, subdividing the 20 miles of river into reaches consistent with changes in velocity, time of passage, and depth of shallows and pool areas—and adjusting k_2 for each reach accordingly. These dissolved oxygen profiles are shown in Figure 4–18. Curve *D* is based on reaeration coefficients computed by the O'Connor–Dobbins equation, curve *C* is based on reaeration coefficients computed by the Churchill equation, and curve *A* is the profile computed by the rational method. Profile *D* (O'Connor–Dobbins k_2), developed by taking into account channel changes, depresses the critical dissolved oxygen value from the previous 6 ppm (Figure 4–17) to 3.5 ppm; curve *C* (Churchill k_2) depresses the critical value from the previous 4.5 ppm (Figure 4–17) to 2.3 ppm. Curve *A*, derived by the rational method, is intermediate between *D* and *C*. Although there are radical differences between the k_2 values computed by the O'Connor–Dobbins and the Churchill equations, the resultant profiles are quite similar in general shape and the critical dissolved oxygen occurs at the same location. It would appear that the O'Connor–Dobbins reaeration coefficient is too high for the given velocities and depths, whereas the Churchill k_2 values are unusually low for the deep, slow reaches.

As a further illustration of the advantage of the rational method and the risk of applying the simplified sag equation, reference is made to Figure 4–19 for River C, which affords opportunity for comparison with observed dissolved oxygen profiles. Curve *A*, the profile computed by the rational method, taking into account channel changes based on good cross-section sounding, is in good agreement with the observed river

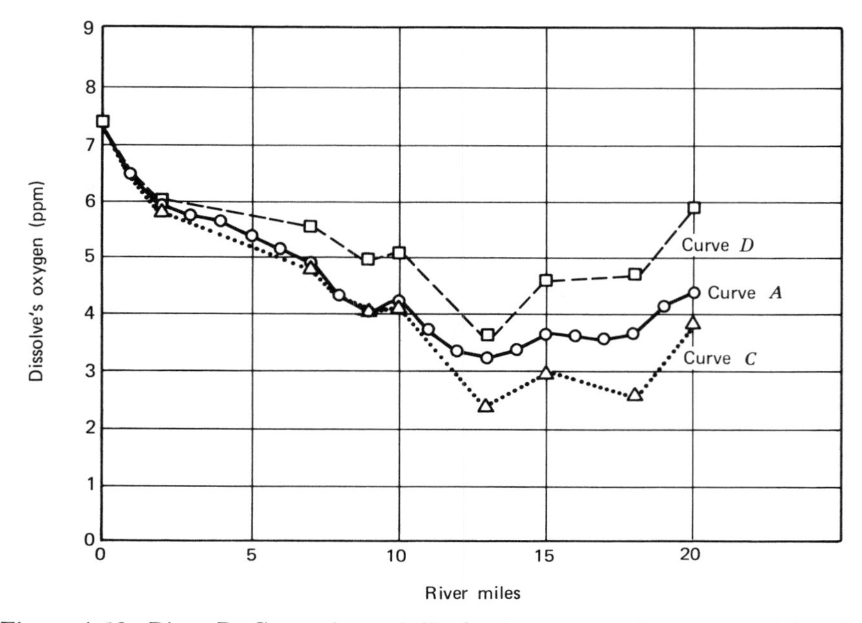

Figure 4–18 River B. Comparison of dissolved oxygen profiles computed by the rational method (curve A) and by the sag equation, adjusting for changes in channel characteristics along the river course. Curve C: Churchill k_2; curve D: O'Connor–Dobbins k_2.

sampling results. The profiles of curves D and C are computed for the same conditions as those of profile A, with the exception that velocity and depth are taken in a single reach (miles 229–226) and applied along the entire course. Adjusted to 1000 cfs, the velocity of 0.336 fps and the mean depth of 12.0 ft applied in the simplified sag equations produce curve D for the O'Connor–Dobbins equation and curve C for the Churchill equation. These profiles are not in agreement with curve A, and the values are decidedly below the observed river sampling results.

It should be recognized that in using the sag equation, where k_2 is changed along the channel in accordance with velocity and depth, the direct solutions for t_c and D_c cannot be applied. In fact a river of such uniform channel is too seldom encountered to ever warrant the use of equations for t_c and D_c. In the proper application of the simplified sag equation we therefore quickly lose the advantage of its simplicity and soon are involved in extensive computations reach by reach. The rational method is less burdensome and more flexible in taking full advantage of detailed characteristics of the waste loads and of the chan-

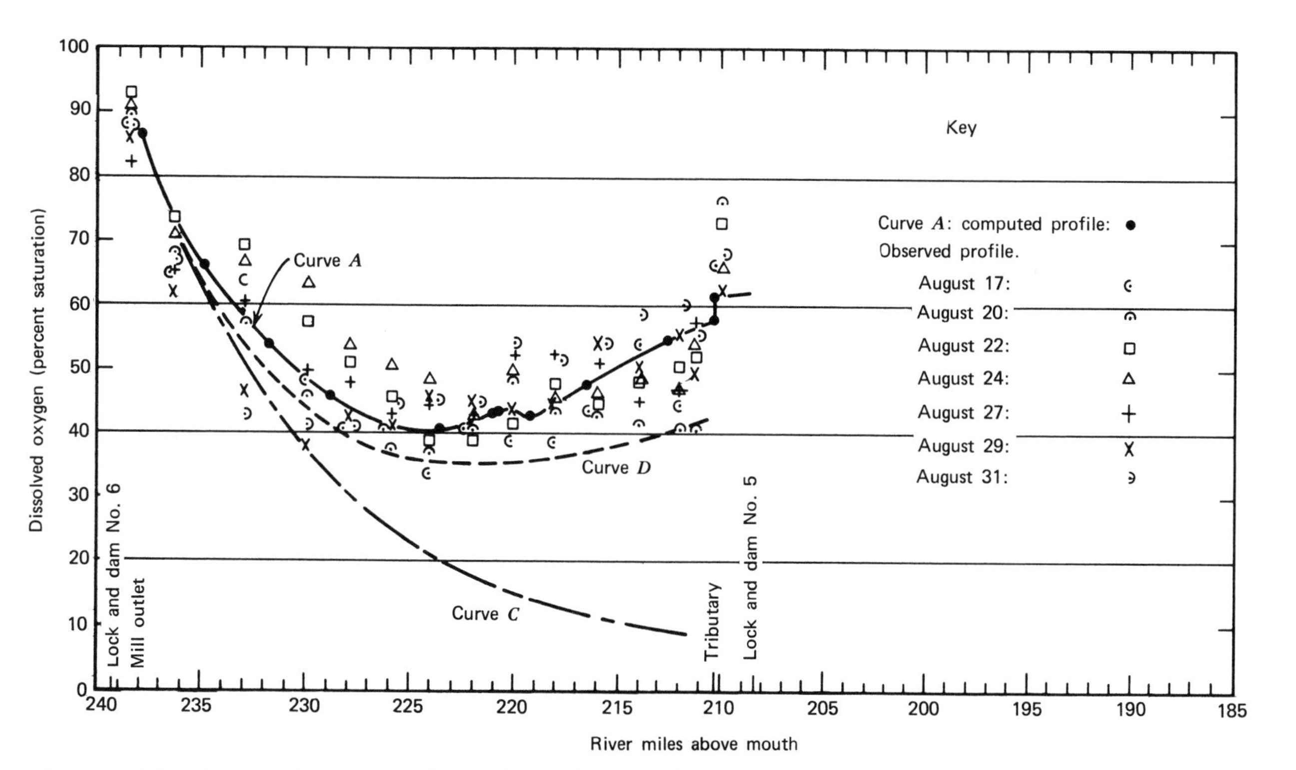

Figure 4–19 River C. Comparison of the observed dissolved oxygen profile (August 1956) and those computed by the rational method (curve A) and by the sag equation. Curve C: Churchill k_2; curve D: O'Connor-Dobbins k_2. Temperature 32°C, runoff 1000 cfs, BOD load 203,000 PE.

nel. Streeter and Phelps in initially proposing the integrated sag equation emphasized the overriding influence of stream depth, time of passage, and the necessity of adjusting k_2 reach by reach.

VERIFICATION

Comparison of Computed and Observed Dissolved Oxygen Profiles

A supporting phase in the rational method (or any method) is verification of the self-purification factors by comparing the computed and observed dissolved oxygen profiles under identical conditions. This comparison is feasible only if existing waste loads are sufficient to depress the dissolved oxygen significantly, but not to the extent of complete depletion. The function of such verification is to confirm *normal* self-purification or to identify potential *abnormalities,* such as sludge deposit, and biological extraction and accumulation.

If not properly identified and taken into account, sludge deposits, the most common abnormality encountered, can lead to a misconception of the nature of self-purification. An observed high removal of BOD in a reach of river may be interpreted falsely as a high rate of satisfaction, when actually it may be a combination of normal rate of satisfaction and a significant deposit of settleable solids BOD. As already discussed, the onshore survey of waste sources, differentiating between the fraction of BOD in colloidal and dissolved form and that in the form of settleable solids, is the first guide to the possibility of sludge deposit. The time of passage curves based on detailed channel cross-section soundings identify potential locations along the channel at which deposit can occur under certain ranges of expected stream runoff. The use of such data in the rational method makes it possible to compute independent dissolved oxygen profiles, taking into account the abnormality of sludge deposit, for verification against reliable observed BOD and dissolved oxygen profiles.

For reliable verification purposes observed stream conditions entail executing a carefully designed river sampling survey under conditions of known waste loads and stable river hydrology (Chapter 10). Intensive sampling over a short period of known stable conditions is superior to grab samples taken over long periods in which hydrology or loads may vary. Even under supposedly stable conditions there is considerable variation among samples; hence a sufficient body of data should be obtained to ensure statistically significant observed conditions. Sampling should also be sufficiently extensive to describe the entire profile, the decline, the critical reach, and the recovery. In making comparisons it is more significant to consider the overall *shape* of the dissolved oxygen profile along extensive reaches of river than to make precise verification at a particular sampling station or a limited reach.

Often it is not feasible to undertake an intensive stream sampling program because hydrologic conditions do not stabilize; then existing river data from limited surveys made from time to time can be utilized. Such surveys, although less than desired, are valuable if selections can be made for stable hydrologic conditions and if the then prevailing waste loadings can be adequately reconstructed.

A number of river cases are illustrated to indicate the variations involved and methods of interpretation used in the verification of normal self-purification, abnormality of sludge deposit, and biological extraction and accumulation. In each case river hydrology data were available, and channel cross-section soundings provided time of passage curves and the necessary channel parameters to permit application of the rational method of computation.

Verification of Normal Self-Purification

Figures 4–19 (curve *A*) and 4–20 illustrate verification of normal self-purification in a relatively large river (River C) below a major industrial source of waste effluent, free of settleable solids subject to deposit entering at mile 238, and a relatively minor municipal and industrial load of 21,500 PE at mile 206. Excellent channel cross-section soundings and a number of river sampling traverses covering the reach mile 238 to mile 210 were available. A period of stable, low flow (1000 cfs) and fairly steady waste effluent loadings prevailed during August, with a high natural river water temperature of 32°C. The observed dissolved oxygen profiles based on seven river sampling traverses are shown in Figure 4–19 by key; the independently computed dissolved oxygen profile for these conditions, based on normal BOD satisfaction adjusted to 32°C, is shown as curve *A*, which is in close agreement with the observed profile. Figure 4–20 represents a verification under winter conditions of 12°C and a monthly average runoff of 1258 cfs. Two river-sampling traverses at this runoff and temperature are shown; the independently computed dissolved oxygen profile for these conditions, based on normal BOD satisfaction adjusted to 12°C, is shown as curve *A*, which is again in good agreement with the observed values. Although no intensive river sampling survey was available in this case, the comparisons based on limited river sampling strongly confirm normal self-purification.

Of equal significance is the agreement between the general *shape* of the observed and computed dissolved oxygen profiles for summer and winter conditions (Figures 4–19 and 4–20, respectively) if the decided change in water temperature is taken into account.

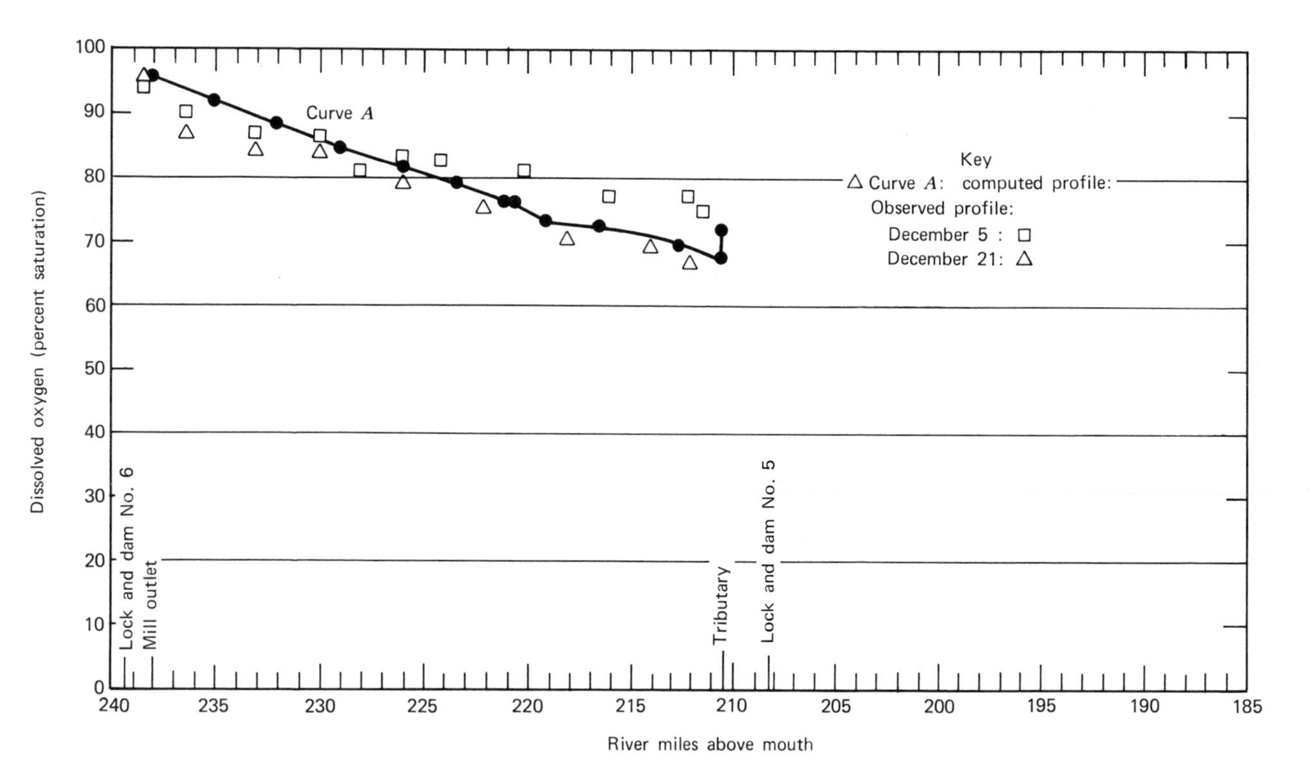

Figure 4–20 River C. Verification of normal self-purification—comparison of computed and observed dissolved oxygen profiles under winter conditions. Temperature 12°C, runoff 1258 cfs, BOD load 208,200 PE.

Verification of Abnormality of Sludge Deposits

The case of River D illustrates difficulties that may be encountered in executing an intensive river survey and in verifying sludge deposits from settleable solids in untreated or inadequately treated municipal and industrial wastes. The raw waste discharges were concentrated in a single zone above deposit areas. A carefully designed intensive river sampling program was prepared for execution during August and September, when steady, low runoff is normally expected. Daily reports of river discharge received from the U.S. Geological Survey and weather forecasts afforded an opportunity to anticipate a steady runoff pattern, undisturbed by a preceding freshet. Unfortunately runoff remained high and was punctuated with freshet flows; finally the intensive sampling program was undertaken at the end of the season, September 13 to 17 (with runoff ranging from 2290 to 1790 cfs and averaging 1930 cfs), after the freshet on September 8. Algae were known to be involved, and dissolved oxygen samples were taken at 2-hr intervals throughout 24-hr cycles. A systematic daily cycle in observed dissolved oxygen developed from which an integrated mean value (balancing daytime oxygen production with nighttime respiration) for the 4-day sampling period was obtained, as shown in Figure 4–21. The mean water temperature was 20°C. The prevailing BOD waste loading from municipal and industrial sources aggregated 622,000 PE, of which 25 percent was estimated to be in the form of suspended solids and hence subject to deposit. From analyses of time of passage curves four areas of potential sludge deposit were identified within and below the zone of waste discharges, as indicated in Figure 4–21. Critical velocities (0.6 fps) for sludge deposit in these areas were 1900, 1300, and 1550 cfs for areas I, II and III, and IV, respectively. The freshet of September 8 was sufficient to scour all areas; hence with the prevailing high runoff deposit is delicately balanced, and only a short period existed for sludge to accumulate in area I as well as along shallow reaches and in eddy areas.

Curve *A* of Figure 4–21 represents the computed dissolved oxygen profile at 20°C at normal BOD satisfaction on the assumption that the entire waste load remained in the flowing stream and no sludge deposits occurred. This profile is above the observed values represented by the integrated means. Curve *B* represents the profile computed by assuming sludge deposit and accumulation for 5 days after the freshet. This depresses the profile to a position in closer agreement with the observed dissolved oxygen level. If the settleable solids fraction of the BOD load is increased from 25 to 30 percent, which subsequent analysis indicated to be more likely, the agreement will be quite close. Considering

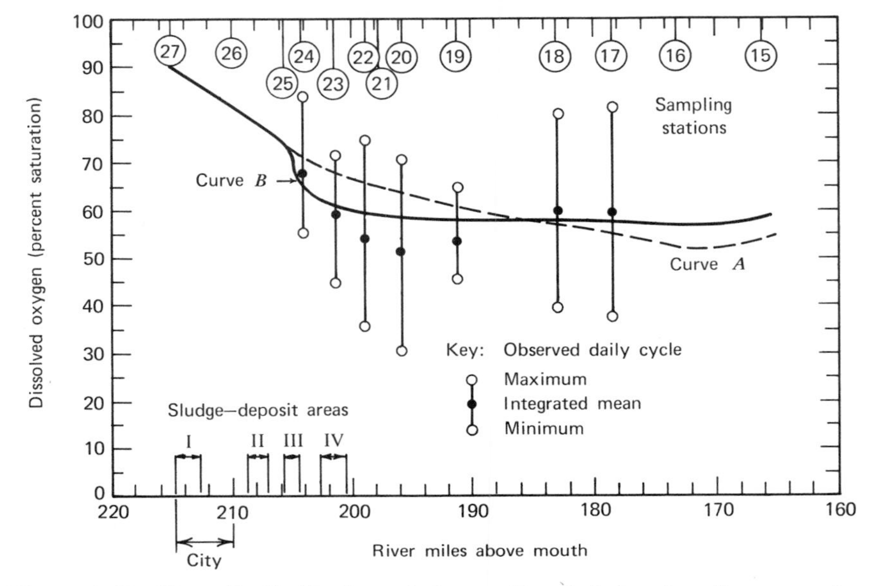

Figure 4–21 River D. Verification of abnormality of sludge deposits—comparison of computed and observed dissolved oxygen profiles. Curve *A*: computed profile—no depositing and no sludge accumulations *B*: computed profile—depositing and 5 days' accumulation.

the great variation in sampling data due to algae and the slight dissolved oxygen sag at this high runoff, there is satisfactory agreement between the computed (curve *B*) and the observed results, with indication that sludge deposit is involved.

To portray what would be likely to occur if runoff prevailed for a protracted period below the critical level permitting deposit and accumulation, Figure 4–22 was developed. Curve *A* is the computed dissolved oxygen profile at 20°C and 1930 cfs, the mean for the period of intensive river sampling, allowing for no depositing and with no sludge accumulation (as previously shown by curve *A* of Figure 4–21). Curve *B* for the same waste load (30 percent as settleable solids) is at a runoff of 1400 cfs, a drought flow as weekly average expected once in 3 years, and a water temperature of 23°C, considering deposition in area I, bypass of areas II and III, and deposition in area IV, with prolonged accumulation approaching equilibrium. Curve *C* is for a runoff of 1000 cfs at 23°C; deposits are found in all areas at equilibrium accumulation. The sharp drop in dissolved oxygen in curves *B* and *C* indicates the significance of heavy demand from sludge accumulation.

An opportunity to verify sludge deposit with significant accumulation periods was afforded in two river surveys made 9 years earlier. Figure 4–23 shows the results of observed river samplings as averages of hourly samples through a 24-hr period taken at various stations during the period July 26 to August 1. Algae effects were present, but in the 24-hr averages the daytime production of dissolved oxygen balanced against the nighttime respiration. The BOD waste loading was substantially lower because of reduced industrial production and was reported as 224,800 PE, of which 30 percent was estimated as settleable solids subject to deposit. Runoff averaged 1400 cfs before and during sampling, with a freshet peaking on July 2; thus opportunity for sludge accumulation ranged from 2 to 3 weeks preceding the respective river sampling days, with sludge distributed among the deposit areas. River water temperature was high at 30°C. Curve *A* represents the computed dissolved oxygen profile for these conditions, allowing for mean sludge accumulation of 20 days. There is substantial agreement between the computed profile and the partial river survey data, definitely confirming sludge deposit of partial accumulation.

Figure 4–24 is a comparison of computed and observed dissolved oxygen

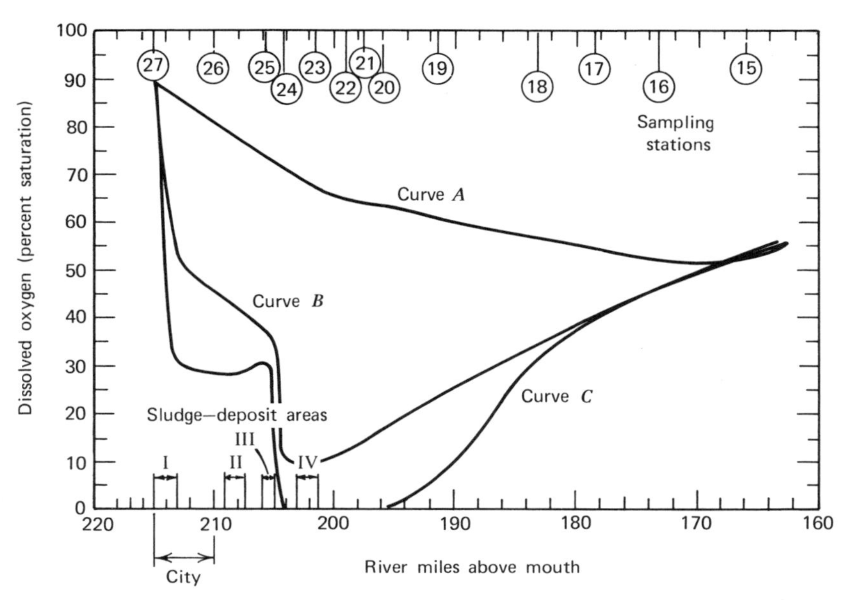

Figure 4–22 River D. Influence of runoff on dissolved oxygen profile. Sludge deposits at equilibrium. Curve *A*: runoff 1930 cfs, no depositing and no sludge accumulation; curve *B*: runoff 1400 cfs, depositing and sludge at equilibrium accumulation; curve *C*: runoff 1000 cfs, depositing and sludge at equilibrium accumulation.

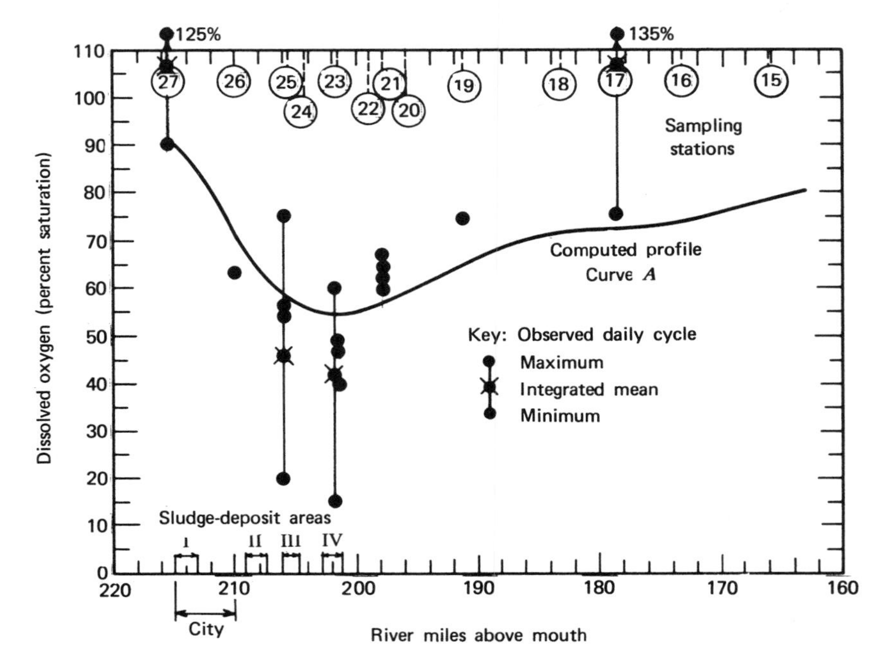

Figure 4–23 River D. Comparison of computed and observed dissolved oxygen profiles; 20-day sludge accumulation.

values in the fall of the same year, September 26 and 27, and October 3 and 4. Runoff on sampling days ranged from 860 to 900 cfs, with an average of 880 cfs, and was preceded by almost 3 months of low flow, permitting deposit and accumulation without interruption by a scouring freshet. This survey, although of only limited sampling, reflects a severe prolonged drought with a cool temperature of 18°C. The prevailing raw waste load was reported as 374,000 PE. Curve *A* represents the computed profile for these conditions, assuming sludge at equilibrium accumulation level and, in view of the low flow, concentrated in area I. The sharp drop in the computed and observed dissolved oxygen profiles, the general agreement at the critical reach, and the subsequent recovery provide strong confirmation of sludge deposit as the dominant abnormality.

The series of comparisons for River D taken as a whole confirms the abnormality of sludge deposit and accumulation, and verifies the self-purification factors adopted in the application of the rational method in this case.

Verification of Abnormality of Biological Extraction and Accumulation

In the case of River E an abnormal BOD removal was known to occur, but the waste load from the single large industry did not contain appreciable settleable solids subject to deposit. Some investigators ascribed the high removal to a high rate of BOD satisfaction. A detailed analysis confirmed the abnormality of biological extraction and accumulation, and verified that the high BOD removal rate was not a high rate of satisfaction. The analyses illustrate how in the process of verification use can be made of old river sampling data. It was necessary, however, to supplement these data by channel cross-section soundings along the course of the river to develop time of passage curves and channel parameters for ranges of stream runoff to permit application of the rational method.

The records from industry provided data on intermittent river samplings and associated waste effluent loadings over a period of years for a variety of hydrologic conditions. Single sample traverses taken under a range of waste loads and river runoff indicated consistently high removal of BOD in the reach from mile 192 to mile 173. Within this

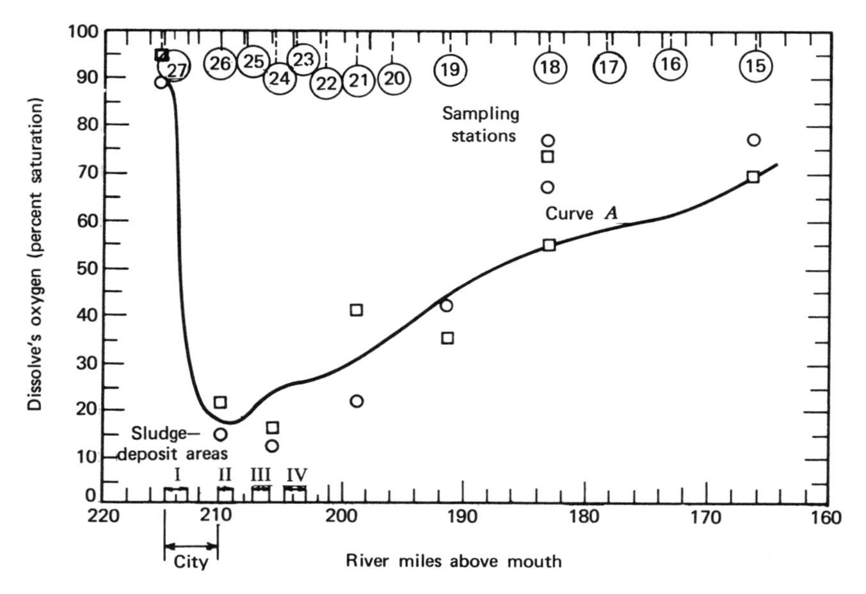

Figure 4–24 River D. Comparison of computed and observed dissolved oxygen profiles under extremely prolonged drought flow. Curve A: computed profile; ⊙ observed by the USPHS on September 26 and 27; □ observed by the USPHS on October 3 and 4.

reach biological floc (predominantly *Sphaerotilus*) was visually verified as heavy at the time of channel cross sectioning; below this reach the floc rapidly disappeared. A critical analysis of these limited data, taking into consideration mill waste loads and the time of passage associated with the runoff prevailing during river sampling, disclosed a rather consistent biological extraction rate b of 0.7 through the active reach (miles 192–173), below which BOD removal reverted to normal. During channel cross sectioning excess biological floc was observed to deposit along the channel banks and in eddy areas of the active zone. On this basis a series of dissolved oxygen profiles was computed by the rational method for comparison with the dissolved oxygen observed in river sampling, taking into account extraction and sludge accumulation of the excess floc.

Figure 4–25 reflects a protracted period of sludge accumulation; the circles represent the dissolved oxygen values observed during an early survey as single samples on 5 days preceded by a scouring freshet and followed by low runoff, providing approximately 40 days' opportunity for

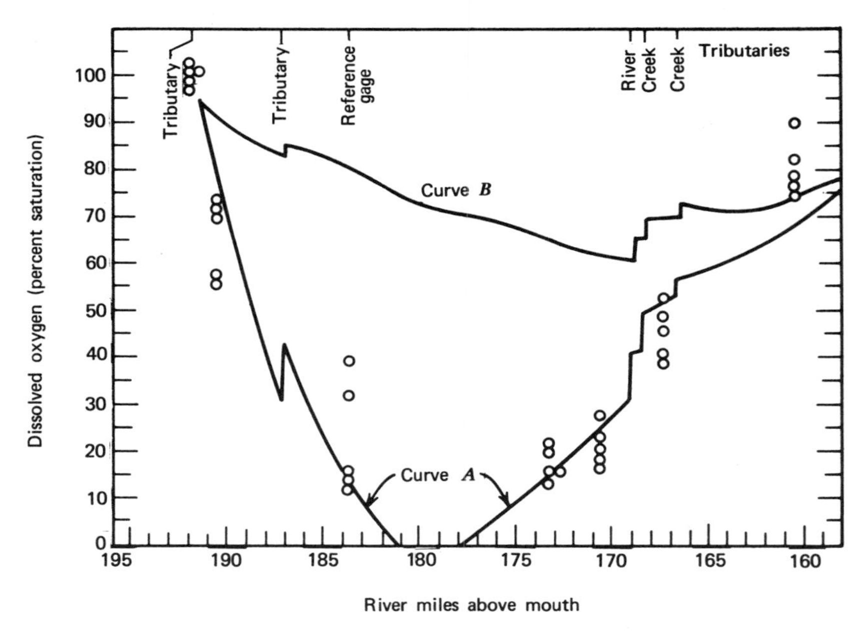

Figure 4–25 River E. Vertification of abnormality of biological extraction and accumulation—comparison of computed (curves *A* and *B*) and observed (○) dissolved oxygen profiles. Curve *A*: 40 days' sludge accumulation; curve *B*: normal self-purification, no sludge. The observed values represent single samples taken on July 22, 26, 27, 28, and 29. Runoff 370 cfs (at reference gage), temperature 22°C, mill-load BOD_5 40,500 lb per day.

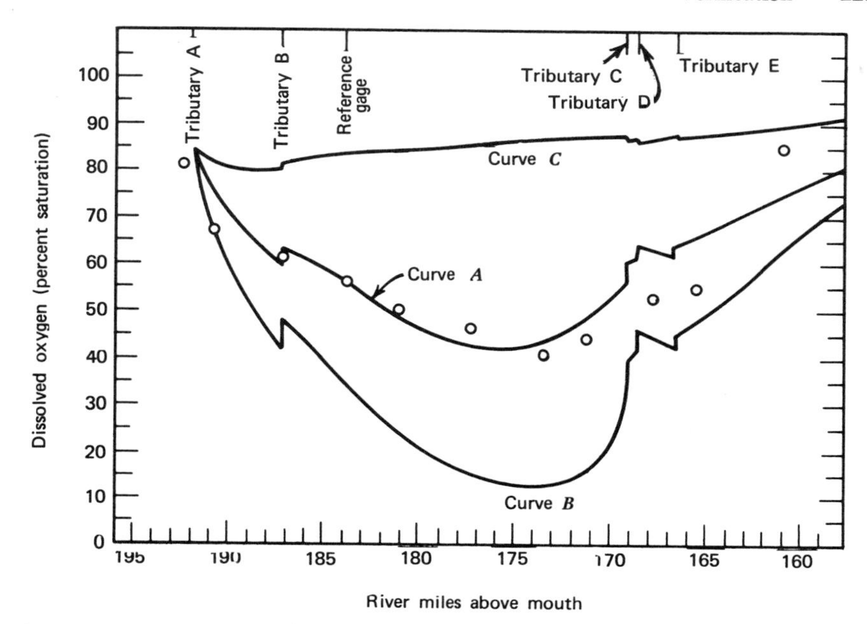

Figure 4–26 River E. Comparison of computed (curves *A, B,* and *C*) and observed (○) dissolved oxygen profiles (during 4th of July shutdown). Runoff 416 cfs (at reference gage), temperature 20.5°C, mill-load BOD_5 4780 lb per day. Curve *A*: 15 days' sludge accumulation; curve *B*: sludge accumulation at equilibrium prior to shutdown; curve *C*: normal self-purification, no sludge accumulation.

sludge to deposit and accumulate prior to the sampling period. Curve *A* is the independently computed profile for these conditions, showing a sharp drop in dissolved oxygen through the zone of deposit to complete depletion, followed by rapid recovery. The limited observed sampling results show a quite similar profile; no samples were taken in the zone of oxygen depletion, but the observed decline and recovery strongly point toward a zone of depletion similar to that of curve *A*. Curve *B* represents the computed dissolved oxygen profile expected if self-purification were normal and no biological extraction and sludge deposit were involved. The marked divergence of curve *B* from the observed profile and from curve *A* is strong verification of the abnormality of biological extraction and sludge accumulation.

That the high BOD removal involves sludge deposit and is not a high rate of BOD satisfaction is verified by a river survey taken some years later on July 6 during a partial shutdown of the industry associated with the July 4 holiday. Curve *A* of Figure 4–26 is the computed dissolved oxygen profile for July 6, based on the reduced mill waste load,

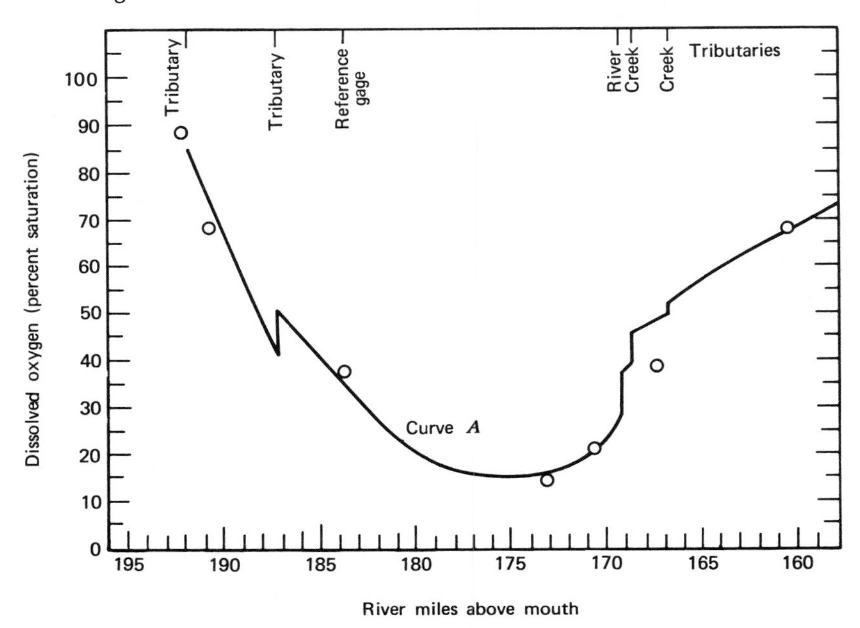

Figure 4–27 River E. Twenty-day biological extraction and accumulation—comparison of computed (curve *A*) and observed (○) dissolved oxygen profiles. Runoff 570 cfs (at reference gage), temperature 22.4°C, mill-load BOD_5 49,400 lb per day.

but allowing for sludge accumulation (after a scouring freshet) of 15 days under normal mill operation prior to the shutdown. The circles represent the observed sample results of July 6, and again there is substantial agreement between the computed and observed profiles. Curve *B* indicates the profile computed for the July 6 condition by assuming opportunity for prolonged sludge accumulation without scour prior to the shutdown and prior to sampling; curve *C* is the profile computed for the reduced mill shutdown load and based on assumption of normal self-purification and no prior sludge accumulation. If the high BOD removal observed were a high rate of satisfaction and not extraction and sludge accumulation, the observed dissolved oxygen profile on July 6 would have been in closer agreement with curve *C* than with curve *A*. The computed profile cannot be brought into agreement with the observed profile without taking into account sludge deposit and accumulation from the prior period of full-scale mill operation.

Figure 4–27 for a warm weather higher runoff condition and Figure 4–28 for a cold weather condition, with allowance respectively for 20 and 25 days of sludge accumulation before sampling, further confirm the abnormality of biological extraction and sludge accumulation.

The consistent agreement between observed and computed dissolved oxygen profiles for a wide range of runoff, temperature, and waste loadings, such as illustrated in Figures 4–25 through 4–28, strengthens verification of the self-purification factors adopted.

A similar confirmation of biological extraction and accumulation, and verification that observed high BOD removal is not a high rate of satisfaction is dramatically illustrated by comparing the computed and observed dissolved oxygen values for River F. Biological flocculation associated with *Sphaerotilus* was visually evident in the reach from mile 25 to mile 0, with a BOD removal rate b of approximately 0.4, below which growth disappeared and removal reverted to normal satisfaction. River sampling data for quite similar waste loads and stream runoff, preceded by a protracted opportunity for excess floc accumulation and for a short period of 10 days' accumulation, are shown in Figure 4–29a and b, respec-

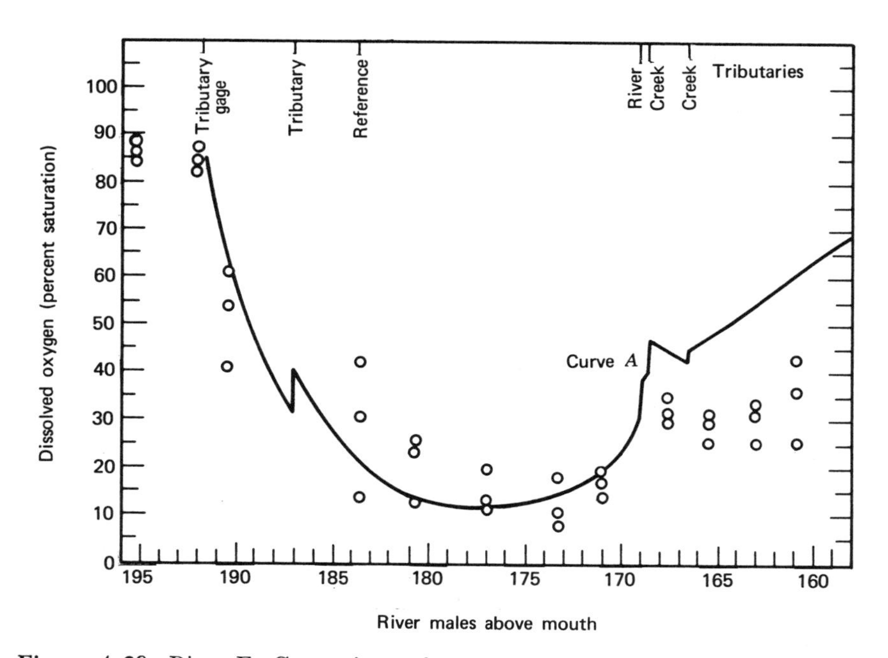

Figure 4–28 River E. Comparison of computed (curve A, 25 days' sludge accumulation) and observed (○) dissolved oxygen profiles. Runoff 290 cfs (at reference gage), temperature 11°C, mill-load BOD_5 42,500 lb per day. Observed values from samples taken on October 23, 28, and November 4.

tively. Curve *A*, the dissolved oxygen profile computed under these conditions by taking into account the prolonged accumulation, is in good agreement with the observed values. Curve *B*, the profile computed by taking into account the short period of accumulation, is also in good agreement with observation. The BOD removal was nearly the same in both survey periods. If the observed BOD removals were truly a high rate of BOD satisfaction in the time of passage along the river course the observed profiles (curves *A* and *B*) should be quite similar regardless of when sampling was taken in relation to a preceding opportunity for sludge accumulation. Actually the contrast in the observed profiles is dramatic: in curve *A* critical dissolved oxygen extends practically to complete depletion, whereas in curve *B* (the short period for accumulation) the critical dissolved oxygen is approximately 50 percent of saturation.

The close agreement between the observed profiles and the computed profiles, which took into account the difference in period of opportunity to accumulate the excess biological sludge, confirms biological extraction and accumulation and endorses the use of the self-purification factors adopted.

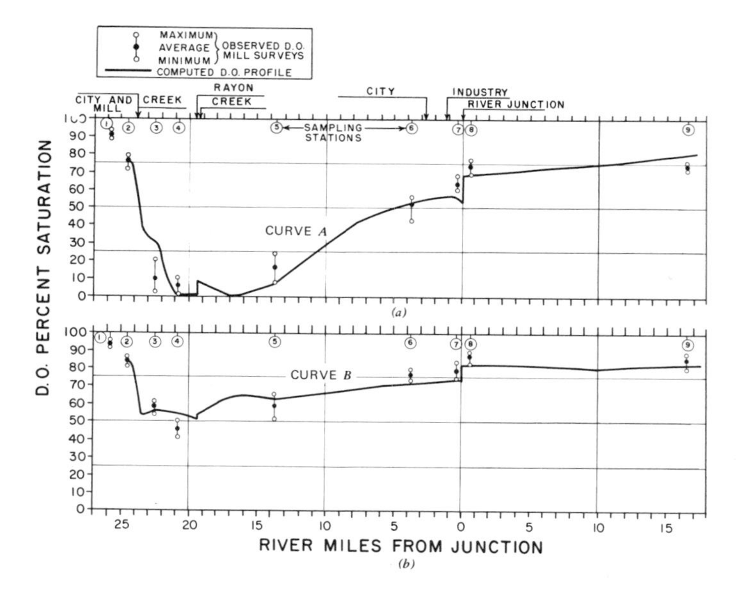

Figure 4–29 River F. Verification of abnormality of biological extraction and accumulation—comparison of computed (curves *A* and *B*) and observed dissolved oxygen profiles: (*a*) prolonged period of biological extraction and accumulation; (*b*) short period of biological extraction and accumulation.

REFERENCES

[1] Phelps, E. B., *Stream Sanitation,* Wiley, New York, 1944, p. 69.
[2] *Standard Methods for the Examination of Water and Wastewater,* 12th ed., APHA, Inc., New York, 1965.
[3] Gannon, J. J., and Downs, T. D., Professional Paper, Department of Environmental Health, University of Michigan, 1964.
[4] Lee, J. D., *Sew. Ind. Wastes,* **23,** No. 2, 164 (February 1951).
[5] Courchaine, R. J., "The Significance of Nitrification in Stream Analysis—Effects on the Oxygen Balance," in *Proc. Purdue Ind. Waste Conf.;* "Report of Oxygen Relationships of Grand River," Michigan Water Resources Commission, May 1962.
[6] Streeter, H. W., *Sew. Works J.,* **7,** 251 (1935).
[7] Velz, C. J., *Sew. Works J.,* **20** (4), 607 (1948).
[8] Black, W. M., and Phelps, E. B., Report to the Board of Estimate and Apportionment, New York City, Appendix, "Determination of the Rate of Absorption of Atmospheric Oxygen by Water," pp. 88–100, March 23, 1911.
[9] Adeney, W. E., and Becker, H. G., *Phil. Mag.,* **38,** 317(1919); **39,** 385 (1920).
[10] Streeter, H. W., Wright, C. T., and Kehr, R. W., *Sew. Works J.,* **8,** 282 (1936).
[11] Velz, C. J., *Trans. ASCE,* **104,** 560 (1939).
[12] Velz, C. J., *Sew. Works J.,* **19,** 629 (1947).
[13] Gannon, J. J., and Downs, T. D., *Water & Sew. Works,* **110,** 3, 114 (1963).
[14] Deininger, R. A., *Some Notes on the Reaeration of Rivers,* Professional Paper. Department of Environment Health, University of Michigan, July 1964.
[15] Streeter, H. W., *Sew. Works J.,* **7,** 535 (1935).
[16] Streeter, H. W., and Phelps, E. B., *Pub. Health Bull.,* **146** (1925).
[17] Phelps, E. B., *Stream Sanitation,* Wiley, New York, 1944.
[18] Streeter, H. W., Wright, C. T., and Kehr, R. W., *Sew. Works J.,* **8,** 2, (1936).
[19] O'Connor, D. J., and Dobbins, W. E., *J. ASCE,* **SA6,** 1115 (1956).
[20] Churchill, M. A., Elmore, H. L., and Buckingham, R. A., *J. ASCE,* **SA4.** 3199 (1962).

5

MICROBIAL SELF-PURIFICATION

When the desiderata are clean streams as an environment favorable to man's well-being, all types of wastes reaching the watercourse are of public health significance. There is, however, special public health concern with the disposal and ultimate fate of *microbial* wastes contained in municipal sewage, especially those of intestinal origin that cause waterborne diseases. These are principally pathogenic bacteria and viruses, and the cysts and ova of parasitic worms. Man's three major contact with water are drinking and domestic uses, consumption of shellfish, and bathing and other recreational activities.

In the United States today serious waterborne disease, such as typhoid, is uncommon. However, the scourge of enteric disease transmitted by the contaminated water supplies of American cities prior to the institution of filtration and chlorination was appalling; in many parts of the world it still remains a major threat to life. The number of pathogenic microorganisms contained in municipal sewage is dependent on the prevalence of the diseases in the population. Although prevalence in American cities is low, this is not to say that the hazards have been eliminated. There remains the potential from persistent carriers or isolated cases, and waters polluted by raw or inadequately treated sewage are always suspect. The number of organisms contributed per day by a single infected person is so great that reliance on dilution and self-purification cannot be depended on to the extent that pertains in dealing with other types of wastes. Success in the control of waterborne diseases is built on the principle of *multiple barriers;* the major lines of defense are treatment of sewage before return of wastewater to the stream and treatment of the domestic water supply again before distribution for use. However, as with all treatment systems, there remains a residual that must be handled by the waste assimilation capacity of the stream, and hence microbial self-purification is a vital factor.

PATHOGENIC MICROORGANISMS OF SEWAGE ORIGIN

A variety of pathogenic bacteria have been isolated from human feces, municipal sewage, and sewage polluted waters. There is an extensive literature on occurrences of devastating epidemics of typhoid fever, caused by the organism *Salmonella typhi,* traced to sewage contamination. Numerous but less serious outbreaks of dysentery and diarrhea are associated with *Shigella* and *Salmonella* organisms; of the latter, forms that are pathogenic to man usually are also pathogenic to animals. Of some 800 or more types of *Salmonella,* the more significant are *Salmonella typhi, Salmonella paratyphi, Salmonella shottmuelleri, Salmonella hirschfeldi,* and *Salmonella enteridis. Vibrio comma,* the causative agent of the dreaded cholera, is of intestinal origin. The municipal sewage of large cities and wastes of institutions treating tuberculosis patients almost always contain *Mycobacterium tuberculosis,* which may also be present in the wastes of dairies and slaughterhouses.

Several species of amoeba have been found in the intestinal tract of man; of these, *Entamoeba histolytica* is known to have caused epidemics of amoebic dysentery. Transmission is by heavy fecal contamination from sewage direct to water supply rather than by general stream pollution.

The tularemia pathogen, *Pasteurella tularensis,* has been isolated from natural waters where the disease is endemic among animals. The route of contamination is through the infected animals, rodents, and rabbits.

Schistosoma and other parasitic worms that undergo an aquatic stage are of major concern in water-contact transmission to humans in tropical countries.

Of increasing concern is the danger of transmission of virus diseases through contaminated water, particularly infectious hepatitis. More than 70 viruses have been recognized in human feces and are thus likely to be present in sewage. Strains of Coxsackie, ECHO, and polio viruses have been found in raw sewage and in various stages of conventional sewage treatment systems. Proven outbreaks of waterborne enteric virus diseases have been so far limited to infectious hepatitis; however, waterborne transmission of other forms is possible, and is of increasing concern and study.

INDICES OF CONTAMINATION

In addition to the pathogens that originate in the intestinal tracts of infected humans, enteric nonpathogens are normally present in uninfected humans and are regularly found in large numbers in the excreta of man and other warm-blooded animals. The two groups that are found most consistently and most abundantly in municipal sewage are the coliform and the enteroccocus. The direct transmission of disease is dependent on the presence of pathogens; this, in turn is a function of the prevalence of specific diseases in the community. Nevertheless the absence of pathogens in water consigned for human use and contact is not a satisfactory criterion of acceptability. The manifold possibilities of the presence of the many known, and of as yet unknown, pathogens preclude reliance on direct examinations for specified pathogens even if reliable laboratory procedures were available. As a practical matter, therefore, it is held that all pollution by human excreta, whether pathogen free or not, is potentially dangerous. Therefore control agencies have evolved criteria of bacterial sewage pollution based on quantitative measure of the normally present *nonpathogenic bacteria.*

The Coliform Group and Fecal Coliform

The first indicator of fecal contamination stems from the early work of Escherich in the identification of *"Bacillus coli."* From this origin the current coliform group has been enlarged to include numerous microorganisms of diverse characteristics, 16 groups composed of 256 types, which by physical and biochemical laboratory procedures are found to fit the coliform classification. In this enlargement of the group difficulty stems from the fact that in addition to *Escherichia coli* originating from the intestinal tracts of humans and warm-blooded animals various types classified in the coliform group are derived from nonfecal vegetable and soil sources that have not been contaminated by sewage. The inclusiveness of the coliform group, drawing microbes from various sources, has led to differences of opinion among public health experts about the validity of this test for water quality. Nevertheless the standard test [1] for detecting sewage pollution still remains the presence of the coliform group as a whole, although it is generally recognized that its inclusive nature may be too severe and that it cannot be considered fully satisfactory as a critical index of *fecal* contamination.

Bacteriologists are engaged in the development of specific tests of increasing diagnostic validity in differentiating between fecal and nonfecal sources, and between fecal organisms of human origin and those from other warm-blooded animals. So far no satisfactory distinction has been made between human and other animal fecal sources; considerable progress has, however, been made in distinguishing between fecal and nonfecal coliforms.

The tentative standard method for differentiating the coliform group into *E. coli* (fecal), *Aerobacter aerogenes* (nonfecal), and *E. freundii* (intermediates) is a combination of four tests (indole, methyl red, Voges–Proskauer, and sodium citrate), commonly referred to as the IMViC test. Many authorities also consider the elevated temperature test with EC medium and the boric acid lactose broth test as satisfactory bases for distinguishing fecal from nonfecal coliforms when used as confirmatory test procedures.

Fecal coliforms are considered to be indicators of recent fecal pollution; the presence of the intermediate group may be indicative of relatively less recent fecal pollution or of landwash runoff. Since all types of coliform organisms (fecal, nonfecal, and intermediate) are found in feces, it is argued by some that the coliform group taken as a whole should remain as the conservative standard test.

Fecal Streptococci

The other group of nonpathogenic organisms proposed as indicators of fecal contamination are the fecal streptococci. The varieties considered strictly as fecal in origin are: *S. faecalis, S. faecalis* var. *liquefaciens, S. faecalis* var. *zymogenes, S. durans, S. faecium, S. bovis,* and *S. equinus.* Large numbers of these nonpathogens occur normally in feces, their abundance being of the same order of magnitude as that of the coliforms. Bacteriological methods of enumeration are available, with tentative standard test procedures included in *Standard Methods* [1].

It has been thought that the fecal streptococcus group occurs only in the feces of humans and other warm-blooded animals and therefore constitutes a more specific test for fecal contamination than the coliform group. However, recent studies indicate that streptococci similar to the fecal streptococci may also be found on certain plants, plant products, and in wastes from food processing plants. The streptococcus group, although not replacing the coliform group as the standard, is considered to be a confirmation that coliform organisms found in water samples are of fecal origin.

The streptococci are more resistant than the coliforms to chlorine and survive longer in sea and brackish waters; hence this group is considered to be a more sensitive test for swimming pools and bathing waters. Moreover, since the streptococci die out rapidly in soil whereas the coliforms may persist for long periods, the streptococcal group is considered to be a more sensitive indicator of recent fecal contamination.

The Sanitary Field Survey and Surveillance

Beyond the results of bacteriological examination of water samples in the laboratory, the interpretation of the quality of water destined for human use and contact depends to a great extent on detailed sanitary field surveys and surveillance of the area tributary to the watercourse. Raw sewage of a single family discharged in close proximity to a point of water use may be more dangerous than the effluent from the sewage treatment works of an entire community at a more distant location. A detailed field survey locating tributary sewers and drains of potential contamination, public and private, is essential in evaluating the risk of contamination and in interpreting laboratory tests. In the control of reservoirs for public water supply and natural bathing areas a continual and systematic surveillance is essential to detect any changes in the existing sources of bacterial pollution.

ENUMERATION

If it is important to select an appropriate group of organisms as an index of sewage contamination, it is equally important to use a meaningful method of reporting the numbers of organisms and appropriate techniques for interpreting the enumeration. Enumeration of bacterial density in samples is most commonly made by the *dilution* or by the *membrane filter* methods. The conventional reporting of bacterial density of samples is in terms of *most probable number* (MPN) or preferably in terms of *mean density,* including an expression of reliability. A detailed discussion of the statistical significance and methods of evaluation is presented in Appendix B.

Enumeration by the Dilution Method

Any laboratory method of bacteriological enumeration is limited to a small sample relative to the large bodies of water from which the sample is drawn. The dilution method furthermore requires examination of a number of *portions* of a sample usually diluted in a manner to obtain a combination of portions which, when incubated in selected media, show both positive and negative results. It is assumed that (a) no bacteria are present in the portions showing negative test results, and (b) one (or more) bacteria are present in the portions showing positive results. By virtue of the necessity of dealing with small samples, and inherently in the test procedure, the enumerations are subject to wide variation in individual sample results.

The two most common test procedures employed are examination of (a) a sample composed of multiple portions of a single dilution and (b) a sample composed of multiple portions of three or more decimal dilutions. For water that is relatively low in bacterial density, such as potable supplies, the single dilution multiple portion sample is applicable; wastewaters and river waters receiving sewage effluents with relatively high bacterial densities require multiple portions in each of several dilutions in order to ensure a significant test result.

Enumeration by the Membrane Filter Method

The membrane filter (MF) method [1] of enumeration of bacteria involves passing a known volume of water sample through a membrane capable of retaining bacteria, placing the membrane on an absorbent pad containing selected nutrient media for the bacteria and incubating for a specified period. Each colony is taken to represent a single organism in the original sample. The size of the sample is governed by the expected density of bacteria; in finished waters low in bacteria large volumes may be filtered; in more contaminated waters it is desirable to employ samples that produce 50 to 200 colonies of all types. The turbidity of the water is a limiting factor in the size of a sample; if it is excessive, the membrane filter method may not be applicable. The bacterial density per milliliter of sample is simply the number of typical colonies divided by the volume in milliliters of the sample filtered.

PER CAPITA CONTRIBUTION AND SEASONAL VARIATION

The daily per capita discharge of nonpathogenic bacteria in municipal sewage is extremely high. In his studies Streeter [2] considered that on the average for large American cities the *B. coli* (coliform group) contribution is roughly 200 billion per capita per day. Intensive studies of the Ohio and Illinois Rivers indicate that the average number of *B. coli* from the metropolitan areas of Cincinnati, Louisville, Chicago, and Peoria ranges from 190 billion to 360 billion per capita daily [3]. Undoubtedly there are variations among communities of different sizes and composition; a large animal population (e.g., stock yards and meat-processing plants) would alter the coliform contribution drastically. In the absence of more extensive data the value suggested by Streeter of 200 billion coliform organisms may be considered as an *annual average* per capita daily contribution in American cities. It is well to consider, however, the seasonal variation.

Figure 5–1 is the seasonal pattern of daily per capita coliform bacteria contributions based on reanalysis of the Frost–Streeter Ohio River data, most probable monthly mean values developed from 1 year of extensive daily sampling data of the influent of the Buffalo sewage treatment works,* and monthly mean values based upon 8 years of MPN determinations of Detroit River water at the Wyandotte intake.† Although

* Data courtesy of George Flynn, Chief Chemist, Buffalo Sewage Authority.
† Data courtesy of George Hazey, Chief Operator, Wyandotte Filtration Plant.

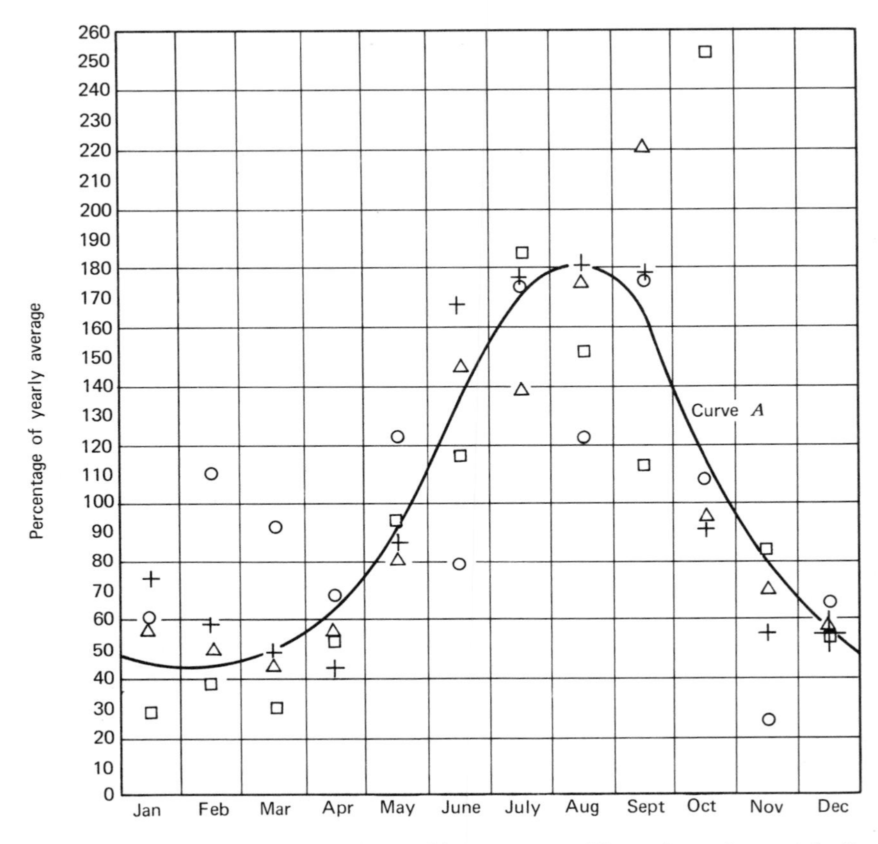

Figure 5–1 Seasonal pattern of monthly average coliform bacteria contribution per capita per day as percentage of yearly average (values are monthly means expressed as percentage of annual average). Curve *A*: general seasonal pattern.

there is considerable scatter, a reasonably orderly seasonal pattern emerges. Of significance is the general agreement between the river data and the seasonal distribution reflected by direct measurements of the Buffalo sewage samples. As reflected by curve *A*, the seasonal low monthly average (January) is 45 percent of the annual average, and the seasonal peak monthly average (August) is 182 percent of the annual average. If this seasonal pattern is applied to Streeter's annual average of 200 billion per capita per day, the range in coliform bacteria contribution as listed in Table 5–1 is significant and must be taken into account in the practical evaluation of stream sanitation; of special concern is

Table 5–1 Expected Seasonal Pattern of Monthly Average Coliform Contribution Based on an Annual Average of 200 Billion per Capita per Day

Month	Monthly Average (Percentage of Annual Average)	Monthly Average Coliform Contribution per Capita per Day ($\times 10^9$)
January	45	90
February	46	92
March	53	106
April	65	130
May	92	184
June	135	270
July	172	344
August	182	364
September	165	330
October	110	220
November	80	160
December	55	110
Annual average	100	200

the increase during the warm summer season, when swimming and other water contact recreation are also at their peak.

Phelps [4] identified such marked seasonal differential in bacterial pollution between the winter and summer seasons, and raised the question as to the cause of this difference:

> A quite unexpected and hitherto unnoted phenomenon has been shown, namely, a great increase in the bacterial evidence of pollution in the warmer months. This effect is shown so consistently in the work of the several laboratories, and upon various rivers, that there can be no doubt of its reality. It is hardly to be believed that there is actual multiplication of the intestinal organisms in the streams themselves, although this possibility cannot, with our present knowledge, be entirely eliminated. It is more probable that the bacterial content of the sewage shows a seasonal variation. Whether this be traceable to actual multiplication of intestinal bacteria within the sewers or to a greater per capita discharge of these organisms in the summer months cannot be stated.

The agreement between the Buffalo sewage seasonal pattern and that of the Ohio and Detroit Rivers is evidence that the seasonal difference is not associated with multiplication of coliforms in the streams. Certainly the pattern is evidenced in the sewers, but it is still not known whether the increase takes place there or is due to a seasonal pattern originating in the intestinal tract of man.

Other normal intestinal nonpathogens are believed to be equivalent to the coliform group in numerical per capita daily contribution and are likely to have a similar seasonal pattern. It can likewise be reasoned that the pathogens of intestinal origin parallel the nonpathogens, but, since they are not the normal organisms, per capita contribution on a community population basis would be quite different, depending on the prevalence of intestinal diseases in the community. The evidence points to the critical period of maximum contribution as the warm weather season.

DEATH RATE AND SURVIVAL IN THE STREAM ENVIRONMENT

Chick's Law

Chick [5] in her early investigation of the fate of bacteria in an unfavorable environment, generalized that bacteria die at a *constant rate;* that is, *a given percentage of the residual population dies during each successive time unit.* This, it is recognized, is the same as the rule that governs the stabilization of organic matter, radioactive decay, and so many natural phenomena. The mathematical formulation (using Phelps' terminology) is as follows:

$$\frac{dB}{dt} = KB$$

integrated to

$$\log_e \frac{B}{B_0} = -Kt$$

or, in common logarithms,

$$\log \frac{B}{B_0} = -kt,$$

where B_0 is the initial number, B is the residual after any time t, and k is the reaction, or death rate. The term B/B_0 represents the proportion of organisms that survive and $[1 - (B/B_0)]$ is the proportion of bacteria that die.

Chick's law is represented graphically on semilog paper as a straight-line decline in numbers. This is a convenient form for integrating multiple sources of bacterial contamination along the course of a stream, as will be illustrated later.

A substantial body of laboratory and field investigation strongly indicates that the survival of pathogens and nonpathogens of special interest in stream sanitation approximates Chick's law.

Self-Purification in the Stream Environment

Modifying Factors

In the complex environment of the stream survival of the coliform group approximates Chick's law, particularly in the early stage, and accounts for the major portion of the decline in numbers. Other modifying factors, however, give rise to variation in the constant k, and in instances may distort the straight-line form. Primary among such factors are temperature, pH, nutrient, sedimentation and adsorption, and competitive life.

The effect of temperature in the usual natural ranges is to stimulate biological activity, and in an otherwise unfavorable environment an increase in temperature increases the death rate. However, under stream conditions, it is possible to distinguish only between two broad temperature ranges, cool weather and warm weather. Although under controlled laboratory conditions a more specific temperature relation can be defined, under field conditions the effect of temperature on death rate is masked by other variables and in some instances is hardly distinguishable even in the broad classifications of "cool" and "warm."

Both acidity and alkalinity increase the bacterial death rate, but under stream conditions the specific contribution of pH is not definable except when there is pronounced deviation from neutrality.

The dividing line between starvation and nutrient for multiplication is ill defined in the stream environment. Seldom is stream pollution of such character and magnitude as to stimulate sufficient growth for a net change in death rate to be detectable.

Sedimentation and adsorption undoubtedly increase the death rate. However, turbid conditions associated with high stream runoff usually show a net increase in bacterial concentration by virtue of contamination flushed by surface wash from the drainage area. Subsequent sedimentation on the receding hydrograph may induce a decrease in bacterial concentration to a point below that expected for the time factor in the death rate. Small, shallow streams afford greater biological contact opportunity for bacterial adsorption than large, deep streams, and they usually have higher death rates.

Perhaps one of the most potent factors, beyond the element of time, in the death of coliform bacteria in the stream environment is the presence of competitive life. The natural biological life of polluted streams is much too rugged for the survival of organisms whose normal habitat is the shelter of the intestinal tract of man and other warm-blooded animals. In some instances coliform bacteria appear to multiply to some extent immediately after discharge into the stream, although as already discussed, there is some doubt about the reality of the apparent increase. That competition from other biological life plays a significant role is evidenced in the greatly retarded death rate of coliform bacteria placed in sterile water in comparison with the death rate of bacteria placed in natural freshwater or seawater.

Although it is known that these, and undoubtedly other, factors affect the survival rate of coliform and pathogenic organisms of intestinal origin, unfortunately their individual or combined effects are not sufficiently understood to permit prediction of a specific survival constant k for a given set of conditions. Reliance must be placed on field survey, observation of the survival of coliform bacteria based on intensive sampling along the course of the stream; on detailed and precise definition of the hydrologic parameters; and on the data which quantify sources of pollution. Decision as to the applicable survival rate is then a matter of professional judgment guided by the field survey of the particular river and a background of experience from other rivers.

Observed Survival Rate of Coliform Bacteria

The many factors in varying combination that effect the bacterial death rate in the stream environment make for deviations from the Chick form of the survival curve and for variations in the constant k.

Deviation in Form

Two principal deviations in form of the survival curve are commonly reported: an initial lag phase preceding onset of the logarithmic decline, and an afterphase of apparent decreasing death rate. In some instances during the lag phase an apparent multiplication in bacterial numbers is observed. This phenomenon is dealt with by considering the initial bacterial load as the peak and shifting the origin of the time scale in determining the death rate to the position of the peak. The afterphase of apparent decline in death rate is shown when the semilog plot bends upward at the end of the depletion curve.

It is not yet known whether these apparent deviations in form of the survival curve are real or are merely reflections of such factors as sampling and enumeration errors, physical disintegration, inaccuracy of time of passage, and extraneous unaccounted for sources of pollution from landwash and drainage along the course of the stream. The general

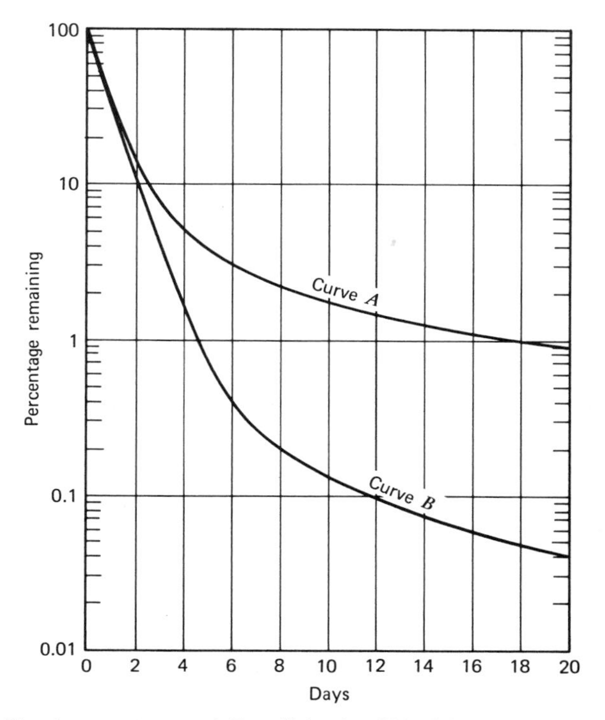

Figure 5–2 Death rate curves of *E. coli* in the Ohio River. Curve A: cool-weather conditions; curve B: warm-weather conditions.

conclusions of the extensive studies of the Ohio River by Frost and Streeter [3] and others lead to the form of survival curve shown in Figure 5–2, curve A reflecting cool weather conditions, and curve B warm weather ones.

A number of mathematical formulations have been proposed in attempts to define the observed deviation in form of the survival curve. One such formulation is the two-stage equation

$$B = B_0(10^{-kt}) + B_0'(10^{-k't}).$$

This calculation is predicated on the assumption that the residual number of bacteria at any time consists of two fractions, one resulting from the application of a death rate constant k to an initial population B_0 and another from similar application of the rate k' to the initial population B_0'. Phelps, although not fully subscribing to this concept as the true representation, determined values of the fractions B_0 and B_0', the corresponding constants k and the half-lives of *E. coli* based on the Ohio River data as follows:

Parameter	Warm Weather	Cool Weather
B_0 (percent)	99.51	97
k (day)	0.467	0.506
Half-life (day)	0.64	0.59
B_0' (percent)	0.49	3.0
k' (day)	0.0581	0.026
Half-life (days)	5.16	11.5

At first glance at the graphical representation (Figure 5–2) it appears that the survival curve deviates substantially from the straight-line Chick form. We see, however, from the tabulation that the coliform group (in summer) is composed of the *large* fraction of 99.51 percent with a half-life of 0.64 day, and a *small* fraction of 0.49 percent of more resistant organisms with a half-life of over 5 days. For practical purposes this means that 99.51 percent of the coliform bacteria dieoff can be represented by the simple Chick straight-line form. Less than 0.5 percent are involved in the apparent deviation reflected in the upward turn of the survival curve.

Our primary concern in practical application of microbial self-purification analysis in streams is in the determination of the applicable reaction constant k, representing the dominant decline in the early phase defined by Chick's law.

Observed Death Rates

In addition to the natural forces that may modify the death rate, inaccurate field observations may skew the rate curve and affect the (presumed) observed reaction rate. As previously emphasized, enumeration of bacteria in the stream involves an intensive sampling program under stable conditions of bacterial loading and river hydrology before and during sampling. In turn the time of passage between sampling stations depends on detailed characteristics of the channel and the prevailing stream runoff. At best, part of the variation in the observed death rate is associated with limitations of field measurement of one sort or another, and therefore a degree of variation must be tolerated in practical application.

Analysis of a number of river surveys carried out by workers aware of the difficulties of field observation discloses the observed death rates shown in Table 5–2 and Figure 5–3. The variation in warm weather

Table 5–2 Coliform Death Rates k Observed in Rivers
[Chick's law: $\log (B/B_0) = -kt$]

River	Reaction Rate k (day) Warm Weather	Cool Weather	Authority for Survey Data	Remarks
Ohio	0.50	0.45	Frost, Streeter et al.	Generalized results of analysis of extensive data
Upper Illinois	0.90	0.32	Hoskins et al.	1-day decline
	0.67	0.29		2-day decline
Scioto	0.96	0.46	Kehr et al.	
Hudson	0.80		Hall, Riddick, Phelps	Freshwater reach below Albany
Upper Miami	0.80		Velz, Gannon, Kinney	Mean through reach above Dayton
Tennessee	0.46		Kittrell	1- and 2-day declines, below Knoxville
Tennessee	0.60		Kittrell	1-day decline
	0.57			2-day decline (below Knoxville)
Sacramento	0.77		Kittrell	1-day decline
	0.65			2-day decline (below Sacramento)
Missouri		0.30	Kittrell	1-day decline
		0.26		2-day decline (below Kansas City)

reaction rate k from 0.46 to 0.96, based on these observed coliform bacteria declines, is not as great as might be expected from more limited surveys reported in the literature. The range in k is further narrowed if a broad classification is made between large and moderate-size rivers, 0.5 ± 0.15 and 0.8 ± 0.2, respectively.

The rates under cool weather conditions are slightly lower than those for warm weather in areas of temperate climate, decreasing substantially in areas of more severe winter.

It will be noted from Figure 5–3 that for summer conditions only a small percentage of coliforms survives after 2 days' time of passage. For a large stream ($k = 0.5$) only 10 percent survive after 2 days; for a moderate-size stream ($k = 0.8$) less than 3 percent survive after 2 days. However, although these survival percentages are small, it must be remembered that, applied to the summer peak coliform bacteria load-

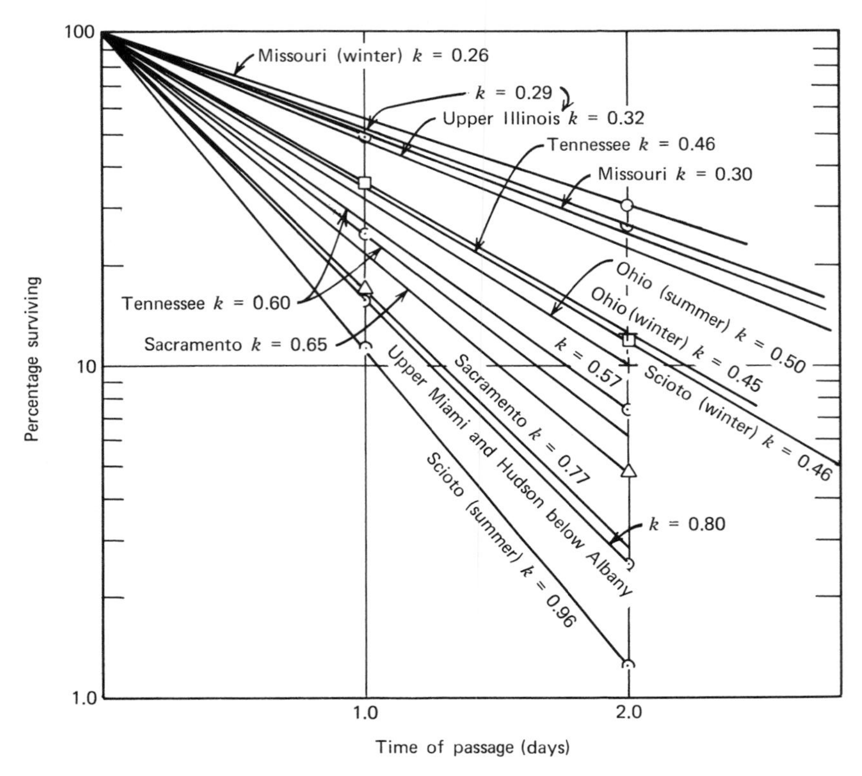

Figure 5–3 Observed coliform bacteria death rate in the stream environment.

ing of 364 billion per capita per day, these declines may result in large absolute numbers of survivors, which, converted to concentrations in the diluting streamflow, may infringe the required criteria of water quality. Thus, although bacterial stream self-purification is great, it cannot be relied on as the sole defense.

Fortunately, with the advent of two-stage chlorination as a part of conventional treatment, an economical means is now available to greatly reduce the bacterial load (by 99.99 percent or more) before discharge to the stream. However, the time of passage factor in bacterial stream self-purification remains an important line of defense; the longer the time of passage from the point of discharge to the point of water use, the more formidable this line of defense. An example illustrates the application of the rational method of stream analysis to the control of bacterial contamination.

Illustrated Example

Selections from a detailed study of pollution of River G, a moderate-size Midwestern river, illustrates the application of the rational method of stream analysis in dealing with bacterial wastes. Five sources discharge inadequately treated municipal waste effluent along a 45-mile reach of the river, which serves as the source of raw water supply for one community and three water recreation areas. Good hydrologic data and river channel cross-sections at frequent intervals are available, and from these streamflow and the time of passage along the course of the river are obtained. An intensive river sampling survey has been made (October 6–8) under stable river hydrology; this includes observed coliform bacteria concentrations at points of water use from which mean density is determined.

Table 5–3 presents the basic data for integrating the survival of coliform bacteria along the course of the river under the conditions that prevailed during the stream survey. The integration of survival expressed in quantity units of population is made at $k = 0.8$ for a river of moderate size, shown as curve A on Figure 5–4, with results summarized in Table 5–4 for comparison with the independently observed coliform bacteria population equivalents determined from the river sampling.

The observed concentrations of coliform bacteria in mean density per 100 ml (column 4) are converted to population equivalents (column 5)

Table 5–3 River G—Data for Coliform Integration[a]

Waste Sources and Water Use Areas	Contributing Population	Time of Passage (days)	Streamflow (cfs)
Municipality No. 1	11,490	0.057	
(A) Water supply intake		1.024	150
Municipality No. 2	17,450	1.921	
(B) Recreation area		2.962	166
Municipality No. 3	10,660	3.111	
Municipality No. 4	3,300	3.469	
(C) Recreation area		4.093	203
Municipality No. 5	930	4.229	
(D) Sampling station		4.749	
(E) Recreation area		5.701	211

[a] October 6–8 conditions.

Table 5–4 River G—Comparison of Computed and Observed Coliform Densities[a]

Location (1)	Computed Residual Population Equivalents from Integration Graph (2)	Runoff at the Station (cfs) (3)	Observed Coliform from River Sampling: Mean Density per 100 ml (4)	Observed Coliform from River Sampling: Population Equivalents (5)
(A) Water supply intake	1950	150	96,000	1360
(B) Recreation area	2700	166	140,000	2190
(C) Recreation area	3150	203	173,000	3310
(D) Sampling station	1300	206	160,000	3110
(E) Recreation area	222	211	17,000	338

[a] October 6–8 conditions.

based on 260 billion coliform bacteria per capita per day as that expected for early October, based on the seasonal curve of Figure 5–1.

Convenient conversions are given by

$$\text{coliforms per 100 ml} = \frac{\text{PE} \times C \times 100}{\text{cfs}},$$

where conversion factor C is

$$\frac{\text{coliforms per capita per day}}{2.447 \times 10^9}.$$

The observed maximum and minimum MPN population equivalents and mean density determined from log-normal probability plots of the sample results are also shown in Figure 5–4. As is to be expected, the ranges in individual sample results show a considerable spread, but the distributions develop reasonably good straight-line fits on log-normal probability paper. As verification of the bacteriological technique and sampling, 19 replicate samples taken consecutively at one river location developed a good log-normal distribution, with a slope $\sigma_{\log}$ that closely approximates that expected for samples of three portions in three decimal dilutions employed (Appendix B).

Considering the many variables involved, there is satisfactory agreement between the observed survival rates of coliform bacteria along the course of the stream and the independently computed values. Of

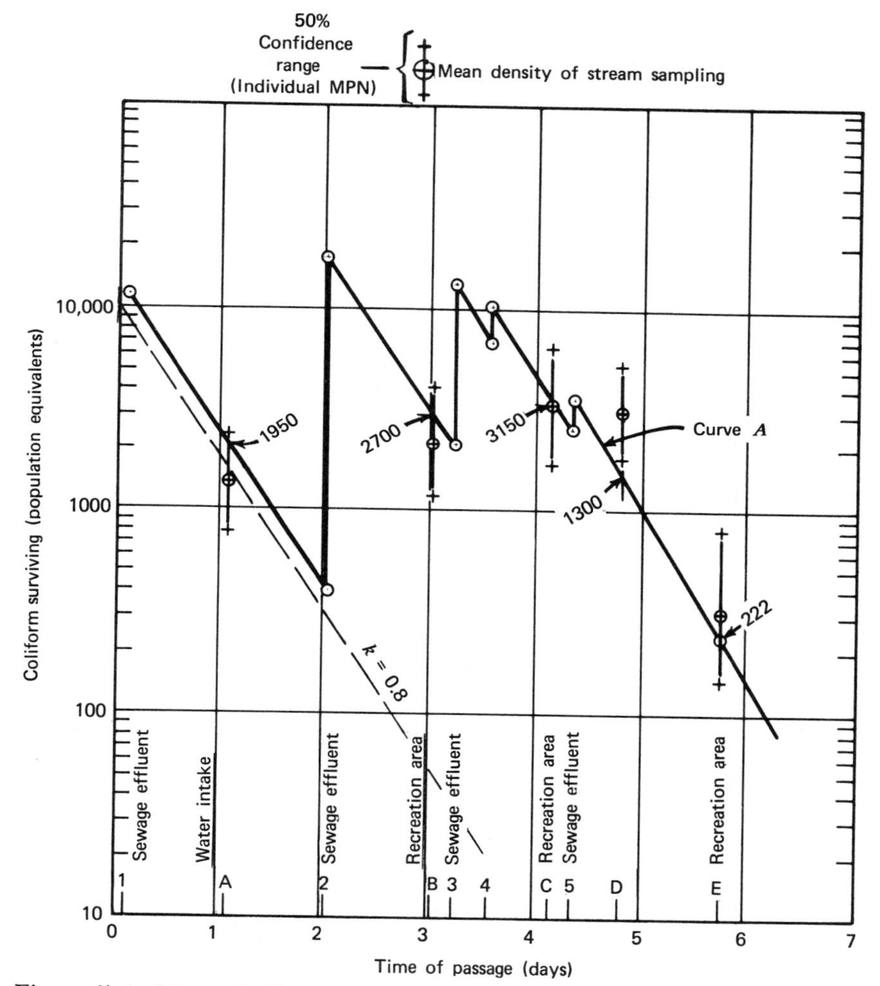

Figure 5–4 River G. Integration of coliform bacteria survival from multiple sources (curve *A*).

particular significance is the close agreement toward the downstream end of the integration; the exception at sampling station D is partly due to uncertainty in determination of the entire municipal load No. 5. The overall agreement is confirmation of the adopted reaction rate $k = 0.8$ and the application of Chick's law.

The water quality is far from satisfactory for the existing uses of the river. Of further interest is the determination of expected water

Table 5–5 River G—Projected Summer Peak Coliform Concentrations at Areas of River Use

Location Areas	Computed Residual Population Equivalents from Integration $k = 0.8$	Projected River Coliforms per 100 ml during Summer Peak
A	1950	193,000
B	2700	242,000
C	3150	230,000
D	1300	93,800
E	222	15,600

quality at the peak of the summer per capita coliform contribution under the same runoff conditions that prevailed on October 6–8. This projection is readily made by applying the peak summer conversion factor to the integrated population equivalents of column 2 in Table 5–4. The summer peak from Figure 5–1 is 364 billion coliform bacteria per capita per day, and the concentration is obtained from

$$\text{coliforms per 100 ml} = \frac{\text{PE} \times 14{,}850}{\text{cfs}}.$$

These projected coliform bacteria concentrations at the areas of water use are shown in Table 5–5, from which it is seen that an even less acceptable quality results under the peak summer conditions.

It is evident that in regard to bacterial contamination satisfactory water quality in this case cannot be obtained by reliance on stream self-purification alone; further reduction in the coliform bacteria load at the source is required before the effluents are discharged to the river. Improvements in existing treatment works and installation of prechlorination equipment were estimated to reduce coliforms by 98.6 percent. The integration of this reduced loading at a k of 0.8 for the same prescribed runoff and the conversion of the residual population equivalents at areas of use to concentration for peak summer per capita contribution yield the improvement shown in Table 5–6.

The water use area A is now acceptable as a raw water supply, but the recreation areas B and C are not acceptable for water contact sports, because the coliform bacteria concentration still exceeds 1000 per 100 ml, the usual criterion for satisfactory protection.

It is apparent from these analyses that, if reasonable safety for water-

Table 5–6 River G—Effect of Single-Stage Chlorination

Location (1)	Reduced Coliform Load (Population Equivalents) (2)	Residual Population Equivalents (Graphical Integration) (3)	Projected River Coliforms per 100 ml (Summer Peak) (4)
No. 1	161		
A		27	2680
No. 2	244		
B		43	3850
No. 3	149		
No. 4	46		
C		45	3290
No. 5	13		
E		3.1	218

contact recreation is to be provided on River G, two-stage chlorination (efficiencies in excess of 99.99 percent under good operating control) is necessary during the recreation season.

Some further points of significance are evidenced by the analysis. Since the population equivalents of columns 2 and 3 of Table 5–6 are the equivalent of raw sewage contribution from such populations at these locations on the river, it illustrates dramatically how the raw sewage of a small number of people drastically contaminates a river. It also emphasizes the necessity of constant vigilance in control by sanitary field survey of areas immediately adjacent to points of use. The analysis also emphasizes the greater protection afforded by long detention in off-channel reservoirs as compared with the shorter time of passage afforded in the natural river channel.

REFERENCES

[1] *Standard Methods for the Examination of Water and Wastewater,* 12th ed., APHA, New York, 1965, Part VII, pp. 567–628.

[2] Streeter, H. W., *Sewage-Polluted Surface Waters as a Source of Water Supply,* Reprint No. 1232 from *Public Health Reports,* June 15, 1928, pp. 1498–1522.

[3] Frost, W. H., and Streeter, H. W., *Pub. Health Bull.,* **143,** U.S. Public Health Service, Washington, D.C., 1924.

[4] Phelps, E. B., *Stream Sanitation,* Wiley, New York, 1944, p. 209.

[5] Chick, H., *J. Hyg.,* **8,** 92 (1908); **10** 237 (1910).

6

RADIOACTIVE AND INORGANIC SELF-PURIFICATION

In this chapter we deal with radioactive wastes subject to decay over time and with inorganic wastes that are stable in character.

RADIOACTIVE WASTES

Types of Radiation

Liquid radioactive wastes are generally classified by the level of radioactivity they contain—for the purposes of this discussion, low (less than 10^{-4} μCi/ml) high (more than 10 μCi/ml), and intermediate (the range between). The radioactivity is classified further as to type, namely, alpha (α), beta (β), and gamma (γ).

The alpha particle has a positive electric charge that is equal in magnitude to twice the electron charge, and its mass is that of a helium nucleus (^{4}He). Alpha particles emitted by radioactive nuclei have slight penetration ability and are completely stopped by a few sheets of paper or a few centimeters of air.

The beta particle has a single negative charge and a mass equal to that of the electron. Beta particles are high-speed electrons ejected from the nucleus and have a greater penetrating range than the alpha particles, a few meters in air or several millimeters in an absorber, such as aluminum.

Gamma rays (not particles) are electrically neutral electromagnetic radiations. They are very penetrating, being only partially absorbed by several centimeters of aluminum or lead.

Magnitude of the Radioactive Waste Load

It was estimated by the Federal Power Commission [1] that by 1980 nuclear fuel steam electric power generating capacity will aggregate nearly 70 million kw; more recently the Atomic Energy Commission revised this estimate upward to 110 million kw by 1980 [2].

As capital costs decline for larger units and costs of nuclear fuel and insurance are lowered, nuclear power plants will become increasingly competitive with fossil fuel plants. Large nuclear power plants will become a significant source of radioactive wastes widely distributed on large rivers, tidal estuaries, and reservoirs. Current nuclear power plants are predominantly of the thermal type; technology soon should make practical the fast breeder type, in which the production of nuclear fuel exceeds consumption.

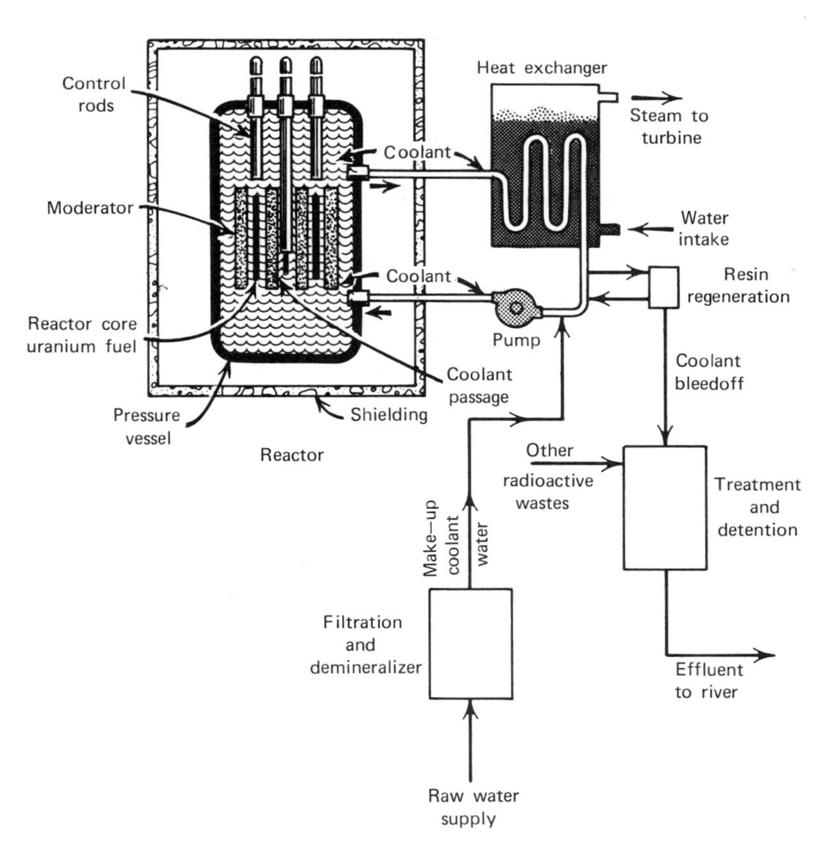

Figure 6–1 Principal components of a nuclear reactor.

The processing of nuclear reactor fuel results in wastes of extremely high activity but of low volume. Such processing plants are relatively few and are rigidly controlled. The wastes are stored in large underground tanks and are not released to the environment. Our primary concern is with the routine release of radioactive wastes to streams from power reactor operations.

Figure 6–1, a simplified schematic diagram, depicts the principal components of a nuclear power reactor of the thermal type. The principal source of routine radioactive waste stems from neutron activation of impurities in the coolant; activation of the coolant itself; corrosion, abrasion, and scale from the equipment; and escape of fuel and fission products into the coolant from defective fuel elements. Various types of coolants, liquid and gaseous, are feasible, but the most common is water in the *pressurized-water* and the *boiling-water* reactors. The water for coolant is filtered and demineralized before and during use to reduce formation of neutron activation products. A portion of the coolant is bled from the system and discharged to waste treatment facilities or detention tanks from which a controlled waste effluent is released to the river. In addition to coolant effluent there are a number of associated sources of radioactive wastes: rinse water, resin and incinerator slurry, shower and laundry wastewater, and decontamination liquids.

At this stage of the practice few data are available concerning unit volume and strength of radioactive wastes expected per unit kilowatt of plant capacity; each plant has unique features. The Shippingport, Pennsylvania, pressurized-water type of power reactor, net capacity 60,000 kw, is reported to produce waste as follows [3]:

Type of Waste and Source	Average Volume per Month (gal)	Average Radioactivity (μCi/ml)
Reactor plant effluent	23,292	4.5×10^{-5}
Cold laundry, etc.	170,000	1×10^{-6}
Hot laundry, etc.	89,000	1×10^{-4}
Decontamination chemicals	7,700	0.27

If the small volume of decontamination chemicals is not released, the daily average volume of radioactive waste from the 60,000-kw nuclear reactor is 9410 gal, a daily quantity of radioactivity of 1270 μCi. If it is assumed that waste volume and radioactivity are proportional to plant capacity, a pressurized-water reactor waste based on Shippingport performance would be approximately 15,700 gal per day and 2120 μCi per day per 100,000-kw capacity.

Since the daily waste volume is small, it is feasible and expedient to provide a very high degree of treatment before release of residual waste effluent to the stream.

Public Health Significance

From the beginning of geologic time man has lived with low levels of natural radiation. Overexposure to radiation (as to any highly toxic agent) may be lethal. Like with many other known and unknown toxic agents encountered by man in his environment, the effects of sublethal doses are not immediately felt, clinically detectable, or fully understood. The experience of Hiroshima and fear of atomic war have engendered a hysterical attitude toward any radiation exposure. Consequently the ultraconservative assumption is made that no threshold exists, and dose–effect relations are often assumed to be linear to infinitely low; therefore it is believed that any amount of radiation, regardless of how small, is not without deleterious effects. Absolute safety in this context is unattainable, and some degree of risk is involved in any adopted standard of exposure level. As has been pointed out by Sterner [4], these assumptions have led to the adoption of standards of radiation protection in the peaceful applications of atomic energy far beyond those imposed on exposure to other public contaminants. In the final analysis any standards of protection hinge on the principle of expediency (see Chapter 12).

Radioactive Decay

The same basic law that governs the decay of organic and microbial wastes applies to radioactive wastes. The decay, or disintegration, of radioactive atoms by spontaneous transformation takes place at a constant rate. This is expressed in the familiar differential form as

$$-\frac{dN}{dt} = \lambda N, \qquad (6\text{–}1)$$

which can be integrated to

$$\ln\frac{N_t}{N} = -\lambda t \qquad (6\text{–}2)$$

and

$$\frac{N_t}{N} = e^{-\lambda t} \qquad (6\text{–}3)$$

or, in common logarithms,

$$\frac{N_t}{N} = 10^{-0.434\lambda t}, \tag{6–4}$$

where N is the initial number of radioactive atoms at time zero, N_t is the corresponding number at time t, N_t/N represents the fraction remaining, and $[1 - (N_t/N)]$ is the fraction that has disintegrated. The rate of reaction, the decay constant λ is the rate at which the reaction proceeds. The common expression of this rate is in terms of *half-life* T, the time t at which 50 percent of the activity is lost by decay and 50 percent remains. In each succeeding half-life interval the remaining activity is reduced by 50 percent so that at the end of two half-life intervals ($2T$) 25 percent remains, at the end of three half-life intervals ($3T$) 12.5 percent remains, and so forth.

Unlike organic decay or microbial death, the rate of radioactive decay is unaffected by temperature, pressure, chemical state, or physical environment; the rate for each specific radioactive species is unique and unchangeable.

A constant rate of decrease, it is recalled, is graphically represented as a straight line on semilog paper. A typical radioisotope decay curve is therefore represented by Figure 6–2, expressing activity in percentage

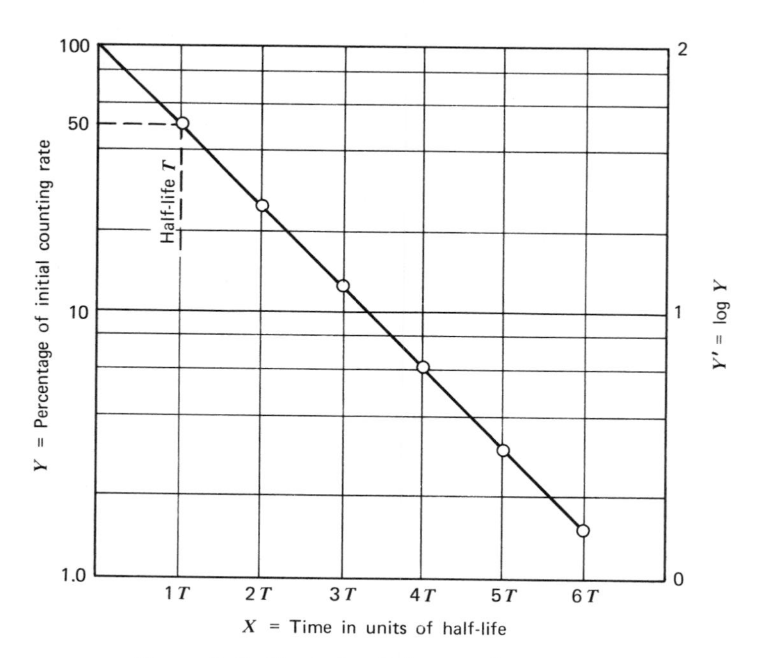

Figure 6–2 Typical radioisotope decay curve.

of the initial counting rate and time t in units of half-life T; the slope $\Delta Y'/\Delta T$ provides the measure of the reaction rate.

The relation between the decay constant λ and half-life T is obtained by substituting $\frac{1}{2}$ for N_t/N and T for t in equation 6–3.

$$\tfrac{1}{2} = e^{-\lambda T} \quad \text{or} \quad \ln 2 = \lambda T,$$

whence

$$\lambda T = 0.693. \tag{6–5}$$

The time unit in which λ and T are expressed may be conveniently stated in years, days, hours, minutes, or seconds, as appropriate for the problem.

Radioactivity Measurements

Measure of Activity

The unit measure of activity is the curie (Ci). One curie of any radioactive nuclide is the quantity that undergoes 3.7×10^{10} disintegrations per second. As with other types of wastes, the radioactive waste load should be measured at its source and expressed in quantity units, or curies per day. This necessitates measurement of both concentration and flow. Depending on strength of the waste, concentration per milliliter may be expressed in curies, millicuries (mCi, 10^{-3} Ci), microcuries (μCi, 10^{-6} Ci), or megacuries (MCi, 10^{6} Ci); flow is expressed in millions of gallons per day (mgd) or cubic feet per second. If concentration is in microcuries per milliliter and flow is in millions of gallons per day, the waste load in curies per day is

$$\text{Ci/day} = \mu\text{Ci/ml} \times \text{mgd} \times 3.7848 \times 10^{3}; \tag{6–6}$$

if flow is in cubic feet per second,

$$\text{Ci/day} = \mu\text{Ci/ml} \times \text{cfs} \times 2.4466 \times 10^{3}. \tag{6–7}$$

The activity of a specified radioisotope or mixture of radioisotopes in a unit quantity of matter is referred to as *specific activity* (μCi/g) in solids and *concentration* (μCi/ml) in liquids. Concentrations in liquid wastes are measured by instruments that detect and record ionizing events in a given period of time, usually expressed as a rate, counts per minute (cpm). In turn counts per minute of the sample are converted to microcuries per milliliter by correction and conversion factors.

Accuracy

The accuracy of radioactivity measurement is influenced by two types of variation, one associated with sampling, the other with counting. Variation in field sampling depends on stability of conditions during the sampling period and a design to ensure representative samples. This is discussed more fully in Chapter 10.

Although analytical errors in the laboratory are more amenable to control than are field errors, a source of variation is inherent in the random manner in which radioactive nuclei disintegrate. The probability that any one among a large number of atoms will disintegrate in a given time is small and constant. The probability of observing such radioactive counts, or rates, is best represented by the Poisson series, as discussed for bacterial enumeration (Appendix B).

An inherent difficulty in all radiation counting instrumentation is the inevitable background due to cosmic rays, to contamination of surrounding materials, and to small amounts of radioactive isotopes in the materials of the instruments. The background count can be controlled to a degree, but it cannot be completely eliminated. The observed counting rate of a sample, therefore, is composed of the sum of two rates, the background and the sample. Consequently the sample rate is obtained as the difference in two rates

$$R_S = R_T - R_B,$$

where R_S is the net count rate, R_B is the background rate, and R_T is the recorded total count rate. The standard deviation of the net count rate R_S is used to define the counting error and the lower limit of detectability. The standard deviation of a difference is the square root of the sum of the squares of the standard deviation of each (see Appendix A). In terms of count rate (cpm) the standard deviation of the net count rate of the sample R_S then is

$$\sigma_{(R_T - R_B)} = \sqrt{\sigma_{R_T}^2 + \sigma_{R_B}^2} = \sqrt{(R_T/t_T) + (R_S/t_B)}, \qquad (6\text{–}8)$$

where t_T and t_B are the counting time for the sample and for the background, respectively. For example, if a 10-min background count (without sample) produces an R_B of 200, or 20 cpm, and a 5-min total count (sample plus background) produces an R_T of 4500, or 900 cpm, then

$$R_S = 900 - 20 = 880 \text{ cpm}.$$

The standard deviation for R_S by equation 6–8 is

$$\sigma_{R_S} = \sqrt{(900/5) + (20/10)} = 13.5 \text{ cpm},$$

and the standard error of the net sample rate expressed in terms of the coefficient of variation is

$$\text{C.V.} = \frac{13.5}{880} = 0.0153,$$

which is to say that the standard deviation is 1.53 percent of the count. Expressed in other terms it can be expected that of other counts taken under similar conditions 68.26 percent will fall within 880 ± 13.5 cpm, 95.46 percent within $880 \pm 2 \times 13.5$ cpm, and 99.73 percent within 880 $\pm 3 \times 13.5$ cpm.

At a 95-percent confidence level the error is 13.5×1.96, or about 26 cpm, which constitutes an error of about 3 percent of the count.

If it is desired to increase the accuracy of the net sample count (assuming the background remains the same), obviously it is necessary to increase the counting time of the sample. The necessary increase can be approximated from the first count, assuming that the total count rate of 900 cpm is the most probable estimate of the true count rate. If it is desired to reduce the error on the net count rate from 26 cpm to, say, 18 cpm at the 95-percent confidence level (from 3 to approximately 2 percent), 1.96 σ_{R_S} is then 18 and

$$\sigma_{RS} = 9.18 = \left(\frac{900}{t} + \frac{20}{10}\right)^{1/2} \quad \text{or} \quad t = 11 \text{ min.}$$

To reduce the error further, to, say, 9 cpm at the 95-percent confidence level (approximately 1 percent error) would require increasing the counting time to 47 min. A balance between acceptable error and counting time can thus be determined, keeping in mind also the errors involved in stream and waste sampling.

The Decay Curve of Mixed Radioactive Isotopes

In dealing with unstable organic matter the rates of decay among specific constituents do not range widely; hence the heterogeneous mixture found in municipal and industrial wastewaters follows quite closely the basic law of constant rate. The parallel does not hold for radioisotopes. Although each radioactive species has a unique rate of decay and adheres precisely to the law of constant rate, the range among radioisotopes is enormous, with half-lives from 10^{-6} sec to 10^{11} years; consequently mixtures of two or more diverse isotopes will not follow the usual exponential law. The plot on semilog paper is usually curvilinear, as illustrated in Figure 6–3. Knowledge of the source and of the waste treatment before discharge usually provides an adequate guide to predominant isotopes in the stream. Furthermore, though complicated and time consuming, radiochemical analyses can further identify specific isotopes.

If the mixture is not too diverse, a distinction can be made between the short-lived and the long-lived components; this then may be reduced to linear forms, and specific decay rates are definable. This analysis

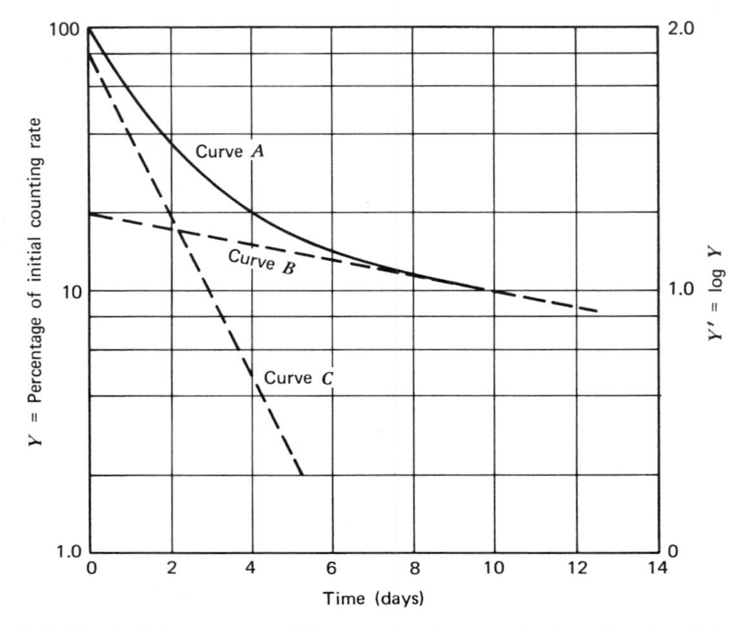

Figure 6–3 Typical decay curve. Mixture of a short- and a long-lived radioiosotope.

can be made graphically on the decay curve, as shown in Figure 6–3. If the measurements of activity are continued long enough, the short-lived component will decay to the point where its contribution is negligible and the curve will approach linearity. A line, curve B, drawn tangential to the tail of curve A, extrapolated to zero time, then defines the activity of the long-lived component; subtracting the values of curve B from the corresponding values on curve A provides the activity due to the short-lived component represented by curve C.

The half-life T for the long-lived component is readily obtained graphically by transposing to a log cycle (100 percent) a line parallel to curve B, and at the intercept with 50 percent T is determined on the time scale as 10 days; from equation 6–5 λ is 0.0693 per day.* Similarly the half-life of the short-lived component, curve C, is 1.0 day and λ is 0.693 per day. If curve C should be curvilinear, the "peel-off"

* Alternatively λ may be derived from the slope; $\Delta Y'$ in common logarithms converted to natural ln and dividing by ΔX, which for the full range of curve C is

$$\frac{(1.301 - 0.940) \times 2.303}{12} = 0.0693;$$

and from equation 6–5 $T = 0.693/0.0693 = 10$ days.

process can be extended to three or more components if the original data are sufficiently accurate.

Application to Stream Conditions

In applying the basic principles of radioactive decay to stream conditions the two primary factors involved in self-purification from radioactive wastes are time of passage along the course of the stream and stream runoff, or dilution. (These two factors and their interrelation are dealt with in Chapter 3.) For radioactive wastes with long half-lives of months or years self-purification by decay is obviously insignificant and the primary factor is dilution by tributary runoff along the course. Both dilution and decay are taken into account in dealing with short half-life isotopes.

The curvilinear nature of the decay curves of mixed radioisotopes precludes direct application of the rational method of stream analysis as it is applied to self-purification from organic and microbial wastes. However, if the curvilinear decay curve of the waste can be broken down into linear components, each component can be dealt with as a separate source (expressed in curies per day) and integrated at its specific linear rate of decay from the point of discharge down the course of the stream at the time of passage of the prevailing runoff regime in a manner similar to that for organic BOD. The composite residual activity at any specified location along the course is the summation of the residuals of the respective components at that location. This residual is then converted to a concentration, such as microcuries per milliliter,* by taking into account the stream runoff (cubic feet per second) tributary to that location.

The residual composite activity and the concentration will be specific for each stream runoff regime. At high stream runoff the time of passage is shortened, but the dilution factor is increased; conversely, at the low runoff expected under drought conditions the time of passage is lengthened, but the available dilution is decreased. The relative effect of these opposing influences (as with bacterial survival) depends on the characteristics of the time of passage curve and the location of downstream positions relative to the point of the radioactive waste discharge. Integration over the range of expected runoff provides a frame of reference of expected residual concentrations along the entire course of the river.

If the radioactivity decay curve of the waste cannot adequately be resolved into linear components, the application of the rational method

* $$\mu\text{Ci/ml} = \frac{\text{Ci/day}}{\text{cfs} \times 2.4466 \times 10^3}.$$

of projecting expected stream conditions may not be possible. An intensive and extensive stream sampling program is then necessary.

Biological Accumulation and Sedimentation

Aquatic organisms serve as *accumulators* of radioactivity, and concentrations found in most forms exceed many times that of the river water. The pattern of accumulation follows the food chain of the biological life, commencing with the plankton and filamentous forms, which absorb radioactive nutrients directly from the water, then the bottom organisms that feed on these, then the juvenile fish that feed principally on these organisms, and at the end of the chain the adult fish that feed on the juveniles. This appears to be the general pattern, but there are deviations, depending on the food habits of a given species and of course on the type and decay constant of the radioactive waste.

The plankton, algae, and filamentous forms usually concentrate the largest amount of radioactivity, and, since there is direct absorption from the water, the level of concentration is directly dependent on that of the river water. The river water concentration varies with the seasonal runoff and at any location relative to the point of discharge is a function of intervening self-purification.

Concentrations in the higher forms of aquatic animals are related to the concentration in their food supply and in addition are dependent on feeding habits and metabolic rates. The primary route of radioactive concentration of most isotopes, except tritium (hydrogen-3), in the higher forms of aquatic animals appears to be through feeding rather than by absorption through the skin or gills. Feeding patterns and metabolism are related to seasonal water temperature, increasing or decreasing with the corresponding change in temperature. Some species of fish virtually stop feeding at low water temperatures, with consequent decline in radioactivity concentration; some continue to feed to some extent, and in these radioactivity is less variable; most species are stimulated by the natural seasonal rise in temperature, and in these radioactivity parallels the temperature pattern. Induced temperature rise from waste heat (to a level at which adverse effects of physiological stresses occur) may stimulate feeding. Some of the anadromous fish—such as adult salmon, which do not feed while in their spawning rivers—contain little or no activity.

Accumulation of radioactivity also takes place in the bottom muds of streams, principally from radioactive suspended solids or colloidal matter that may be contained in the wastewater, and from living and dead aquatic organisms contributed to the deposit. As with organic sludge deposits, stream velocity at which deposit or scour may occur is related to runoff level; accordingly it is expected that variation in concentration of deposits at any river location will be erratic and may be subject to redistribution, depending on river channel characteristics and the flashy or stable character of the runoff hydrograph.

The degree to which biological accumulation and sedimentation alter the expected self-purification decay rate depends on the specific isotopes contained in the waste effluent and their nutrient value and on the density of the aquatic population. In small streams the population density may be high; in large streams density may be low or negligible; and the density is highly responsive to seasonal variations. The accumulation is over the life cycle, and, although the accumulated concentration may be many times that of the river water, the biological population in terms of weight or volume in relation to that of the riverflow is very small. Hence the amount of radioactivity removed is small.

As is to be expected, the nutrient phosphorus (phosphorus-32) is the radioisotope that is most selectively assimilated by the aquatic chain and usually constitutes the largest accumulation found in the populations. Since phosphorus-32 generally constitutes a small percentage of reactor waste effluent, the biological accumulation does not normally have a great effect on the overall self-purification decay rate.*

In view of its erratic nature, biological accumulation and sedimentation cannot be depended on for removal of radioactivity from streams. The conservative approach is to determine downstream residuals on the basis of the radioactivity decay rate. The primary concern with biological accumulation is the effect on aquatic life and on humans who consume fish and shellfish.

Hydrologic Factor in Accident Protection

A frame of reference for the expected radioactivity along the course of the stream is one benefit of the rational method (or of stream sampling); a second benefit is that it provides a thorough definition of the hydrologic characteristics of the river basin (Chapter 3)—a vital element in protection in the event of accidental discharge of an unusually heavy radioactive waste load. Again the two elements in protection are the time of passage and tributary runoff along the course of the stream.

A series of time of passage curves covering the range of expected runoff, including the high as well as the low stages, should be prepared and readily available to the major producers of actual or potential radioactive wastes and to major downstream users of the river water, particu-

* Where phosphorus-32 is employed as a tracer in streams the effect of biological accumulation reduces the residual phosphorus-32 activity at locations downstream substantially below that expected from its normal rate in the time of passage.

larly for municipal and industrial supplies. Such time of passage curves would enable the downstream users to make immediate prediction of the arrival of the accidental slug. A daily reporting of stream runoff at the key stream gaging station to which the time of passage curves are referenced provides the appropriate curve to apply in determining the time of passage to any location. Radioactive waste producers would undoubtedly be aware of an accident and could immediately notify downstream water users. Such forewarning time may be crucial in preparation to bypass use at downstream locations and to take other precautionary action. One serious episode of this type without such hydrologic data and analyses immediately available would be inexcusable neglect. On some rivers they are available but perhaps not in readily usable form; on others the basic hydrologic data are meager or lacking. Although so far no serious accidental releases have been reported, with the increased use of atomic fuel for electric-power generation the potential risk becomes yearly greater.

Application to Reservoirs

Radioactive waste discharged to a reservoir, unlike that to a flowing stream, may result in buildup of activity many times that of the daily waste load. In a closed system (no advection) an equilibrium accumulation is attained when the aggregate daily decay of the accumulated activity equals the activity of the daily waste load contributed to the system. (This is akin to the equilibrium accumulation of BOD attained in organic sludge deposits.)

For a decay curve of normal linear form the equilibrium accumulation is

$$E = \frac{1}{\lambda}, \tag{6–9}$$

or in terms of half-life from equation 6–5,

$$E = 1.44\,T, \tag{6–10}$$

where E is equilibrium accumulation expressed in units of daily radioactive waste load. For example, if the half-life of the waste is 1 month (30 days), the equilibrium accumulation is 43.2 times that of the daily waste addition to the closed system. If the waste comprises two or more linear components, its equilibrium accumulation is the sum of the individual components. Thus in a closed system a substantial buildup of activity is possible with daily discharge of long-lived radioactive wastes.

In large reservoirs, such as are required for condenser circulating water and waste heat dissipation of nuclear fuel steam electric power generation, streamflow must be maintained below the reservoir; the result is a daily loss of radioactivity by advection. The amount of advection loss depends to a large degree on the distribution of wasted discharge in the reservoir, a product of many factors. As with waste heat load (Chapter 7), it is necessary to assume a distribution, and from this distribution an approximate advection loss is computed.

One simplification is to assume a uniform distribution of radioactive waste throughout the reservoir and to consider the daily advection loss as the ratio of the daily reservoir release to the total reservoir volume. On the basis of monthly average conditions, the advection loss then is at a constant rate for the month and may be incorporated into the decay rate for determination of accumulation of activity. As the reservoir release and volume change from month to month, accumulation is integrated over a period sufficient to approach an equilibrium accumulation. Again, as with sludge deposits, partial accumulation may be integrated graphically or analytically by substituting λ for K' in equation 4–10.*

$$E_p = \frac{1}{\lambda}(1 - 10^{-0.434\lambda t}). \tag{6–11}$$

The following example illustrates the procedure:

Radioctive waste:	
Half-life, days	30
Decay rate, $\lambda_D(0.693/30)$	0.0231
Advection:	
Reservoir release, percent per day	1.0
Equivalent λ_A; [log (100 − 1) = 2.0 − 0.434 λ_A]; λ_A	0.0092
Combined decay and advection, $\lambda_{(D+A)}$	0.0323
Equilibrium accumulation	
E in units of daily waste load (1/0.0323)	31
Partial accumulation at end of first month	
$31(1 - 10^{-0.434 \times 0.0323 \times 30}$	
(31 × 0.62) units of daily waste load (E_p)	19.2
Second month:	
Average advection, percent per day	0.5
Equivalent λ_A; [log (100 − 0.5) = 2.0 − 0.434 λ_A]; λ_A	0.00461
Combined decay and advection $\lambda_{(D+A)}$	0.0277
Equilibrium accumulation	
E in units of daily waste load (1/0.0277)	36
First month's accumulation in terms of second month's equilibrium (19.2/36)	0.533
Equivalent time t; [log (100 − 0.533) = $10^{-0.434 \times 0.0277t}$]	28.3
Partial accumulation at end of second month	
Period (28.3 + 30 days)	58.3
$36(1 - 10^{-0.434 \times 0.0277 \times 58.3})$	
36 × 0.7985 units of daily waste load (E_p)	28.8

* If data on daily reservior contents and release are available, a simulation program can be written for solution by computer. However, since it is necessary to make an assumption as to the initial distributon of waste, the graphical solution as an approximation is adequate.

Since accumulation in the 2 months attained approximately 80 percent of equilibrium, the process is extended through the low runoff season (low advection rate) until a maximum accumulation is attained approaching equilibrium. As runoff increases, advection rate increases and the level of equilibrium accumulation is lowered. Thus the critical period is the low runoff season when there is least advection and the reservoir volume also diminishes. The analytical or graphical integration should be made for critical drought periods as determined from statistical analysis of drought probability. The maximum average concentration of radioactivity in the reservoir (and reservoir release to downstream) is then determined as the maximum seasonal activity accumulation divided by the corresponding reservoir volume.

The assumption of uniform distribution of activity in the entire reservoir is not always valid. Appropriate adjustments to effective reservoir volume may be necessary, such as elimination of isolated arms. Also a degree of variation within the reservoir is inevitable, depending on configuration and size in relation to the location and design of the radioactive waste discharge system and to the location of the advection outlet. Where hydro pumped storage is an adjunct to nuclear fuel steam electric power generation, good mixing is provided over wide reaches of the reservoir (see Chapter 11).

Radioactivity accumulation approaching equilibrium levels does not necessarily imply a high concentration of activity. The large reservoir volume is available for dilution; however, as the reservoir volume is increased, the advection rate decreases and accumulation increases. Computations made for a range of likely conditions specific for each reservoir development plan provide a reference frame of expected activity accumulation and concentration.

INORGANIC SELF-PURIFICATION

Inorganic wastes constitute the stable portion found in various sources, principally as dissolved salts and inert suspended matter. Since this type of waste does not decompose, decay, or dissipate, the dissolved fraction is cumulative in amount in progression from source to source; the only significant self-purification factor that can reduce concentration is the increase in tributary stream runoff along the course of the stream. The critical period is the low runoff season. The inorganic salts may become concentrated enough to interfere with water uses and may contain end products that are toxic to animals and man.

The inert suspended fraction, silt from erosion and other inorganic suspended matter, although not degradable, is subject to deposit along the course of the stream. Thus the concentration that remains in the flowing stream at any location is dependent on two factors—namely, tributary dilution and removal by deposition. Removal by deposition depends on stream velocity and the critical velocity for scour; the former is readily defined by the time of passage curves and the latter by the scour equation, as discussed in Chapter 3.

Unlike organic sludge deposit, which reaches an equilibrium accumulation limit, the inorganic deposits can accumulate progressively, limited only by scour velocity. However, as the channel cross-section is progressively diminished by accumulated deposit, the velocity increases and ultimately the critical velocity at which scour takes place is attained. Except in large quiescent pools and reservoirs, freshets and floods also provide effective scour to limit accumulation. Depending on the channel characteristics as reflected by the time of passage curves, redistribution of deposit may occur from location to location at various critical runoff regimes.

Where suspended matter of waste includes both organic and inorganic fractions, the heavier inorganics may entrap the lighter organic ones, resulting in increased organic sludge deposit. Usually the most damaging effect of inorganic deposit is the destruction of fish spawning areas and other habitat by blanketing the streambed with silt.

REFERENCES

[1] *National Power Survey,* Federal Power Commission, U.S. Government Printing Office, Washington, D.C., 1964, pp. 77–92.

[2] *Engr. News Record,* November 10, 1966, p. 18.

[3] Straub, C. P., *Low-Level Radioactive Wastes, Their Handling, Treatment, and Disposal.* USAEC, Division of Technical Information, U.S. Government Printing Office, Washington, D.C., 1964, p. 32.

[4] Sterner, J. H., Paper No. P/288, *Proc. Third International Conference on Peaceful Uses of Atomic Energy,* Vol. 14, 1964, p. 147.

7

HEAT DISSIPATION IN STREAMS, PONDS, AND RESERVOIRS

Disposal and dissipation of heat released in streams, estuaries, cooling ponds, and reservoirs from industrial and utility operations is of increasing significance. Large quantities of condenser and cooling water used in modern thermal electric power generating stations, nuclear reactors, and industries place a heavy demand on water resources. The total electrical energy consumption in the United States for 1960 was reported as 846 billion kw-hr. The National Power Survey [1] projects consumption of approximately 2800 billion kw-hr by 1980, or slightly more than three times the amount generated in 1960. The installed generating capacity necessary to meet this demand is estimated at 525 million kw, compared with about 200 million kw of capacity in existence at the end of 1963. In regions where fossil fuel is not readily available, nuclear fuel will be employed increasingly.

In addition, the trend is definitely toward concentration of generating capacity at large stations. The entire flow of a stream may be circulated through the condensers during drought periods. The enormous waste heat loads returned to watercourses may raise the temperature of the water substantially above normal levels. Fortunately the waste heat is dissipated in a predictable profile, and the water temperature is gradually restored to normal. However, an elevated temperature profile interferes with subsequent downstream water uses, affects the capacity of the watercourse to assimilate other waste products, and may affect aquatic life.

IMPACT OF WASTE HEAT ON RECEIVING WATERS

The influence of waste heat on other waste assimilation factors and on aquatic life should be briefly summarized.

Influence on Waste Assimilation Capacity

As was shown in Chapter 4, the temperature of the water is a major factor in organic waste assimilation capacity because it plays a triple role by affecting the rate of stabilization, the dissolved oxygen saturation capacity, and the rate of reaeration. As the water temperature rises, the rate of oxidation of organic matter increases, placing a greater demand on the dissolved oxygen. However, as the temperature rises, the dissolved oxygen saturation capacity of water declines. In reaeration temperature plays opposing roles; an increase in temperature increases the rate of oxygen diffusion within the body of water, but the increase in temperature restricts solution from the atmosphere by reducing the saturation capacity. The net effect is usually adverse. However, where heat loads melt the ice cover, reaeration can take place; warm discharges are an asset to waste assimilation in severe climates.

Where organic sludge deposits are involved, their rate of decomposition is increased with rise in temperature. A secondary effect is the influence on the equilibrium accumulation of BOD attained in deposited sludge. As temperature rises the equilibrium accumulation declines; conversely, as temperature declines, BOD accumulates to higher equilibrium levels. Thus a sudden rise in temperature from an excessive heat load following a protracted period of accumulation established a combination of high rate of satisfaction acting on an excess BOD accumulation, which can induce an abnormally sharp decline in dissolved oxygen. Treatment of wastes prior to discharge should eliminate sludge deposits, but, if they occur within the zone of heat loading, a sudden rise in waste heat can aggravate conditions.

In self-purification from bacterial contamination the effects of temperature on the survival of coliform and pathogenic bacteria are less precisely established (see Chapter 5). In broad terms the death rates are increased with increase in temperature and hence survival is lowered. Offsetting this higher mortality is the presumed increase in initial aftergrowth associated with increase in temperature. Research has not yet clearly established whether seasonal temperature changes are responsible or whether increases in temperature from waste heat are a critical factor.

Influence on Aquatic Life

Various species of fish and the aquatic plants and animals on which they feed are affected by the temperature of their water environment, and each has its tolerance range. A moderate temperature rise may not be injurious, and for some organisms may actually result in increased population. However, a sudden increase in temperature or an increase beyond the tolerance range may kill aquatic life or adversely affect reproduction. Elimination of one species in the food chain may upset the entire ecological balance. In addition to survival are the more elusive effects on sustained population structure, activity and movement, reproduction, feeding, and development. So many other environmental factors enter that biologists are not in agreement on these effects except in very general terms.

Fish and aquatic organisms differ greatly in their tolerance to changes in temperature. Many are able to acclimate themselves to higher temperatures for short periods as a physiological stress response but cannot survive for long beyond their specific optimal range. With respect to fish population, streams, as suggested by Wurtz [2,3] may be roughly grouped into three temperature classes: *cold water, intermediate,* and *warm water.*

Cold water streams associated with the highest type of sport fish, the trout and salmon, are rated with a summer water temperature range between 52 and 68°F. At temperatures above 75°F it is reported that the trout population declines to 10 percent or less, whereas in the cold range trout represent 90 percent or more of the population. Also at temperatures above 75°F salmonides are rarely observed. Although temperatures above the optimal range may not kill trout, good production cannot be sustained except within the optimal cold water range.

Intermediate streams in the summer water temperature range of 70 to 80°F support such sport fish as small-mouth bass and walleye pike. Although too warm for trout, such streams provide good angling. When their temperature is raised above 85°F, intermediate streams tend to shift toward a coarse fish population.

In southern warm water streams in the summer range 80 to 90°F the fish population and the supporting aquatic organisms tend to adjust to a somewhat higher temperature tolerance. Large-mouth bass predominate, with a variety of other desirable warm water fish populations. If the temperature of such streams is raised and maintained above 95°F, the population tends to shift toward coarse fish.

The upper tolerance limit of temperature beyond which most intermediate and warm water fish cannot survive, except for very short periods, is in the range 100 to 105°F.

A factor that is not clearly defined by stream biologists is the extent of the reach of river through which elevated temperatures must prevail in order to significantly affect the fish population of the stream. Temperature decline is almost logarithmic in character, and hence a superelevation is rapidly brought toward normal. If superelevation at a single waste heat source is not beyond the lethal level, thereby producing a temperature block, migration and movement may not be hindered and only a short reach of stream may be involved in the shift in natural fish and aquatic populations. However, multiple sources of waste heat along the course or excessive superelevation can result in long reaches of stream being maintained at unfavorable temperature levels, with consequent drastic changes in the characteristics of the stream life.

In addition to the direct effects of elevated temperature, two significant indirect impacts are associated with waste heat discharges to streams: changes in dissolved oxygen and increased sensitivity to toxic compounds. Some aquatic forms are tolerant to wide ranges in dissolved oxygen concentration; others are more sensitive; many adjust or adapt to changes in the level of concentration.

Increase in water temperature affects the oxygen requirements and minimum concentration for survival. Generally rise in temperature stimulates respiratory and metabolic rates, thereby increasing oxygen utilization. As temperature rises the saturation capacity of water declines, and, if organic wastes are also involved, the combination may result in critical oxygen conditions downstream from waste heat discharges.

What has been stated with respect to direct temperature effects also applies to dissolved oxygen tolerance variations, but sensitivity to dissolved oxygen is less acute. The cold water streams naturally carry high dissolved oxygen concentrations because of high saturation capacity. Biologists have not determined whether temperature or oxygen is the dominant factor. Cold water species appear to thrive in reaches and depths of wide variation in dissolved oxygen. The normal species of intermediate and warm water streams appear to tolerate much lower oxygen concentrations. Lethal limits are quite low, but the optimal oxygen ranges for various species of fish are rather ill defined and too frequently are arbitrarily established.

The tolerance limits of substances toxic to fish and aquatic organisms are usually lowered by an increase in temperature. This is particularly true if the temperature stress response is excessive. Differences in reported experimental results, however, are difficult to interpret because of the wide variety of conditions employed in testing. Seldom are controlled laboratory results comparable to the complex natural environment. A precise definition of lowered tolerance in relation to temperature elevation cannot be made at present; however, there is general agreement among biologists that elevated temperatures render fish more susceptible to toxicity.

Some aspects of the effect of waste heat on aquatic life can be forecast quite confidently, but often under apparently similar circumstances quite different end results develop. Further field study and laboratory research are needed for a better understanding of biological thermodynamics.

MAGNITUDE OF WASTE HEAT AND COOLING WATER REQUIREMENTS

Quantitative data concerning waste heat loads and the proportion assignable to utility, industry, and municipal sources generally are not available. With the trend toward large power stations, condenser circulating water systems of steam electric power generating plants are likely to remain the largest single source of waste heat. The heat loads from specific installations vary widely, depending principally on the size of unit and the fuel source, fossil or nuclear.

Fossil Fuel Plants

Power engineering experience with fossil fuel plants employing two-pass condensers indicates a direct relationship between generating capacity, circulating water requirements, and waste heat load, as shown in Figure 7–1. Curve *A* shows the relation between the unit capacity in kilowatts and the circulating water requirements; curve *B* shows the relation between the unit capacity in kilowatts and the waste heat discharge. Capacity refers to the size of the individual units of which the total plant capacity is composed. Circulating water requirements refer to summer conditions. As the size of a generating unit decreases, the heat load and the circulating water requirements increase sharply. For large capacity (over 200,000 kw) modern units waste heat load levels off at about 4400 Btu/kw-hr, and circulating-water requirements at about 0.55 gpm/kw. During the winter period the water requirements have been reported to be about 60 percent of the summer requirements.

The current trend is toward large generating units of 350- to 750-MW rating. A few units with capacities in excess of 750 MW are in operation or under construction. Some reduction in net heat rate (the amount of fuel required to generate 1 kw-hr) is obtained with larger units. In 1946 the size of largest unit design was 140 MW, with a net heat rate of 10,600 Btu/kw-hr; in 1962 unit size increased with a maximum of up to 1000 MW and corresponding net heat rate below 8400 Btu/kw-hr.

Of the net heat fuel input, about 40 percent is converted to electrical energy, about 50 percent is waste heat discharged in the condenser cir-

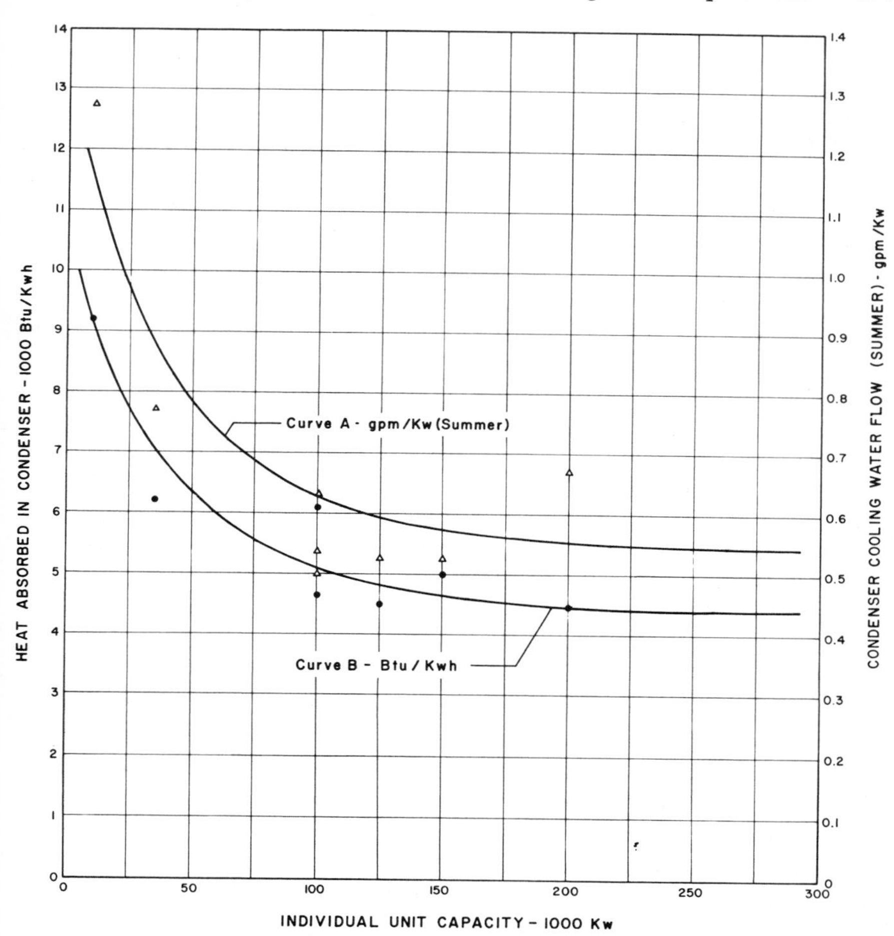

Figure 7–1 Characteristics of fossil fuel steam power plant condensers (two-pass condensers).

culating water, and about 10 percent is stack loss discharged to the atmosphere. Thus 4400 Btu/kw-hr is a reasonable waste heat rejection rate to the watercourse expected from modern fossil fuel steam electric power generating plants.

The generalized graph is in fairly close agreement with recent reported operating data. For specific installations the waste heat load can be determined from the operating records of cooling water pumping rate, power load, and temperature rise through the condensers.

Nuclear Fuel Plants

The trend in power generation is toward nuclear fuel. The record of safety, reliability, and economy has made nuclear fuel competitive with fossil fuel. In large generating units (above 800,000 kw) nuclear fuel has the cost advantage. Nuclear fuel technology is developing rapidly, and the prospect is for continued reduction in fuel costs. Coal mining is by contrast a mature industry; the dramatic lowering of cost that automation and the unit train innovation brought about early in the 1960s is not likely to be repeated. Another advantage of nuclear power plants is their excellent safety record, which doubtless will be reflected in lower insurance rates. These factors point toward nuclear fuel as the dominant energy source for the electric utility industry by the early 1970s.

Waste heat rejection from the condensers of nuclear power plants is greater per unit energy production than it is for conventional fossil fuel plants. The difference is due to the lower efficiency of the nuclear plants; moreover heat discharged from fossil fuel as hot gases to the atmosphere must be removed from nuclear plants by condenser flow. Thus condenser circulating-water flow and waste heat rejection associated with nuclear plants are as much as 50 percent greater than those of fossil fuel plants.

The efficiency of nuclear power plants under the present state of technology (generally boiling water or pressurized water) is on the order of 32 to 34 percent. Efficiencies of 40 percent (approaching the efficiency of modern fossil fuel plants) are possible with gas cooled reactors, some of which are now under construction.

The amount of cooling water flow for boiling-water reactors can be calculated from a formula proposed by the General Electric Company as follows:

$$P = \frac{1000(1 - e)}{2.45e\,\Delta T},$$

where P is the condenser circulating water flow in gallons per kilowatt-hour, e is the efficiency of the steam cycle, and ΔT is the temperature rise across the condenser in degrees Fahrenheit.

Thus for a plant of 33-percent efficiency and ΔT of 19°F, P is computed to be 43.6 gal/kw-hr, or 0.727 gpm/kw. Under these conditions the waste heat rejection is 6900 Btu/kw-hr. Referring to Figure 7–1, it will be noted that both the condenser water rate and the waste heat rejection rate of boiling water nuclear plants are actually about 50-percent greater than those for conventional fossil fuel plants.

As reported by the Westinghouse Electric Company, pressurized water type reactors have similar condenser water requirements and heat rejection rates.

Variation in Waste Heat Loading

Thermal electric power plants, whether utilizing fossil or nuclear fuel, do not necessarily operate at continuous and uniform load. The nature of the electrical load served is such that great variation in demand occurs on a seasonal, weekly, and daily basis. In general thermal units tend to follow this load pattern, and waste heat rejection varies accordingly.

During the useful life of a thermal electric unit a definite pattern of operation is followed, based on the age of the equipment. New thermal units normally are operated on the base of the system load curve for a few years, resulting in a high rate of use, often approaching 7000 hr/yr. Traditionally new thermal units are more efficient; their lower production costs lead management to operate them at continuous load.

However, these units are themselves moved from the base load operation as still newer units are added, with the result that their operation becomes less and less continuous as the years progress. Ultimately the unit reaches a point at which it is operated strictly on a peaking basis, involving only a few hundred hours per year. In general waste heat loading follows the above pattern of operation, with greatest effect in the early years. However, maximum system loads occur during summer hot spells, doubtless a result of air conditioning requirements. Thus, even if a given thermal station is not operating continuously on base load, it contributes a sizable waste heat load during the critical temperature period.

The efficiency of the steam plant cycle has a significant effect on the production of waste heat. Condenser design also influences temperature superelevation of discharged cooling waters. The discharge water of the two-pass condenser is about 10°F warmer than that of the one-pass condenser, the former involves a smaller flow of water. The higher temperature of discharge water increases the heat transfer from the water surface and is thus preferable for cooling pond duty. Single-pass condensers with greater water flow and less temperature elevation are more applicable to once-through cooling, and are preferable for locations on large rivers. The temperature of intake cooling water to steam plant condensers significantly affects turbine power and performance. A lower condenser water temperature results in a minimum of backpressure on the turbine exhaust, thereby ensuring maximal power and efficiency.

Concentration at Large Power Stations

As population and economy expand, good plant sites become increasingly difficult to find; therefore the trend is to develop the largest generating plant at a single site that is technically and economically feasible. It is therefore to be expected that total capacity concentration of 3000 to 5000 MW at a single generating station will be the practice.

The magnitudes of waste-heat load and condenser water requirements at a single plant of 4000-MW capacity (fossil fueled) based on 4400 Btu/kw-hr and 0.55 gpm/kw (Figure 7–1) are 17.6 billion Btu/hr and 2.2 million gpm, or 4900 cfs. The rise in temperature of the 4900 cfs of condenser water for this heat load would be 16°F. The condenser water rate of 4900 cfs constitutes a very large river; during drought severities of, let us say, 0.1 cfs per square mile yield this would require a tributary drainage area of 49,000 square miles. If the natural summer water temperature under such drought conditions is 80°F and the entire riverflow of 4900 cfs passes through the condensers, the temperature of the river at the plant discharge would be raised to 96°F. If the plant uses nuclear fuel, the waste heat load will be greater.

The magnitude of the probable summer waste heat load and the very large riverflows essential to avoid excessive temperature superelevation can be seen clearly. There are not many river systems that support a natural drought flow on the order of 5000 cfs; hence water is a key factor in the location of a steam electric plant. Drought control by storing of floodwaters and releasing them during the drought season is increasingly essential as society's demands lead to progressively larger power plants.

In watersheds that cannot ensure adequate streamflow, either as natural or as augmented runoff, other means of dealing with waste heat must be employed; off-stream cooling ponds, large reservoirs, or various types of cooling towers all are alternatives to normal stream dissipation. All these alternatives, unfortunately, entail a loss of water to the atmosphere by evaporation. Well-designed cooling towers require make-up water to compensate for a water loss on the order of 15 to 20 cfs per 1000 MW; cooling pond loss by induced evaporation (exclusive of natural evaporation from the pond area and gain by rainfall) may range from 5 to 15 cfs per 1000 MW; and induced evaporation losses in a flowing stream may be on the order of 10 cfs per 1000 MW, or about one-half that for cooling towers. These are relative values; actual evaporation losses depend on many local factors, including climatologic conditions and specific design arrangements.

QUANTITATIVE DETERMINATION OF HEAT DISSIPATION

Empirical Approximations

One of the early attempts to establish a numerical basis for predicting the cooling effects of ponds was that proposed by Lima [4]. His relationships are empirical, established from data collected at existing installations, and centered around a coefficient K related to air temperature under standard conditions of a 6-mph wind. Another method for predicting the cooling effect of ponds was proposed by Throne [5], who employed a modified energy budget relationship based on observations made in Colorado.

LeBosquet's [6] pioneering work presents a mathematical basis for predicting heat loss in a flowing stream. His formulation is based on the simplified assumption that the rate of heat transfer is proportional to the temperature differential between air and water. In turn he assumes that the rate of decrease of temperature differential between air and water, when the temperature of the air remains constant, is a constant proportion of the remaining temperature differential. This is expressed in differential form as

$$-\frac{dT}{dt} = KT, \tag{7-1}$$

which integrates to

$$\log_e \frac{T_1}{T_0} = -Kt \tag{7-2}$$

or, in common logarithms,

$$\log \frac{T_1}{T_0} = -0.434\, Kt, \tag{7-3}$$

where T_1 is the temperature differential between water and air at any time t, T_0 is the initial temperature differential, and K is a constant depending on specific situation and units.

If referred to the usual units of heat transfer U (Btu per hour per square foot per degree Fahrenheit),

$$K = \frac{0.434UA}{G}$$

and

$$\log \frac{T_1}{T_0} = \frac{-0.434UAt}{G}. \tag{7-4}$$

For a river of uniform-cross section and no increments of runoff along its course, equation 7–4 becomes

$$\log \frac{T_1}{T_0} = -U \frac{(0.0102wD)}{Q}, \tag{7–5}$$

where A is the surface area of the stream in square feet, D is the distance along the river course in miles, w is the mean width of the stream channel in feet, G is the weight of water in pounds in the channel between location D_0 and D_1, and Q is the runoff of the stream in cubic feet per second.

The coefficient of heat transfer U will depend on the physical characteristics of the river and meteorologic factors other than air temperature, and must be determined for each particular stream or reach of stream. If the temperature differential is observed at two points, D_0 and D_1, U is obtained by transposing equation 7–5

$$U = \frac{Q}{0.0102wD} (\log T_{D0} - \log T_{D1}). \tag{7–6}$$

The following example illustrates the application of the method in determining a specific heat transfer coefficient for the Ocmulgee River below a steam electric power station upstream from Macon, Georgia:

Cooling water employed	367 cfs
Temperature rise of cooling water	17°F
Runoff below power station	955 cfs
Temperature of river above power station	86°F
Temperature of river below station (mile 215)	90.23°F
Temperature of river at Macon (mile 205.7)	88.7°F
Air temperature (daily mean) at Macon	85.0°F
Mean river width	145 ft

Substituting in equation 7–6, we obtain

$$U = \frac{955}{0.0102 \times 145 \times 9.3} \log (90.25 - 85) - \log (88.7 - 85)$$

$$= 10.4 \text{ Btu/hr-ft}^2\text{-°F}.$$

LeBosquet's formulation, it is recognized, is the familiar semilogarithmic form, which also expresses BOD amortization, bacterial survival, and radioactive decay. If the temperature differential is plotted as the ordinate on the logarithmic scale and distance in river miles as the abscissa on the linear scale, a straight-line relation emerges. If U is determined and w and Q remain constant along the river reach, the slope of the line is given by $(0.0102Uw/Q)$. From such a plot the residual temperature differential can readily be determined for any location along the course, and in turn the profile of the river water temperature can be constructed.

The limitation of LeBosquet's formulation is its dependence on a measured temperature profile under controlled conditions to determine the heat loss coefficient U; relating heat loss only to the temperature differential between water and air, he neglects other meteorologic factors involved in the heat balance. The convenience of the semilogarithmic form, however, warrants its application where approximations are sufficient. The heat loss coefficient U varies widely from river to river—according to LeBosquet from 6 to 18 Btu/hr-ft^2-°F. The value for the Ocmulgee River as computed above (10.4) and the value computed for the Tittabawassee River (8.75), fall within this range. A heat loss coefficient determined by British workers for the river Lea is 5.4 Btu/hr-ft^2-°F.

The Energy Budget Formulation

A more fundamental approach to heat dissipation is based on energy budget relationships. A detailed investigation of the various terms of the energy budget of water as related to meteorologic parameters was undertaken at Lake Hefner, Oklahoma, by a federal task force [7]. Although the primary purpose was investigation of evaporation from water surfaces, basic data were obtained on heat loss. The micrometeorologic measurements employed, however, preclude general applications based on routine U.S. Weather Bureau records.

Langhaar [8] has proposed a method for predicting heat loss from cooling ponds based on energy budget relations and conventional U.S. Weather Bureau records; his method permits direct computation without dependence on measurements of water temperature. Modification of Langhaar's basic approach as developed below permits determination of probabilities of occurrence from the long-term meteorologic records employed.

In the development of basic equations four principal surface mechanisms are involved in heat transfer and temperature change as a body of water is exposed to the atmosphere. Heat is lost through evaporation, convection, and radiation—and gained through solar radiation. Heat may also be transferred to the enclosing earth bottom by contact and infiltration, but in most flowing streams and in well-sealed ponds and reservoirs this small loss may be neglected. In ponds and reservoirs net heat transfer by advection may also be involved. The heat balance is represented as

$$H = H_e + H_c + H_r - H_s - H_a, \tag{7–7}$$

where H is net heat loss, H_e is heat loss by evaporation, H_c is heat loss by convection, H_r is heat loss by radiation, H_s is heat gain by solar radiation, and H_a is net heat transfer by advection.

The three heat losses—evaporation (H_e), convection (H_c), and radiation (H_r)—can be computed from meteorologic measurements of air temperature, wind velocity, and vapor pressure. Net heat transfer by advection (H_a) is the balance associated with inflow and outflow of water.

Evaporation Loss

Heat loss by evaporation (H_e) is readily calculated by Meyer's formula [9], which has wide acceptance in engineering practice and is particularly applicable because it is based on routine U.S. Weather Bureau observations. Meyer's formula for evaporation, expressed in inches per month, from natural bodies of water is

$$I = C_1(1 + 0.1W)(V_w - V_a), \tag{7-8}$$

where W is mean wind velocity in miles per hour measured about 25 ft above the water surface or above the surrounding land area; V_w is water vapor pressure in inches of mercury corresponding to the surface water temperature taken about 1 ft below the surface; V_a is the mean absolute water vapor pressure prevailing in the overlying atmosphere 25 ft above the water surface or above the surrounding land area; and C_1 is a constant, ranging from 10 to 15, depending on the depth and exposure of the body of water and frequency of meteorologic measurements. For large deep lakes and reservoirs C_1 is taken near the lower value, and for shallow ponds and surface accumulations C_1 is taken as 15. For flowing streams of moderate depth and velocity C_1 may be taken as 14. Using latent heat of vaporization H_v for the given water temperature, I in inches per month is converted to H_e, heat loss in Btu per hour per square foot of water surface by

$$H_e = 0.00722H_vC_1(1 + 0.1W)(V_w - V_a). \tag{7-9}$$

Abridged tabulations of the latent heat of vaporization H_v and saturated water vapor pressure V_a are presented in Tables 7–1 and 7–2, respectively.

Convection Loss

Heat loss by convection depends on the temperature differential between water and air, and on the wind velocity at the surface of the water. In still air convection loss from a flat surface ranges from 0.5 Btu/hr-ft^2-°F for a temperature differential of a few degrees to

Table 7–1 Latent Heat of Vaporization H_v[a]

T (°F)	H_v (Btu/lb)	T (°F)	H_v (Btu/lb)
32	1075.8	100	1037.2
35	1074.1	105	1034.3
40	1071.3	110	1031.6
45	1068.4	115	1028.7
50	1065.6	120	1025.8
55	1062.7	125	1022.9
60	1059.9	130	1020.0
65	1057.1	135	1017.0
70	1054.3	140	1014.1
75	1051.5	145	1011.2
80	1048.6	150	1008.2
85	1045.8	155	1005.2
90	1042.9	160	1002.3
95	1040.1	165	999.3

[a] After Keenan and Keyes.

1.0 Btu/hr-ft^2-°F for temperature differentials of 50 to 100°F; a generally applicable value may be taken as 0.8 Btu/hr-ft^2-°F. Air movement at the surface is reported to increase this value by a coefficient C_2 of 0.16 to 0.32 for each mile per hour of wind velocity. For a reasonably quiescent body of water Langhaar suggests an average value of 0.24. For flowing streams of moderate velocity a higher value, approaching 0.32, would be reasonable. The greatest factor of uncertainty is the conversion of nearby U.S. Weather Bureau wind record to velocity at the surface of the water. Wind velocity may vary widely with elevation of the recording instruments above the surface. Various relations are proposed for correcting to surface level. Of great significance is the relative exposure, which is a highly variable factor and is difficult to evaluate. In practical application surface velocity can be taken as half that recorded at official Weather Bureau stations.

Convection loss for a water surface then is given by

$$H_c = \left(0.8 + C_2 \frac{W}{2}\right)(T_w - T_a), \tag{7–10}$$

where T_w is the surface water temperature and W and T_a are Weather Bureau measurements of mean wind velocity in miles per hour and mean air temperature in degrees Fahrenheit, respectively.

Table 7–2 Saturated Water Vapor Pressure V_a[a] in Inches Hg

Air T (°F)	Vapor Press. (in. Hg)	Air T (°F)	Vapor Press. (in. Hg)	Air T (°F)	Vapor Press. (in. Hg)	Air T (°F)	Vapor Press. (in. Hg)
30	0.164	60	0.517	90	1.408	120	3.425
31	0.172	61	0.536	91	1.453	121	3.522
32	0.180	62	0.555	92	1.499	122	3.621
33	0.187	63	0.575	93	1.546	123	3.723
34	0.195	64	0.595	94	1.595	124	3.827
35	0.203	65	0.616	95	1.645	125	3.933
36	0.211	66	0.638	96	1.696	126	4.042
37	0.219	67	0.661	97	1.749	127	4.154
38	0.228	68	0.684	98	1.803	128	4.268
39	0.237	69	0.707	99	1.859	129	4.385
40	0.247	70	0.732	100	1.916	130	4.504
41	0.256	71	0.757	101	1.975	131	4.627
42	0.266	72	0.783	102	2.035	132	4.752
43	0.277	73	0.810	103	2.097	133	4.880
44	0.287	74	0.838	104	2.160	134	5.011
45	0.298	75	0.866	105	2.225	135	5.145
46	0.310	76	0.896	106	2.292	136	5.282
47	0.322	77	0.926	107	2.360	137	5.422
48	0.334	78	0.957	108	2.431	138	5.565
49	0.347	79	0.989	109	2.503	139	5.712
50	0.360	80	1.022	110	2.576		
51	0.373	81	1.056	111	2.652		
52	0.387	82	1.091	112	2.730		
53	0.402	83	1.127	113	2.810		
54	0.417	84	1.163	114	2.891		
55	0.432	85	1.201	115	2.975		
56	0.448	86	1.241	116	3.061		
57	0.465	87	1.281	117	3.148		
58	0.482	88	1.322	118	3.239		
59	0.499	89	1.364	119	3.331		

[a] After Meyer [9].

Radiation Loss

Heat loss by radiation from the water surface is practically equivalent to that of a blackbody. The radiation loss in Btu/ft^2-hr can be computed by the Stephan–Boltzmann law from $H_r = 0.173 \times 10^{-8}(T_w + 460)^4$. Counter to this heat loss, water receives radiation in varying amounts from the atmosphere and from the terrain. Assuming that the surrounding objects are at the same temperature as the air, Langhaar suggests a linear approximation of the net radiation loss for ordinary temperature as simply the differential in temperature between the water surface and the air:

$$H_r = (T_w - T_a). \tag{7–11}$$

Solar Radiation Gain

Heat gain from solar radiation, direct and diffuse, cannot be computed from meteorologic factors and must either be estimated or, preferably, measured. As measured it includes both the direct and the diffuse, or scattered, sky radiation received per day on a horizontal surface. The diffuse radiation in relation to the total varies radically, depending on cloud cover, water vapor, air pollution, and shade from overhanging foliage. At high elevations total radiation is high, and the diffuse radiation from cloudless skies may be as low as 5 percent of the total; in cloudy regions at low elevations total radiation is reduced, and the diffuse fraction may constitute over 40 percent; in highly industrialized areas and urban centers atmospheric pollution at times virtually eliminates direct radiation and the major component is the diffuse fraction. Along small streams shade trees may reduce the direct fraction, and total radiation may be substantially below that for large streams and open bodies of water.

Hence estimating H_s from maximum possible sunshine leads to values that are too high, as local sky cover radically reduces the direct radiation reaching the water surface. The absorption coefficient of water surface is high, exceeding 95 percent; consequently the controlling factor is local sky cover. The solar radiation as measured at nearby Weather Bureau stations is considered directly applicable if exposures are similar; for some locations it may be necessary to apply a correction factor to adjust for difference in exposure. The adjustment to apply is a matter of judgment and may be as much as 25 percent upward or downward. A fuller discussion of variation in solar radiation based on recorded measurements at key Weather Bureau stations is presented in Chapter 3.

Equilibrium Water Temperature

A body of water exposed to a given set of meteorologic conditions will come to equilibrium in time: the heat losses by evaporation, convection, and radiation will balance the heat gain from solar radiation. Under equilibrium conditions net heat loss is zero ($H = 0$), and from equation 7–7 (assuming no net advection),

$$H_e + H_c + H_r = H_s. \tag{7–12}$$

Under these conditions the temperature of the water will also approach equilibrium. This limiting temperature is referred to as the *equilibrium temperature* E and can be computed from the measurements of the prevailing meteorologic conditions as follows:

Substituting the three expressions of heat loss in equation 7–7 gives

$$H = 0.00722H_vC_1(1 + 0.1W)(V_w - V_a) + \left(0.8 + C_2\frac{W}{2}\right)(T_w - T_a) + (T_w - T_a) - H_s. \tag{7–13}$$

At equilibrium the temperature of the water is derived

$$T_w = E \quad \text{and} \quad V_w = V_E.$$

Substituting and combining gives an equation for E and V_E in terms of the measurable meteorologic factors:

$$\left(1.8 + C_2\frac{W}{2}\right)E + 0.00722H_vC_1(1 + 0.1W)V_E = \left(1.8 + C_2\frac{W}{2}\right)T_a + 0.00722H_vC_1(1 + 0.1W)V_a + H_s. \tag{7–14}$$

A direct solution for E cannot be made by equation 7–14 since V_E is also involved, but by assuming values for E, successive approximations can quickly approach a balance in the equation, which thus determines the proper value of E, as illustrated in the following example for a flowing stream:

Air temperature T_a	61.5°F
Wind velocity W	10.3 mph
Relative humidity	71.0%
Water vapor pressure $V_a(0.545 \times 0.71)$	0.387 in. Hg
Solar radiation recorded	331 langleys
Heat gain from solar radiation $H_s(331 \times 0.1535)$	50.8 Btu/hr-ft²-°F

First assumption, $E = 65°F$:

H_v at 61.5°F by Table 7–1 = 1059.1, $C_1 = 14$, $C_2 = 32$, V_E at 65°F by Table 7–2 = 0.616.

By equation 7-14:

$$(1.8 + 0.16 \times 10.3)65 + 0.00722 \times 1059.1 \times 14(1 + 0.1 \times 10.3)0.616 = 358,$$

$$3.45 \times 65 + 217.2 \times 0.616,$$

$$(1.8 + 0.16 \times 10.3)61.5 + 0.0072 \times 1059.1 \times 14(1 + 1.03)0.387 + 50.8 = 347.$$

$358 > 347$; $\therefore$ E assumed too high.
Second assumption, $E = 63°F$:

$$3.45 \times 63 + 217.2 \times 0.575 = 342.3.$$

$342.3 < 347$; $\therefore$ E assumed too low.
Third assumption, $E = 63.5°F$:

$$3.45 \times 63.5 + 217.2 \times 0.585 = 346.$$

$346 < 347$; $\therefore$ E assumed too low.
Fourth assumption, $E = 63.6°F$:

$$3.45 \times 63.6 + 217.2 \times 0.587 = 346.8.$$

$346.8 \approx 347$; $\therefore$ $E = 63.6°F$.

The Water Temperature Profile

With given sustained meteorologic conditions, theoretically an infinite time of exposure is required for water to attain equilibrium. Moreover, an infinite surface area would be required for warm water introduced into a river or basin to cool to the equilibrium temperature. But because temperature decline is nearly logarithmic, a close approach to equilibrium temperature is attained within practical limitations of time, basin area, or river reach.

The relation between the water temperature profile and required surface cooling area involves these concepts of net heat loss H based on observed meteorologic conditions as follows. An increment of water temperature decline dT_w per increment of time dt is equal to the heat loss from the corresponding increment of surface area divided by the weight of the underlying water of depth b.

$$\frac{dT_w}{dt} = -\frac{H}{62.4b} \tag{7-15}$$

which integrates to

$$\int_{T_1}^{T_2} dT_w = -\frac{H}{62.4b}\int_{t_1}^{t_2} dt. \tag{7-16}$$

Expressing t in hours, in terms of area A in square feet per cfs of streamflow, we obtain

$$t = \frac{bA}{3600},$$

Solving for A gives

$$A = -224{,}640 \int_{T_1}^{T_2} \frac{dT_w}{H}. \tag{7-17}$$

From equation 7–12, when $T_w = E$, $H_s = H_e + H_c + H_r$, from which

$$H_s = 0.00722 H_v C_1 (1 + 0.1W)(V_E - V_a) + \left(1.8 + C_2 \frac{W}{2}\right)(E - T_a). \tag{7-18}$$

Substituting into equation 7–13 gives

$$H = 0.00722 H_v C_1 (1 + 0.1W)(V_w - V_E) + \left(1.8 + C_2 \frac{W}{2}\right)(T_w - E). \tag{7-19}$$

Combining equation 7–19 with equation 7–17 and substituting α for $0.00722\, H_v C_1 (1 + 0.1W)$ and β for $[1.8 + C_2(W/2)]$ gives

$$A = -224{,}640 \int_{T_1}^{T_2} \frac{dT_w}{\alpha(V_w - V_E) + \beta(T_w - E)}. \tag{7-20}$$

Although equation 7–20 cannot be integrated directly, it is possible to reach solutions without resorting to empirical procedures by dividing the temperature differential $(T_1 - T_2)$ into a number of increments (ΔT_w) and performing the integration by successive summations, taking T_w and its corresponding V_w as the mean for each successive increment in accordance with

$$A = -224{,}640 \sum_{T_1}^{T_2} \frac{\Delta T_w}{\alpha(V_w - V_E) + \beta(T_w - E)}. \tag{7-21}$$

Summation of four to eight increments is usually adequate, and, since the complete temperature profile is desirable, the successive increments provide the intermediate points between the initial or entering water temperature T_1 and the final temperature T_2.

Where long-term weather records are available, therefore, it is possible by equation 7–14 to compute the equilibrium water temperature expected from any probability of occurrence of meteorologic conditions. In turn cooling to any level of temperature above equilibrium can be computed by equation 7–21. Knowing the cumulative surface area along the course of a stream (or outward from a point of heat discharge into a pond or reservoir) makes possible the construction of a reliable temperature profile.

Surface Water Temperature and Vertical Gradients in Deep Reservoirs

The phenomenon of thermal stratification in deep lakes and reservoirs in temperate climates results in a seasonal succession of vertical temperature gradients ranging from uniform surface-to-bottom temperature to characteristic profiles of marked temperature differences with depth.

The process begins after the spring turnover, when the body of water is effectively mixed at a uniform low temperature. The minimum temperature experienced at the time of spring turnover is dependent on local climate but is generally above 39.2°F, the temperature of maximum density of water. Above 39.2°F the density of water decreases with increase in temperature, and the rate of decrease increases as temperature rises. These temperature–density relations give rise to patterns of circulation and stratification through the seasons.

If the water at the time of turnover is of uniform transparency and is quite undisturbed, the increment of solar radiation would be absorbed exponentially with depth and would heat the water at a rate to produce an exponential temperature gradient. Two principal factors prevent such a process from taking place. Heat loss to the atmosphere will cool the surface layers and set up convection currents, and wind will disturb the surface and generate turbulent motion, with downward transport of heat.

The resultant temperature distribution takes a characteristic form. In lakes and reservoirs of sufficient depth the water divides into an upper region of uniformly warm, circulating, turbulent water termed the *epilimnion* and into a deep, cool, and relatively undisturbed region termed the *hypolimnion*. The region of rapid decrease in temperature separating the two layers is referred to as the *thermocline*. This characteristic form is accentuated as the warm season progresses and recedes through the cool season. As the surface water cools, the denser upper layers sink. With this turnover the cycle goes to the completion, and the body of water reverts to uniform temperature throughout its depth.

Under given climatologic conditions, the size and depth of the body of water are the most significant factors in development of thermal stratification, but many factors may alter the characteristic shape and progression of the resultant vertical temperature profiles. The principal factors include reservoir configuration and terrain, area–volume–stage relations, orientation of prevailing winds, and hydrologic (or induced) characteristics of inflow and outflow.

Large steam electric power generating plants take advantage of thermal stratification by withdrawing condenser circulating water from the cool hypolimnion and discharging the heated condenser water into the epilimnion. It is necessary in practical application, therefore, to predict the succession of vertical temperature gradients through the seasons.

Methods of such prediction have been either highly theoretical mathematical formulations that cannot adequately take into account the many local factors involved or have consisted of merely superimposing arbitrary characteristic temperature gradients in their entirety and balancing total heat content with a computed heat content. The Reynolds procedure [10] is a compromise between these two extremes; it adheres to fundamental concepts but tailors the predictions to local conditions without depending on the rigid relationships that must be assumed in the strictly mathematical treatment.

In dealing with a dynamic situation influenced by many factors and variations in climatology it is expected that any method based on conditions averaged over a period of time is subject to lag effects and deviations induced by abnormal variations over short periods. The prediction method based on monthly average conditions, satisfactory for most practical application, is illustrated in the procedure described below.

Hydrology

The first step in the procedure is to establish the reservoir hydrology for the design year or series of years. Statistical analysis of monthly average runoff yield applicable to the tributary drainage area defines probability of inflow. Sequential analysis of inflow and outflow establishes the reservoir operating conditions and stage over the period of record and identifies the normal and critical years. Precipitation and evaporation may be considered separately or, if of the same order of magnitude, may be combined. The temperature of the inflow water for each month is either based on past records or is computed as equilibrium values from climatologic data. If inflow is not natural runoff and constitutes a withdrawal from an upstream reservoir, the temperature may differ substantially from that of the natural streamflow. Similarly the outflow temperature from the reservoir depends on the depth at which withdrawal is taken. These variables determine the advection of heat to and from the reservoir for the inflow and outflow.

Meteorology and Net Heat Flux

When hydrology has been established, it is necessary to determine for each month of the climatologic year in question the relationships for net heat flux according to monthly average reservoir surface temperatures. The net heat flux is the monthly average solar radiation (H_s) recorded for the specific month of the design year less the total surface heat loss H_e, H_c, and H_r computed from the specific monthly average climatology by equations 7–9, 7–10, and 7–11, employing coefficients for large reservoirs appropriate to the local conditions. From these equations it is apparent that heat loss can be expressed as a function of surface-water temperature after substitution for the given climatologic conditions has been made. Hence it is possible to plot for each month of the design year the net heat flux expressed in energy units, such as acre-feet-°F versus surface water temperature, as shown in Figure 7–2*a*. This is a convenient form for interpolation, because frequent reference to this evaluation is made in balancing the heat budget in successive months through the design year.

For example, the net heat flux curve for January for the critical year is computed for four selected surface water temperatures T_w expected to encompass the January range. The critical January reservoir mean surface area is 14,150 acres, and the January mean recorded meteorologic conditions are as follows:

T_a	46.5°F
Relative humidity	75%
V_a	0.237 in. Hg
W	7.8 mph
H_s	34.4 Btu/ft²-hr

Selecting 40°F as a first value for T_w:

H_e by equation 7–9 with $C_1 = 10$ is 1.38,
H_c by equation 7–10 with $C_2 = 24$ is 11.3,
H_r by equation 7–11 is 6.5.

Thus $H_e + H_c + H_r - H_s$ is 50.82 Btu/ft²-hr, or a net heat gain for January for the entire reservior of 2.33×10^{13} Btu $(50.82 \times 43{,}560 \times 14{,}150 \times 24 \times 31)$. The heat required to raise an acre-ft of water 1°F is 2.72×10^6 Btu $(43{,}560 \times 62.5)$; hence the net heat flux for the critical January for a T_w of 40°F is 8.566×10^6 acre-ft-°F $[2.33/(2.72 \times 10^7)]$. Similarly net heat fluxes for a T_w of 45, 50, and 55°F, are 5.079×10^6, 1.322×10^6, and -2.61×10^6 acre-ft-°F, respectively.

The four values describe the range of net heat flux for the January curve, as shown in Figure 7–2*a*. Curves for the other months of the critical year are developed in a similar manner.

Thermal Conductivity in the Hypolimnion

Before proceeding with the method of vertical distribution of net heat, consideration must be given to thermal conductivity in the hypolimnion. In a water column of a deep lake or reservoir, if H_z is the total heat that passes through a unit area at any plane at depth z and T_z is the temperature at z, then

$$\frac{dH_z}{dt} = -A_z \frac{dT_z}{dz}, \tag{7–22}$$

where A_z is referred to as the coefficient of eddy conductivity.

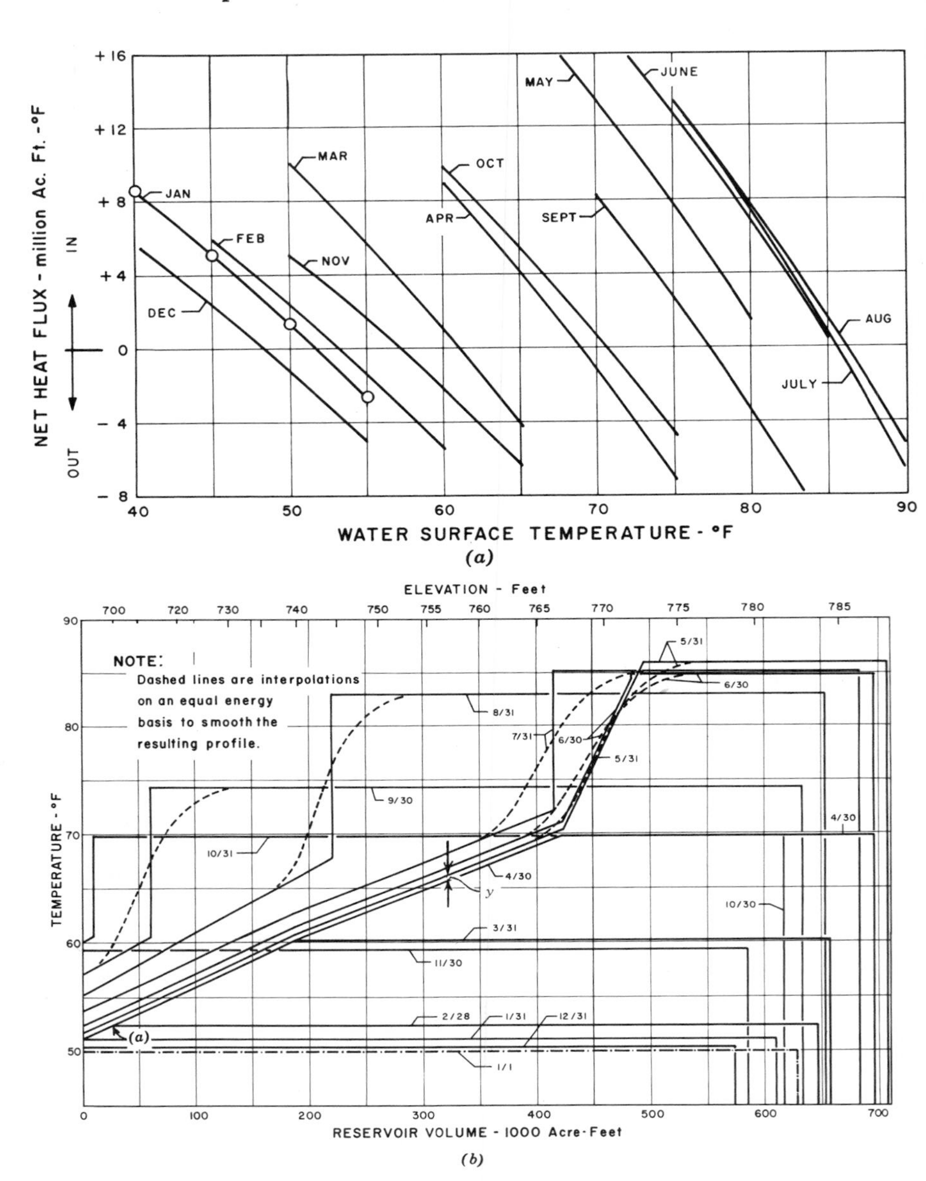

Figure 7–2 Sample diagrams: (*a*) heat flux and (*b*) temperature–volume relationships.

Equation 7–22 may be applied by computing the rate of increase in the heat content of a column of water of unit area below a certain depth over a designated period and dividing by the value of the mean thermal gradient at the depth under consideration. Though the results are highly variable, the procedure is acceptable for practical application and provides suitable guidelines. Values of A_z for large reservoirs in temperate climates computed by equation 7–22 generally fall between 0.05 and 0.15 g/cm-sec. For example, in Reservoir H, referred to later, A_z based on t of 61 days, depth of 110 ft (3350 cm), average temperature gradient of 0.5°F/ft, and average temperature rise of 2°F is computed as follows:

$$A_z = 3350\left(\frac{1}{61}\right) \times 2\left(\frac{1}{0.5}\right) \times 30.48\left(\frac{1}{86400}\right) = 0.078,$$

which is within the range expected for large reservoirs.

Under present knowledge of hypolimnion heating a further simplification is practicable. If it is assumed that reservoir basins are similar and that in a given region the temperature gradients are within the error of estimating the eddy conductivity, a direct shift in temperature may be made on the temperature–volume diagram month by month at different depths, as illustrated in Figure 7–2*b*.

The general procedure is illustrated in application to a proposed reservoir K in a southeastern river basin at latitude 34°54′N, under natural exposure without induced heat loading. Depths of mix and conductivity in the hypolimnion as monthly averages are estimated from observations made at an existing adjacent reservoir H. The reservoir stage and volume for the critical year and the depth of mix are shown in Table 7–3. The resultant computed seasonal surface water temperature and natural vertical gradients are shown in Figures 7–2*b* and 7–12*c*, and are summarized in Table 7–4.

Figure 7–2*b* represents the temperature–volume working diagram of the reservoir on which the heat balance is developed. The energy content of the reservoir expressed in acre-feet-°F is represented by the area under the working diagram for each month. The reservoir volume and corresponding water surface stage for the last day of each month of the design year, as determined from the area–volume–stage diagram, are marked on the volume scale (to which is also affixed the corresponding reservoir stage scale). Measured from the water surface stage at the end of each month, the predetermined depths of mix are marked off.

In determining the heat balance it is necessary to assume a surface-water temperature, and by successive approximations the measured heat

Table 7–3 Reservoir K—Stage, Volume, and Depth of Mix for the Critical Year

Month	Stage Elevation (End of Month) (ft)	Volume (End of Month) (acre-ft)	Mix Depth (ft)
December	782.5	628	Full depth
January	781.2	610	Full depth
February	783.8	646	85
March	784.4	657	45
April	786.9	696	20
May	787.6	708	15
June	786.9	696	15
July	786.1	683	20
August	784.1	653	40
September	782.8	633	70
October	781.8	617	95
November	779.5	585	Full depth
December	778.7	574	Full depth

content represented by the enclosed area of the temperature–volume diagram for the month is balanced with the computed heat gain for that month. The surface temperature in balance at the end of the month becomes the temperature at the beginning of the next month; an assumption is again made for the end of this month and a new balance obtained, and so through the year.

The computed heat added or lost in the reservoir during each month is determined from the monthly average climatology and hydrology for each month of the design year as shown in Table 7–4. Since reservoir K is a deep reservoir, the surface temperature at the beginning of January is likely to be below the equilibrium temperature and is taken as 50°F (column 2); the temperature at the end of January as a first assumption is taken as 50.8°F; the average for January is therefore 50.4°F (column 3). From Figure 7–2*a* the net heat flux for January at 50.4°F is 970×10^3 acre-ft-°F (column 5). Advection from natural drainage is the January tributary runoff 10,680 acre-ft at the mean January surface runoff temperature of 46.5°F, or a gain of 497×10^3 acre-ft-°F (column 6). Other advection (column 7) amounting to a net loss of 1412×10^3 acre-ft-°F is the difference between that received from reservoir J (22,000 acre-ft $\times$ 50.4°F); and that released from reservoir K (50(000 acre-ft $\times$ 50.4°F); the temperature at the withdrawal levels with no stratification in January is 50.4°F, the assumed value.

Table 7–4 Computed Heat Balance for Reservoir K—Critical Year, Natural Exposure

	Surface Temperature (°F)			1000 Acre-ft-°F							
									Measured Heat Content		
Month (1)	Initial (2)	Average (3)	Final (4)	Heat Flux (5)	Natural Drainage (6)	Other Advection (7)	Heat Content Previous Month (8)	Summation Calculated Heat (9)	Above 50°F (10)	Below 50°F (11)	Total (12)
January	50.0	50.4	50.2	970	497	−1412	31,400	31,455	488	30,500	30,988
		50.6	51.2	840	497	−1415	31,450	31,322	732	30,500	31,232
February	51.2	51.8	52.4	1020	3152	−1605	31,232	33,799	1,560	32,300	33,860
March	52.4	56.2	60.0	4520	2123	−1627	33,860	38,876	5,750	32,850	38,600
		56.3	60.2	4400	2123	−1630	33,860	38,753	5,890	32,850	38,740
April	60.2	65.0	69.8	4000	4063	−1871	38,740	44,932	10,040	34,800	44,840
May	69.8	78.0	86.2	4020	2769	−1729	44,840	49,900	14,620	35,400	50,020
		77.9	86.0	4150	2769	−1729	44,840	50,030	14,580	35,400	49,980
June	86.0	85.2	84.4	180	1385	−2095	49,980	49,450	14,170	34,800	48,970
		85.4	84.8	−110	1385	−2095	49,980	49,160	14,260	34,800	49,060
July	84.8	85.2	85.6	400	1974	−2620	49,060	48,814	15,120	34,150	49,270
		85.0	85.2	700	1974	−2620	49,060	49,114	15,020	34,150	49,170
August	85.2	84.2	83.2	2500	1011	−3325	49,170	49,356	16,950	32,650	49,600
		84.1	83.0	2650	1011	−3320	49,170	49,511	16,870	32,650	49,520
September	83.0	78.6	74.2	−1650	1142	−2830	49,520	46,182	14,420	31,650	46,070
		78.7	74.4	−1780	1142	−2830	49,520	46,052	14,305	31,650	45,955
October	74.4	72.1	69.8	−1660	1002	−2235	45,955	43,062	12,180	39,850	43,030
November	69.8	64.6	59.4	−6000	775	−3040	43,030	34,765			34,780
December	59.4	54.9	50.4	−4910	1220	−2140	34,780	28,950			28,980

The total heat content at the beginning of January with reservoir volume at 628×10^3 acre-ft and uniform temperature throughout the depth of 50°F is $31{,}400 \times 10^3$ acre-ft-°F (column 8). The total heat content at the end of January (column 9) is therefore $31{,}455 \times 10^3$ acre-ft-°F, $(970 + 497 + 31{,}400 - 1412)$.

The measured heat content on the temperature–volume diagram to the end of January at the assumed surface temperature of 50.8°F with the reservoir volume at 610×10^3 acre-ft is 488×10^3 acre-ft-°F above 50°F (column 10) and $30{,}500 \times 10^3$ acre-ft-°F below 50°F (column 11) (temperature base below 50°F is arbitrarily set at 0°F). The measured total, $30{,}988 \times 10^3$ acre-ft-°F (column 12) is less than the computed $31{,}455 \times 10^3$ acre-ft-°F (column 9); hence the assumed surface temperature 50.8°F is too low. A second assumption of 51.2°F produces a close balance, which then becomes the initial temperature for the beginning of February.

When the mixing depth at the end of the month is less than the total depth, the heat distribution is considered linear through the month and the temperature–volume diagram is closed as a straight line between the beginning and the end of the month. For example, at the end of February from Table 7–3 with reservoir stage at 783.8 and mixing depth 85 ft the elevation of the mixing depth is 698.8 located at (a) in Figure 7–2b; the temperature–volume diagram is closed from (a) at 2/28 to full depth at 1/31. During the warming season a relatively small amount of heat is added to the hypolimnion by eddy conductivity as a direct shift in temperature increment based on a predetermined temperature rise; the remainder of the heat is distributed in the epilimnion linearly on the temperature–volume diagram. For example, this shift commences in May, is represented by the vertical shift (y) of 0.5°F between 4/30 and 5/31, as shown in Figure 7–2b, and increases through the warm season until mixing extends again to full depth in November. Thus proceeding through the year the surface temperature at the end of December returns to 50.4°F, which is in close agreement with the initial assumed temperature of 50.0°F, indicating that the heat gain and loss through the year is about in balance.

The heat gain (except for the amount assigned to hypolimnion heating) is always added in the epilimnion from the mixing depth of the previous month to the mixing depth of the month in question. In some cases, however, during rapid cooling mixing depth may necessarily be below that specified due to natural convection of the cooler surface waters below this point. Also, in cases where water is directly advected into or out of the reservoir at depth, the thermal profile must be adjusted to correspond with the change in volume at the proper temperature.

In dealing with deep reservoirs used as a cooling device for large steam electric power generating installations the procedure for computing surface water temperatures and vertical gradients is similar to that for natural exposure, taking into account the necessary modifications involved with the induced thermal loading and condenser water intake and discharge. The depth of mix, except as changed by regression of the temperature profile, is assumed initially to be constant, generally 10 to 15 ft. This assumption reflects the strong tendency of the heat discharge to stratify in the epilimnion in the cool season as well as in the warm season, in contrast to the deep mixing that occurs in these seasons under natural exposure. Thermal conductivity in the hypolimnion is insignificant relative to the rapid lowering of the surface layers by the removal of condenser intake water from the deep, cooler waters.

PRACTICAL APPLICATION CONSIDERATIONS

Variation in Meteorologic and Hydrologic Factors

In applying the basic equations to a practical situation judgment in the application of meteorologic and hydrologic factors is at least as important as the adoption of equations and coefficients. Air temperature, wind velocity, water vapor pressure of the atmosphere, solar radiation, and streamflow are dynamic variables. Although these factors vary radically with time, within the natural setting the variation is orderly and can be defined by statistical methods applied to long-term records of observations reported by the U.S. Weather Bureau and by the U.S. Geological Survey at or near the location under investigation (see Chapter 3). As shown later, the long-term records of selected nearby official Weather Bureau stations constitute a guide to local meteorologic variation.

In addition, to relate a water temperature profile in a stream to physical locations below a point of introduced heat load it is essential to define the channel characteristics of width, depth, surface area, channel volume, and time of passage along the course. Each runoff regime produces its corresponding channel characteristics (see Chapter 3). Similarly in reservoirs stage–area–volume relations must be established.

The statistical nature of meteorologic change induces a dynamic rather than an equilibrium state. Water lags behind meteorologic change, but constantly the shift is toward equilibrium temperature. Stable conditions for a month would result in temperatures closely approaching equilibrium. In shallow streams and ponds deviations from the general seasonal temperature trends can be expected with diurnal cycles; in large bodies of water these deviations are less noticeable. A sustained meteorologic condition for a period of a week may induce a 3 to 5°F temporary deviation from the general seasonal trend. In large, deep reservoirs where stratification takes place wind action induces mixing to varying depths—depending on intensity, direction, and duration—which can cause temporary deviations in normal temperature patterns. Conductivity and circulation due to density differences between warmer upper and cooler lower layers result in deviations and lags from normal. Hence interpretation of observed or computed water temperatures should be made in the light of the actual dynamics of the situation with due regard to possibilities of short-term deviations. Since many of the refinements of theory fall well within the frame of meteorologic and hydrologic variation, theoretical calculations must be tempered by practical considerations; the prudent analyst is deliberate in his selection of a probability basis, for example, occurrence expected once in 5, 10, or 20 years.

Expected Seasonal Pattern of Natural Equilibrium Water Temperature

Since there are few long-term records of observed water temperature on which probability studies can be made directly, it is usually necessary to develop the seasonal pattern from computed equilibrium values employing equation 7–14. Ideally equilibrium water temperature should be computed for each month over the period of years for which simultaneous records of air temperature, wind velocity, vapor pressure, and solar radiation are available. However, the short-term records of solar radiation that are available for most areas limit this approach unless it is feasible to generate an extension synthetically. Not all situations, of course, warrant the time and labor involved in computing equilibrium water temperatures on a monthly basis.

Short and Long Methods of Computation

A review of the observed patterns of occurrence of meteorologic factors suggests a compromise short approach in developing the expected seasonal pattern of water temperature, retaining the concepts of probability. The most probable equilibrium water temperature expected for each month of the year is computed from the most probable values of air temperature, wind velocity, vapor pressure, and solar radiation determined from the respective seasonal patterns. The range within which equilibrium water temperature is expected to fall for various degrees of confidence (e.g., for 80 percent of the years) is computed for the upper limit on the combination of once-in-10-years high air temperature, vapor pressure, and solar radiation, with the once-in-10-years low wind velocity; and for the lower limit from the combination of once-in-10-years low air temperature, vapor pressure, and solar radiation, with once-in-10-years high wind velocity.

Figure 7–3 shows the seasonal pattern of equilibrium water temperature as monthly average based on the compromise approach, applied to the Tittabawassee River in the vicinity of Midland, Michigan. Curve *A*, the most probable, may be taken as representative of natural seasonal water temperature expected normally, and the shaded belt is the 80-percent confidence range. As a verification of this approach, individual monthly equilibrium water temperature was computed for the 14-year

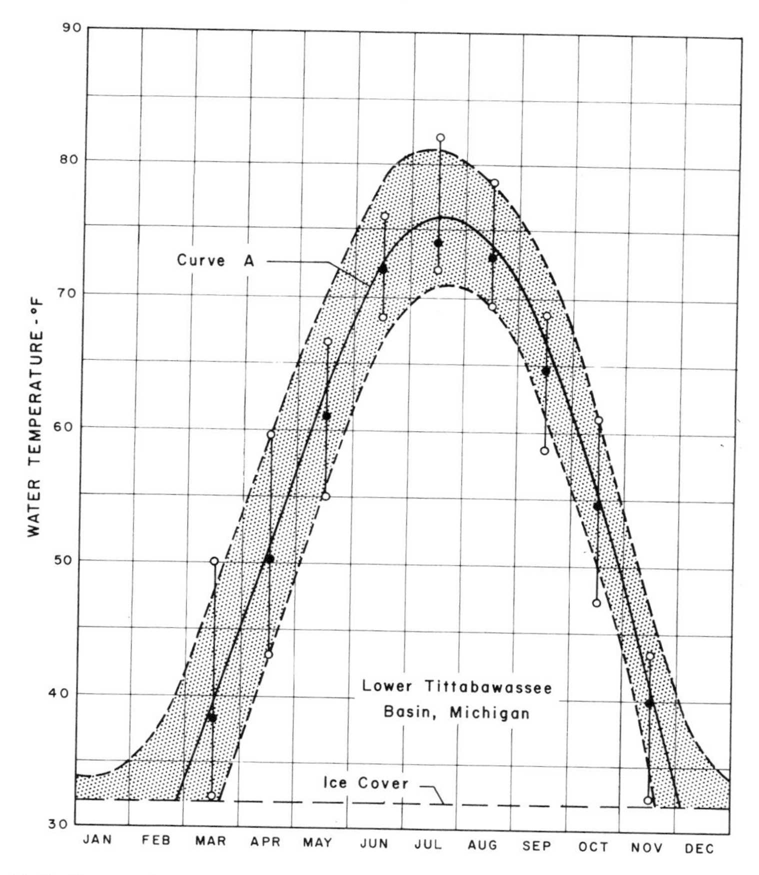

Figure 7–3 Seasonal pattern of equilibrium water temperature. Comparison of results of short and long methods of computation.

period for which simultaneous meteorologic observations were available at the East Lansing, Michigan, Weather Bureau station. The results of these computations are also shown in Figure 7–3, summarized for each month of the year, the mean indicated by solid circles and the maximum and minimum by open circles. There is close agreement between the 14-year monthly means and the most probable (curve *A*); in addition the minimum and maximum computed monthly values of the 14-year period fall reasonably within the 80-percent confidence range. This close agreement warrants application of the short compromise approach in developing expected seasonal patterns of equilibrium water temperature, particularly where only short-term records of solar radiation are available.

Verification of Use of Nearby Weather Bureau Records

In connection with studies for a wastewater storage and cooling pond, opportunity was afforded to verify the employment of nearby long-term Weather Bureau records with records obtained at the site after the pond was placed in operation. Figure 7–4 shows the results of the statistical

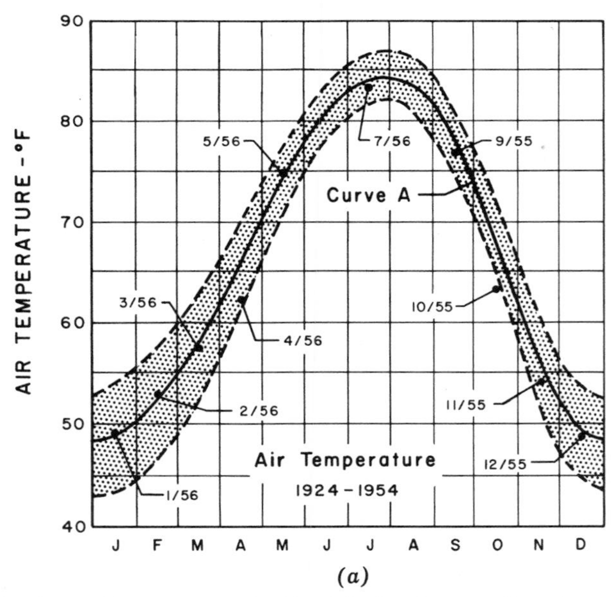

Figure 7–4 Seasonal patterns of climatologic factors (based on Shreveport, Louisiana, records): (*a*) air temperature, (*b*) water vapor pressure, (*c*) wind velocity. Curves *A*: most probable monthly averages; shaded areas: range within which monthly averages can be expected for 80 percent of years; ● 7/56: observed monthly average at the pond site.

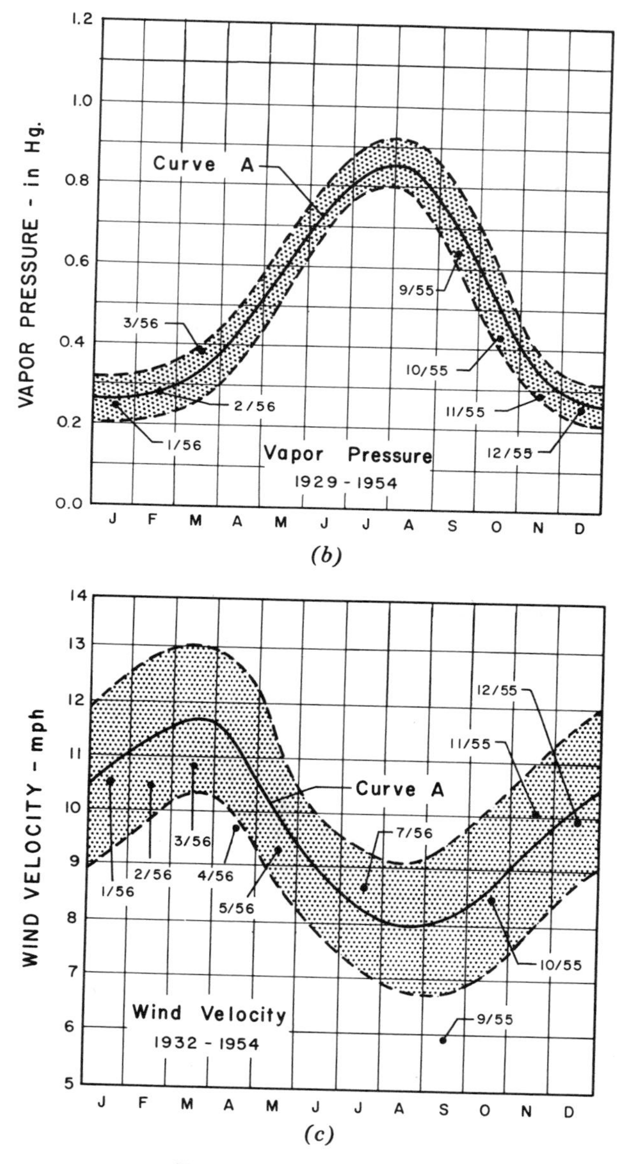

(*b*)

(*c*)

Figure 7–4 (**Continued**)

analyses of the long-term monthly average meteorological records at the Shreveport, Louisiana, U.S. Weather Bureau station, located approximately 40 miles from the cooling pond site. Figures 7–4*a*, *b*, and *c* show the expected seasonal patterns of air temperature, water vapor pressure, and wind velocity, respectively. Curves *A* show the most probable monthly average values, and the shaded bands show the 80-percent confidence ranges. The monthly average values of air temperature, water vapor pressure, and wind velocity observed at the weather station located at the cooling pond are also indicated in Figures 7–4*a*, *b*, and *c*. It will be noted that, with the exception of the September 1955 wind velocity, the observed data at the pond site fit well within the expected seasonal pattern based on the Shreveport readings. With due regard to topographic conditions, employment of long-term meteorologic data at nearby official Weather Bureau stations is a responsible procedure. The long-term records afford a basis for determining the normal patterns and the variations from normal expected for any desired probability. Further verification is also afforded in the comparison of computed and observed equilibrium water temperatures and temperature profiles induced by waste heat load.

Deviation Induced by Unusual Hydrology

In some river basins the water temperature deviates from the equilibrium temperature one would expect under the prevailing meteorology. This deviation is primarily associated with an unusual hydrologic setting of the basin where stream runoff is fed by melting snow accumulated in mountain regions or by contributions from large springs and extensive groundwater sources. A dramatic illustration of such variation from an expected norm is that found in the Deschutes River basin in Oregon, as shown in Figure 7–5. Curve *A* is the most probable monthly average seasonal pattern of equilibrium water temperature, and the shaded band is the 90-percent confidence range expected from meteorologic factors. The actual monthly average river water temperatures recorded at the Madras gage for the period March 1952 to September 1952 are shown as curve *B* in Figure 7–5 and are projected as a normal seasonal pattern. The general stability of the river water temperature is reflected by curve *B* and can be compared with the off-stream equilibrium temperature as reflected by curve *A*. The river water remains distinctly below the equilibrium temperature during the summer peak, with a normal differential of 9°F; during the winter the river water remains above the equilibrium temperature, with a normal differential of 10°F. The western boundary of the Deschutes River basin is formed by the Cascade Mountains, with Mount Hood as the northern terminus. Melting snows and

Curve A - Most probable seasonal equilibrium water temperature.

Range within which monthly average equilibrium water temperature can be expected for 90 percent of the years.

Curve B - Projected normal seasonal river water temperature.

● — Observed monthly average water temperature - Deschutes River at Madras gage, March 1952 - September 1952.

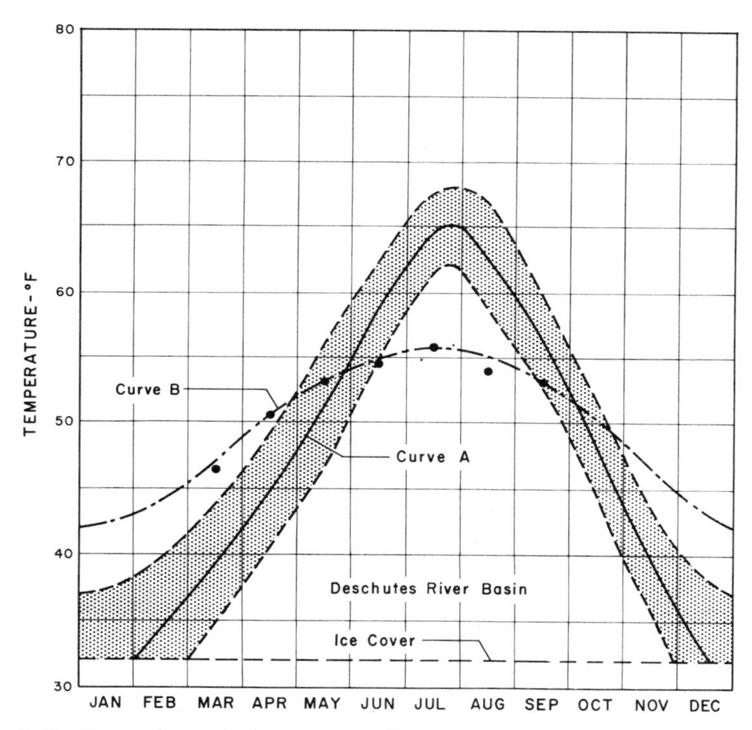

Figure 7–5 Comparison of off-stream equilibrium water temperature and observed river water temperature. Curve *A:* most probable seasonal equilibrium water temperature; curve *B*: projected normal seasonal river water temperature; shaded area: range within which monthly average equilibrium water temperature can be expected for 90 percent of years; ●: observed monthly average water temperature for the Deschutes River at the Madras gage March 1952–September 1952.

chilled groundwater feed characterize the Deschutes as an unusually cold stream.

Other upland streams, fed by cool groundwater, show similar patterns of deviation from expected off-stream equilibrium water temperatures, but usually less dramatically than does the Deschutes. Where dense vegetative cover in the drainage basin shades small streams and tributaries from direct solar radiation there is also a pattern of lowered seasonal peak water temperatures. These differentials in temperature are usually dissipated rather quickly after the cold tributary reaches a larger exposed main stream, and the water temperature tends again to approach equilibrium conditions.

Computed and Observed Induced Temperatures in Ponds and Streams

As a verification of the basic equations and of the methods of practical application, comparisons of computed with observed conditions for a cooling pond and for a river under waste heat loading are presented. Both situations involve relatively shallow water depths with isothermal conditions; deep water reservoirs involving temperature stratification require special consideration and are dealt with later.

A Cooling Pond

Following operation of the wastewater cooling pond near Shreveport, Louisiana, temperatures of the return water for reuse were observed daily at the return intake, affording a direct comparison with the previously computed temperatures based on the heat loading and the Shreveport Weather Bureau meteorological records. Figure 7–6*a* is the computed seasonal pattern of equilibrium water temperature for the pond without heat loading, which represents the maximum cooling possible.

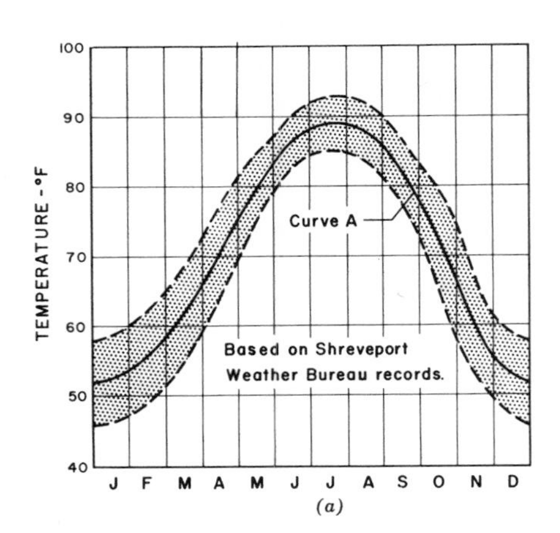

Figure 7–6 Cooling pond water temperatures: (*a*) expected seasonal pattern of equilibrium temperature; (*b*) expected temperature profile through pond; (*c*) comparison of computed and observed monthly average temperatures at outlet. Curves *A*: most probable monthly averages; shaded areas: range within which monthly averages can be expected for 80 percent of years.

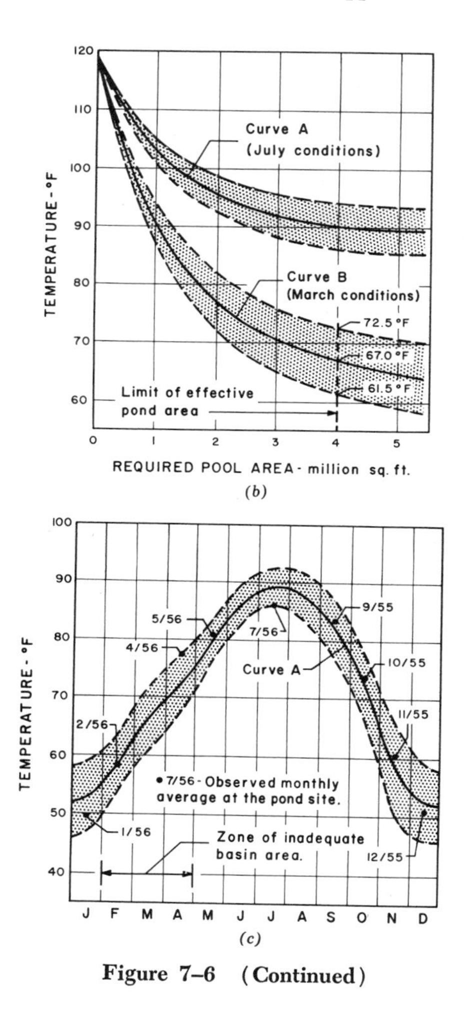

Figure 7–6 (Continued)

Figure 7–6*b* shows the computed temperature profiles expected through the cooling pond from the inlet of the hot wastewater to the outlet for the return. The wastewater inflow averages 25,000 gpm, or 23.2 cfs, at 120°F. For the month of July, the height of the warm weather season, curve *A* represents the most probable temperature profile; the shaded band about curve *A* represents the 80-percent confidence range. During this season, with the stage of the lagoon high, excess surface area is available and water temperature at the outlet approaches equilibrium, which for the most probable meteorologic conditions is 89°F and for the 80-percent range is 85 to 93°F.

Curve B represents the most probable temperature profile and the 80-percent range for cool weather conditions in March, a season when the wastewater pond is at low stage and surface area is limited. For most probable March meteorologic conditions the cooling limit is to 67°F. that is, 5°F above equilibrium level. The limiting cooling temperature range expected for 80 percent of the years is 61.5 to 72.5°F, 5.5 and 3.5°F above the equilibrium level, respectively. From such computed temperature profiles Figure 7–6c was constructed; it shows the expected seasonal pattern of water temperature of the return wastewater to be reused as condenser water. Curve A is the most probable temperature profile, and the shaded band is the 80-percent confidence range. This graph also shows the monthly average observed water temperatures of the return water recorded after the pond was placed in operation, which, it will be noted, are in close agreement with the computed temperatures.

A River Temperature Profile

For a verification of a computed river temperature profile it is essential that meteorologic and hydrologic conditions remain relatively stable for a reasonable period before and during the time of the river temperature survey. Likewise the introduced heat load inducing the temperature profile must be known and remain relatively stable for the same period. Such conditions were met for the survey on the Tittabawassee River, Michigan, during September 9–18, 1953. The prevailing meteorologic conditions for this period as reflected at the nearby East Lansing Weather Bureau station were as follows:

Air temperature T_a	61.5°F
Wind velocity W	10.3 mph
Water vapor pressure V_a	0.387 in. Hg
Solar radiation H_s	50.8 Btu/ft^2-hr

From these data the equilibrium water temperature above the introduced heat load, by equation 7–14, is computed as 63.6°F. The measured stream runoff (corrected for diversions) at the point of heat load averaged 394 cfs, with only minor increments added along the course. Channel characteristics for this runoff were available from detailed cross-section soundings at 500-ft intervals along the course. On the basis of these conditions and a prevailing heat load of 1.312 billion Btu/hr, the river-temperature profile computed by equation 7–21 is shown as curve A of Figure 7–7. It will be noted from curve A that the heat load induces a temperature rise of 14.8°F, from 63.6 to 78.4°F, which is followed by a rapid heat dissipation, with temperature declining to 72.5°F at mile 19.24 and to 64.4°F at the river mouth, again approaching equilib-

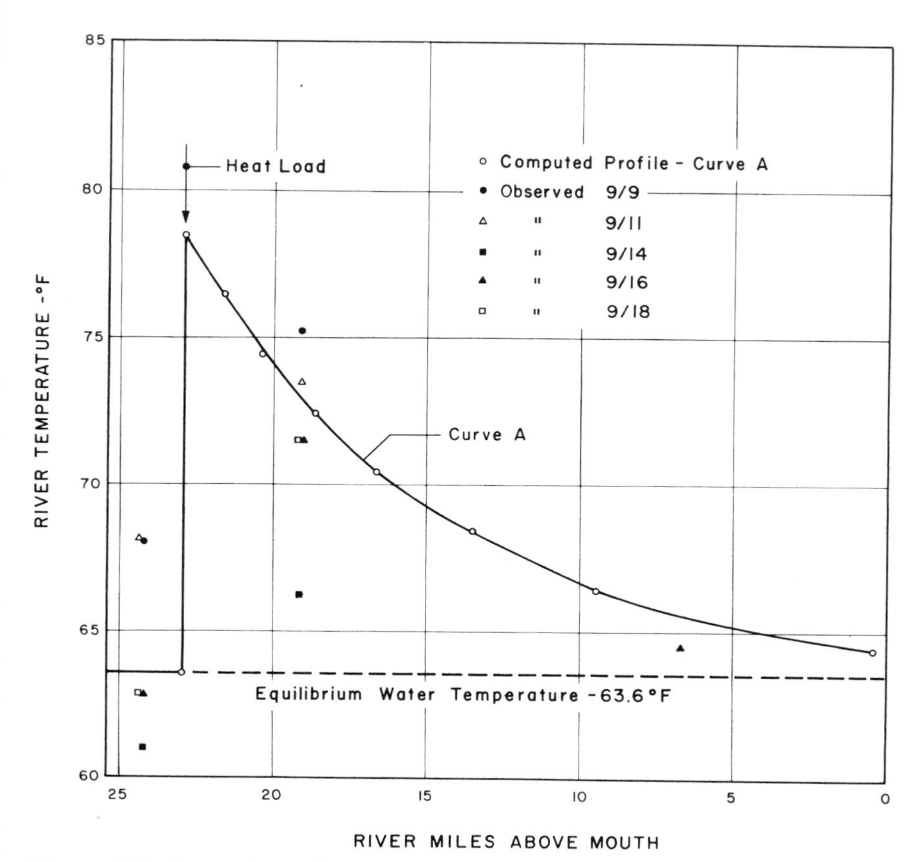

Figure 7–7 Comparison of computed and observed river temperature profiles—Tittabawassee River.

rium within 0.8°F. It will also be noted that the limited observed temperature measurements are in reasonable agreement with the computed temperature profile along the lower reach; likewise the observed river water temperatures at mile 24.2 above the point of heat load are in reasonable agreement with the computed equilibrium water temperature, 63.6°F.

Recognizing the sensitivity and dynamic character of the temperature profile, these verifications warrant employment of the basic equations in forecasting river water temperature profiles expected under other natural or regulated streamflows and heat loadings.

Example Computation of River Temperature Profile. The following computations are the basis for Figure 7–7 and illustrate the procedure in the development of the river temperature profile:

Runoff	394 cfs
Waste-heat load	1.312×10^9 Btu/hr
Equilibrium temperature (E)	63.6°F
V_E at 63.6°F	0.587 in. Hg
$T_{w_0} - T_E = \dfrac{1.312 \times 10^9}{394 \times 3600 \times 62.4}$	14.8°F
T_{w_0} (63.6 + 14.8)	78.4°F
T_a	61.5°F
W	10.3 mph
V_a	0.387 in. Hg
H_s	50.8 Btu/ft²-hr
C_1	14
C_2	32

By equation 7–21:

Since $T_{w_0} - T_E = 14.8$°F, consider seven increments of ΔT_w of 2°F, or a decline of 14°F.

	T_w (°F)	ΔT_w (°F)	T_w (°F) (average)	V_w (in. Hg) (average)
T_{w_0}	78.4			
		2	77.4	0.939
T_{w_1}	76.4			
		2	75.4	0.878
T_{w_2}	74.4			
		2	73.4	0.821
T_{w_3}	72.4			
		2	71.4	0.767
T_{w_4}	70.4			
		2	69.4	0.717
T_{w_5}	68.4			
		2	67.4	0.670
T_{w_6}	66.4			
		2	65.4	0.625
T_{w_7}	64.4			

$$\alpha = 0.00722 \times 1059.1 \times 14(1 + 0.1 \times 10.3) = 217.2,$$

$$\beta = [1.8 + (0.16 \times 10.3)] = 3.45,$$

$$A\,(\text{ft}^2) = \frac{224{,}640 \times 2 \times 394}{\alpha(V_w - V_E) + \beta(T_w - E)} = \frac{177 \times 10^6}{217.2(V_w - V_E) + 3.45(T_w - E)}.$$

	Required Incremental Area ($ft^2 \times 10^6$)	Σ Area ($ft^2 \times 10^6$)
T_{w_0} to T_{w_1}: $A_1 = \dfrac{177 \times 10^6}{217.2(0.939 - 0.587) + 3.45(77.4 - 63.6)} =$	1.428	1.428
T_{w_1} to T_{w_2}: $A_2 = \dfrac{177 \times 10^6}{217.2(0.878 - 0.587) + 3.45(75.4 - 63.6)} =$	1.704	3.132
T_{w_2} to T_{w_3}: $A_3 = \dfrac{177 \times 10^6}{217.2(0.821 - 0.587) + 3.45(73.4 - 63.6)} =$	2.092	5.224
T_{w_3} to T_{w_4}: $A_4 = \dfrac{177 \times 10^6}{217.2(0.767 - 0.587) + 3.45(71.4 - 63.6)} =$	2.682	7.906
T_{w_4} to T_{w_5}: $A_5 = \dfrac{177 \times 10^6}{217.2(0.717 - 0.587) + 3.45(69.4 - 63.6)} =$	3.672	11.578
T_{w_5} to T_{w_6}: $A_6 = \dfrac{177 \times 10^6}{217.2(0.670 - 0.587) + 3.45(67.4 - 63.6)} =$	5.691	17.269
T_{w_6} to T_{w_7}: $A_7 = \dfrac{177 \times 10^6}{217.2(0.625 - 0.587) + 3.45(65.4 - 63.6)} =$	12.292	29.561

The river surface area along the course is computed from the channel cross-sections and plotted as cumulative area against river milage from the waste heat outlet to the river mouth. The corresponding coordinates for the river water temperature profile (based on the computed surface area required) are interpolated from the cumulative plot as follows:

Required River Surface Area ($ft^2 \times 10^6$)	River Mile above Mouth	Computed River Water Temperature (°F)
0 (waste heat outlet)	23.0	78.4
1.428	21.7	76.4
3.132	20.4	74.4
5.224	18.7	72.4
7.906	16.7	70.4
11.578	13.6	68.4
17.269	9.5	66.4
29.561	0.4	64.4

If Figure 7–7 is replotted on semilog paper substituting the residual temperature differentials above equilibrium as the ordinate (log scale) against the required surface areas as the abscissa scale, the curve approximates a straight line. The slope of this line defines the heat loss coefficient. The slope for the total decline, 14.8°F at mile 23 to 0.8°F at mile 0.4, for the total required surface area of 29.561×10^6 ft² is

$$\frac{\Delta Y'}{\Delta X} = \frac{\log 14.8 - \log 0.8}{29.561 \times 10^6},$$

from which U is derived (equation 7–12) as

$$U = \frac{394}{0.0102 \times 247 \times 22.6} (\log 14.8 - \log 0.8) = 8.75 \text{ Btu/hr-ft}^2\text{-°F}.$$

This is in agreement with values reported by LeBosquet [6] and others previously referred to.

Waste Heat Discharge Distribution

Modifying Factors in Rivers

As with all types of pollution, the distribution of waste heat discharge into the streamflow is seldom uniform throughout the channel cross section. The completeness of intermixing depends on the design of the outlet, the channel configuration, width, and depth, and the hydraulic characteristics of velocity, turbulence, and relative discharge rates. The tendency toward unequal distribution with waste heat is accentuated because the heated condenser water has a lower density than the cooler streamflow. Hence it is necessary to evaluate the expected distribution pattern for specific locations. Where markedly unequal distribution occurs, adjustments in parameters of the heat dissipation equations may be necessary. In some streams the waste heat distribution is nearly complete throughout the channel cross-section, vertically and laterally, in a relatively short distance below the discharge outlet works; in other locations hot discharge may be channeled along the bank of the outlet, intermixing very little with the streamflow through extensive reaches along the course; in still other situations lateral distribution may be quite uniform, whereas vertical stratification is pronounced, warmer water overlying the colder stream runoff.

The critical period is the low runoff warm weather season, because then maximal temperature superelevation occurs. During cooler seasons of higher runoff inequality of distribution is likely to be more pronounced, but then it is of less consequence. During the drought period all or most of the runoff is used as circulating water, stream velocity is lower, depth and width are less, and channeling opportunity is less.

In very wide streams the degree of channeling depends to some extent on the design of the waste heat discharge outlet. High velocity discharge, guide vanes, multiple outlets, and appropriate placement in the natural channel configuration are aids to lateral distribution. If the temperature superelevation of the circulating water is moderate, channeling may actually be an advantage, providing natural streamflow along the opposite shore as a bypass for fish and aquatic organisms. If such channeling is reasonably well defined, it can be dealt with as a stream within a stream. Similarly, if good lateral distribution is achieved but vertical stratification occurs in a deep stream, the higher initial surface temperature superelevation accelerates heat dissipation, and lower water temperature results at a downstream reach after vertical stratification has been broken up.

Modifying Factors in Ponds

The formula (equation 7–21) that describes the decrease of temperature with increasing area is basically applicable to a cooling pond of the flowthrough type, ideally a long pond with a clearly identified flow pattern moving from the outlet of the condenser water discharge to a distant intake point. Under these conditions the pond behaves essentially like a river; the factors governing lateral and vertical mixing in the pond can be taken as for a flowing river. Since shallow ponds are particularly sensitive to wind action, however, unpredictable currents and flow patterns may occur. These irregularities may be minimized by guiding the flow by means of appropriate structures, such as dikes, or by subdividing the impoundment into individual parts—a primary hot pond receiving the condenser discharge, followed by a secondary pond leading to the thermal plant condenser water intake. Water from the hot pond is discharged to the cooler part by means of an inverted weir, thus ensuring that the cooling takes place in a well-defined way. Another type, which might be called the completely mixed pond, is one in which the temperature drop takes place in a relatively short distance. Wind action and internal density currents keep this type fairly well mixed, producing a more uniform temperature over the entire pond. In such a pond the temperature will be at a level above equilibrium such that the net heat addition from the power plant approximately equals the net heat transfer to the atmosphere.

If one compares both types of ponds on the basis of the intake water temperature, it can be shown that to tolerate the same thermal intake the mixed pond would require a larger surface area. This is because a flowthrough pond dissipates more heat in the first few acres than the mixed pond does.

Most cooling ponds will lie somewhere between these two types unless special precautions are taken to ensure a streamlike flow through the pond.

Modifying Factors in Reservoirs

Although the distribution pattern of waste heat from thermal plants is usually two-dimensional, lateral and longitudinal, in deep reservoirs the vertical dimension may be involved. The basic mechanism for predicting the temperature distribution and vertical gradients as outlined above provides the *average* condition expected. There are a number of modifying factors that potentially may induce deviations from the normal average. These are the topography and shape of the reservoir, wind action, the influent–effluent system, internal density currents, and thermal power plant arrangements.

Theoretical models for predicting waste heat distribution from thermal plants located on reservoirs are generally more reliable for reservoirs of regular shape and uniform depth. Reservoirs that consist of many arms and branches of unequal depth present greater uncertainty, since some portions may be isolated and virtually ineffective for cooling purposes, particularly during periods of severe drawdown in reservoir stage.

Wind action over the surface of a reservoir exerts a shearing stress at the air–water interface that induces surface currents on the order of 1 to 3 percent of the wind velocity. Heavy wave action also increases the depth of vertical mixing and exchange between the epilimnion and the hypolimnion. A steady wind along a major axis of a reservoir produces a slight elevation of the water at one end and a lowering at the other end. This differential in elevation induces currents that with time will come to equilibrium and result in a returning underflow. In a stratified reservoir underflow may cause a temporary depression of the thermocline at one end and an elevation at the other. When the wind subsides, the reservoir and the vertical temperature gradients will gradually return to normal in a series of oscillating movements. If such a wind occurs on a reservoir where the intake to the thermal plant is not sufficiently deep, a considerable variation in intake water temperature may result. Depending on the relative location of the thermal plant condenser water intake and discharge—and the direction, duration, and intensity of the wind—short-circuiting between heated condenser discharge and the intake may occur. Modifications associated with wind action are usually of short duration; the reservoir reverts quite quickly to the normal seasonal temperature distribution patterns.

If large amounts of water are added to a reservoir at levels that do not have a corresponding temperature or turbidity, strong internal density currents will occur and will alter the normal stratification patterns. In hydro reservoirs likewise analysis must take account of the turbulent energy associated with sudden large withdrawals or additions that may induce strong oscillations and, depending on magnitude relative to reservoir capacity and the location of the advections, tend to break up normal stratification. In Chapter 11 it is shown that the design of pumped storage facilities, whether for flow augmentation or for peak power demands, can be so arranged as to ensure complete mixing; such facilities can therefore be either beneficial or detrimental to heat dissipation, depending on how they are arranged.

In large, deep reservoirs used for heat dissipation from thermal plants the objective is to withdraw cool condenser circulating water from the hypolimnion and discharge the heated water to the epilimnion. The success of such a two-layer system in avoiding short-circuiting of heated condenser discharge water to the plant intake depends on many hydraulic factors. Where intake and discharge are in close proximity and reservoir depth is limited, skimmer walls are installed to prevent drawdown of the heated surface water. Basically two-layer stratification is expected by virtue of the difference in density of the cooler intake water and the heated discharge water. If some short-circuiting does occur, the condenser discharge temperature cannot increase indefinitely as increasing density differential ultimately brings about separation.

The predominant design features and local factors that influence potential short-circuiting are the temperature differential across the condensers, the depth of the intake, the condenser circulating water flow rate, the intake velocity, and importantly the geometry of the local reservoir in the approaches to the intake. The latter determines the restriction of the approach flow pattern, whether confined to a narrow two-dimensional approach or a completely unrestricted three-dimensional 360° radial pattern. Usually some degree of restriction of approach flow pattern between two- and three-dimensional idealized forms is inherent in the local geometry. Similarly the vertical temperature gradient in a deep reservoir in the vicinity of the condenser water intake and discharge and intervening channels is influenced by local geometry, reservoir stage, and operating patterns of associated hydro facilities. The temperature gradient usually lies between a completely stratified two-layer system and a continuous gradient from surface to bottom. This complexity precludes precise prediction of the conditions under which short-circuiting will or will not occur. Some theoretical approaches and laboratory model studies developed for idealized conditions afford a general guide.

Debler's Formulation. Debler's formulation [11] is based on a two-dimensional flow pattern and a continuous temperature gradient. His experimental work has shown that when the Froude number is less than

0.28 the flow pattern to a bottom line sink is divided into two regions, an essentially stagnant upper region and an active lower zone that contributes the entire flow. At Froude numbers greater than 0.28 the upper layer increasingly contributes until the entire depth is involved in active flow. The Froude number as used by Debler is derived from

$$F = \frac{q}{d^2} \sqrt{h\rho_0/g\Delta\rho},$$

where F is the densimetric Froude number, d is the depth of the flowing fluid, h is the entire depth of fluid, ρ_0 is the density at bottom of tank, $\Delta\rho$ is the density difference between surface and bottom, g is the acceleration of gravity, and q is the flow per unit of width.

This formula can be used to predict the thickness of a withdrawal layer from any reservoir that approaches the idealized criteria of geometry and depth essential to a two-dimensional flow pattern and which has the inlet at the bottom. Rearranged, the equation can be written as

$$d^4 = \frac{q^2 h}{(0.28)^2 g'},$$

where $g' = g(\Delta\rho/\rho_0)$.

Harleman's Formulation. Harleman's formulation [12] is based on a three-dimensional flow pattern into a point sink and a two-layer stratified fluid. The maximum withdrawal depth from the lower layer without simultaneous withdrawal from the upper layer is determined for a 360° radial approach by the equation

$$d^5 = \frac{16Q^2}{4.2\pi^2 g'},$$

where Q is the total flow into the discharge pipe. For flows approaching from an angle of, say, 90° the flow Q must be multiplied by a factor of 4; for flows approaching from an angle of 60° Q must be multiplied by 6.

The Debler [11] and Harleman [12] theoretical formulations are developed for two opposite idealized extremes of flow pattern and temperature gradients; they therefore, in a sense, define a range within which to expect actual conditions. However, local conditions and reservoir geometry are difficult to relate to rigid mathematical formulations—the theoretical predictions must be tempered with the professional judgment that comes with practical experience.

Natural and Induced Temperatures in Deep Reservoirs

The determination of the effect of thermal discharge in a deep reservoir remains an inexact science. In addition to natural climatologic variables there is a complex of other modifying factors that are less precisely measurable. Although a body of scientific knowledge is available to evaluate idealized situations, it must be recognized that in evaluating any individual practical situation it is necessary to make certain judgments and assumptions about the influence of a number of these less precisely measured factors. At the present state of the science and art no single direct solution can be made; a number of computations must be made under rational assumptions to define a framework of resultant conditions within which judgments can be made.

An example will demonstrate the role of judgment in the face of complexities. The data here are extracted from a study made for a proposed large thermal electric power development to be located on a deep reservoir in a southeastern river basin, and to be operated in conjunction with pumped storage and conventional hydroelectric power facilities in a chain of reservoirs.

A Proposed Development

Figure 7–8 shows the proposed developments, which include reservoir K, on which is located a 3000-MW nuclear fuel steam electric power generating station, and a conventional 140-MW hydroelectric power station; to the north, reservoir J, on which is located a 610-MW pumped storage facility utilizing reservoir J as the storage pool and reservoir K as the lower pool; reservoir H is an existing development with a conventional hydroelectric power plant at the lower end. Reservoir K at full pool has 911,000 acre-ft with 18,120 acres of water surface area. Figure 7–9 is the area–volume relation, and Figure 7–10 is the reservoir operational pattern under the prescribed rules and natural hydrology and meteorology for the period 1951–1960.

Computed Equilibrium Water Temperatures. Equilibrium water temperatures are computed by equation 7–14 (using the coefficients appropriate for deep reservoirs) month by month, based on the historical meteorologic records of air temperature, wind velocity, and vapor pressure observed at the nearby Weather Bureau station and the corresponding monthly average solar radiation recorded at the nearest available station, Atlanta, Georgia. These results are summarized in Table 7–5.

Since not all months of a given year will experience conditions of the same probability of occurrence, for the selection of a sequence representative of normal conditions and of a critical condition such as may

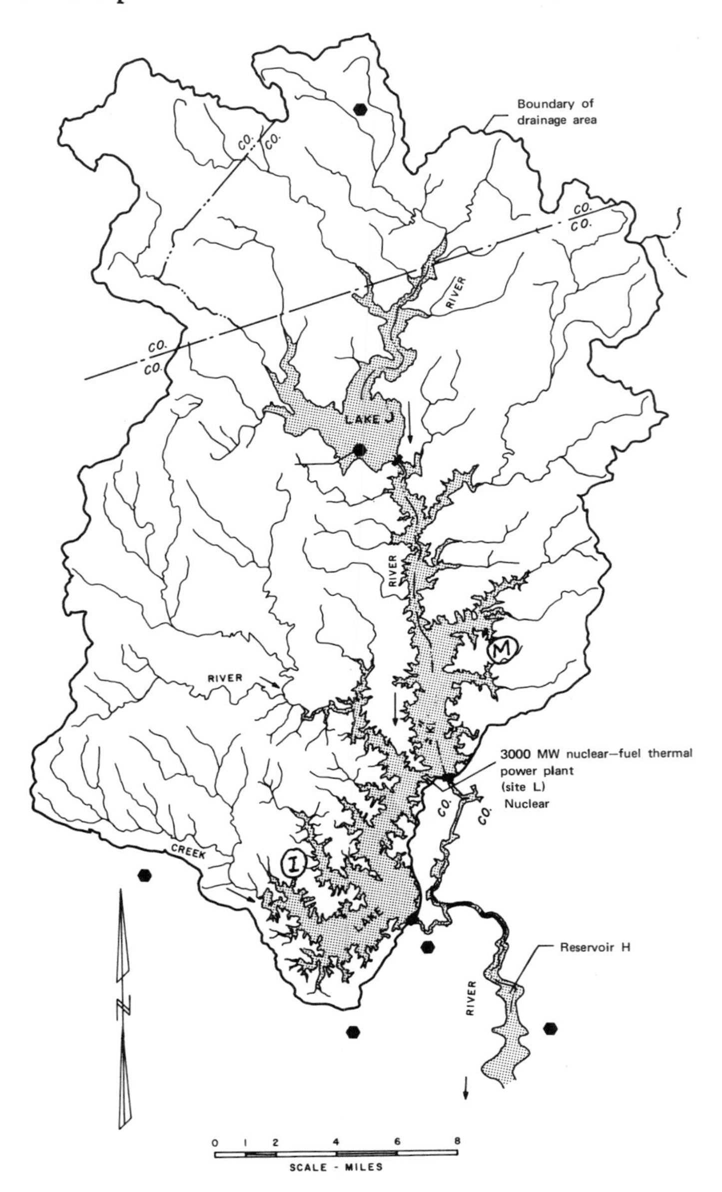

Figure 7–8a Proposed large thermal electric power development—vicinity map.

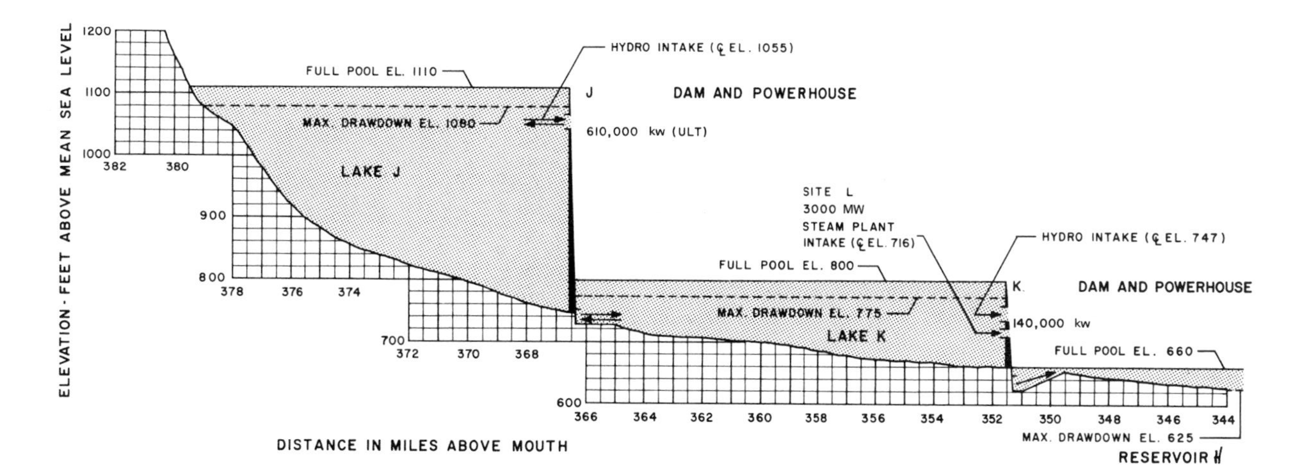

Figure 7–8b Proposed large thermal electric power development—cross section.

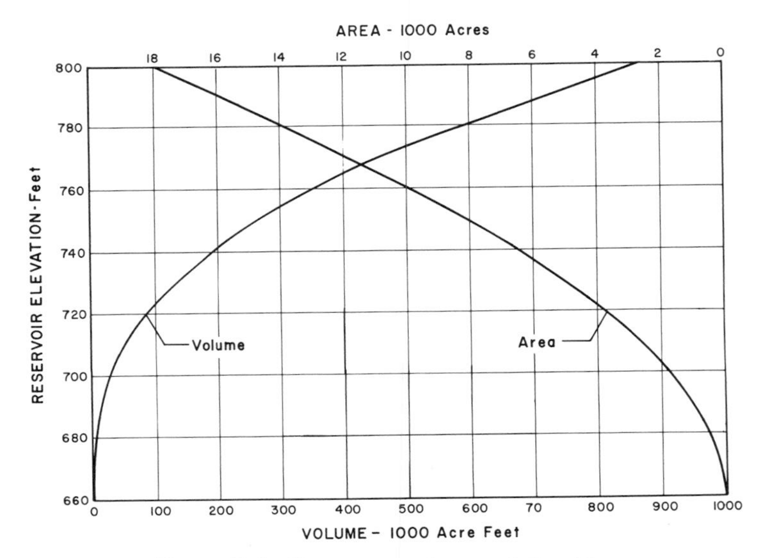

Figure 7–9 Stage–area–volume relationships.

be expected on the average once in 20 years, statistical analyses of the average annual equilibrium water temperatures are made of the values as listed in Table 7–5. The average annual values describe a quite normal distribution, as shown in Figure 7–11.

The year that most nearly approximates the most probable annual

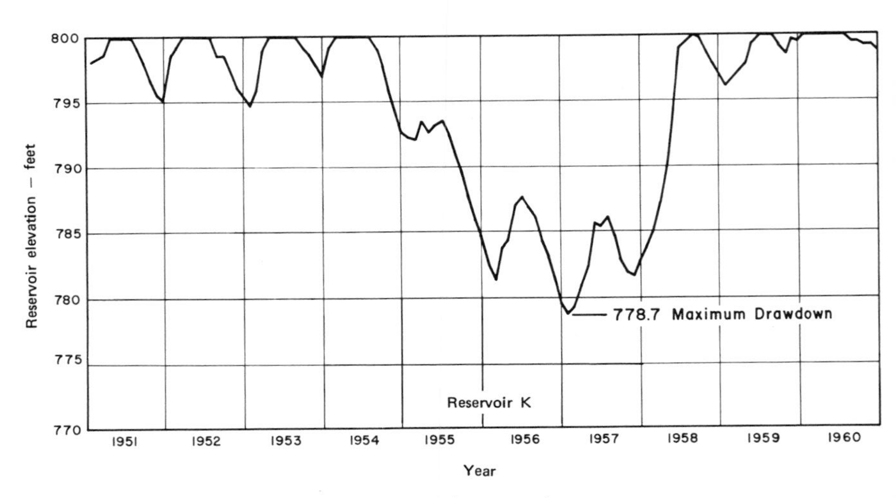

Figure 7–10 Reservoir operation pattern.

Table 7–5 Computed Monthly Average Equilibrium Water Temperatures

Year	January	February	March	April	May	June	July	August	September	October	November	December	Average Annual
1965	48.8	51.0	55.4	70.4	82.4	79.8							
1964	45.7	47.1	59.5	68.1	77.6	85.3	83.8	83.3	78.1	64.5	60.0	49.3	66.86
1963	45.4	47.8	63.4	69.3	77.4	81.8	85.8	83.6	75.9	70.5	56.9	42.8	66.71
1962	45.5	55.5	56.9	(a)	83.6	85.8	87.9	86.3	78.4	(a)	55.0	46.0	68.02
1961	45.6	52.2	60.3	63.9	73.6	80.3	86.0	83.5	80.3	68.1	59.6	(a)	66.69
1960	47.4	49.3	48.4	67.9	73.4	82.8	84.6	85.6	78.6	70.4	57.6	44.6	65.88
1959	(a)	50.5	56.8	68.5	78.5	82.5	85.1	86.0	77.8	66.9	55.4	(a)	66.85
1958	42.9	46.2	54.8	65.8	77.7	82.0	86.2	84.6	76.5	65.7	58.5	45.7	65.55
1957	48.0	55.3	57.1	71.3	77.5	83.8	85.0	82.9	76.5	62.1	56.1	48.6	67.02
1956	44.5	52.0	57.7	65.4	75.6	80.9	(a)	(a)	(a)	65.8	54.7	56.6	66.72
1955	46.7	51.3	59.1	70.2	76.1	77.7	84.1	84.2	75.4	65.1	55.0	44.0	65.74
1954	49.6	54.5	60.2	72.9	73.8	83.9	85.5	84.7	79.1	67.9	53.0	45.8	67.57
1953	51.7	53.2	60.9	68.8	81.1	85.3	85.5	86.1	77.3	70.5	57.1	48.3	68.82
1952	53.1	52.1	58.8	67.7	76.4	88.5	86.1	82.4	77.7	65.1	56.6	49.2	67.81

[a] Records incomplete, substitute most probable value for determination of the average annual.

value as identified from Figure 7–11 is 1964, and the extreme expected on the average once in 20 years is identified as 1953. The monthly historical meteorologic conditions at the Weather Bureau station with Atlanta solar radiation as recorded for the months of 1964 are therefore considered as the normal year, and for the months of 1953 as the critical year. The equilibrium water temperatures computed for the sequence of months for these 2 years, the normal and the critical, are found in Table 7–5.

These seasonal patterns represent the equilibrium water temperatures expected at the site of reservoir K. The natural surface water temperature and the vertical temperature gradients in deep reservoirs, such as the proposed development, will deviate somewhat from these patterns, as shown in the following computations.

Natural Surface Water Temperatures and Vertical Gradients. The natural water temperatures expected in reservoir K (without waste heat loads) are computed through the months of the year in accordance with the method described above. Seasonal patterns are developed for two conditions, the expected normal year of hydrology, meteorology, and stage, and for the critical once-in-20-years climatology and critical 1956 stage.

In establishing the criteria for the development of natural stratification in reservoir K reliance is placed on occasional temperature measurements made in reservoir H during 1964 and on experience at other deep reser-

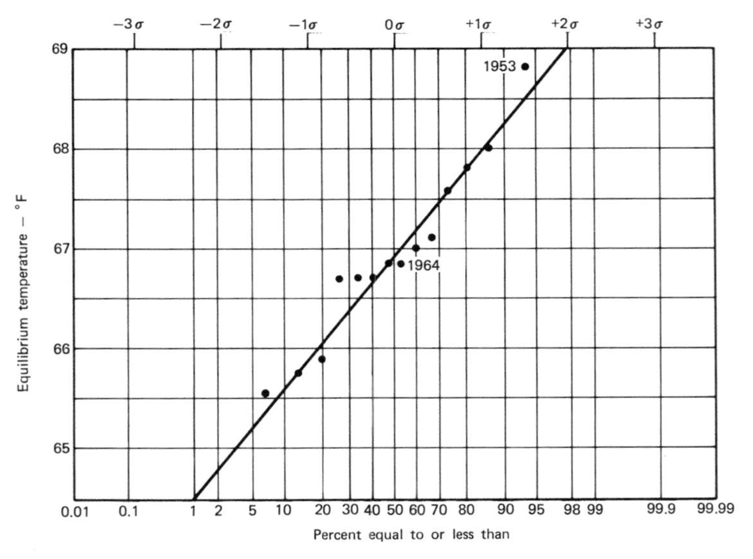

Figure 7–11 Distribution of average annual equilibrium water temperatures.

voirs. Mixing depths were quite closely defined in the adjacent reservoir H at depths ranging gradually from 15 ft in May and June, to 95 ft in October, to full depth November through January, and then gradually decreasing to 20 ft in April.

The normal seasonal water temperatures expected under natural exposure in reservoir K were computed for successive months through the year based on the monthly average historic hydrology. Figure 7–12*a* shows the computed vertical water temperature gradients at the end of each month. Commencing on January 1, 1964, the stage is at elevation 794.7 ft and the surface water temperature is 51°F and homogeneous throughout the entire depth. From this base the surface temperature and vertical gradients are computed in succession: stratification commences in March, changes to marked thermoclines throughout the summer–fall season, and again approaches a homogeneous temperature in October, ending December 31, at a reservoir stage of 799.3 ft and a temperature of 52.4°F.

A number of water temperature vertical gradients on specific days during 1964 were recorded in reservoir H, as shown in Figure 7–12*b*. These compare quite well with the computed gradients (Figure 7–12*a*) expected in reservoir K under 1964 conditions, making allowance for the forced variation in hypolimnion heating.

Some deviation is to be expected between an observed gradient measured on a specific day and that computed for the end of each month based on monthly average meteorology and hydrology. Short-term climatologic conditions may combine in such a manner as to cause abnormal rates of heat transfer at the surface, and winds may cause mixing and shifting of the respective water layers (seiches) on a short-term basis. Water temperature measurements at 6-in. depth in reservoir H during April and May 1965 indicate that at such depth the temperature may vary more than 5°F over a period of 1 day. Temperatures close to the surface are most sensitive to abnormal conditions during the periods of very small mixing depths because the epilimnion volume that must respond is much smaller than during other periods of the year.

Allowing for the dynamic, cyclic, and day-to-day variations, the close agreement between computed and observed temperatures is verification of the methods of computation adopted.

The *critical* seasonal pattern of water temperatures expected under natural exposure in reservoir K was computed in a similar manner; it was based on monthly average historic meteorology of 1953 and the 1956 hydrology, as illustrated in Table 7–4 and Figure 7–2*b*. These severe conditions resulted in a 22-ft drawdown in combination with once-in-20-years climatology, and produced high water temperatures. Figure 7–12*c* shows the computed vertical water temperature gradients expected; a similar seasonal pattern develops as for the normal year but as expected the gradients and surface temperatures shift somewhat to warmer temperatures.

As a further verification of the computational methods, a comparison is made of the computed surface temperatures expected in reservoir K under natural exposure of 1964 with surface temperatures measured in reservoir H on specific days during 1964, as shown in Figure 7–13. Curve A represents the seasonal pattern of surface water temperature computed for reservoir K as monthly average 1964 conditions; curve B represents the computed equilibrium surface water temperature as

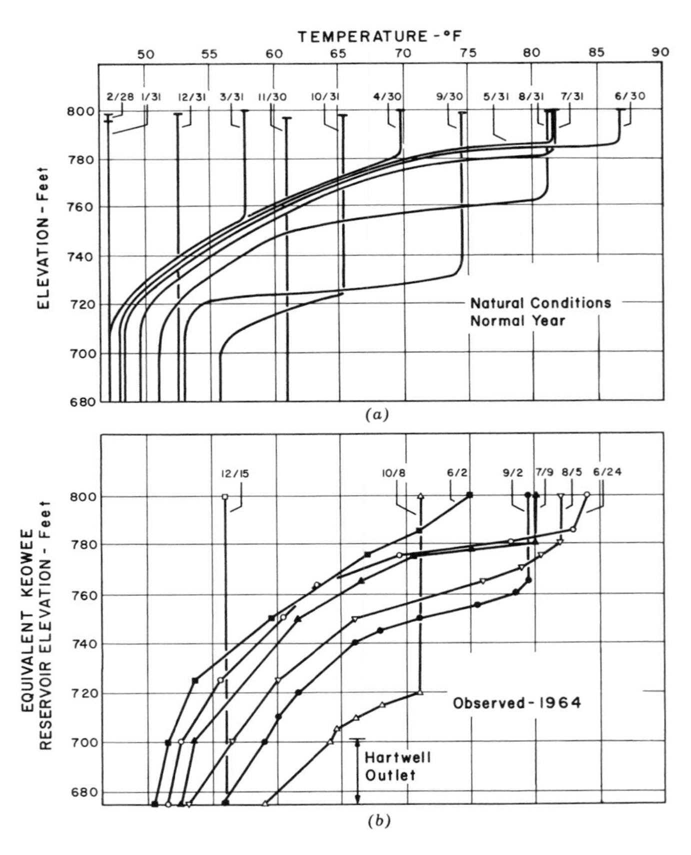

Figure 7–12 Computed natural monthly water temperature and comparison with observed gradients: (a) vertical gradients expected in reservoir K; (b) observed gradients in reservoir H; (c) vertical gradients expected in reservoir K—natural conditions critical year.

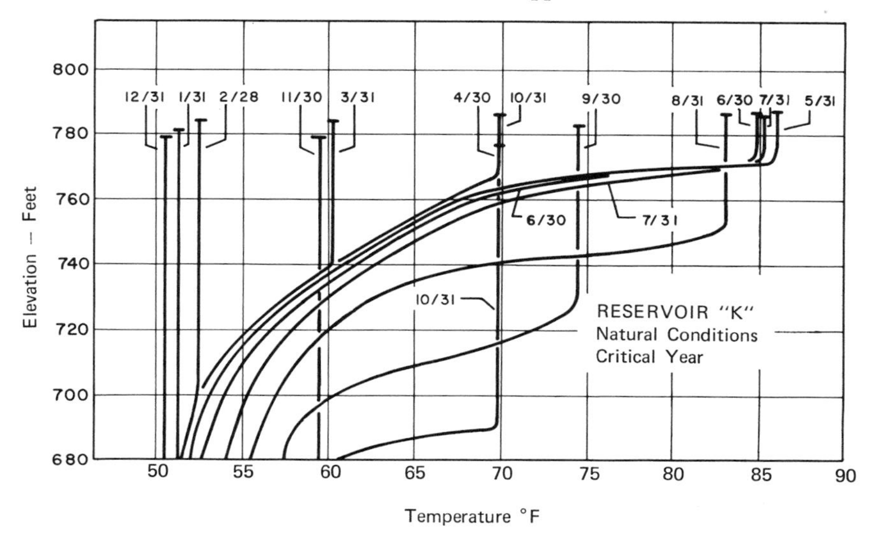

Figure 7–12 (Continued)

monthly averages for 1964 at the reservoir site; and the solid squares are the recorded surface water temperatures measured on specific days during 1964 on reservoir H. Again recognizing that individual daily measurements may deviate from monthly average conditions, there is good agreement between the computed monthly average pattern and the limited daily observed values at reservoir H.

The deviations significantly below the computed surface water temperature on June 2 and July 9 can be accounted for in part from observed meteorologic conditions on preceding days. The solar radiation on June 1 was 452 langleys, compared with monthly averages of 557 and 593 langleys for May and June, respectively. The average solar radiation for the 4 days before the measurement on June 2 was only 301 langleys. Similarly the solar radiation reported on July 8 was only 216 langleys, compared with the monthly average for July of 474 langleys. The other days for which water temperatures are reported experience conditions closer to average; however, it must be emphasized that a number of circumstances may affect an instantaneous surface temperature condition.

Also on Figure 7–13 are plotted the temperatures measured at a depth of 6 in. in reservoir H at intervals during April and May of 1965. These measurements, converted to approximate monthly average values, are indicated by the solid triangles; the computed equilibrium surface-water

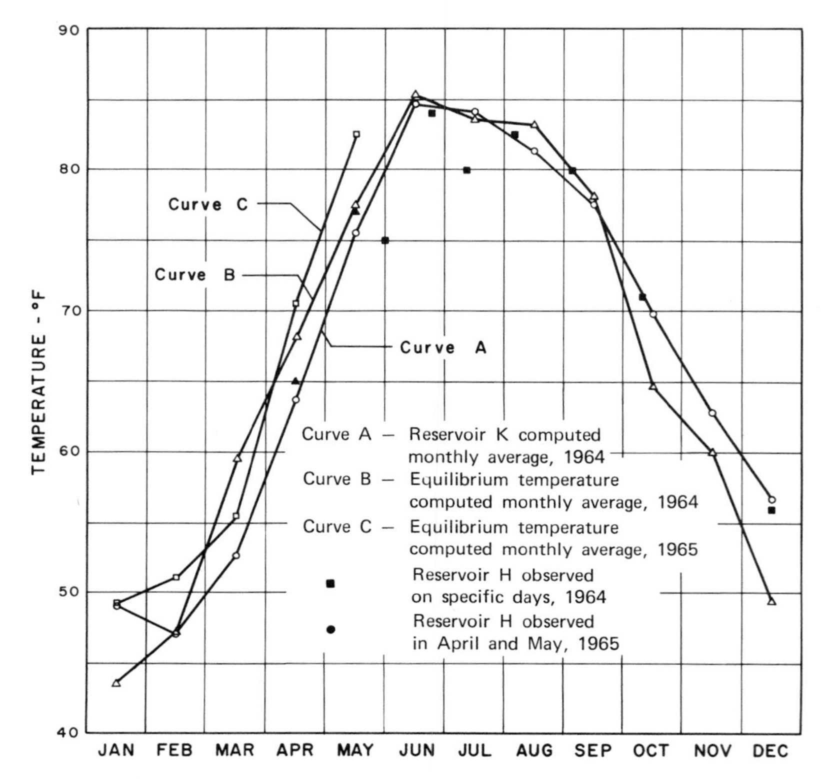

Figure 7–13 Comparison of reservoir K computed natural surface water temperatures and reservoir H observed surface temperatures for 1964 and 1965.

temperatures based on the climatology of the early months of 1965 are shown as curve C.

In comparing curve A, the surface water temperature of reservoir K, with curve B (the equilibrium water temperature pattern) it is normally expected that a large, deep reservoir will lag behind climatologic seasonal changes with a delayed rise in water temperature during spring and early summer and a delayed decline in water temperature in the fall season. The seasonal surface temperature peak and the equilibrium peak (both of them consistent with reservoir H observed values) are in good agreement at approximately 85°F. The relative deviation between curve C and the reported surface temperatures for the spring period of 1965 provide a further check on the computation procedures.

Induced Water Temperatures and Vertical Gradients. The waste heat load from the 3000-MW nuclear fuel steam electric power generating plant on reservoir K (Figure 7–8) on an average annual basis is 16.2×10^9 Btu/hr at an 80-percent load factor. This rating requires a condenser circulating water flow of 3800 cfs and produces a 19°F rise in the condenser water discharge. The condenser water intake is from the hypolimnion at elevation 710 ft (bottom of the intake 90 ft below the surface at full pool), and the condenser discharge is near the surface into the epilimnion.

In order to establish a range of expected temperatures under normal and extreme conditions and to indicate the sensitivity to assumptions as to reservoir effectiveness, induced water temperature gradients were computed for this waste heat loading for four conditions: (a) normal year climatology and reservoir stage, entire reservoir effective; (b) critical year climatology and stage, entire reservoir effective; (c) critical year, entire reservoir effective but complete vertical mixing; (d) critical year, 50 percent of reservoir effective.

The computation procedure in determining the temperature expected in reservoir K with the induced heat loading follows that set forth for deep reservoirs, taking into account two essential modifications. Since approximately 230,000 acre-ft per month are withdrawn from the hypolimnion for cooling purposes and discharged to the epilimnion, the temperature gradients are thereby lowered at a rate that overshadows any natural eddy conductivity in the hypolimnion. Therefore adjustments for eddy conductivity as applied in deep reservoirs under natural exposure are not made where such condenser water withdrawals of such magnitude are involved. Also, since the heated condenser water is discharged at the surface, thermal stratification occurs at shallow depths early in the season and persists throughout the year. Therefore, instead of varying the mixing depth through the seasons, a constant minimum average mixing depth of 10 ft is assumed throughout the year.

The most logical assumption is that the entire reservoir is effective in heat dissipation. For the normal year, entire reservoir effective, Figure 7–14*a* shows the computed water temperature vertical gradients at the end of each month of the year based on the monthly average climatologic conditions. Commencing on January 1, with homogeneous temperature throughout the reservoir depth, and applying the monthly heat loads, vertical temperature gradients and thermoclines are induced, rapidly increasing the temperatures in the epilimnion and gradually increasing the temperatures in the hypolimnion until August, when maximum temperatures are attained and the gradient is least from surface to bottom. There follows in the fall a rapid decrease in water temperature, homogeneous throughout the depths by the end of September, and continuing through the fall to the end of December, returning to approximately

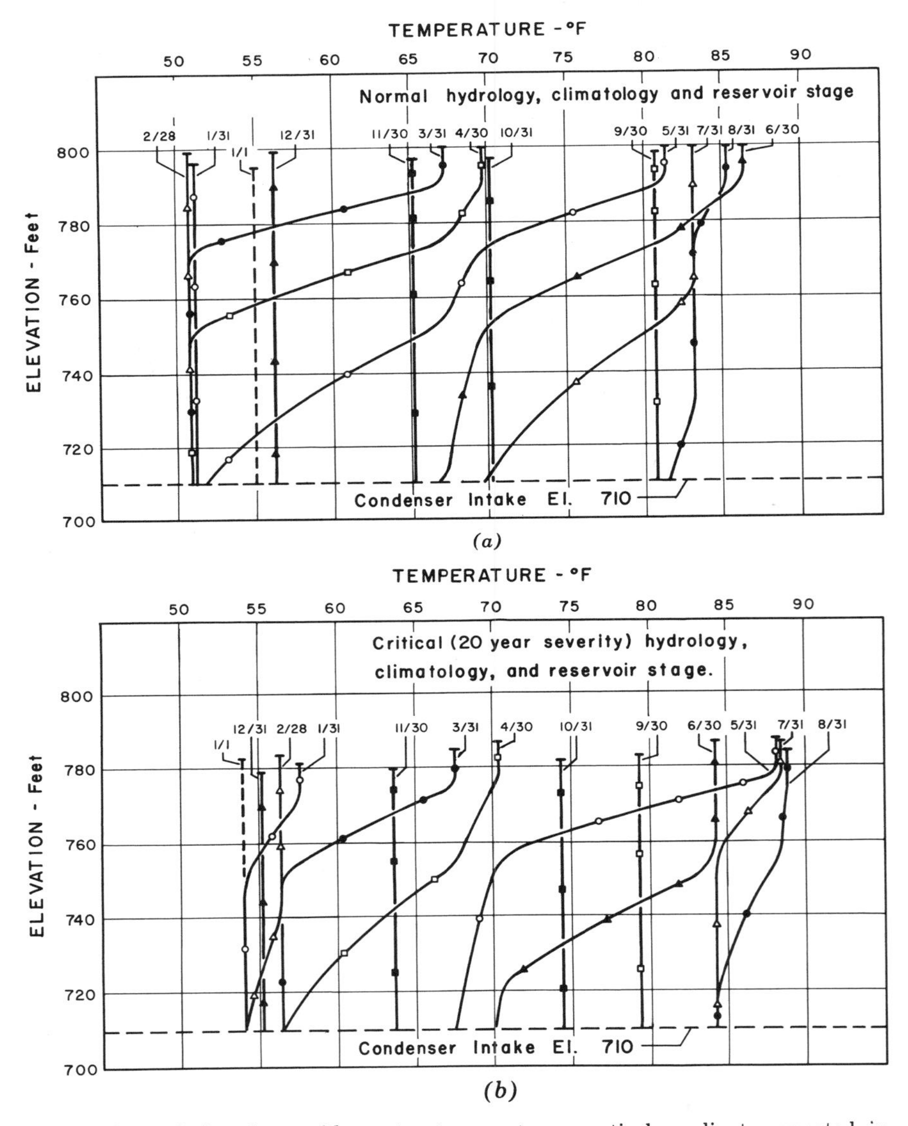

Figure 7–14 Induced monthly water temperature—vertical gradients expected in reservoir K: (*a*) normal hydrology, climatology, and reservoir stage; (*b*) critical (once in 20 years) hydrology, climatology, and reservoir stage. Entire reservoir effective. (Symbols on curves are for identification purposes only and do not represent data points.)

1°F above that at which the year commenced. Since the initial temperature is approximately 5°F above that expected without the heat load, the reservoir can be considered to contain 5°F of undissipated heat; however, cooling back to that level is ensured and no significant buildup of heat is likely over successive years.

Comparing Figure 7–14*a* (the temperature gradients induced by the heat load) with Figure 7–12*a* (the natural temperature gradients), a marked difference is seen in the increase in temperature in the deeper portions of the reservoir through the summer season as a result of the heat load.

From the seasonal vertical temperature gradients of Figure 7–14*a* the monthly average intake and discharge temperatures of the circulating condenser water of the nuclear fuel steam electric power plant are determined as listed in Table 7–6. The monthly average intake temperature corresponds to the average temperature of the reservoir for that month taken at the centerline elevation of the intake (elevation 716 ft). The maximum monthly average intake water temperature from the lower depths of reservoir K is 81.2°F (September) and the discharge temperature from the condensers of the steam plant to the surface of the reservoir is 100.2°F. The seasonal peak at the end of August is 81.8°F for the intake and 100.8°F for the condenser discharge.

Table 7–6 Computed Monthly Mean Intake and Discharge Temperatures of Circulating Condenser Water, Normal Climatology

Month	Intake (°F)	Discharge (°F)
January	53.1	72.1
February	51.0	70.0
March	50.8	69.8
April	50.8	69.8
May	51.9	70.9
June	60.2	79.2
July	69.0	88.0
August	76.2	95.2
September	81.2	100.2
October	75.4	94.4
November	67.6	86.6
December	60.7	79.7
Peak, end of August	81.8	100.8

The monthly water temperature vertical gradients were computed similarly for the once in 20 year severity of climatologic conditions expected and the critical 1956 drawdown of reservoir K. The seasonal pattern of vertical gradients through the year is shown in Figure 7–14*b*. The trend through the critical year is similar to that for the normal year (Figure 7–14*a*), except that the critical year, with lowered reservoir level and less favorable heat dissipation conditions, shows a shift toward warmer water temperatures. Considering the substantial differences in climatology and reservoir stage, the shift is moderate.

As an extreme assumed condition in distribution of waste heat load to the reservoir, computations were made on the basis of complete mixing throughout the depth of the reservoir through all months of the year. It is evident that such a condition is highly improbable because the lower density water discharged from the condensers will resist deep mixing. The results of this calculation, however, when compared with the more likely assumptions outlined earlier, indicate how sensitive a matter the distribution of heat through depth can be. Under the assumption of complete mixing the net heat loss to the atmosphere is less because the surface temperature is reduced. As the warm season progresses, with

Table 7–7 Computed Seasonal Water Temperature of Reservoir K at Critical Climatology under Assumption of Complete Mixing of Reservoir

Month	Reservoir at End of Month (°F)	Monthly Mean Intake (°F)	Monthly Mean Discharge (°F)
January	56.6	55.8	74.8
February	57.2	56.9	75.9
March	64.0	60.6	79.6
April	71.2	67.6	86.6
May	83.6	77.4	96.4
June	88.2	85.9	104.9
July	88.2	88.2	107.2
August	89.2	88.7	107.7
September	80.2	84.7	103.7
October	74.6	77.4	96.4
November	64.2	69.4	88.4
December	55.8	60.0	79.0
Peak, end of August	89.2	89.2	108.2

greater retention of the heat load, the entire body of water from top to bottom attains a maximum temperature superelevation; the advantage of lower temperature intake water at lower depths is no longer available. Since vertical temperature gradients are eliminated, the seasonal temperatures computed are as listed in Table 7–7.

As shown in Figure 7–8, reservoir K is extremely irregular and has several relatively narrow arms extending outward from the main channel. It is conceivable that the area and volume of some of these arms may not at all times be effective in the complex interrelations of circulation and heat dissipation.

Since it is not possible to determine precisely what portions of the proposed reservoir K may potentially be inactive because of the combination of local factors, an arbitrary reduction of 50 percent is adopted, under conditions of critical climatology and reservoir stage. As will be seen later, this reduction is consistent with the area required for the surface water temperature profile to approach the equilibrium temperature. In developing the reduction no attempt is made to eliminate particular arms of the reservoir, and the active portion involves both branches of reservoir K outward from the power plant site.

Figure 7–15 shows the vertical temperature gradients computed as before by the methods outlined earlier. The seasonal pattern is similar

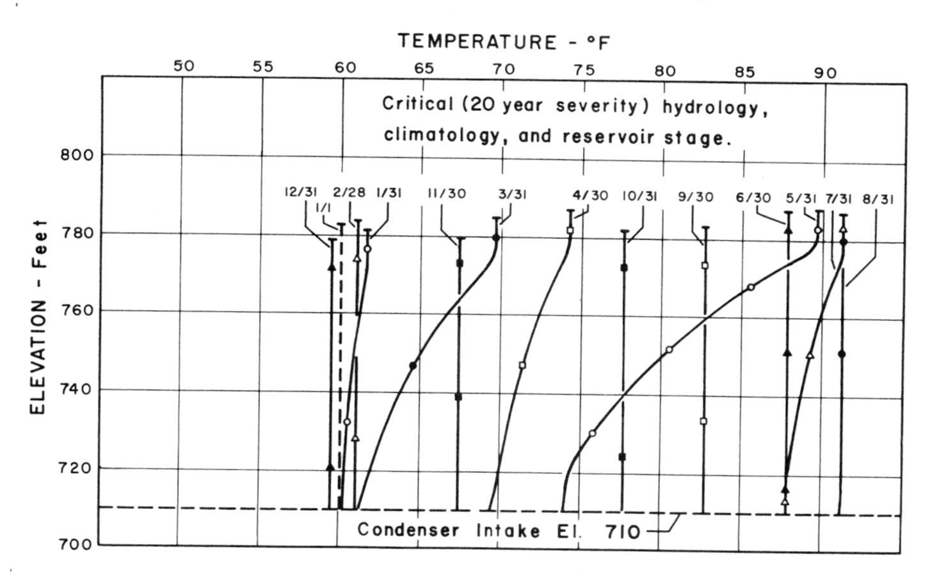

Figure 7–15 Potential induced monthly water temperature—vertical gradients expected in reservoir K. Reduced effective reservoir capacity. (Symbols on curves are for identification purposes only and do not represent data points.)

Table 7–8 Comparative Induced Peak Temperatures of Condenser Water Intake and Discharge

Climatology and Reservoir Stage	Assumed Reservoir Conditions	Condenser Water Temperature (°F)	
		Intake	Discharge
Normal year	Entire reservoir effective and stratified	81.8	100.8
Critical year	Entire reservoir effective and stratified	84.2	103.2
Critical year	Entire reservoir effective, completely mixed	89.2	108.2
Critical year	50 percent of reservoir effective and stratified	91.2	110.2

to that computed for the same climatology and stage (Figure 7–14*b*) where the entire reservoir is considered active. However, as would be expected, there is a shift toward warmer temperature in the lower depths.

In each computed vertical temperature gradient induced by the waste heat load from the nuclear fuel steam electric power plant (Figures 7–14*a*,*b*, Table 7–7, and Figure 7–15) there is an evident seasonal peak temperature at the end of August. These peak conditions are summarized for comparative purposes in Table 7–8 in terms of expected condenser circulating water intake and discharge temperatures.

Thus during a critical year, such as expected on the average once in 20 years, the deep condenser water intake withdraws water from the reservoir at the seasonal peak temperature in the range 84.2 to 91.2°F, and the condenser water at the surface discharge to the reservoir is in the range 103.2 to 110.2°F. The temperatures are more likely to be toward the lower values of these ranges because the computed values are for the rather unlikely extreme conditions of complete mixing and one-half of the reservoir ineffective in heat dissipation.

Surface Water Temperature Profile from Point of Waste Heat Discharge. The vertical temperature gradients established in the depths of the reservoir, from which the condenser intake water temperature is determined, represent a *mean* reservoir condition developed through the seasons from which a critical seasonal peak intake temperature is established. In this process the heat load is essentially distributed uniformly over the entire reservoir. The heat, and therefore the temperature distribution from which the average is derived, is actually not uniform but for any given short period is in the form of a declining profile with maximum superelevation at the point of discharge of the condenser water to the upper layers of the reservoir. The profile extends outward, declining at a nearly logarithmic rate and approaches the equilibrium temperature toward the remote areas of the reservoir. The profile is defined by equation 7–21 and represents the upper seasonal limit of surface temperature distribution. Any entrainment of surrounding water and successive lowering of surface layers, such as employed in the development of the seasonal gradients, will lower the surface water profile.

For reservoir K the seasonal peak intake water temperature drawn from the lower depths to the condensers under critical climatologic conditions considering the entire reservoir to be effective is 84.2F° in August; the surface discharge from the condensers with the 19°F rise in temperature is 103.2°F at the outlet. The temperature profile outward from the source as defined by equation 7–21 is shown as curve A in Figure 7–16. The reservoir surface area required to cool to prescribed temperature levels is designated in acres outward from the waste heat source. The surface water temperature is expected to decline from 103.2 to 94°F within an area of approximately 1800 acres. At the critical drawdown

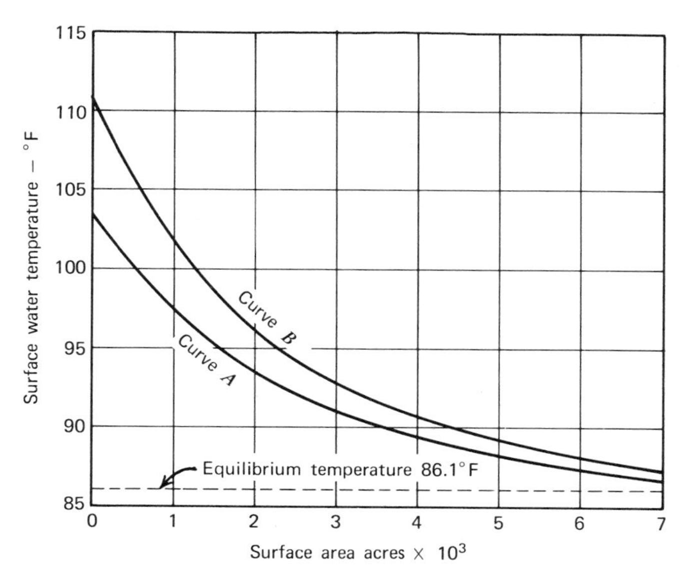

Figure 7–16 Expected surface water temperature profile—August of critical year. Curve A: entire reservoir effective; curve B: reduced effective volume.

stage the total reservoir area is 14,600 acres, and hence the affected area of temperature superelevation above 94°F constitutes approximately 12.3 percent of the total reservoir. Equilibrium water temperature for the critical August meteorological conditions is 86.1°F. Surface temperature cooling to 1°F above equilibrium by curve *A* is achieved in approximately 6400 acres, 43.8 percent of the total reservoir area available.

Similarly the surface water temperature profile computed for the discharge condenser water superelevation of 110.2°F, when only 50 percent of the reservoir volume is operative in heat dissipation, is shown as curve *B*. The area required to cool to 94°F in this case is 2550 acres, or 17.5 percent of the entire surface area of the reservoir; cooling to 1°F above equilibrium requires approximately 7500 acres, or 51.4 percent of the entire reservoir area. This is in close agreement with the 50-percent arbitrary reduction in effective reservoir employed in computing the vertical temperature gradients on which the 110.2°F superelevation was based.

It should be recognized that the temperature profiles and affected areas represent the upper limits of surface water seasonal peak conditions; and that the temperatures at the intake depth remain as projected through the seasons at 84.2°F (entire reservoir effective) and 91.2°F (50 percent of the reservoir effective), respectively.

The superelevation in temperature in the immediate path of the condenser water discharge (103.2–110.2°F) could be lethal to many types of sport fish. However, this superelevation is expected only under severe reservoir drawdown and climatology, and only during the peak of the warm season. Also it should be noted that only 12.3 to 17.5 percent of the reservoir area is above 94°F; and furthermore the superelevated water is of shallow depth, overriding a larger body of much cooler water below. Thus the waste heat load from the 3000-MW nuclear fuel thermal plant should not seriously interfere with fish life, and in fact may enhance it. The impoundment of small upland streams may change the type of sport fish from trout to walleyes and small-mouth bass or to large-mouth bass, but the aggregate sport fishing and recreational opportunity is greatly enhanced by the large reservoir area. Waste heat dissipation and sport fishing and recreation are not necessarily mutually inimical; under good engineering design and scientific management these two water resource assets are compatible.

Considerations of Two or More Waste Heat Sources

As the trend toward concentration of large thermal electric power developments continues, advantage will be taken of large, deep reservoirs for condenser water and waste heat dissipation, and for peaking hydroelectric power, such as illustrated (Figure 7–8) and discussed above. In such developments it can be expected that two or more thermal electric power plants will be located on a single reservoir or interconnected system. For example, referring to reservoir K, on the basis of surface cooling theoretically the computations indicate that the induced peak surface water temperature superelevation from the 3000-MW nuclear fuel plant can be expected to approach within a degree or two the natural equilibrium temperature within an area outward from the condenser discharge of approximately one-half that of the total reservoir area available. Hence reserve capacity for dissipation of additional waste heat load is indicated, particularly if advantage is taken of remote locations to avoid serious overlap of surface temperature profiles.

To illustrate such a possibility two additional fossil fuel steam electric power generating plants of 2000-MW capacity each are located at sites M and I (Figure 7–8), with surface condenser water intake from one arm and discharge to the other. These two additional plants, operating at 80-percent load factor with the 3000-MW nuclear fuel plant constitute a combined waste heat load of approximately 3×10^{10} Btu/hr. Under conditions of critical climatology and reservoir stage, and considering the entire reservoir effective, the surface water temperatures and vertical gradients were computed through the months of the year from which seasonal peak temperatures of intake and discharge condenser water were determined. For the fossil fuel plants peak intake and discharge temperatures are 91.5 and 110.5°F, respectively; the deep intake temperature to the nuclear fuel plant is 86.7°F and the temperature of the discharge is 105.7°F. Referring to the previous computations for the nuclear fuel plant alone, under similar conditions the intake and discharge temperatures are 84.2 and 103.2°F, respectively. Considering the magnitude of the extra heat loading, the increase in temperature at the nuclear power plant is moderate; however, the surface water temperatures of the remote areas of the reservoir are substantially higher than those resulting from operation of the nuclear power plant alone. Obviously the plants have an effect on each other.

Again it is stressed that in practical applications it is not possible to predict precisely the heat distribution, the relative effectiveness of various reservoir reaches, and potential short-circuiting because of irregular geometry and operational patterns of associated hydroelectric plants. Although the effect of multiple waste heat loadings can be theoretically computed for various assumptions of heat distribution and effective reservoir area, there is no reliable method of determining which assumption will be in best agreement with the actual conditions until some of the thermal units are actually in operation and field measurements can be made. Large installations of the type illustrated are usually scheduled for construction over a period of years, and opportunity is thus afforded of verifying the influences of local factors by monitoring reservoir temperatures and flow patterns.

As part of such a monitoring program, at least one weather station should be established at an appropriate location at the site for recording air temperature, relative humidity, and wind velocity. Longitudinal, vertical traversing of the reservoir for temperature profiles and gradients should be taken under stable conditions through the seasons, before as well as during operation of the early thermal units. Infrared flights over the reservoir also afford a sensitive means of delineating surface water temperature boundaries and profiles. Pumped storage and conventional hydroelectric power operation oriented to peaking generation and off-peak pumping involve large volume exchanges in short periods of each day, whose influence is difficult to anticipate without field observations. As field monitoring data become available, application of theoretical computations can be refined, narrowing assumptions and improving precision. Such studies, combining field monitoring and theoretical analyses, then would form the basis for further thermal plant site selection and capacity limitations of additional thermal electric power generation stations.

The Alternative of Small Reservoir and Cooling Towers

Practical considerations and economics dictate alternative solutions to waste heat disposal associated with large mine-mouth electric power generating stations. The abundance of low cost high grade bituminous coal within range of high voltage transmission to centers of energy demand make it economically attractive to locate large generating stations at the site of coal fields. Since such deposits of coal are usually in the headwaters of drainage basins, only small streams are available, of insufficient summer dry weather flow to meet satisfactorily condenser water demands and waste heat disposal by conventional on-stream methods. An alternative is the combination of large cooling towers and a small storage reservoir. Sufficient drainage area is usually available to provide storage capacity for *make-up* water to compensate for evaporation and other water losses. The large cooling towers, usually of the hyperbolic type, provide dissipation of heat to the atmosphere, principally by evaporation, and with make-up water from the reservoir permit continuous recycling of condenser water.

An example of such a mine-mouth development is the large Keystone project [13,14] constructed in the coal fields northwest of Johnstown, Pennsylvania, straddling the watershed divide between the Allegheny and Susquehanna rivers (Figure 7–17). The plant consists of two large generators providing a total capacity of 1800 MW. Practically all of the coal comes from within a 15-mile radius of the plant, and 60 percent of it supplied direct from the mines by conveyor belts. Electrical energy is distributed into an interconnected 500,000-volt transmission system. Four large cooling towers of the hyperbolic type, 325 ft high and 247 ft in diameter at the base, receive the condenser water at a rate of 560,000 gpm (1250 cfs). Make-up water under summer conditions is estimated at 35 cfs, of which 30 cfs replaces evaporation loss.

The Keystone plant site is on a small stream, with a tributary drainage area of 191 square miles. The average annual flow is 300 cfs, with a critical monthly average of 12 cfs and a minimum recorded of 2.4 cfs. The make-up storage reservoir located about 5 miles north of the plant site on a small tributary provides storage capacity of about 27,000 acre-ft. The storage capacity, in addition to supplying make-up and other plant needs, is employed to maintain a minimum flow of 30 cfs below the plant site through the critical drought season to enhance downstream water quality. Studies of streamflow augmentation and plant consumption indicate that drawdown of the reservoir level prior to Sep-

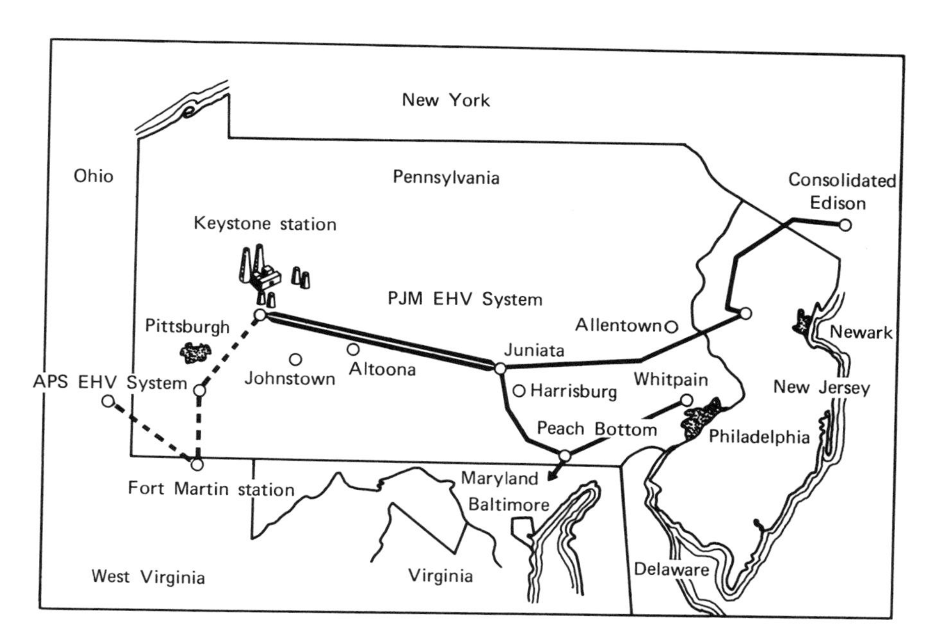

Figure 7–17a Keystone mine-mouth generating station. Schematic diagram of project.

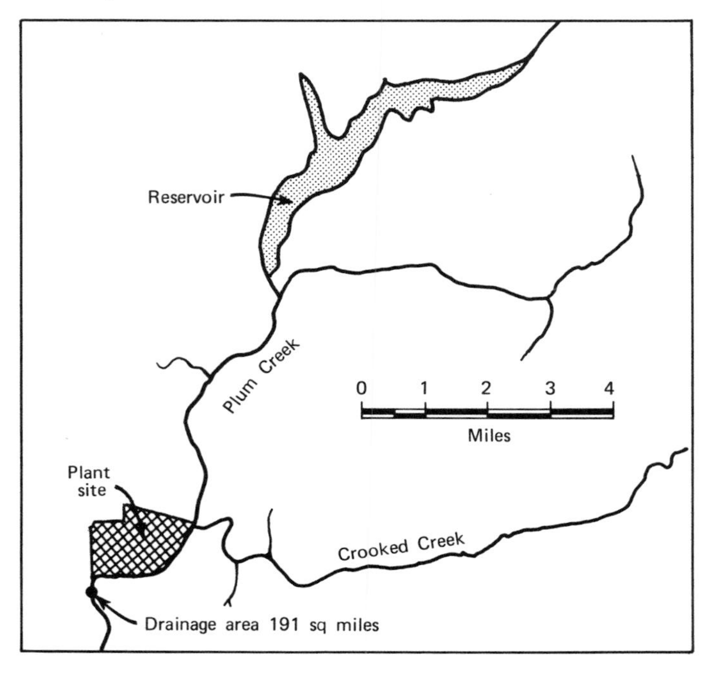

Figure 7–17b Keystone plant site and reservoir.

tember will not be substantial even in poor water years. Hence the reservoir can also be used for recreational purposes. Keystone is (on a limited scale) an example of multiple use of a water resource.

Of interest is the economic analysis by the designing engineer [13,14] in comparing the cost of a conventional plant located on the Allegheny River (using the river for condenser circulating water and heat dissipation) and the mine-mouth plant (using the combination cooling towers and make-up reservoir). Initial investment savings favor the mine-mouth plant by $1.00 per kilowatt of capacity; however, this is at a sacrifice of operating costs associated with higher condenser pressure, increased pumping power, and loss in capability. The big operating savings favoring the mine-mouth plant are in the low cost of fuel.

A conventional river plant of 1800-MW capacity, based on the generalized values of Figure 7–1, would require circulating condenser water of 2210 cfs with a temperature rise of 16°F, as compared with the mine-mouth cooling tower plant with lower circulating water recycling flow of 1250 cfs but with a temperature increase of about 28°F. The optimum cooling range for cooling towers is high, between 23 and 30°F. A factor in conventional river plants is the limit of superelevation of the riverflow induced by the rejected waste heat of the condenser water discharge, usually in the range 10 to 20°F.

It appears therefore that construction of large steam electric power generating plants is technically feasible not only on large rivers or reservoirs but also in small drainage basins (if there is adequate reservoir capacity for make-up water), and is economically feasible if cheap fuel and high voltage transmission are accessible.

Cooling towers, however, have limitations and inherent disadvantages that usually preclude their use in or near urban areas. Higher capital and operating costs are a liability; another is that the vapor plume exhausted from the towers induces a local fog. Under certain meteorologic conditions fog may blanket extensive areas and seriously add to air pollution smog. Freezing of vapor may also create icing conditions in the vicinity of the towers. Also, although the hyperbolic shape is a pleasing esthetic form, the massive scale of the towers is overpowering in an urban setting and may be considered unacceptable in a scenic natural setting: a single tower will be 25 to 35 stories high, and its diameter will equal two city blocks—and two to four such towers are required for a large generating station. A final and serious liability is that cooling towers consume greater quantities of water than other cooling methods. Although cooling towers are effective in heat dissipation, they ought not necessarily to be preferred over reservoir or streamflow methods of cooling.

An alternative method of heat dissipation that incorporates broad multiple water use benefits is the adaptation of pumped storage, discussed in Chapter 11. This avoids the disadvantages of cooling towers (a costly, primarily single purpose investment) and can provide such additional benefits as nearby community and industrial water supply, water based recreation, peaking electric energy capacity, and low flow augmentation for stream quality improvement.

REFERENCES

[1] *National Power Survey,* Part 1, Federal Power Commission, U.S. Government Printing Office, Washington, D.C., 1964.
[2] Wurtz, C. B., Paper No. 61-WA-142, *ASME,* August 1961.
[3] Wurtz, C. B., Pub. No. 3, Division of Sanitary Engineering, Pennsylvania Department of Health, January 1962.
[4] Lima, D. O., *Power,* **80,** 142 (1936).
[5] Throne, R. F., *Power,* **95**, 86 (1951).
[6] LeBosquet, M., *J. New England Water Works Assn.,* **60,** 2, 13, 755 (1959).
[7] Anderson, E. R., Technical Report, U.S. Geological Survey circular 229, 1952 and Professional Paper 269, 1954.

[8] Langhaar, J. W., *Chem. Engr.,* **60,** 194 (1953).
[9] Meyer, A. F., *Evaporation from Lakes and Reservoirs,* Minnesota Resources Commission, June 1942.
[10] Velz, C. J., Calvert, J. D., Deininger, R. A., Heilman, W. L., and Reynolds, J. Z., *Report on Waste Heat Dissipation,* prepared for the Department of Interior, April 1966.
[11] Debler, W. R., *Trans. ASCE,* Paper No. 3157 (1959).
[12] Harleman, D. R. F., in *Handbook of Fluid Dynamics,* Streeter, ed., McGraw-Hill, New York.
[13] Smith, A. F., III, and Bovier, R. E., *Proc. Amer. Power Conference,* Vol. 25, 1963, p. 406.
[14] Estrada, H., and Smith, A. F., III, *Proc. Amer. Power Conference,* Vol. 26, 1964, p. 381.

8

SELF-PURIFICATION IN ESTUARIES

APPROACHES TO EVALUATION OF SELF-PURIFICATION IN ESTUARIES

The complexity of estuaries has led to several approaches to evaluation of self-purification [1–17]. These can be roughly grouped into five schools: (a) tidal prism, (b) mathematical model, (c) experimental model, (d) statistical, and (e) rational. The early work of engineers centered on the belief that the waters of the tidal prism are completely available for dilution and that wastes are flushed to sea on each ebb tide, not to be returned on the following flood. On this fallacy it was postulated that estuaries possessed great capacity to assimilate wastes. However, except in unusual situations where sources of pollution are located near an estuary outlet to the ocean and where there are strong lateral ocean currents, this theory does not hold. Deterioration of harbors has disproved the tidal prism theory, and it is now well established that in the majority of polluted estuaries wastes carried toward the sea on the ebb tide actually return on the flood tide, oscillating back and forth with but slow net movement toward the sea.

Divergence among schools currently centers on different approaches to determining mixing or diffusion of wastes, accumulation patterns, and mean retention time within the estuary.

With the advent of computers, a great deal of effort is being devoted to solutions through elaborate mathematical models that apply to tidal phenomena hydrodynamic theory developed for atmospheric diffusion. A great many independent variables are involved only a few of which are measurable, and in the end, a number of quite arbitrary assumptions must be made in application of the mathematical models to a specific estuary.

In an effort to verify theoretical models and define the most relevant variables, experimental models of specific estuaries are sometimes constructed and tested. The limitation of these hydraulic models is the difficulty of incorporating the features of the specific prototype so as to test simultaneously the complexities associated with tides, the hydrodynamic diffusion, and the fate of unstable wastes, such as are associated with bacterial survival, deoxygenation, and reaeration. So far experimental model studies have not successfully developed methods for simultaneous testing of hydrodynamic, hydrologic, and biologic elements. The usual procedure is to employ stable dyes. Although such investigations are useful in predicting the fate of stable, nondegradable waste products, there is risk in projecting these findings to the fate of unstable, degradable bacterial, organic, and radioactive wastes.

Limitations of the mathematical and experimental model approaches have given rise to the statistical school of thought. In this approach no attempt is made to define and formulate the fundamental relationships involved and reliance is placed on statistical analyses of observed water quality end results from a given waste loading, correlating with recorded variables, such as landwater runoff, seasonal water temperature, and tide. Obviously such procedures based on observed end results develop regression equations that check these particular end results. The risk is in extrapolating these equations to conditions other than those that prevailed during the period of observation. The statistical approach is helpful in interpolation within the frame of observed results, but, since it is not based on self-purification fundamentals, it cannot be employed with any sense of security in analyses of other conditions, such as interception of waste loads for treatment with new locations of effluent discharge, or such alterations in hydrologic regime as channel changes, storage of upland runoff, or low flow augmentation. For variables of this kind some other method of analysis must be applied.

In the current state of knowledge we are left with the *rational* approach—that is, breaking down the complex problem into rational parts and applying to each its primary relevant factors. The fundamental principles of self-purification developed for application to inland rivers will apply to estuaries. It is necessary only to consider modifications associated with tidal translation and seawater intrusion.

In the rational approach seawater intrusion is the critical factor; estuaries are classified by this factor. At locations well above the brackish reach, such as at Albany on the Hudson River or at Portland on the Columbia–Willamette rivers, modifications are minor since only ebb and flood translation need be taken into account; at downstream locations within the brackish reach, such as the New York metropolitan area tributary to New York Harbor or the urban-industrial complex tributary to Savannah Harbor, more significant modifications are necessary, as both tidal translation and seawater exchange associated with the salinity gradient are involved.

A fuller appreciation of the complexities encountered in dealing with estuarial problems is gained from a brief review of the nature of tides and tidal currents. However, formulations that attempt to incorporate all of these vagaries quickly lead to an impenetrable labyrinth of confusion. Much of this confusion can be avoided, without loss of significance, by reducing analyses to simple *mean tide* conditions.

TIDES AND CURRENTS IN ESTUARIES*

Tide Producing Forces

The alternate rise and fall of the level of the sea is a matter of gravity, which, it will be recalled, varies directly with the mass and inversely with the square of the distance from the attracting mass. The attraction for the solid earth as a whole varies with the distance to the center of the earth. But the waters of the oceans are on the surface of the earth and on one side or the other are closer or farther away from the attracting mass, the sun and moon, than is the center of the solid earth. This difference in distance results in imbalance in attraction, establishing the forces that cause tides as the ocean waters move relative to the solid earth.

The relative proximity of the moon to the earth more than compensates for the greater mass of the sun, and thus the moon plays the dominant role in tidal fluctuation. The combination of influences of the moon and the sun set up periodic variations in tide producing forces, which with varying configuration of estuaries, terrestrial features, and wind action make for wide variation in tide from place to place and at different times of the day, month, or year at any particular place.

Based on the periodic forces, predicted tide tables are prepared annually for many geographical locations by the U.S. Coast and Geodetic Survey. In all there are 20 or 30 short- and long-term periodic forces that are usually taken into consideration in predicting and preparation of the tide tables. Since these tide tables, as well as current tables, are readily available from the U.S. Coast and Geodetic Survey, we need

* For a fuller discussion see references 18–20.

be concerned, for most estuarial problems, with only three primary tide cycles: the *semidaily*, the *moon's phase*, and the *moon's declination.*

Periodic Variation in Tides

Semidaily Tide Cycle

Since the moon in its apparent movement around the earth has both an upper and a lower meridian passage in 1 lunar day, there results for most areas of the world two complete cycles of rising and falling tide, with two high and two low waters in each tidal day. The variation in tide through a lunar day takes the form of a sine curve, as shown in Figure 8–1. The time of occurrence of high and low water referred to the instant at which the moon crosses the meridian is very nearly constant. However, since the lunar, or tidal, day is 24.841 hr, the moon passes a given meridian about 50 min later each solar day; hence high water and low water likewise occur about 50 min later each day. Assuming for the moment that the two tidal cycles in a lunar day are of equal range and duration, the interval for a complete rise and fall in one cycle would be 12.42 hr. Thus in observing a given location at the

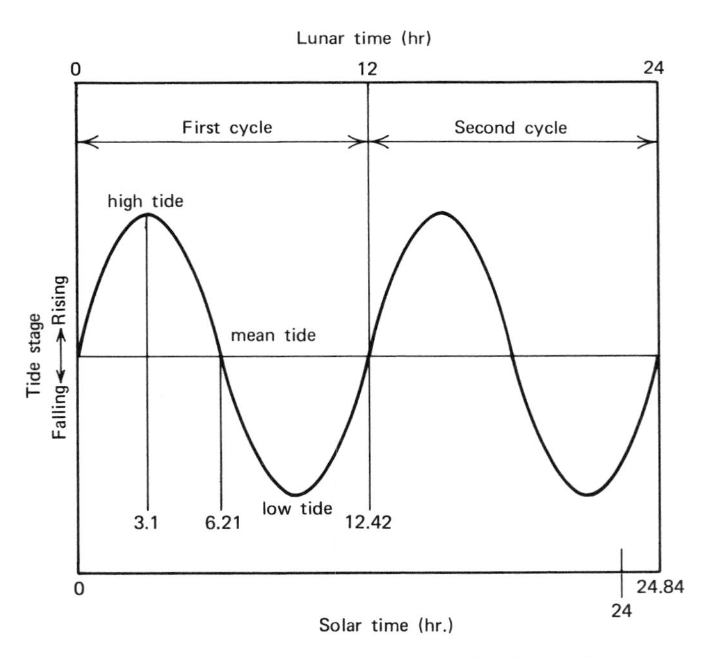

Figure 8–1 Variation in tide through a lunar day.

same hour each day (assuming no other forces are involved) all phases of a complete cycle of rising and falling tide are reflected in 12.42/0.841, or approximately 15 days.

Moon's Phase Cycle

There are also day-to-day variations, depending on the position of the moon relative to the earth and the sun. The most noticeable variation from day to day is that related to the moon's phase, which has a cycle of 29½ days. Twice in the lunar month spring tides occur, at times of new and full moon, when the earth, moon, and sun are in direct alignment; then high water rises higher and low water falls lower than usual. Neap tides occur when the moon is in its first and third quarters (moon at right angles to the sun as seen from the earth); then tide does not rise as high or fall as low as usual.

Moon's Declination Cycle

The moon in its orbit about the earth moves in a plane inclined to the plane of the equator of the earth. Hence its declination is constantly changing during the month, being minimal when the moon is on the equator and maximal when it is in either of the tropics. It proceeds through the cycle (the tropic month) in approximately 27⅓ days. When the moon is on, or close to, the equator (twice in the tropic month) the two high waters and likewise the two low waters of a day are nearly similar, with minimal difference between morning and afternoon tides. As the declination of the moon increases, the difference between the morning and afternoon tides increases, and when the moon attains its maximal declination at or near either of the tropic zones, the difference between morning and afternoon tides is at a maximum. These differences between morning and afternoon tides in some geographical areas are substantial and are referred to as *diurnal inequality*.

Nonperiodic Variables

Although all the periodic variables that influence tides are felt all over the world, geographical location determines to a great extent which factor or factors will dominate. Hence the specific characteristics of tide vary radically from place to place.

Terrestrial features also radically affect tides. The dynamic characteristics of tidal currents approaching or confined in constricted formations induce local tidal ranges that are quite different from that of the adjacent open sea. Similarly underwater configuration and water depth affect local tides. In addition a labyrinth of interconnected estuaries with multiple inlets to the open sea shows wide variations and peculiarities due to differences of timing of tidal currents approaching through different inlets.

The most significant nonperiodic factor that affects self-purification in estuaries, as will be shown later, is *landwater runoff*. Where a landwater stream discharges into the ocean, the lower reach of the river experiences intrusion of seawater far beyond the range of the flood current, forming a salinity gradient—a mixture of seawater and landwater, called *brackish water*.

Tidal Currents

The periodic vertical rise and fall of the tides induces corresponding horizontal movements of the water, known as tidal currents—flood with rising tide and ebb with falling tide. At the end of each flood or ebb, in the process of reversing direction, there occurs a brief period of relative quiescence, referred to as slack water.

The velocity of the current is usually expressed in knots (nautical miles per hour).* The tidal currents exhibit periodic changes in strength of current corresponding closely with the periodic changes in range of tides. As related to the moon's changing phases, the variation in current strength from day to day is approximately proportional to the corresponding changes in the range of tide. With the moon's change in declination, however, the tide and current do not change alike, and generally the diurnal inequality in tide is greater than the associated diurnal inequality of current.

The velocity of the current is generally referred to the surface; however, because of frictional resistance, it varies throughout the channel cross-section, being near the bottom approximately two-thirds that at the surface, greatest at the center of the channel, and decreasing toward the shores.

The average velocity in the entire cross-section is approximately three-fourths of the central surface velocity. These relationships apply to a uniform straight channel; irregularities of shape and direction set up quite different paths and distributions.†

Flood and Ebb Translation

In simple wave motion associated with a true periodic tide the time elements of the tidal currents bear a definite relation to the time of occurrence of high and low water. This relationship can be expressed in the form of a sine curve. Strength of current occurs midway between

* A nautical mile is 6080 ft; knots $\times$ 1.15 = statute mile/hr.

† Current charts are available for selected harbors from the U.S. Coast and Geodetic Survey.

high and low water; slack occurs at high and at low water; duration of the flood is equal to duration of the ebb (6.21 hr); and strength occurs 3.1 hr after high and low water. Under these conditions the distance traveled on the ebb equals that on the flood, and the body of water therefore simply oscillates back and forth, without any net movement in either direction. Based on the sine curve, the distance traveled in flood or ebb may be estimated from the maximum velocity at strength from

$$D = 0.637 \times V_s \times 6.21,$$

where D is horizontal distance traveled in miles and V_s is velocity at strength in miles per hour.

In actual practice, however, these simplified theoretical conditions seldom exist since the occurrences and durations are unequal and are either retarded or accelerated. Such variations are pronounced where diurnal inequality exists and particularly where the nontidal factor of landwater runoff is involved.

Landwater runoff not only increases the strength of ebb and decreases the strength of flood, but it also changes the time of occurrence of slack water. Slack before flood now comes later, whereas before ebb it comes earlier, resulting in increased duration of the ebb and decreased duration of the flood. In heavy landwater runoff these tendencies are magnified, and conversely during the drought season the differential is minimal. Because of these inequalities, the distance of horizontal travel on the ebb current is greater than that on the flood, and consequently there is a net displacement of the entire body of water toward the sea. Landwater runoff is therefore crucial to self-purification of estuaries; we must examine its effect in further detail.

NET TIME OF PASSAGE

Channel Volume Displacement Concept

In dealing with inland stream self-purification we have seen that the time of passage down the course of the stream from reach to reach is expressed as occupied channel volume displaced by runoff and that this determination is confirmed by dye tracer measurement and by computed and observed water quality profiles. This expression of the time of passage implies blocklike movement of the body of water, with practically complete vertical and transverse mixing throughout the cross sections but no appreciable longitudinal mixing. This simplified concept is equally applicable to most estuaries with but minor modification.

Blocklike movement of the water mass is characteristic of estuaries, and extensive longitudinal mixing does not occur; the steep water quality gradients observed in polluted estuaries confirm these similarities to inland streamflow. If longitudinal mixing dominated, steep gradients would be leveled and the observed water quality would show more uniform conditions over long reaches. Obviously with reversal of tidal currents some longitudinal mixing does occur in short reaches but not with the completeness assumed in some theoretical approaches. The more acceptable concept, therefore, is blocklike movement.

The net movement from reach to reach toward the sea is brought about by displacement of the occupied channel volume by the discharge of landwater into the tidal reach. Since the occupied channel volume at mean tide remains quite fixed, it is axiomatic that with the continuous addition of landwater runoff from the tributary drainage area an equivalent amount must find its way into the ocean; otherwise the channel level would continue to rise. Though a seaward movement takes place in steplike action as net excess of ebb over flood translation, conversion to mean tide is equivalent to elimination of these oscillations, and movement through the estuary may be considered as a continuous forward net time of passage similar to that in a freshwater stream: $T = V/Q$, where V is the occupied channel volume to mean tide, Q is the rate of landwater runoff, and T is the net time of passage.

Modification in Channel Volume

Only minor modification, depending on the configuration of the estuary, is necessary in computing the net time of passage. When a well-defined channel without bays leads directly to the ocean, no modification need be made except to convert channel soundings to mean tide. In other instances more or less isolated bays are interconnected to the main channel, and the portions that interchange with the main channel must be included as part of the total occupied channel volume. Similarly estuaries in their lower reaches commonly have a number of interconnected auxiliary channels. These auxiliary volumes, to the extent of flood or ebb translation, are collapsed into the main channel volume.

As with freshwater streams, channel volumes from section to section are determined from cross-section soundings along the course of the estuary. The net time of passage seaward is then determined as occupied channel volume adjusted to mean tide stage (with modifications where necessary) displaced by the tributary landwater runoff. Since the mean tide volume remains unchanged, the net time of passage for other runoff yields is inversely proportional to runoff: $T_2 = T_1(Q_1/Q_2)$.

INITIAL DISTRIBUTION OF WASTE LOAD BY TIDAL TRANSLATION

Unlike the wastes in the one-directional flow of an inland stream uninfluenced by tides, waste discharges from outlets along the course of an estuary lose identity with respect to their onshore locations because of tidal translation and overlapping distribution from adjacent outlets. The waste contributed in any given reach is not measured as sharp increments at individual outlets but as a continuous increment (expressed as a smooth distribution curve) integrated by ebb and flood translation. The nature of the initial distribution of wastes might best be visualized by considering the tidal estuary as a stationary body of water and the waste outlets as rotating through an arc up and down stream, spreading the waste over a reach of the estuary upstream of the outlet equal to the flood translation and downstream equal to the ebb translation, with overlapping sprays from outlets within their ranges of arc.

Employing this concept of distribution of effluent and the blocklike movement of the water mass, we can illustrate the method of determining the initial distribution of waste loading from multiple outlets by an example, estuary A.

Distribution from a Single Waste Outlet

First consider the initial distribution of waste by tidal translation from a single outlet. Current velocities in estuary A vary with stage of tide, landwater discharge, and geographic position depending on channel configuration. During extreme drought runoff ebb and flood currents are nearly equal, and Figure 8–2 shows the average tidal current cycle. Current velocity at strength approaches 1.8 mph, with a mean through the tide cycle of 1.1 mph, resulting in a translation on an ebb or a flood of about 7 miles. Thus, referred to a fixed onshore position, such as a waste outlet, tidal translation is conceived as a 7-mile block of water moving back and forth before that outlet. Such a movement through a flood and ebb cycle is illustrated in Figure 8–3. Commencing at slack water at the end of the ebb cycle, the upper end of the 7-mile block is directly opposite the onshore outlet. For each successive hour of the ensuing flood cycle the 7-mile block is moving upstream past the outlet, with the lower end of the block positioned opposite the outlet at slack water at the end of the flood; then translation reverses through successive hours of the ebb tide, and the 7-mile block moves downstream until again the upper end is positioned opposite the outlet at slack water at the end of the ebb. Thus, assuming a continuous discharge,

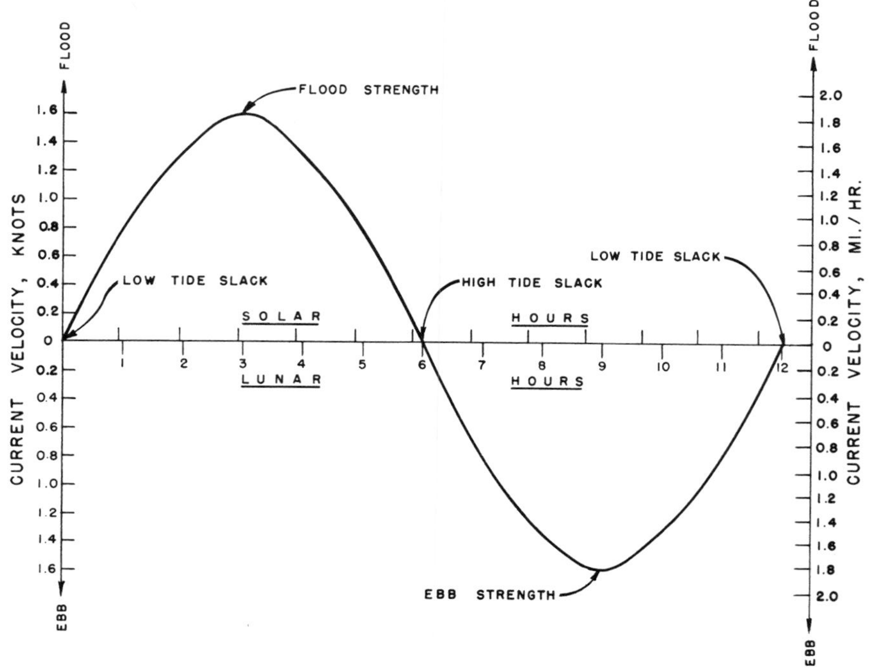

Figure 8–2 Estuary A. Average tidal current cycle (extreme drought).

waste from the fixed outlet is distributed throughout the 7-mile block and at the end of the flood translation is momentarily detectable at a position 7 miles upstream from the outlet and again at the end of the ebb translation is momentarily detectable at a position 7 miles downstream of the outlet. Thus the distribution from the single fixed outlet is identified over a range twice the tidal translation, or in this instance 14 miles.

The amount of waste at any fixed geographical position along estuary A is represented by a mean value through a tidal cycle passing such a fixed channel cross section. The mean value is a function of the proportion of the time the 7-mile block is positioned opposite the section. This places the maximum proportion of the waste discharged at a cross-section position opposite the outlet, with proportionally less at other cross sections within the 14-mile range, symmetrically distributed above and below the outlet in the form of a bell-shaped sine curve, such as the white portion of Figure 8–3 viewed from its side. Commencing at a point 7 miles upstream from the outlet and integrating this curve

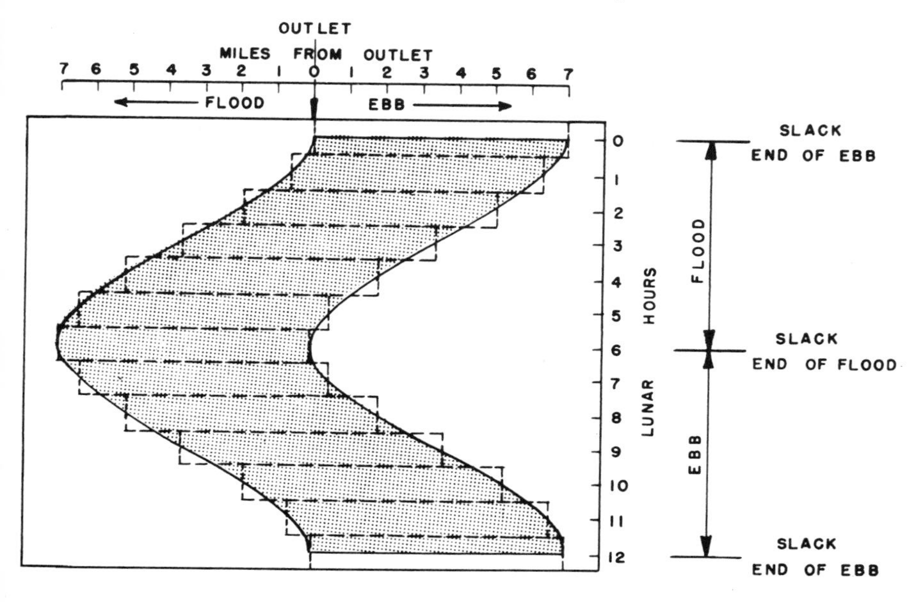

Figure 8–3 Estuary A. Translation of a unit 7-mile block of river water relative to a stationary pollution outlet at successive hours through a flood and ebb tide.

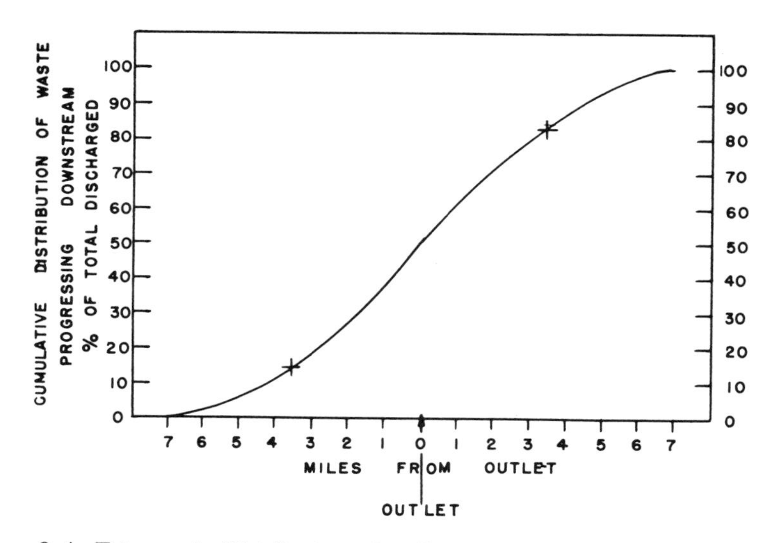

Figure 8–4 Estuary A. Distribution of pollution from an outlet by tide (mean value passing a fixed channel section through a tidal cycle).

progressively downstream gives the cumulative proportion of the total waste discharged at or above any cross section in the 14-mile reach. Such a cumulative integration is illustrated as the mass curve in Figure 8–4. Since the initial assumption is an equal flood and ebb translation, 50 percent of the total is distributed upstream of the outlet and 50 percent downstream. It is further noted that only 15 percent is distributed above a location 3.5 miles upstream of the outlet, and likewise 15 percent is assignable below 3.5 miles downstream, leaving 70 percent within the central 7 miles about the outlet. Through this reach the distribution curve is approximately linear, a convenient feature facilitating integration of wastes from overlays of multiple outlets along the course of the estuary.

Integrated Distribution from Multiple Outlets

The Mass Curve Moving Average Approximation

If analysis of cumulative distribution from a single outlet (developed in Figures 8–3 and 8–4) validates the assumption that linear distribution is a reasonable representation and that a single mean tidal current is essentially representative of the estuary, it is possible greatly to simplify the complex task of integrating the cumulative distribution from overlapping multiple waste outlets. This simplification is to apply a *moving average* to a mass curve of cumulative waste loads along the course of the estuary.

The application of this simplification is illustrated for estuary A in Figure 8–5. The first step is to plot the waste loads as a cumulative mass curve commencing with the upstream residual and progressively adding successive loads at the mile locations of each outlet, shown as curve A in Figure 8–5. With the mean tidal current velocity as 1.1 mph, the ebb and flood translation over which each outlet discharge is distributed, we have seen, is 14 miles. Curve B of Figure 8–5 is the trace of moving average values through the mass curve A. It is produced by applying a 15-point moving average over 14 miles to the ordinates of curve A, commencing 14 miles above mile 21 (i.e., just above the first waste outlet), placing the average of the 15 ordinate values at the midpoint, and regularly at 1-mile intervals dropping an upstream ordinate and adding a downstream ordinate. This smoothed curve B then approximates the *initial integrated distribution* of overlappings of the multiple outlets effected by ebb and flood translation.

Individual sources of waste now lose their identity and are replaced by increments of waste loading mile by mile along the course, not as specific contributions at each outlet but as the differential between any

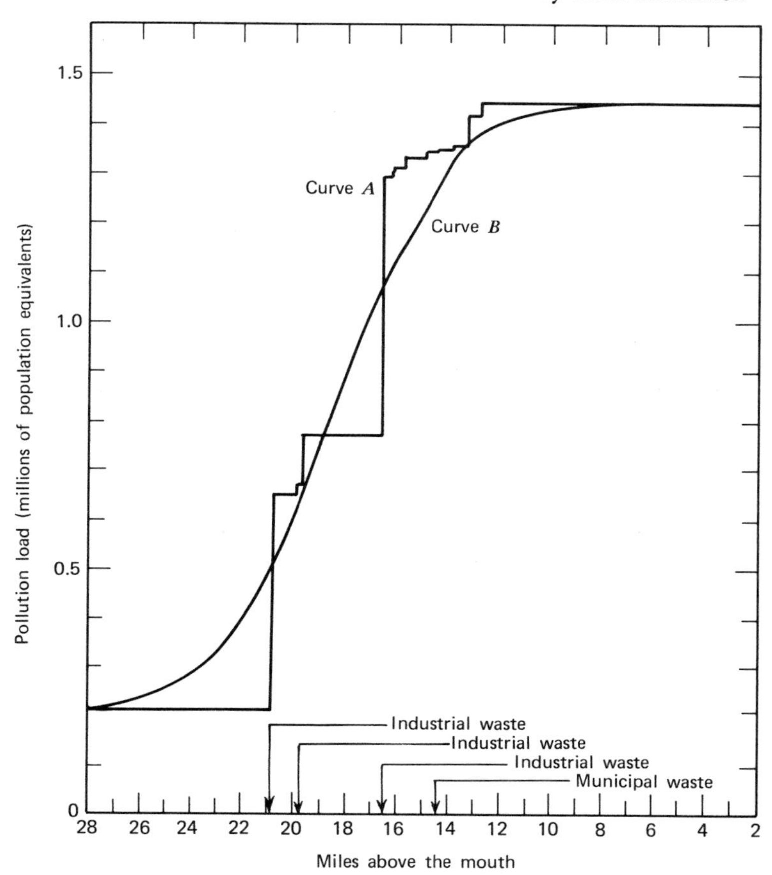

Figure 8–5 Estuary A. Distribution of pollution leads by tides. Curve A: mass curve of sources of pollution without tidal oscillation; curve B: integrated distribution by ebb and flood oscillation.

two successive mile points along the integrated distribution curve B. Differential increments may be taken at shorter or longer unit distances; usually for tidal translation of the magnitude in this illustration differentials at 1-mile intervals are adequate.

These differential increments replace the original sources and are dealt with as if they were the initial waste loading introduced along the estuary. With this modification in the expression of waste loading, deoxygenation can now be computed in the same manner as for inland, nontidal streams. The time of passage to the location of each increment is readily determined from the net time of passage curve for the particular runoff regime. The differential increments of waste load are integrated for this time of passage from increment to increment at the appropriate deoxygenation rate, thus establishing the deoxygenation down the course of the estuary.

Interdistribution among Tributary Branches

In large tidal streams it is not uncommon for tributary drainage systems to join the main estuary along the tidal reach and in turn become tributary estuaries. These may vary from a simple interconnection between two branches to complex interconnections among multiple branches with a common outlet, such as New York Harbor with the Hudson River, the East River, and the Hackensack–Passaic–Newark Bay–Kill Van Kull system, all discharging into Upper New York Bay with a common outlet through the Narrows. In such estuaries interchange takes place among the tributary branches on ebb and flood translation with interdistribution of waste load liabilities and of self-purification assets. Such interdistributions are taken into account in developing loading and water quality profiles for each branch of the estuary. However, in considering a location below the junction of tributary estuaries such interdistribution need not be developed, since the downstream condition may be obtained as the composite summation of each branch integrated separately to the downstream location in question.

Interdistributions are complex, but rational simplifications permit practical solutions. The basis for determining interdistribution usually depends on two factors which are readily measured: the extent of ebb and flood translation and the relative proportion of the total floodflow at the junction that extends up the main estuary to that which extends up the tributary estuary. On the ebb tide the channel contents of the main estuary and its tributary are combined and mixed laterally so that on the flood tide the combined contents are returned and redistributed to each in proportion to their respective tidal flows. The proportion of each estuary's liabilities (or assets) that is thus mixed can be estimated from the length of time during the ebb period that the liability (or asset) from a particular section of one estuary is commingled with that of a corresponding section of the other.

The following example, illustrated in Figure 8–6, outlines the general procedure for a main estuary B and a major tributary estuary C. Ebb and flood translation is determined as approximately 8 miles (42,000 ft) in both estuaries, and the proportion of flood tide flow up B and C is 62 and 38 percent, respectively. Both estuaries receive discharge of wastes at multiple outlets along their respective courses, including locations below the junction. The initial distributions as determined by

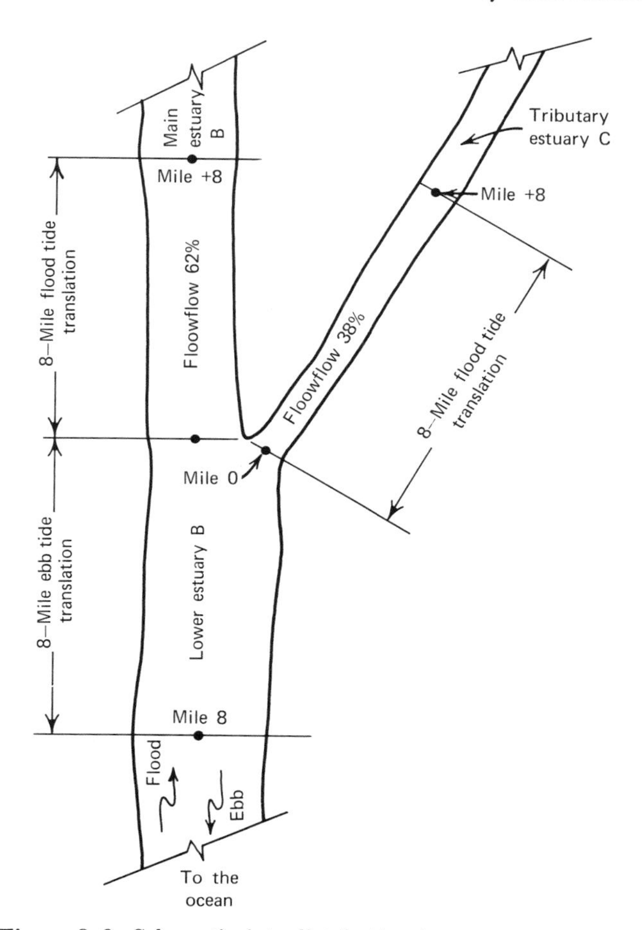

Figure 8–6 Schematic interdistribution between two estuaries.

a 17-point moving average at approximately 1-mile intervals through the mass curves of loading, considering each estuary separately (uninfluenced by interdistribution), are shown in Figure 8–7, curve *A*; curve *B* represents the distribution of waste loading located above the junction for estuaries B and C, respectively, and curve *C* represents the distribution of waste loads located below the junction along lower estuary B. It will be noted that portions of the downstream sources are distributed upstream above the junction and that portions of the upstream sources are distributed below the junction.

If the period of commingling on the ebb tide is taken as inversely proportional to the distance above the junction (a reasonable representa-

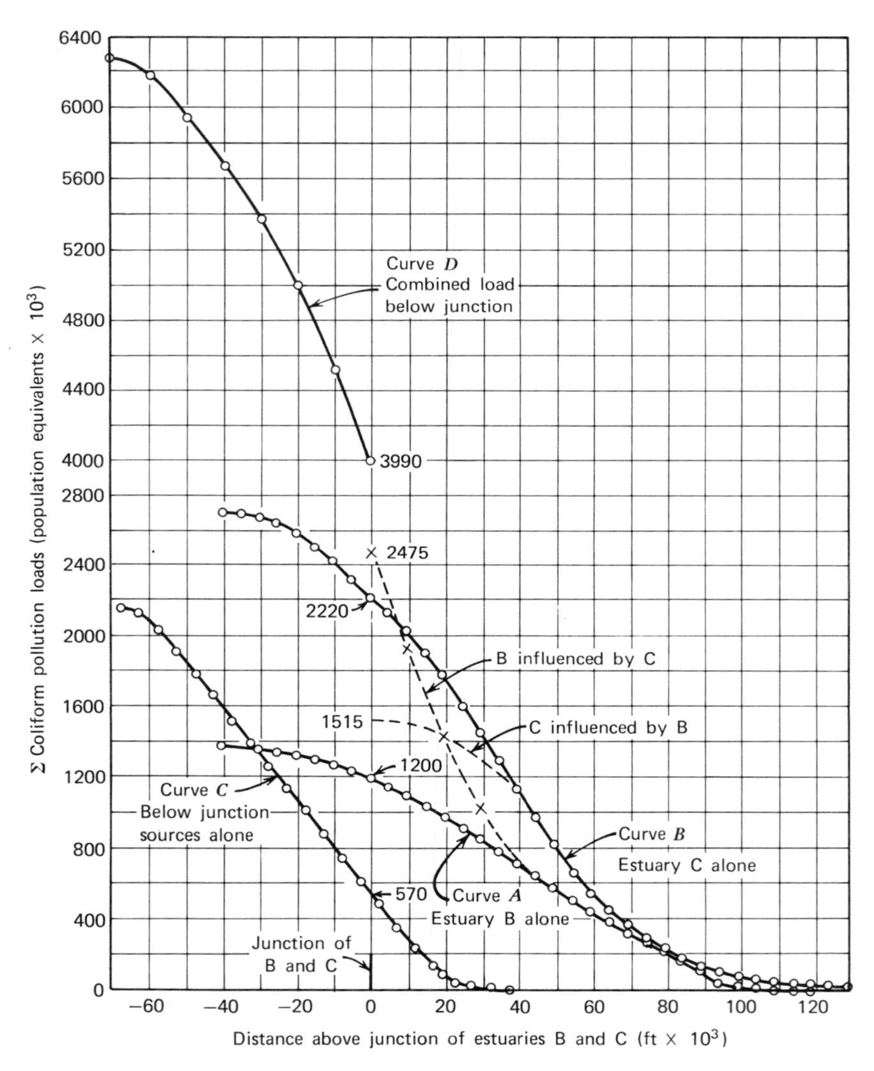

Figure 8–7 Interdistribution between estuary B and tributary estuary C (tidal range 42,000 ft).

tion, as shown in the development of the moving average), interdistribution of the separate distributions of waste loading is readily developed. For the sections at the junction at mile 0 the two estuaries are together for the complete period of the ebb tide and hence are considered completely mixed laterally.

The combined waste loadings of curves *A*, *B*, and *C* (expressed in thousands) at the junction is 3990 PE (1200, 2220, and 570); 62 percent, or 2475 PE, is redistributed to estuary B, and 38 percent, or 1515 PE, to estuary C. At the section 10,000 ft above the junction 10/42 of the time the sections are not commingled on the ebb, and 32/42 of the time the sections are laterally mixed. The separate waste loading at this section for estuary B from curve *A* is 1100 PE, 10/42, or 483 PE, remains unmixed and 32/42, or 1547 PE, is mixed. The combined portion mixed is 2385 PE; 62 percent, or 1479 PE, is redistributed to estuary B, and 38 percent, or 906 PE, to estuary C. These redistributed loadings, added to the respective unmixed portions, give the interdistribution at section 10,000 in estuary B as 1741 PE (1479 + 262), and in estuary C as 1389 PE (906 + 483). These values are the interdistribution that would obtain between estuaries B and C at this section were no waste loads added below the junction. However, the back distribution (from sources below the junction) at the section 10,000 from curve *C* is 310 PE, of which 62 percent, or 192 PE, is assignable to estuary B, increasing the total to 1933 PE; and 38 percent, or 118 PE, is assignable to estuary C, increasing the total to 1507 PE.

In a similar manner interdistribution is obtained for sections further above the junction, decreasing the influence of one estuary on the other until at mile 8 the estuaries remain completely unmixed throughout the ebb; hence waste loadings remain independent at and above mile 8. The interdistributions shown in Figure 8–7 as dashed lines indicate how significantly one estuary can influence another through the reach of the flood tide translation.

Downstream of the junction the three separate initial distributions of loading, ordinates of curves *A*, *B*, and *C*, are combined as a single distribution along the main estuary, shown as curve *D* in Figure 8–7. Differential increments taken from curve *D* then constitute the initial loading on the downstream reach. These differential increments for the downstream reach are substantially greater than that obtained from curve *A* (which represents the case were no waste discharged along the tributary estuary). Thus a polluted tributary estuary influences both the upstream and the downstream reaches of the main estuary.

So far we have considered interdistribution of initial waste loadings. If these are stable nondegradable wastes, the cumulative moving average mass curve represents the loading at any section taking into account the upstream reach beyond the influence of interdistribution. In converting these quantity loadings to concentrations the dilution landwater runoff of the main estuary and the tributary estuary must be also interdistributed. Freshwater assets are dealt with in the same manner as interdistribution of waste liabilities. If seawater intrusion is involved through the brackish reach, replacement seawater (as defined later) is a dilution asset also subject to interdistribution.

In considering unstable degradable wastes, such as bacteria or radioactivity, in which interest centers on the surviving residuals along the course, it is usually more convenient to integrate the initial waste distribution increments of each branch separately at the net time of passage and apply interdistribution to the residuals. Again if concentrations are required, dilution assets are also interdistributed.

Special consideration is required in dealing with organic wastes since reaeration is involved, and interdistribution includes this asset. Reaeration is dealt with by interdistributing the resultant dissolved oxygen profiles, which by successive approximations are balanced with the changes in reaeration induced by interdistribution in the affected reaches. Computer adaptation greatly facilitates this operation.

SEAWATER INTRUSION

Quite apart from its influence on self-purification in estuaries, seawater intrusion affects industrial and municipal water supply, boiler and condenser water for thermal power plants, fish and shellfish harvesting, irrigation, harbor and marine structures and machinery (by corrosion), and channel configuration (through deposition induced by coagulation of suspended and colloidal matter). Consequently much study and investigation continue to be devoted to identification and understanding of the multiplicity of the interrelated factors that determine seawater intrusion. Currently there is no workable theoretical basis for predicting what will occur under the varied and complex conditions encountered for all phases of the tide. Nevertheless rational simplification is possible for practical solutions to special problems.

Salinity Gradient, the Brackish Reach

In any landwater stream discharging into an ocean, seawater intrudes upstream counter to the landwater flow to form a brackish reach that extends well beyond the flood tide translation. The extent of the intrusion depends primarily on the characteristics of the local tide, the configuration of the channel, and the tributary landwater runoff. The tongue of seawater intrusion reduced to mean tide condition usually extends in a well-defined salinity gradient from a location near the mouth to a location of about 10 percent seawater and then diminishes at a more gradual rate to a location of 100 percent landwater. As landwater runoff increases, the salinity gradient is moved seaward; conversely, as runoff declines, the salinity gradient moves upstream, extending the seawater intrusion.

Under a steady hydrologic regime the mean tide salinity gradient comes to an equilibrium position; during freshets or periods of marked change in the rate of runoff the gradient is unstable and erratic: landwater flows near the surface, seawater near the bottom.

Stable equilibrium salinity gradients usually prevail during the low-runoff drought season, when these overruns and underruns are minimal, with quite uniform salinity distribution throughout the channel cross section. The stable condition at this period permits evaluation of the critical waste assimilation capacity of the estuary.

Hydraulic and Channel Factors

Channel configuration—particularly mean cross-section, mean depth, bed profile, and hydraulic gradient—affects the seawater intrusion pattern. The extent of seawater intrusion (other factors being constant) is a function of the difference in specific gravity between the seawater and the landwater, the mean water depth of the estuary, and the mean prevailing hydraulic gradient of the water surface.

Basically seawater intrudes on landwater by virtue of its greater density; the greater the depth, the greater the hydraulic superelevation of landwater required to counterbalance the heavier seawater. For example, if the specific gravity of seawater is taken as 1.03, then for a depth d the difference in specific gravity represents a hydraulic head of 0.03 d. This difference in head, added as a superelevation to the landwater, would produce an equilibrium of pressures at the bottom B ($p' = p$), but would result in an unstable condition at all lesser depths.

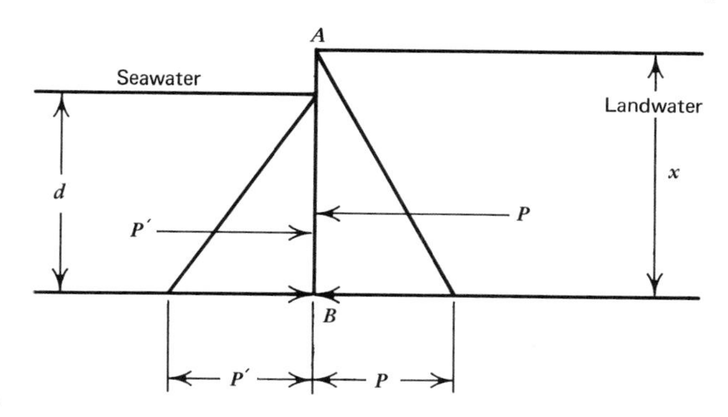

Taking moments about a vertical plane representing a theoretical gate (A,B), the equilibrium of total pressure on the two sides of the gate is obtained when the landwater depth x equals $d\sqrt{1.03}$, or a difference in hydraulic head of 0.015 d. This difference in head will produce hydraulic equilibrium for the section as a whole but will leave unbalanced pressures by virtue of which saltwater will tend to flow upstream on the bottom of the channel and landwater downstream on the top.

If the entire brackish water reach is considered as the gate separating 100 percent seawater from 100 percent landwater, the brackish reach will extend upstream until it acquires a difference in its hydraulic gradient of 0.015 d. Such a hydraulic gradient through the brackish reach will not permit further extension of seawater, but neither will it permit landwater movement toward the sea.

The Influence of Landwater Runoff

When a hydraulic gradient through the brackish reach does not permit further extension of seawater, the resultant condition represents only a momentary hydraulic equilibrium with no landwater flow seaward. There must be superimposed on the brackish water gradient a landwater gradient sufficient to induce net landwater flow to the sea equal to the tributary landwater runoff. It follows, therefore, that, as landwater runoff increases, increasing the hydraulic gradient through the brackish reach, the upper limit of seawater intrusion moves downstream to a new location so that the hydraulic superelevation balancing the seawater again approaches 0.015 d while at the same time the steeper gradient through the brackish reach allows net seaward flow of landwater counter to the seawater intrusion.

This generalization about the conditions of seawater intrusion assumes a reasonable uniformity of channel characteristics. Estuaries with irregular beds may show marked vertical stratification in salinity distribution, and deep pockets may accumulate salinity to a concentration approaching that of static water, with little or no interchange through the tidal cycles with the overlying active water. Estuaries of this type defy generalization; the only recourse is extensive stream sampling over the full range of tide and landwater runoff. In general the commonly encountered estuaries that enter the sea through alluvial coastal plains present more uniform channels; irregular conformation is usually associated with estuaries entering the sea from rugged terrain.

Specific Illustrations

In dealing with the common estuary of relatively uniform channel characteristics it is possible to use the general relations formulated for seawater intrusion to extend expected salinity gradients for a range of landwater runoff, even when field measurements are limited. For any given estuary one needs channel cross sectioning to characterize the channel hydraulic and salinity observations under runoff at which equilibrium conditions prevail. The sampling of salinity must be sufficiently intensive through tidal cycles to permit determination of mean tide values; three or more sampling locations in the brackish reach are necessary to establish the salinity gradient for one or more runoff regimes. Such analysis is best illustrated by application to a specific estuary.

Observed salinity data through the brackish water reach of Savannah Harbor illustrates a number of significant relationships. Three sets of salinity measurements were made in 1951 and 1952 by local interests, and in 1950 by the Army Corps of Engineers. The local surveys were made during fairly steady runoff, with three to nine samples in the cross-section taken daily at regular sampling stations over a period to reflect variations through a tidal cycle. These salinity measurements at each station are oriented to stage of tide at the date and time of sampling, providing a trace through the tidal cycle from which a mean tide value can be integrated.

The salinity measurements made by the Army Corps of Engineers were intensive through a tidal cycle on specific days. Simultaneous current and salinity readings were taken at about hourly intervals through a tidal cycle at 10 to 15 depth positions from surface to bottom, usually at the center of the main channel. From the daily hydrograph for 1950 (Figure 8–8), it will be seen that most measurements were taken during freshets or periods of marked change in stream discharge and that they therefore reflect unstable or erratic salinity conditions with underrun of seawater and overrun of landwater, as shown typically in Figure 8–9. The erratic distribution illustrates the importance of making salinity measurements under a stable hydrologic regime—sufficiently delayed after a preceding freshet to permit the salinity gradient to regain equilibrium. However, the measurements of October 12 were under favorable steady runoff conditions, covering five locations in the lower reach. The upper reach of the brackish water section was measured at three locations at about the same runoff on June 29–30 and August 31–September 1. These intensive data show quite uniform velocity and salinity from top to bottom, as reflected typically by Figure 8–10. This homogeneity is in marked contrast with the stratification observed during periods of unstable hydrograph, as reflected in Figure 8–9. The intensive salinity data at each section, reduced to mean tide through the tide cycle, provide the salinity gradient through the brackish reach.

Figure 8–11 shows the mean tide salinity gradients for these three survey periods; curve *A* represents the observed gradient for the survey period August–September 1951 at the stable landwater runoff of 4730 cfs (Clyo gage); curve *B* is the observed gradient for the May–June

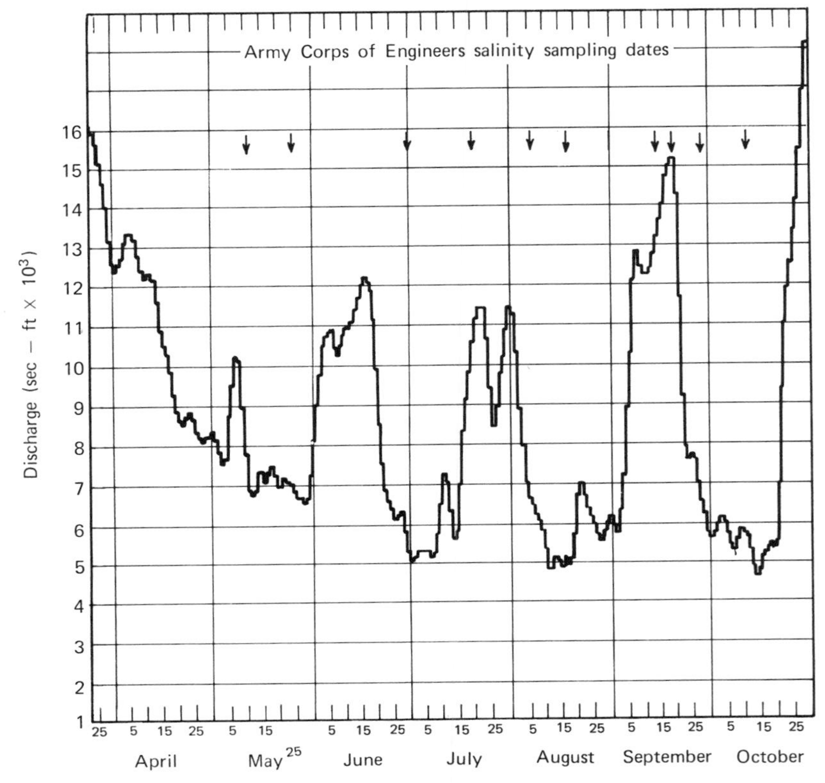

Figure 8–8 Savannah River. Daily hydrograph, 1950 (provisional). U.S. Geological Survey gage near Clyo, Georgia; drainage area 9850 square miles.

1952 survey at a runoff of 6590 cfs (Clyo gage); and curve *C* is the observed gradient for the selected periods of the Army Corps of Engineers 1950 survey at the mean prevailing runoff of 5700 cfs (Clyo gage).

Well defined salinity gradients are developed for each survey, with remarkably linear decline in seawater from a location near the mouth, upstream to the vicinity of Savannah City, then tailing off in a curve approaching 100 percent landwater in the vicinity of mile 20. As landwater runoff declines, the salinity gradient moves upstream; conversely, as runoff increases, the gradient moves downstream, which is consistent with blocklike movement and the generalized formulation outline above. Taking 100 percent seawater as 35,300 ppm total salt, the salinity gradients for the three runoff regimes of 6590, 5700, and 4730 cfs (Clyo gage) are 4.01, 4.21, and 4.48 percent per statute mile, respectively, averaging 4.21 percent. Although the slopes of the gradients are similar,

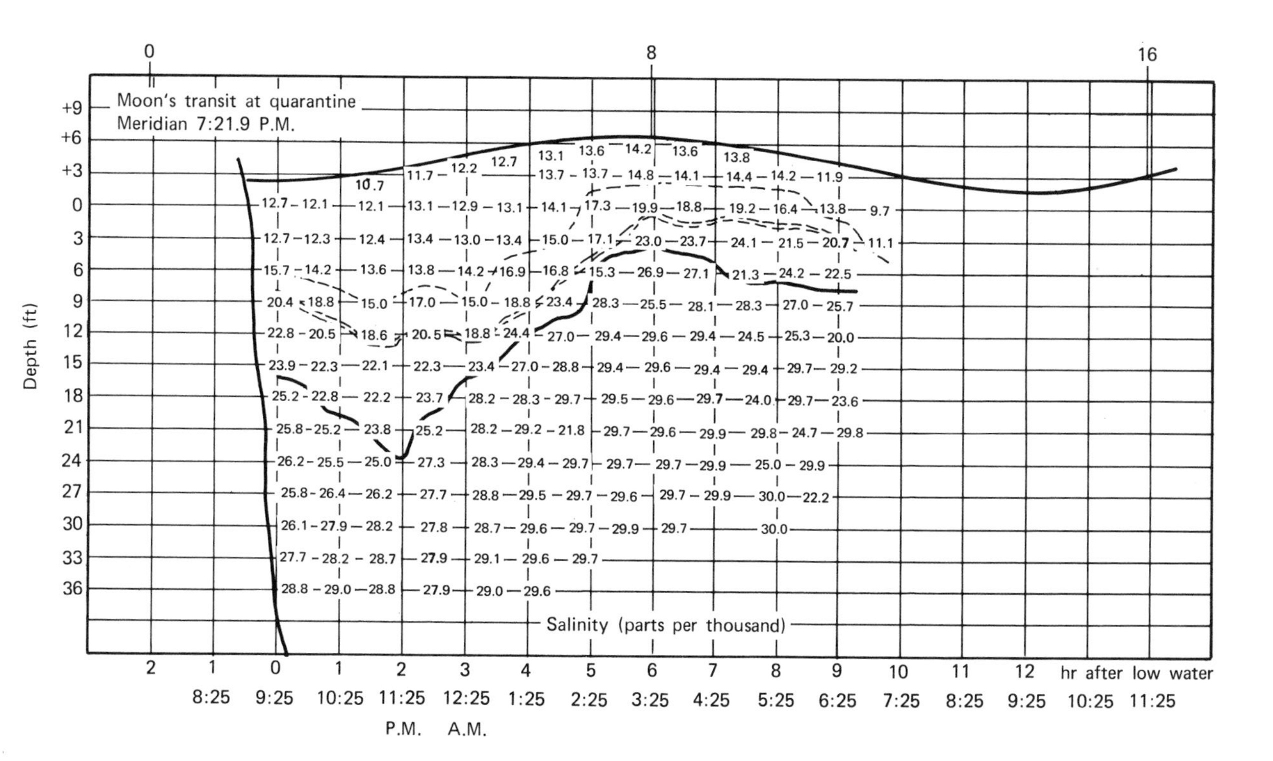

Figure 8–9 Savannah Harbor, station 173, center range; August 7, 1950.

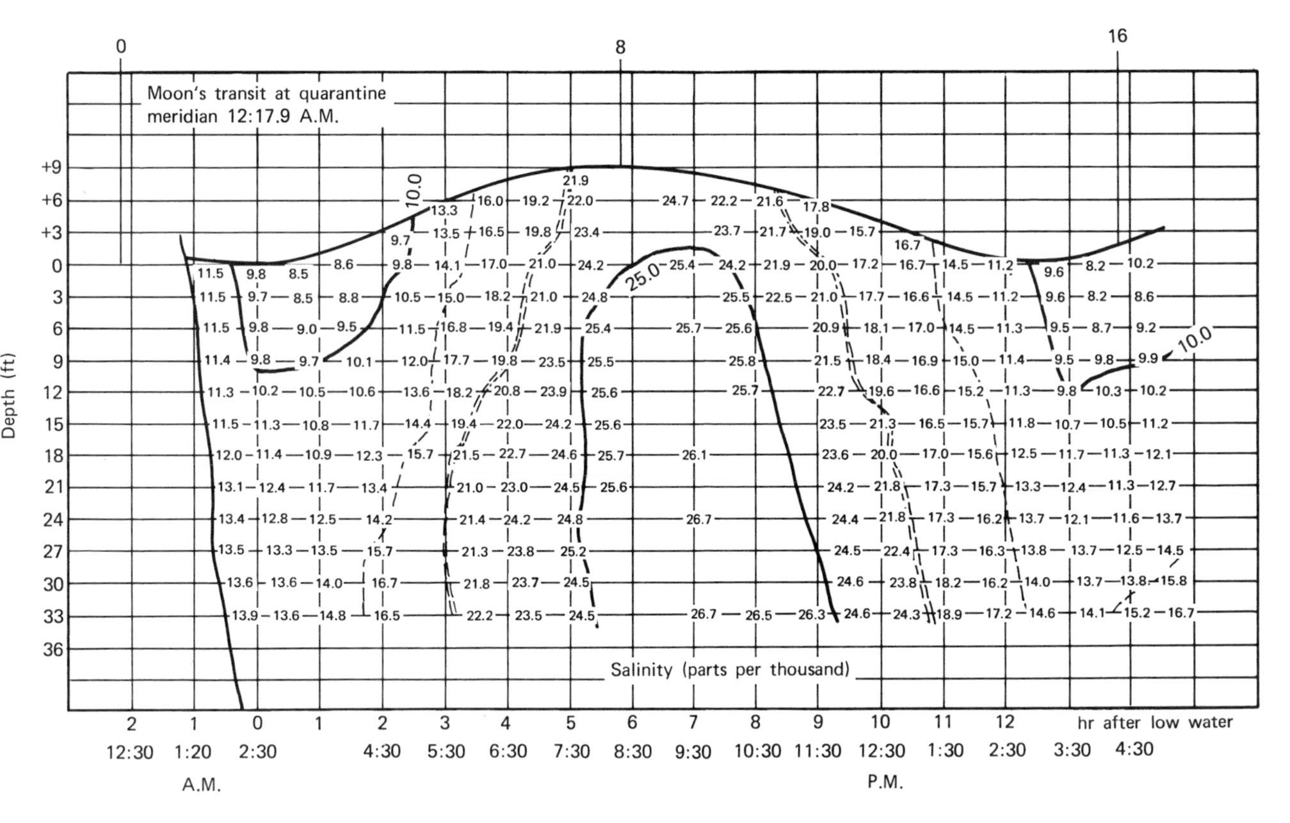

Figure 8–10 Savannah Harbor, station 173, center range; October 12, 1950.

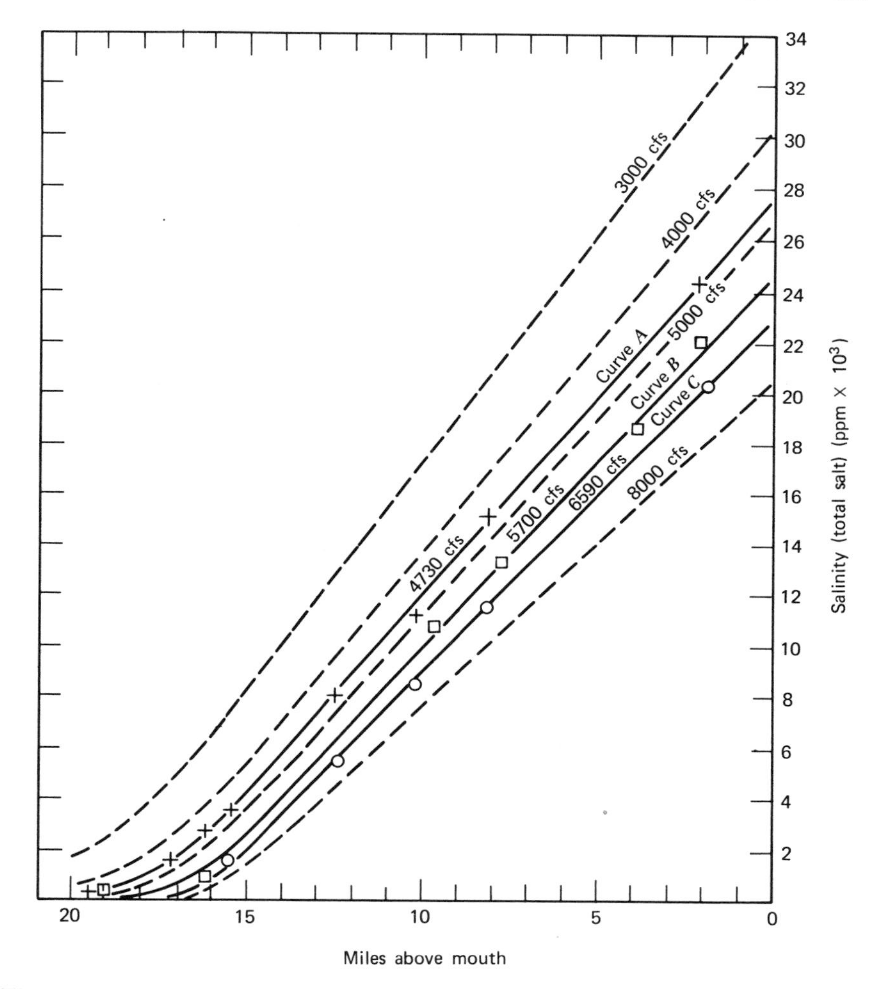

Figure 8–11 Savannah Harbor. Salinity gradient—mean through tidal cycle—at various freshwater runoffs (Clyo gage). Curve *A*: survey, August–September 1951; curve *B*: Army Corps of Engineers Survey, June 29, August 31, October 12, 1950; curve *C*: survey, May–June 1952.

there does appear to be a consistent slight increase in slope with decrease in runoff. After conversion to nautical miles, the average salinity gradient is 4.86 percent per nautical mile.

A detailed study [21] of the extensive salinity data of New York Harbor and Hudson River collected by the Metropolitan Sewerage Commission shows a well defined salinity gradient of about 2 percent per nautical mile from the Narrows upstream to the 10-percent seawater location, beyond which it flattens. This gradient persists through a wide range in landwater runoff through the seasons, moving seaward at high runoff and inland at low runoff, in accord with the generalized formulation.

Precise measurements by the Army Corps of Engineers of the difference in mean tide elevation between a downstream and an upstream location in the brackish reach, taken at a steady landwater runoff regime for a lunar month, confirm the generalized formulation. The superelevation of landwater at equilibrium with seawater with a specific gravity of 1.03 for the mean channel depth of 32 ft based on the general formulation is 0.48 ft, or 0.0096 ft per nautical mile. This is the initial hydraulic gradient necessary to counterbalance the heavier seawater with no seaward movement; to this is added the hydraulic gradient necessary to induce landwater movement through the brackish reach toward the sea. Reduced to mean tide conditions, the Corps of Engineers measurements disclose a water surface gradient of 0.0127 ft per nautical mile. The landwater runoff as a mean for the lunar period was 23,000 cfs. Hence the hydraulic gradient inducing this landwater flow is the observed total gradient minus the equilibrium gradient (0.0127 — 0.0096), or 0.0031 ft per nautical mile, a value predictable for a channel of the mean section and hydraulic radius for the reach involved.

The relation between hydraulic gradient and runoff can now be specifically defined; and at the salinity gradient of 2 percent per nautical mile the geographical position of seawater intrusion (such as to 10 percent seawater) can be determined for specific landwater runoff rate. The positions of salinity gradients computed by this method are found to be in close agreement with observed mean tide salinities developed from estuary sampling undertaken at stable runoff regimes.

As these illustrations suggest, each tidal estuary under a stable hydrologic regime establishes a quite definite mean tide salinity gradient through the brackish reach. The gradient is erratic during and immediately after freshets, but will again approach an equilibrium condition after a period of stable runoff. The mean tide gradient remains reasonably constant at different levels of stable landwater runoff but is shifted seaward as runoff increases and landward as runoff declines. If measurements by intensive field sampling are made during a period of stable runoff, these specific relationships can be defined.

Replacement Seawater Flow, Quantitative Determination

Landwater discharge to the sea is actually a complex process of landwater and seawater interchange through the brackish reach in a succession of tide cycles. By reducing to mean tide condition the process is simplified and the net effect of the interchange may be resolved into two opposite net flows, one equivalent to the landwater moving seaward and the other equivalent to the seawater intruding upstream through the brackish reach, discharging seawater into the downward landwater flow at such intervals and in such amounts as to provide the observed salinity gradient. Thus this hypothetical seawater river is a great flow near the ocean but gradually diminishes in volume as it proceeds upstream, vanishing at the upper end of the brackish reach. The magnitude of the landwater runoff is the key factor since it controls both the position of the salinity gradient and the amount of seawater contribution to maintain the salinity gradient.

Since it has been shown that under a stable hydrologic regime of constant landwater flow seawater intrusion approaches a stable salinity gradient, it follows that there must be a loss of landwater through the brackish reach to the sea, equivalent to the landwater discharge to the brackish reach.

It has also been shown that under such equilibrium conditions of stable runoff and stable salinity gradient landwater and seawater are well intermixed throughout the cross-section. Hence the loss of landwater passing any section in the brackish reach must take with it seawater in proportion to the landwater and seawater mixture at that section. In turn this seawater loss must be replaced by other seawater in equal quantity flowing upstream to maintain the stability of the salinity gradient at that cross-section. This economy results in a large replacement seawater flow near the outlet into the ocean where the proportion of seawater to landwater is great, progressively decreasing upstream along the salinity gradient to zero at the upper end of the brackish reach.

Thus the rate of replacement seawater flow available at any particular cross-section within the brackish reach is dependent upon the rate of landwater runoff and the proportion of seawater at that section defined by the associated salinity gradient. An increase in landwater runoff increases the loss to the ocean, but the seaward shift of the salinity gradient associated with increase in runoff decreases the proportion of seawater. The relative proportion of these opposing influences at different cross sections may produce quite different relationships between the landwater flow and the replacement seawater flow at various sections along the course through the brackish reach. Once the relation between the position of the salinity gradient and landwater runoff is established, the replacement seawater flow is then readily computed along the entire brackish reach for any runoff.

At any location in the brackish reach two kinds of valuable dilution are therefore available: landwater runoff and associated replacement seawater flow. Replacement seawater is a primary factor in the sharp recovery in estuaries, particularly through the lower reach of the brackish section. Its effectiveness also depends on the location of waste outlets in relation to the position of the salinity gradient and the nature of tidal translation affecting the initial distribution of waste loading. It is obvious, therefore, that the placement of waste outlets and the maintenance of adequate landwater runoff are both crucial to the waste assimilation capacity of estuaries.

WASTE-ASSIMILATION CAPACITY OF ESTUARIES

The modifications to self-purification to be applied to estuaries depend on the type of waste involved and whether it is discharged into the brackish reach or above the influence of seawater intrusion.

At this point it is well to reemphasize that in applying to estuaries the rational method of self-purification developed for inland streams a major simplification is effected by reducing tide cycle variations to mean tide condition. All computations are for mean tide with but one exception—namely, the initial distribution of the loading by ebb and flood translation, approximated by applying the moving average through the mass curve. However, since blocklike movement obtains in tidal estuaries, a good approximation of conditions expected at the end of the ebb tide and at the end of the flood is readily obtained from the computed mean tide condition by shifting the mean tide profile seaward one-half the ebb translation and upstream one-half the flood translation, respectively.

Where intensive sampling data reflect stable landwater runoff and minimum diurnal inequality of tide, plots for each sampling station produce a tide cycle pattern from which a mean tide value can be integrated and a value at the end of the ebb and flood can be identified, as illustrated by Figure 10–6. These values from respective sampling stations provide the mean tide profile and the profiles at the end of ebb and of flood. Where the water quality gradients are steep, with sharp differences at fixed sampling stations, it is consistently found that the positions of the end of flood and end of ebb profiles are located close to one-half the tidal translation on either side of mean tide. This evidence not only supports the simplified method for location of end of ebb and end of flood profiles, but more importantly it supports the concept of blocklike movement of the water mass so basic to the rational method of analysis.

Organic Wastes

Above the Brackish Reach

In estuaries well above seawater intrusion, the only modification necessary in calculation of self-purification is the initial distribution of wastes from single or multiple outlets induced by the local ebb and flood translation; of course it may also be necessary to take into account auxiliary channels and bays in determining occupied channel volume, net time of passage, and the interdistribution among branches. After the initial waste load distribution, ebb and flood action can be eliminated in practice by reducing to mean-tide conditions; in effect the estuary is dealt with as a continuously seaward moving stream. Deoxygenation is dealt with as in inland streams. The differential increments of BOD loading taken from the moving average mass curve are integrated at the net time of passage for the particular runoff regime from increment to increment and summarized as usual on Form A.

Similarly reoxygenation requires no special modification except that consideration must be given to shortening the mix interval in reaeration computation because of reversal in current with ebb and flood tide. A relation found to fit most large estuaries is shown as curve *B* of Figure 4–13. The oxygen balance and resultant dissolved oxygen profile are summarized as usual on Form B.

In the Brackish Reach

In dealing with organic wastes in the brackish reach the general procedure requires only two additional modifications, one affecting deoxygenation and the other reoxygenation. Quite apart from settleable solids subject to sludge deposit, organic wastes in contact with a high concentration of seawater have a tendency to coagulate nonsettleable solids and colloidal BOD, particularly in estuaries receiving turbid landwater runoff. In such instances deoxygenation must be modified to include a sludge deposit factor.

Dissolved Oxygen Assets of Replacement Seawater. One other modification in computing organic self-purification in the brackish reach is an allowance for the dissolved oxygen assets of the replacement seawater available along the course. Since replacement seawater flow initially derives from the ocean, it is considered to be, at its source, free of BOD and fully saturated with dissolved oxygen. If this intruding seawater flow were not required to pass through any zones of pollution, it would contribute at any cross section in the brackish reach its full asset of dissolved oxygen. Where organic pollution exists, however, the dissolved oxygen content of the replacement seawater is diminished, and the amount available from seawater at any cross section is a function of deoxygenation and reaeration in the reaches through which the replacement seawater must intrude. The available dissolved oxygen at any point is determined by distributing by small increments from section to section the deoxygenation and reaeration along the course of the brackish reach to the ocean. Commencing at the ocean end the calculation allocates deoxygenation and reaeration to replacement seawater in proportion to the seawater content in each small reach, taking into account the constantly diminishing quantity of replacement seawater flow progressively upstream, as determined by the salinity gradient.

This procedure constitutes no special problem insofar as deoxygenation is concerned, since the BOD satisfied along the entire course to the sea is readily determined from the load distribution integrated in the usual manner in accordance with the net time of passage, accounting for the complete satisfaction of any residual in the last increment into the sea.

A problem arises in assigning reaeration assets—it is necessary to

Table 8–1 Estuary D—Summation of BOD Amortization at 25.7°C and Control Landwater Runoff of 6850 cfs

	BOD Debt Amortized (Population Equivalents)		
Mile Point (1)	Colloidal and Dissolved Fraction (2)	Sludge at Equilibrium (3)	Total (4)
27.31	0	0	0
24.18	21,000	0	21,000
21.87	50,600	0	50,600
20.91	65,750	52,000	117,750
19.75	118,900	104,000	222,900
16.67	231,700	169,150	400,850
14.33	402,000	257,400	659,400
12.58	574,000	280,000	854,000
10.30	715,000	280,000	995,000
7.95	835,200	280,000	1,115,200
5.94	922,200	280,000	1,202,200
3.98	994,200	280,000	1,274,200
0 (Mouth)	1,080,200	280,000	1,360,200

know the reaeration assets for the entire reach to the sea. Since this quantity is not known until the entire dissolved oxygen profile has been computed, it is necessary to utilize the familiar procedure of successive approximations. Initially assume a complete reaeration summation curve through the entire reach to the ocean; solve for the resultant dissolved oxygen asset of the replacement seawater section by section; compute the entire dissolved oxygen profile along the course to the sea; and finally check the computed reaeration summation curve against the assumed curve. Successive approximations are made until what has been assumed checks closely with that computed. This computation is much more difficult than the usual procedure employed in computing reaeration reach by reach because here the entire reaeration is needed in a succession of reaches. Plotting the reaeration asset as a separate summation greatly

Table 8–2 Estuary D—Replacement Seawater Based on Salinity Gradient at Landwater Runoff of 7060 cfs

	Salinity Gradient		
Mile Point (1)	Percentage of Seawater (2)	Percentage of Landwater (3)	Replacement Seawater (cfs) (4)
0 (Mouth)	65.16	34.84	13,204[a]
1	61.19	38.81	11,131
2	57.23	42.77	9,446
3	53.40	46.60	8,090
4	49.29	50.71	6,862
5	45.47	54.53	5,887
6	41.50	58.50	5,008
7	37.54	62.46	4,243
8	33.57	66.43	3,567
9	29.60	70.40	2,968
10	25.78	74.22	2,452
11	21.81	78.19	1,969
12	17.85	82.15	1,533
13	13.88	86.12	1,137
14	9.91	90.09	777
15	5.95	94.05	446
16	2.12	97.88	152
17	0.14	99.86	10
18	0.00	100.00	0

[a] Landwater = 7060 cfs, or 34.84 percent; seawater = (7060/0.3484) × 0.6516 = 13,204 cfs.

aids in evaluating the magnitude and direction of successive approximations; the process is nevertheless tedious and time consuming. With the advent of high-speed computers the task has been greatly facilitated.

The Computed Dissolved Oxygen Profile and Verification against Observed Values. The procedure in computing the dissolved oxygen profile through the brackish reach is best illustrated by an example drawn from a study of estuary D. A waste survey, intensive estuary sampling, and good channel cross-sections established the salinity gradient, the net time of passage, the organic pollution loading, the residual BOD

Table 8–3 Estuary D—Dissolved Oxygen Assets of Replacement Seawater at 25.7°C and a Landwater Runoff of 7060 cfs

			Computed Dissolved Oxygen	
Mile Point (1)	Replacement Seawater (cfs)[a] (2)	Dissolved Oxygen (Population Equivalents at Saturation) (3)	Percentage of Saturation at Section (4)	Population Equivalents at Section (5)
0 (Mouth)	13,204	1,970,037	100.0	1,970,037[b]
1	11,131	1,660,745	99.7	1,653,395
2	9,446	1,409,343	99.2	1,396,990
3	8,090	1,207,028	99.0	1,193,330
4	6,862	1,023,810	99.0	1,012,000
5	5,887	878,340	99.2	870,908
6	5,008	747,194	99.4	740,830
7	4,243	633,056	99.0	626,000
8	3,567	532,196	99.0	526,671
9	2,968	442,826	98.2	435,040
10	2,452	365,838	96.9	354,600
11	1,969	293,775	94.6	278,410
12	1,533	228,724	91.8	210,040
13	1,137	169,640	87.0	147,450
14	777	115,928	81.1	94,160
15	446	66,543	75.1	50,340
16	152	22,678	67.5	15,315
17	10	1,492	61.9	924
18	0	0		

[a] Column 4 of Table 8–2.

[b] Saturation of Seawater at 25.7°C = 6.65 ppm;

$$\frac{6.65 \times 8.34}{0.24 \times 1.547} = 149.2 \text{ PE/cfs of replacement seawater,}$$

13204 cfs × 149.2 = 1,970,037 PE..

from upstream, and observed dissolved oxygen values. The BOD increments of the colloidal and dissolved fraction were determined from the moving average of the mass curve at mile intervals along the course and were integrated at 25.7°C ($k = 0.13$) for the landwater runoff of 7060 cfs prevailing during the estuary sampling; the sludge deposits involved were considered at equilibrium accumulation. Table 8–1 summarizes the amortization of the sludge and the colloidal and dissolved fraction BOD through the brackish reach to the mouth.

Replacement seawater, based on the salinity gradient and landwater runoff of 7060 cfs, was computed at mile intervals as listed in Table 8–2. The dissolved oxygen asset of the replacement seawater at any particular section in the brackish reach is dependent on the replacement seawater flow reaching that section and the deoxygenation and reaeration below that section to the mouth through which it must pass. The saturation values of the replacement seawater flow and the residual dissolved oxygen contained when reaching the section (expressed in population equivalents) are summarized in Table 8–3; the computations are illustrated as follows. As an aid to computing the residual dissolved oxygen

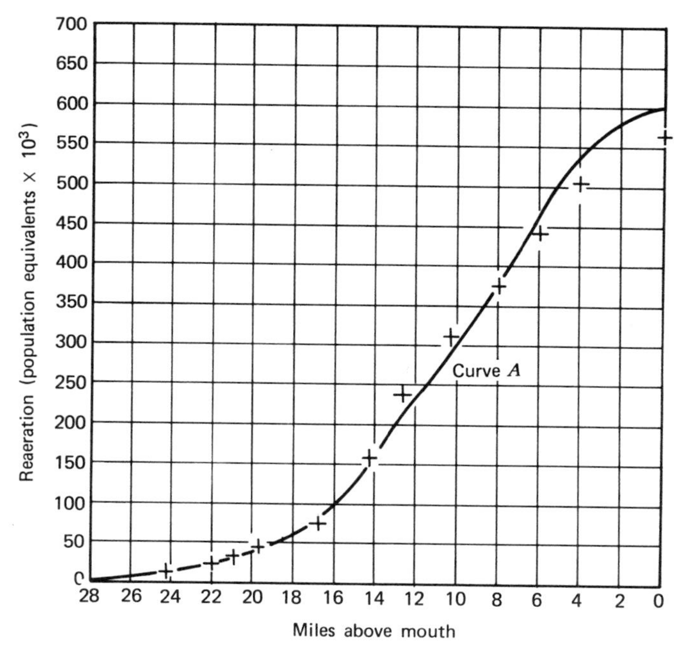

Figure 8–12 Estuary D. First approximation of reaeration. Data from previous computation: —•—; data from column 13 of Table 8–4: +.

of the replacement seawater, the BOD amortization is plotted to permit interpolation of increments between mile points. Since it is necessary to have also a summation plot of reaeration from reach to reach to the mouth, reaeration obtained from previous computations was used as an approximation (shown in Figure 8–12). It was assumed that any residual BOD at the mouth escaped to the open sea and that replacement seawater entering at the mouth was saturated with dissolved oxygen. The computations were made at 1-mile intervals commencing at the mouth, as illustrated for the first 2-mile reaches.

Mile 0 to 1:

Available at mile point 1 at saturation (column 3 of Table 8–3)	1,660,745 PE
BOD amortized in the reach (from plot)	−20,000 PE
Reaeration increment gained in the reach (Figure 8–12)	8,000 PE
Net change	−12,000 PE
Seawater at mile point 1 (column 2 of Table 8–2)	61.19%
Net change assigned to seawater (0.6119 × 12,000)	−7350 PE
Residual dissolved oxygen in seawater at mile point 1 (1,660,745 − 7350)	1,653,395 PE
Percentage of saturation of replacement seawater (1,653,395/1,660,745)	99.7%

Mile 1 to 2:

Replacement seawater available at mile point 2 at 99.7% saturation (0.997 × 1,409,343)	1,401,000 PE
BOD increment amortized in the reach	−21,000 PE
Reaeration increment gained in the reach	14,000
Net change	−7,000 PE
Seawater at mile point 2	57.23%
Net change assigned to seawater (0.5723 × 7000)	−4010 PE
1,401,000	
−4,010	
1,396,990	99.2%
1,396,990/1,409,343	

In a similar manner the dissolved oxygen content of the replacement seawater through the entire brackish reach was computed, as summarized in columns 4 and 5 of Table 8–3.

We are now in a position to compute by the rational method the dissolved oxygen profile through the brackish reach, considering both landwater and seawater from section to section commencing at mile point 27.31 at the first increment of back distribution of sources of pollution and proceeding to the mouth. These computations are summarized in Table 8–4 on Form B as modified to include the replacement seawater; columns 4 and 6 are obtained from the channel cross-section soundings, column 5 is obtained from Figure 4–13, curve *B*, for estuaries, column 7 is the BOD amortization including the sludge deposit (column 4 of Table 8–1), column 10 is the sum of the saturation values of the landwater runoff and the replacement seawater, column 14 is replacement seawater contribution obtained from interpolation of a plot of column 5 of Table 8–3. The reaeration and oxygen balance in the reaches above seawater intrusion are computed in the same manner as illustrated in Chapter 4 for landwater streams. Commencing with the reach from mile point 16.67 to 14.33, replacement seawater is involved and is dealt with as follows:

Mile 16.67 to 14.33:

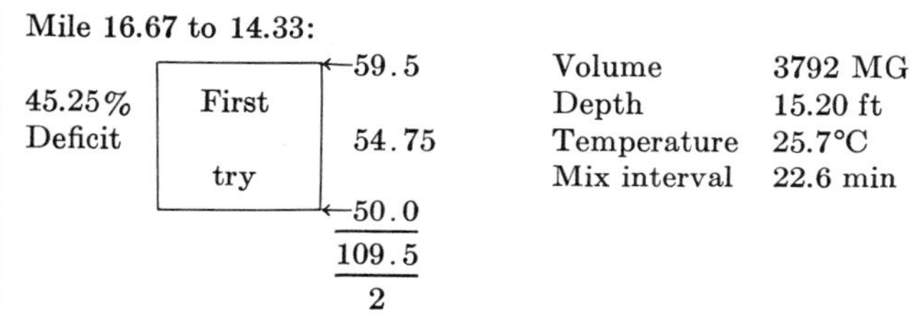

Dissolved oxygen per mix at 100% deficit (Figure 4–12) 0.265%
0.4525 × 0.265 × (1440/22.6) = 7.63% per day at 45.25% deficit
100% seawater taken as 35,300 ppm total salts
Mean percentage of seawater in the reach 7.9%
Saturation at 25.7°C = 8.2 ppm
8.2 × (8.34/0.24) = 285 PE/MG

Reaeration in reach 0.0763 × 285 × 3792 =	82,400 PE
Previous reaeration	77,500 PE
	159,900
Net	420,000
Replacement seawater	82,000
	661,900

Landwater 1,270,000 saturation, seawater 100,000 saturation
1,270,000 + 100,000 = 1,370,000
661,900/1,370,000 = 48.4% (too low)

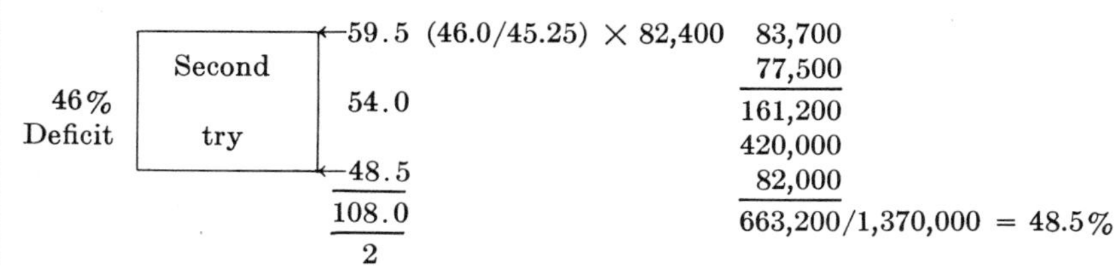

The computation continues in this manner, keeping in mind that, since the mean salinity content of each reach increases as the mouth is approached, the saturation value of dissolved oxygen declines in successive reaches.

The reaeration summation as computed (column 13 of Table 8–4)

Table 8–4 Estuary D

FORM B: REOXYGENATION AND OXYGEN BALANCE

Station (river mile) (1)	Runoff at Station (cfs) (2)	Mean Temperature (°C) (3)	Effective Depth (ft) (4)	Frequecy of Turn-over (min/mix) (5)	River Volume (MG) (6)	Total Demand Satisfied above Station (PE) Table 8-1 (7)	Dissolved Oxygen (Population Equivalents): Added by Runoff between Stations (8)	Total Added by Runoff above Station at 85% Saturation (9)	Total Runoff at Station at Saturation (10)	Net at Station (Column 9 minus Column 7) (11)	Added by Reaeration between Stations (12)	Total Added by Reaeration above Stations (13)	Total Added by Replacement Seawater (14)	Net Oxygen Balance at Station (Column 11 plus Column 13) (15)	Percent Saturation at Station Divided by Column 10 (16)
27.31	6850	25.7				0		1,080,000	1,270,000	1,080,000		0	0	1,080,000	85.0
			12.22	20.0	1070		0				9,900				
24.18						21,000		1,080,000	1,270,000	1,059,000		9,900	0	1,068,900	84.0
			12.80	20.6	1369		0				12,850				
21.87						50,600		1,080,000	1,270,000	1,029,400		22,750	0	1,052,150	82.7
			10.76	18.6	845		0				11,600				
20.91						117,750		1,080,000	1,270,000	962,250		34,350	0	996,600	78.4
			20.6	26.4	1541		0				12,350				
19.75						222,900		1,080,000	1,270,000	857,100		46,700	0	903,800	71.0
			19.73	25.9	2730		0				30,800				
16.67						400,850		1,080,000	1,270,000	679,150		77,500	0	756,660	59.5
			15.20	22.6	3792		0				83,700				
14.33						659,400		1,080,000	(L) 1,270,000 (S) 100,000	420,600		161,200	82,000	663,200	48.5
			15.70	22.8	3240		0				78,000				
12.58	6850					854,000		1,080,000	(L) 1,270,000 (S) 192,000	226,000		239,200	172,000	637,200	43.6
			19.60	25.8	3890		33,200				67,900				
10.30	7060					995,000		1,113,200	(L) 1,310,000 (S) 347,000	118,200		307,100	327,000	752,300	45.4
			19.57	25.8	4054		0				66,200				
7.95						1,115,200		1,113,200	(L) 1,310,000 (S) 541,000	−2,000		373,300	530,000	901,300	48.7
			15.47	22.8	3471		0				68,400				
5.94						1,202,200		1,113,200	(L) 1,310,000 (S) 755,000	−89,000		441,700	755,000	1,107,700	53.5
			18.90	25.3	4383		0				60,500				
3.98						1,274,200		1,113,200	(L) 1,310,000 (S) 1,016,000	−161,000		502,200	1,016,000	1,357,200	58.3
			23.40	27.8	7345		0				61,300				
0.0	7060	25.7				1,360,200		1,113,200	(L) 1,310,000 (S) 1,970,000	−247,000		563,500	1,970,000	2,286,500	69.8

L = landwater; S = replacement seawater.

is plotted in Figure 8–12 for comparison with the assumed reaeration employed in determining the oxygen assets of the replacement seawater. It is to be noted that the computed reaeration in the upper portion of the brackish reach is somewhat above that assumed (curve *A*) and somewhat lower toward the mouth. A refinement can now be made by recomputing the asset of replacement seawater by assuming as a second approximation of reaeration, that computed in Table 8–4; however, this will not significantly alter the final dissolved oxygen profile, and for the purpose of the illustration column 16 of Table 8–4 is acceptable.

For verification of the rational method adapted to estuaries the computed dissolved oxygen profile is compared with the observed values of the estuary sampling taken under the same conditions. Figure 8–13 shows the observed sampling results. Curve *A* is the integrated mean through the tide cycle, and curves *B* and *C* represent the end of the ebb and flood tide taken from smoothed plots of the tide cycle. The range in variation through the tide cycle defined by these profiles is shown by the shaded belt. Each daily observed dissolved oxygen value as the mean of three samples at each station unadjusted for phase of tide is also shown in Figure 8–13. The computed mean tide profile (column 16 of Table 8–4) is shown as curve *D*, which fits well within the observed dissolved oxygen variations and is in good agreement with the observed mean tide profile (curve *A*) through the decline and the critical reach and is 3 to 5 percent of saturation below the observed values through the recovery reach. It will be noted, however, that there is a wide range in the daily observed sample results through the recovery reach, in contrast with the more compact sampling results through the decline in profile. The computed recovery is, however, well within the variance of the observed data.

As an indication of the importance of taking replacement seawater into account, a computation was made without replacement seawater. Although the conditions for this computation were slightly different, it was comparable with that just illustrated, and the profile is shown in Figure 8–13 as curve *E*. Compared with curves *A* and *D*, the dissolved oxygen decline is substantially lower, with the critical dissolved oxygen in the vicinity of mile 8 at 26 percent as compared with 44 percent at mile 12.5; the recovery zone is completely below the observed values and 25 to 30 percent of saturation below the computed and observed mean tide profiles. This discrepancy emphasizes the substantial asset of replacement seawater in inducing the marked recovery shown by the computed profile (curve *D*) and that observed by sampling.

A further validation of the rational method for analysis of estuaries is reflected in the comparison of the computed and the observed dis-

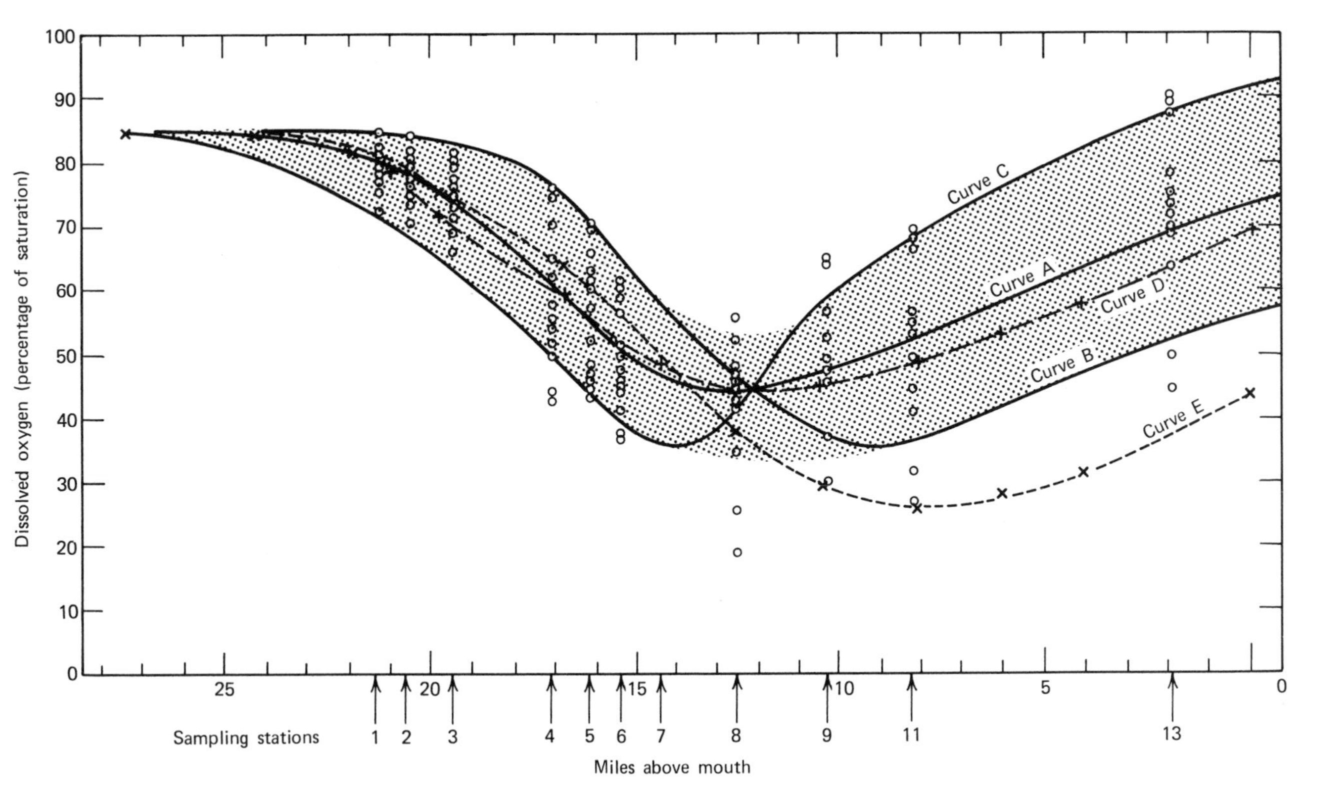

Figure 8–13 Estuary D. Comparison of computed and observed dissolved oxygen profiles, showing range through tidal cycle. Curve A: observed profile integrated mean for tidal cycle; curve B: observed profile, end of ebb tide; curve C: observed profile, end of flood tide; curve D: computed with replacement seawater; curve E: computed without replacement seawater; ○: daily observed dissolved oxygen mean of three samples (center of channel) unadjusted for phase of tide.

solved oxygen profiles for estuary D, as shown in Figure 10–7. These profiles represent conditions that prevailed 5 years after those represented by Figure 8–13, with a landwater runoff of 7500 cfs at the control gage, a somewhat different waste loading, and a substantially lower temperature, 13 to 17°C. The dissolved oxygen profile (curve *D* of Figure 10–7), computed by taking into account replacement seawater, is in good agreement with the observed intensive estuary sampling results.

BACTERIAL CONTAMINATION

For wastes that are degradable but do not demand dissolved oxygen, chiefly bacteria and radioactive material, the residual at any location depends on the survival to that point in the time of passage and is computed in the same manner as for an inland stream. In tidal reaches well above the seawater intrusion the initial distribution of loading is determined as for other wastes by the local ebb and flood translation integrated by the moving average mass curve. The differential increments of loading taken at short distances are then integrated at the net time of passage for the particular runoff from increment to increment at the appropriate death (or decay) rate, resulting in a sawtooth trace of residual survival along the course of the estuary. As in computing for inland streams, it is desirable to express loading in population equivalents, leaving the adjustment for tributary runoff increments until the final step of converting to concentration. At any location, then, the concentration is obtained by converting the survival in population equivalents to numbers of bacteria and dividing by the runoff (in appropriate units) tributary to that location.

If the loading extends into the brackish reach, the replacement seawater becomes an added dilution factor, with residual survival converted to concentration by dividing by the sum of the tributary landwater runoff plus the replacement seawater flow at the particular cross-section. For practical purposes it can be assumed that any increment of survivors carried upstream by replacement seawater may be ignored. In view of the uncertainties in numbers of bacteria contributed per capita and the rather wide variation in death rate in the stream environment, further refinement is not usually warranted in most estuary problems.

However, as with organic waste, a refining computation is to integrate the downstream increments below the brackish water cross-section in question, commencing at the ocean and employing passage time between increments the same as the net seaward time of passage. Survivals are assigned to replacement seawater in proportion to the seawater content defined by the salinity gradient; the number of survivors in each small reach is therefore reduced in accord with the diminishing magnitude of the replacement seawater flow as it progresses along the declining salinity gradient. A direct solution is possible here, obviating the successive approximations that are necessary in organic waste problems. It is evident that with both the high death (or decay) rate and the rapid decline in the flow rate of replacement seawater, the number of survivors thus carried by replacement flow to the upstream sections of the brackish reach is relatively small. Such survival at a particular section is added to that from the downstream integration to that section, and the net concentration is then determined by dividing the combined sum by the sum of landwater runoff and replacement seawater flow.

STABLE WASTES

Like other wastes, nondegradable stable waste products in the reaches of estuaries above seawater intrusion can be analyzed with only minor modification: the initial distribution of waste sources by the ebb and flood translation is integrated by the moving average through the mass curve. After the initial distribution of loading, ebb and flood translation is eliminated by reducing to mean tide condition. Since net time of passage is then a continuous seaward movement and the waste products are stable, the incremental loads along the course are cumulative. The pattern of waste concentration through the zone of initial distribution is therefore obtained by dividing the ordinate (*not* the differential increment) to the moving average mass curve by the tributary runoff expressed in appropriate units.

If the distribution of waste loading extends into the brackish reach, the concentration is obtained by dividing the ordinate by the sum of the tributary landwater and the replacement seawater at the particular section. Again this procedure assumes a minimal back distribution by replacement seawater flow beyond that provided in the initial ebb and flood translation, which for most practical problems is adequate. If further refinement is warranted, a procedure similar to that described above may be employed—except that there is no loss from degradation and the assignment to replacement seawater is in proportion to the seawater content in the small reach. However, there is of course left behind in each small reach an amount accumulated from previous reaches because of the decline in replacement seawater along the salinity gradient; therefore it is necessary to deal with small reaches commencing at the seaward end, employing in this case the differential increments from the moving average mass curve (not the ordinate). Any back-distribution residual contained in the replacement seawater at a particular section in the brackish reach is added to the ordinate at that section, and concentration is determined by dividing by the sum of landwater runoff and replacement seawater flow tributary to that section.

REFERENCES

[1] Parsons, H. de B., *Trans. ASCE,* **76,** 1979 (1913).
[2] Tully, J., *Fish Reserve Bd. Bull.,* **83,** (Canada) (1949).
[3] Ketchum, B. H., *Sew. Ind. Wastes,* **23,** No. 2, 198 (1951).
[4] Arons, A. B., and Stommel, H., *Trans. Am. Geophysical Union,* **32,** 419 (1951).
[5] Pritchard, D. W., *Advances in Geophysics,* **1,** 243 (1952).
[6] Stommel, H., *Sew. Ind. Wastes,* **25** (9), 1065 (1953).
[7] U.S. Corps of Engineering Reports on Delaware River Model Study, 1951–1954.
[8] Preddy, W. S., *J. Marine Biology Assn. U.K.,* **33,** 645 (1954).
[9] Rattray, M., Jr., and Lincoln, J. H., *Trans. Am. Geophysical Union,* **36,** 82, 251 (1955).
[10] Pearson, E. A., Reduced Area Investigation of San Francisco Bay, California State Water Pollution Control Board, July 1958.
[11] O'Connor, D. J., *Proc. ASCE,* **86,** SA 3, 35 (1960).
[12] Kent, R., *Proc. ASCE,* **86,** SA 2, 15 (1960).
[13] O'Connell, R. L., and Walter, C. M., *Proc. ASCE,* **89,** SA 1 (1963).
[14] Thomann, R. V., *Proc. ASCE,* **89,** SA 5 (1963).
[15] Pyatt, E. E., U.S. Geological Survey Water Supply Paper 1586F, 1964.
[16] Gameson, A. L. H., Barrett, M. J., and Preddy, W. S., *Int. J. Air Water Pollution,* **9,** 655 (1965).
[17] O'Connor, D. J., *Proc. ASCE,* **91,** SA 1, 23 (1965).
[18] Marmer, H. A., Special Publication No. 111, U.S. Coast and Geodetic Survey, Washington, D.C.
[19] Marmer, H. A., *The Tide,* Appleton Century-Crofts, 1926.
[20] Strahler, A. N., *The Earth Sciences,* Harper and Row (1963).
[21] Phelps, E. B., and Velz, C. J., *Sew. Works J.,* **5,** No. 1, 117 (1933).

9

IMPACT OF RIVER DEVELOPMENTS ON WASTE ASSIMILATION CAPACITY

DETRIMENTAL AND BENEFICIAL EFFECTS

Chapter 1 emphasized the role of satisfactory disposal of the end products of community and industrial activity as a limiting factor in urban-industrial growth and the importance of the waste assimilation capacity of streams, a key factor in multiple water resource use and development. It is easy to grasp the significance of the impact of waste disposal on other water resource uses, but all too seldom is adequate attention given to the effect of other water uses and river developments on the complex problem of waste disposal. Depending on how planned and operated, other water uses and river developments can be either detrimental or beneficial to the waste assimilation capacity of the stream and hence to water quality. This chapter deals with these detrimental and beneficial impacts.

Modification of Factors That Influence Self-purification

The factors that influence the self-purification of streams and estuaries as identified in Chapters 4 through 8 may be consolidated into three primary elements: (a) streamflow, (b) water temperature, and (c) channel characteristics. The latter involves a number of interrelated parameters—depth, surface area, volume, time of passage, and velocity—that are usually altered through change in depth of channel. As we have seen, every change in streamflow induces change in the channel parameters; thus runoff has a compounding effect. The impact of river developments on self-purification and waste assimilation capacity is in turn readily identified by how these three primary factors are altered.

It is difficult to conceive of any river development that does not change at least one of these primary factors. Therefore it is axiomatic that those responsible for satisfactory waste disposal and pollution and water quality control are concerned with every river development, existing and proposed. Concern is not only with the detrimental impacts but also with alterations in the primary factors that can be incorporated into river developments to enhance the waste assimilation capacity and water quality. The time to consider these potential effects is at the planning stage, and no river development should be undertaken without a careful evaluation of its influence on waste assimilation capacity.

A critical review of typical river developments—hydroelectric plants, thermal electric power plants, navigation, flood control, irrigation, and other diversions—illustrates the nature of the effects and ways of minimizing the detrimental and incorporating the beneficial.

HYDROELECTRIC POWER

Electrical energy cannot be stored in large blocks; it must be generated as demanded. It travels through the distribution system at the speed of light from the points of generation to the points of use. The practice is to carry the base load by large steam electric power generating units and to rely primarily on hydroelectric power installations to meet the peak demands. The unique ability of hydroelectric power generation is to be placed on the line quickly, a feature that is not possible with thermal generation. The amount of energy generated is a function of the rate of flow and the head on the turbines. Hydroelectric developments fall into two general classifications, run-of-the-river and storage type. The former depends on the natural stream runoff, usually with dams at falls to exploit the natural head available, but with little storage in the ponds behind the dams. Storage type developments involve large dams and reservoirs to exploit both head and flow, augmenting the natural runoff from storage. The impacts of hydro power developments are conveniently discussed as the downstream and the upstream effects.

The Downstream Effects

By its very nature hydro power operation induces radical diurnal and weekend pulsations in streamflow downstream; peak flow parallels peak energy demand during set hours of the day on weekdays and minimal flow (or none) corresponds to the low energy demand on weekends. The practice is to save as much water as possible behind the dam on weekends to be utilized for peaking on weekdays. Without equalization, these surges in streamflow persist for many miles downstream, as illustrated in Figure 3–28. A daily and weekly cycle in streamflow is thus established, as reflected in instantaneous and daily hydrographs, also illustrated in Chapter 3.

The impact on self-purification stems directly from change in streamflow, and in turn from corresponding alterations in channel parameters. For most existing hydro power installations, both the run-of-the-river and the storage types, the effect is detrimental. The weekend shutdown frequently reduces streamflow below that of the once in 10-year natural drought expectancy, and thus drastically limits waste assimilation capacity and potential community and industrial growth at downstream locations. Although our interest centers on the effect on waste assimilation capacity, such restriction in streamflow obviously also affects all other downstream water uses.

Two additional downstream impacts usually occur at existing hydropower installations, *lowered temperature* and *lowered dissolved oxygen.* The former may be beneficial, but the latter is frequently seriously detrimental. The impoundment behind the dam, if deep, results in thermal stratification, with the lower depths consisting of cooler water but usually low in dissolved oxygen, in many instances approaching 1 ppm or less. The intakes of penstocks leading to the turbines are usually at a fixed elevation in the lower depths of the reservoir to permit drawdown; thus the cooler water, depleted in dissolved oxygen, is withdrawn and discharged downstream.

Reaeration replenishes dissolved oxygen along the downstream reach, but usually, before there is adequate dissolved oxygen for self-purification, the benefit of cooler water is lost as the temperature approaches the equilibrium temperature of the prevailing meteorologic environment.

Because of its role in dissolved oxygen depletion, a hydro power installation can be considered a pollutor. In many situations the release of oxygen depleted waters is a serious loss of water quality and natural waste assimilation capacity to downstream users located in the affected zone. Typical examples are on the Flambeau River, where a recreational area is located downstream of the hydro power plant, and on the Roanoke River at Roanoke Rapids, with industry immediately downstream. In both situations remedial measures were required to improve the dissolved oxygen content.

The Upstream Effects

The construction of a hydroelectric dam radically changes the channel characteristics. The increase in channel depth from a high dam may extend for miles upstream, replacing the natural channel by a slack-water pool. Time of passage is greatly increased, placing a heavy burden of amortization on a short reach of stream, and with the increase in depth the reaeration rate is greatly reduced. The effect for organic waste disposal is a dissolved oxygen profile sharply contracted from that of the natural channel. The velocity in the slack-water pool is so reduced that any natural or introduced suspended solids are certain to deposit and accumulate as sludge; this adds an additional burden on dissolved oxygen. Unless compensated by low flow augmentation, these detrimental effects on upstream waste assimilation capacity are so drastic as to preclude adjacent industrial and urban growth.

A secondary impact in the reach upstream from a hydroelectric dam is the effect on water quality in the pool area. If upstream pollution is slight and the pool is of moderate depth, water quality may be enhanced. If thermal stratification occurs in deep pools, though the lower depths are certain to be low in dissolved oxygen, the upper layers of the pool are usually high in dissolved oxygen. But if stratification is pronounced and prevails for long periods with the bottom completely devoid of dissolved oxygen, the iron and manganese content of the overlying water may increase significantly, and during periods of seasonal reservoir turnover may be damaging to water uses.

Corrective and Beneficial Measures

Corrective measures are called for at many existing hydroelectric installations, and beneficial features should be incorporated as routine requirements in proposed developments. At existing installations immediate relief can be obtained by change of the rules of release to maintain a suitable downstream flow; this may mean some sacrifice in peaking practice for the protection of downstream interests.

In many locations a reregulation reservoir can be constructed below the existing hydroelectric plant as a permanent solution. In all proposed hydroelectric power developments it should be standard practice to provide a reregulation reservoir below the power installation to iron out diurnal and weekend pulsations in discharge and to equalize the flow. Since most proposed hydroelectric power projects now are planned as large storage type, such reregulation would not only remove the detrimental daily and weekend fluctuations but would ensure passing downstream the benefits of streamflow augmentation released from storage. A great benefit would result in increased waste assimilation capacity and water quality as well as to all other water uses. The relatively small extra cost of reregulation is more than offset by the great downstream benefits ensured and the increased urban-industrial growth potential created. This should be attractive to the electric utility industry. Two typical examples illustrate the implications of reregulation, one on the Savannah River and the other on the Chattahoochee River.

Prompted by the need for assured streamflow for the Savannah River nuclear development below Augusta, Georgia, and for navigation from Augusta to Savannah Harbor, the Corps of Engineers constructed large storage-type projects, Clark Hill–Hartwell, above Augusta, incorporating the Stevens Creek development as reregulation reservoir. A statistical analysis of drought flow based on the record at the Clyo gage above Savannah City prior to the operation of Clark Hill develops the drought flow reference frame shown in Figure 9–1. The once in 10-year to once in 20-year drought flow ranges from 3500 cfs as minimum monthly average to 2500 cfs as a minimum daily average at the Clyo gage, as shown by curves *C* and *D*. With the Clark Hill storage development and Stevens Creek reregulation, the assured flow during such severe droughts as those experienced in 1925 and 1931 was estimated as 6033 cfs and with the additional storage at Hartwell as 6933 cfs (indicated by curves *X* and *Y*, respectively). This shows that reregulation equalizes flow and passes downstream the benefits of upland storage. The Clark Hill–Hartwell–Stevens Creek development eliminates all of the severe natural drought risks and raises the sustained yield above even the most probable ordinary dry weather flow. Experience since the construction and operation of these river developments verifies the high sustained flow at Savannah Harbor.

In contrast is the development by the Corps of Engineers on the Chattahoochee River above Atlanta, Georgia. Initially as planned and constructed, the storage type hydroelectric power project at Buford did not include a reregulation reservoir to equalize flow and pass downstream the benefits of the storage available. The Buford reservoir provides about 1 million acre-ft of usable storage for flow augmentation for peak power generation. During the severe drought years this could provide a flow at Atlanta of 1690 cfs as a minimum monthly average. However, rules of release for operation of Buford for peak power demand specified weekday peak discharges for a few hours of the day of 3000 to 6000 cfs, with off-peak discharges of 500 cfs on weekdays and during weekend shutdown. These rules of release would result in a flow of 550 cfs at Atlanta, the extra 50 cfs constituting tributary runoff in the 48-mile

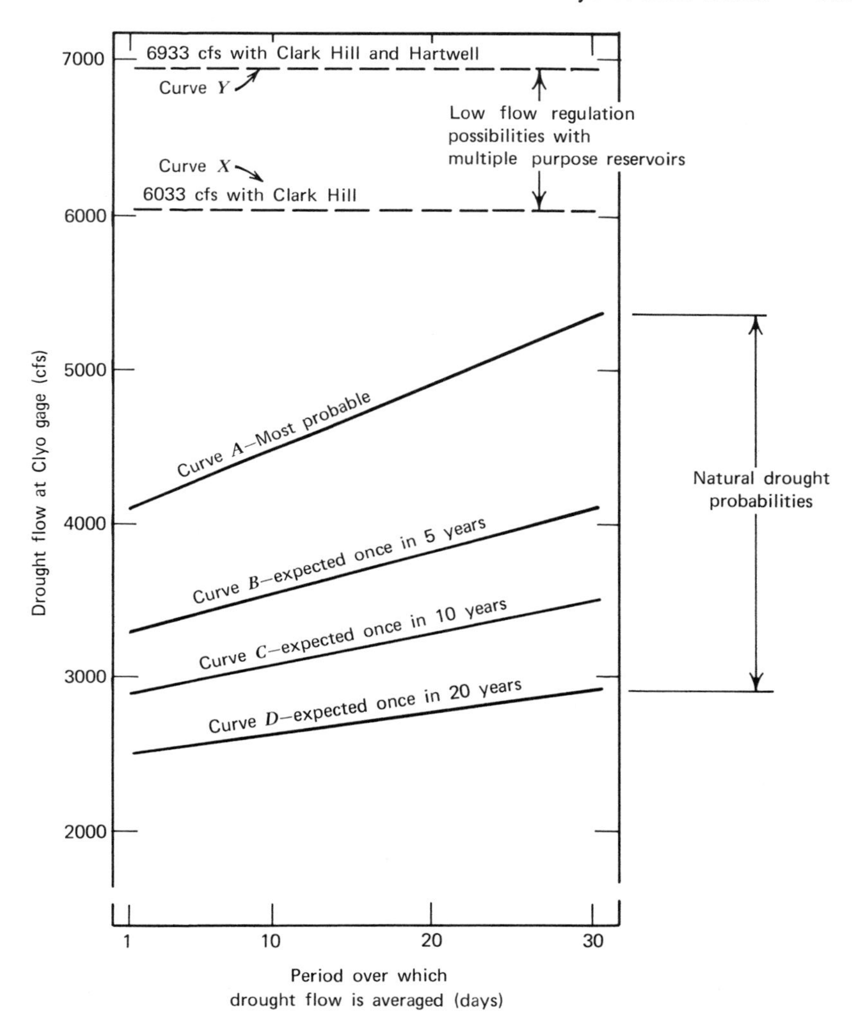

Figure 9–1 Savannah River at the Clyo gage. Relation between drought severity and duration over which flow is averaged with and without proposed regulation.

intervening reach. Thus, although Buford could potentially ensure a uniform flow of 1690 cfs at Atlanta, actually, for hours on weekdays, flow at Atlanta would decline to 550 cfs and on the weekend remain at 550 cfs for 2 to 2.5 days. Water surging down the river during the peak hours of weekdays is no help during the long weekend shutdown.

A statistical analysis of drought flow yields, applying the theory of extreme values to the record before the operation of Buford, discloses a natural drought severity expected on the average once in 10 years at Atlanta of 600 cfs as a minimum monthly average and of 490 cfs as a minimum daily average. Thus the 1 million acre-ft of storage at Buford actually provides very minor increase in flow above the once in 10-year drought. Furthermore the natural drought as a daily minimum is expected on the average to be as low as 490 cfs not more often than once in 10 years, whereas under the rules of release at Buford the flow of 550 cfs would occur every weekend in the dry-weather season, or for about 30 weeks each year, or in 10 years about 300 times. From the standpoint of waste assimilation capacity of the Chattahoochee River, although great potential benefit is available in the 1 million acre-ft of storage, the Buford development as initially conceived is of little value. Reregulation is obviously necessary to realize this potential for any use beyond development of peak electrical energy.

The importance of reregulation of the pulsations from the Buford hydroelectric plant is further emphasized when the interrelation of the various water needs of metropolitan Atlanta is considered. The district, extending well beyond the central city, is currently estimated to have a population of over 1 million and is projected to 2.4 million by 1988.* The community and industrial water supply is dependent principally on the flow of the Chattahoochee River. The intake of the city system is located approximately 48 miles below Buford dam, with the effluent of the wastewater treatment works returned to the river a short distance downstream of the supply intake. The current city maximum daily demand during the warm weather dry season approximates 120 mgd, and it is likely that as growth continues the needs of the entire district will continue to be supplied from this source.

About a mile below the city works are located two large steam electric power stations of the Georgia Power Company, the old Atkinson station of 240,000-kw capacity and the new Jack McDonough station of 490,000-kw capacity. The combined condenser water requirement of these two stations as reported† is 1276 cfs at rated capacity. A third generating station, Yates (550,000-kw capacity) is located approximately 41 river miles farther downstream; its condenser water requirements are reported to be 920 cfs.

In order to meet the needs for increased assured flow, the city of Atlanta and the Georgia Power Company entered into an agreement

* Atlanta Region Metropolitan Planning Commission.

† Atlanta Region Comprehensive Plan: Water and Sewerage, Atlanta Region Metropolitan Planning Commission, February 1969.

to install reregulation capacity of 3200 acre-ft at the Morgan Falls dam of Georgia Power, located about 12 miles above the city water intake. If this reregulation capacity is utilized without power peaking at Morgan Falls, it can ensure approximately 1550 cfs at the city intake and at the steam electric power plants. The cost of the Morgan Falls reregulation is stated to be about $1,000,000, a small sum considering the large federal investment at Buford. Analysis also indicates that, if reregulation capacity of 5000 acre-ft could be provided, it would ensure a uniform flow of 1690 cfs.

The interrelation of the heat loads from the steam power plants in close proximity to the city wastewater effluent discharge places a heavy burden on self-purification; analysis shows that dependence only on wastewater treatment cannot provide satisfactory stream quality in the Chattahoochee and reregulation to provide low flow augmentation is essential.

Considering the future urban-industrial needs and the requirements of steam electric power generation, it is quite evident that ultimately complete reregulation of the peak power releases at Buford is essential; alternatively peaking power practice must be abandoned and Buford must be operated primarily as a low flow augmentation reservoir.

In further consideration of the upstream reaches above hydroelectric dams remedial measures in the reservoir should include elimination of extensive backwater areas, sloughs, and exposed mudflats by construction of dikes. The reservoir area also should be cleared and grubbed to minimize organic decomposition and impairment of recreational uses. Low flow augmentation from upland storage can, to a degree, offset some of the detrimental impact on waste assimilation through the hydro pool, but the channel changes are usually so drastic that assimilation capacity is not adequate for large scale industrial-urban waste effluent loadings. It is preferable to reserve the pool reach for water supply source and for recreational use, and to direct industrial-urban growth to the downstream reach.

There are several methods of correcting oxygen deficiency of the water withdrawn from the hydro pool. Air intake to the turbines adds some dissolved oxygen, but at the expense of generating efficiency. Artificial aeration of the stream theoretically can add dissolved oxygen but is usually not feasible because it necessitates obstructing the channel.

Intake towers with multiple inlets to ensure withdrawal of surface water high in dissolved oxygen are effective but costly, and not always feasible. Submerged weirs to induce surface withdrawal and breakup of stratification can be effective but are costly. Induced circulation by updraft pumping can also break up stratification and increase dissolved oxygen, but to be effective it entails costly installation and operating expense.

The most effective method of improving the quality of the pool water and ensuring release of high oxygen content to the downstream reach is the adoption of pumped storage. If the reregulation reservoir can be employed for the dual purposes of equalizing downstream flow and of serving as the lower pumping pool for recycling a portion of the release back to the upper pool, not only is the potential beneficial effect great but the economics are attractive from the standpoint of peak energy produced. Recycling back to the hydro upper pool on off-peak hours is sufficient to prevent stratification and provides extensive reaches of the pool with water approaching dissolved oxygen saturation throughout the entire depth and at temperatures below the equilibrium level of the prevailing meteorologic environment. Adaptation of pumped storage is dealt with in some detail in Chapter 11.

THERMAL ELECTRIC POWER PLANTS

Temperature and Channel Changes

The principal effect of thermal electric power plants, both fossil and nuclear fuel developments, is stream temperature superelevation. The influence of heat on assimilation of waste products and on the aquatic life of the stream has been discussed in Chapter 7. Suffice it here to reiterate that superelevation of stream temperature is detrimental to organic waste assimilation capacity, primarily in acceleration of the rate of amortization and in reduction of the dissolved oxygen assets of the stream. With the trend toward large concentration of thermal generating capacity at a single power plant, the impact of stream temperature superelevation through urban-industrial reaches is critical.

In addition, some installations necessitate alteration of the stream channel by construction of a low dam to form a backwater pool for the circulation intake system; the adverse effect of power pools applies, except that diurnal pulsations in downstream flow are not as pronounced. However, if the pool is operated as a storage reservoir during the low streamflow season, retention of water during the weekend low electric load may also be detrimental to downstream flow.

Corrective and Beneficial Measures

There are several ways of alleviating the detrimental effects or incorporating features that can result in a net benefit. Since thermal electric power generation requires great quantities of condenser water, upland storage that will augment the streamflow and ensure a high sustained runoff during the critical dry weather season is the best solution in many situations. This not only reduces the initial temperature superelevation to moderate levels but the additional streamflow provides increased waste assimilation capacity to compensate or even to exceed the loss entailed in temperature increase. This is illustrated in the Clark Hill–Hartwell–Stevens Creek development to meet heat load and other water needs at the Savannah River atomic works. It is further illustrated in studies on the Tittabawassee River, discussed in Chapter 7.

A promising approach to meeting water needs and controlling temperature superelevation is the combined development of pumped storage hydroelectric and thermal electric installations. The pumped storage not only provides substantial peak energy but also ensures low flow augmentation with water at cooler than the natural equilibrium temperature. This is illustrated in studies on the Miami River, discussed in Chapter 11.

When low flow augmentation is not feasible, it may be necessary to construct expensive cooling towers or ponds to dissipate the heat load. This results in substantial water loss by evaporation and necessitates large quantities of make-up water. Cooling ponds are usually not possible at locations within the urban-industrial area, since they require a large area.

An alternative for proposed thermal electric power developments is to locate the power station outside the urban-industrial complex and transmit the electrical energy over high voltage lines. This is feasible at mine-mouth sites with access to cheap coal resources. Although this isolates the heat problem to areas where the adverse impact may be less damaging, water remains a critical problem, as usually mine-mouth sites are in the headwaters and only small tributary streams are available. If such isolated streams must be protected against adverse temperature effects on aquatic life or other downstream uses, a combination of cooling towers and upland storage of runoff to provide make-up water and to control temperature superrelevation can be utilized. A good illustration of this approach is the Keystone generating station, discussed in Chapter 7.

NAVIGATION WORKS

Streamflow and Channel Changes

Navigation works are of three types: dredged channel, low flow augmentation, and locks and dams; in some instances combinations of all three are employed. The effects on the waste assimilation capacity of the stream are reflected in changes in streamflow and alterations in channel characteristics. In most current navigation developments the effect on waste assimilation is detrimental.

Dredging deepens the channel and thereby slows the time of passage. It increases amortization of organic burden in shorter reaches of stream and the increased depth and lower velocity reduce the oxygen assets by lowering the reaeration rate. In large rivers and estuaries the adverse effects of dredging may be minor, since channel alterations are insignificant in relation to the total channel cross section; in smaller streams the proportionate channel change is greater, with corresponding significant loss of waste assimilation capacity.

In early works, when shallow draft navigation prevailed, dredging was generally limited to the removal of snags and shoals. Efforts were made to ensure required navigation depths during the dry weather season by increasing the river stage through flow regulation from upland storage reservoirs. Such practices usually more than compensated for loss due to dredging and resulted in a substantial net benefit to waste assimilation capacity and all other water uses.

The recent trend, however, is toward deep navigation by locks and dams, with higher dams replacing older works. The navigation objective is to form deep slack water pools and to eliminate the initial and continual expense of dredging. These river developments are extremely detrimental to the natural waste assimilation capacity of the stream. They are therefore a hindrance to the industrial growth which the navigation improvements are trying to promote and on which the economic justification of the works has been largely based. Many otherwise good industrial sites with respect to high natural waste assimilation capacity have been seriously impaired.

Implications of Multipurpose Developments

A typical example of this trend is the progression of navigation works on the Tombigbee–Warrior River system. Figure 9–2 shows the existing and proposed navigation works between Mobile Bay and Birmingham, Alabama. Initially, navigation improvements consisted of dredging of shoals; then construction of a system of 17 locks and dams provided an 8-ft controlling navigation depth; now under way is a plan to replace the 17 low dams by three high dams, Jackson, Demopolis, and Warrior, forming three deep slack water pools. Figure 9–3 shows the profile of the lower river to Demopolis. Curve *A* is an approximation of the natural low water profile; the first series of locks and dams create slack water pools between No. 1 and No. 2 dams at elevation 13, and successive

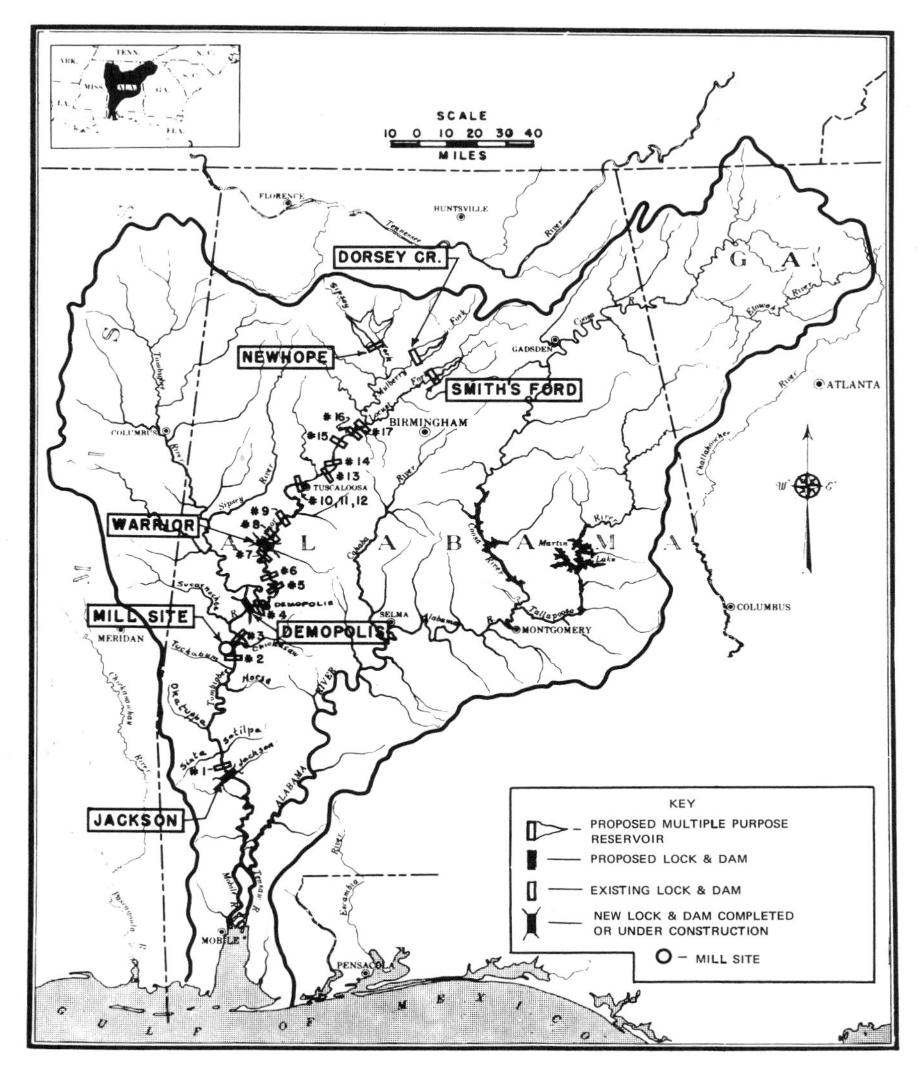

Figure 9–2 Alabama–Tombigbee–Warrior sytsem—river developments.

pools at 10-ft rises, to elevation 23, 33, and to elevation 43 above No. 4 at Demopolis. The current plan replaces No. 1, No. 2, and No. 3 dams with a new lock and dam at Jackson (mile 100.7). This raises the pool 20 ft to elevation 33, with backwater extending 117.4 miles upstream to the new Demopolis lock and dam. In turn the Demopolis

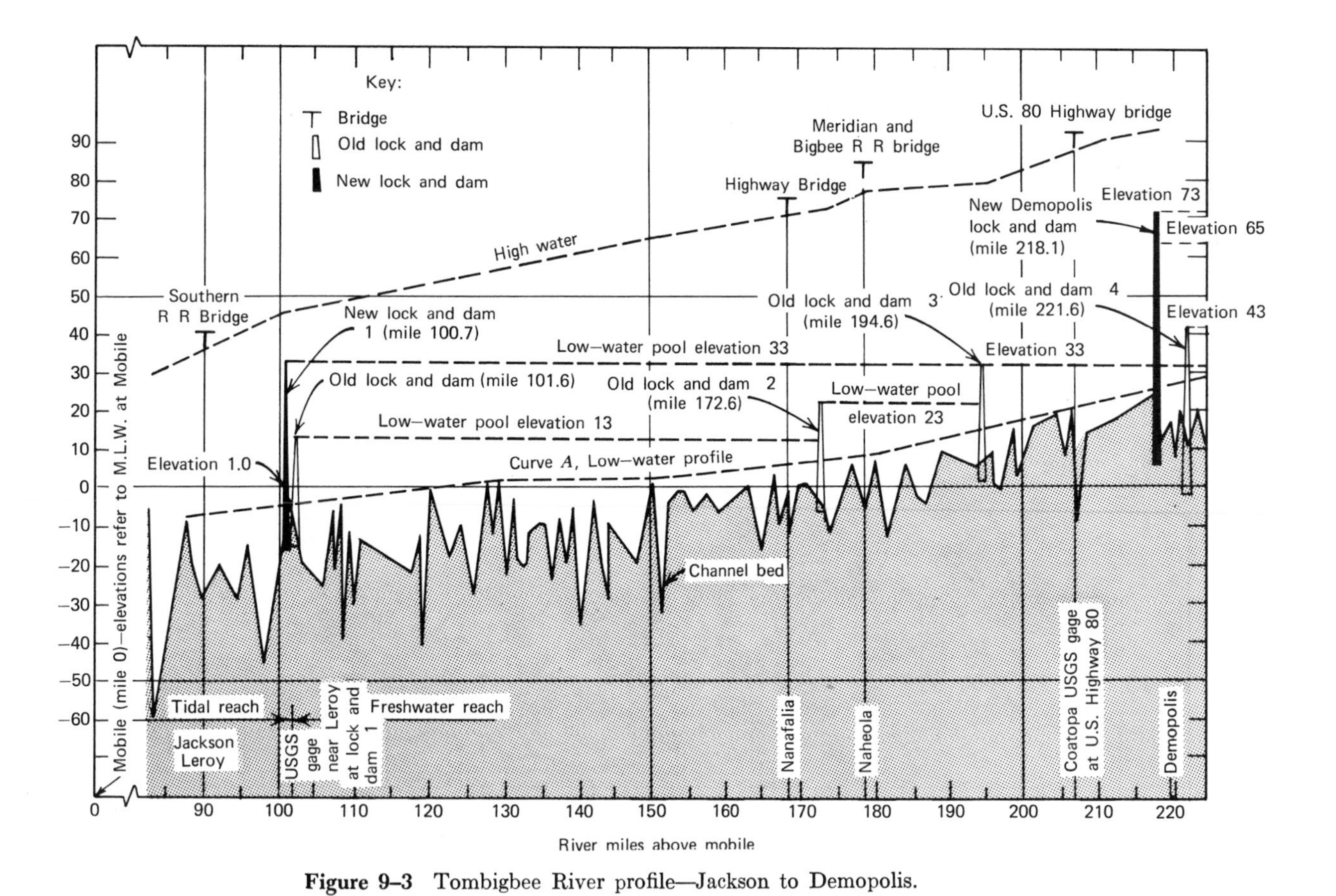

Figure 9–3 Tombigbee River profile—Jackson to Demopolis.

dam replaces dams 4, 5, 6, and 7, raising the pool 30 ft to elevation 73.

The effect of these navigation works on organic waste assimilation capacity is illustrated in computations for an industrial site at mile 178 between lock and dam nos. 2 and 3. The once-in-10-years minimum consecutive 5-day average drought flow at mile 178 is 816 cfs, coincident with the peak warm-weather temperature of 30°C. The allowable waste load, free of suspended solids subject to deposit, under these drought conditions prior to alteration of the natural stream channel is computed as 172,000 population equivalents to maintain dissolved oxygen of 50 percent of saturation (3.8 ppm at 30°C) in the critical reach. Under the same drought conditions but with channel alteration associated with lock and dam No. 2 with pool elevation 23, the allowable waste load to maintain dissolved oxygen of 50 percent saturation in the critical reach is computed as 62,000 population equivalents, a decrease in waste assimilation capacity of 64 percent.

The Jackson lock and dam raises the pool an additional 10 ft, which results in a further reduction in waste assimilation capacity. The dam, in altering the natural channel, increases the time of passage and depth, and thereby results in heavy oxygen demand in a short reach of the river and poor reaeration in the deep, slow moving pool. Not only is the waste assimilation capacity decreased, but the dissolved oxygen profile is steeper and more sensitive to variation in waste load. The location of the critical dissolved oxygen of 50-percent saturation for the natural channel is about 12 miles downstream, whereas for the navigation pool at elevation 23 the critical dissolved oxygen is only 4 miles below the industrial outlet. A high degree of waste effluent control would be essential to equalize normal variations. Also the very low velocity in the slack water pool increases the possibility of deposit and accumulation of organic sludge. A very high degree of settleable solids removal would be essential since any release in the waste effluent would be certain to form sludge deposits, which would further limit the waste assimilation capacity. As pools become deeper, they are also subject to the adverse effects of streaking and stratification, with the lower depths devoid of dissolved oxygen.

It should be recognized that the slow time of passage in a pool reach results in almost complete satisfaction of the organic waste load within the pool, and in this sense the pool is acting as a treatment works within the river, protecting downstream reaches. If the pool reach were allowed to degenerate in the interest of protecting downstream reaches, the waste loading on the pool as a treatment system could be substantially higher than indicated in the above computations. However, this is not acceptable in present American practice, although some day it may be necessary, as European practice has demonstrated.

A further element to be considered is the effect on streamflow that is associated with the operation of navigation locks. This can be beneficial or detrimental. Streamflow is subject to pulsations since locks are filled and emptied at irregular intervals, depending on traffic. During the drought period there is a tendency to conserve water when traffic is light; this may result in no release from one pool to another for significant periods. Where traffic is heavy and drought flow is inadequate for lockage, low flow augmentation is necessary. This can be supplied by single purpose or multipurpose upland storage. Figure 9–2 shows three proposed upland multipurpose developments on the Warrior. The storage in these proposed developments, if operated with due consideration of low flow augmentation, could increase drought flow from the natural once in 10-year severity of 816 cfs to about 2500 cfs at the downstream site. This potential increase in flow would not be realized as a continuous assured flow unless reregulation is provided below the proposed upland hydroelectric plants, and operation of successive locks passed the benefits along the course.

FLOOD-CONTROL WORKS

Ordinarily flood control is not thought of as influencing the waste disposal problem; in most situations it does not. The real significance stems from failure to capitalize on the potential benefit inherent in relating flood control to drought control. This beneficial aspect is dealt with in Chapter 11. Here we shall briefly consider the detrimental effects.

Flood control works can be grouped into three classes—reservoir developments, levees, and local protection works. What has been stated with respect to the upstream effects of hydroelectric dams applies equally to flood control reservoirs. In single purpose developments there is a difference in that the floodwaters are retained only until the downstream channel capacity permits release. The reservoir is thus emptied or drawn down in readiness for the next storm. An example of this type of control is the system of five single purpose retention reservoirs of the Miami Conservancy District in Ohio. The reservoirs are designed without gated outlet works to automatically discharge so as to empty in a matter of days after a storm; hence the upstream channel reverts to its original characteristics in a short time. In a sense this single purpose development (although effective in flood control) is a misappropriation of good reservoir sites and a waste of water resources needed to maintain dry weather flow along the highly developed urban-industrial downstream reach.

Levees and local protection works have an effect on channel characteristics. The channel changes induced by levee construction are not necessarily immediately experienced but may over time change the location, width, and depth. Local protection works, particularly channel straightening and dredging, have an immediate effect, usually detrimental to waste assimilation capacity. The increased channel cross-section increases the time of passage and may also increase depth during the low flow season. An example of this is the local protection works proposed on the Kalamazoo River, involving channel dredging and a submerged-weir type of dam to maintain the low water profile. The low-water pool results in a longer time of passage through the critical reach of 1.5 days during the low runoff period. Computations show that the loss in waste assimilation capacity would result in a decrease of dissolved oxygen to the levels that obtained before the installation of costly municipal and industrial waste treatment works. There is little gain when one river development undoes what another is trying to achieve.

IRRIGATION AND OTHER DIVERSIONS

Most uses of water result in return of the spent water to the stream. Diversions without return or compensation from storage are a direct impact on streamflow and, depending on the magnitude and the season, may be seriously detrimental to waste assimilation capacity.

In arid regions where the principle of prior appropriation is applied, the entire natural streamflow is diverted to agricultural irrigation during the warm weather growing season. Most of the water is consumed by transpiration and evaporation, and little, if any, is returned through seepage; any that is returned by drainage is likely to be of poor quality. The effect of such irrigation practice on streamflow is illustrated in the analyses of drought flows in the upper Deschutes basin. The recorded flow at Benham Falls, Oregon, located below storage reservoirs but above irrigation diversion canals, (corrected for change in reservoir contents) provides a basis for estimating the natural drought yield. Statistical analyses of the corrected flow, employing the theory of extreme values, give a most probable natural drought yield of 0.739 cfs per square mile as a monthly average, and 0.495 cfs per square mile as the yield expected on the average once in 10 years. The stream gage at Bend, Oregon, located below the irrigation diversion canals shows the effect of the diversion. From analyses of this record the minimum monthly average drought flows expected at Bend under the prevailing diversion practice are 145 and 31 cfs as the most probable and the once in 10-year severity, respectively. The natural drought yields applied to the tributary drainage area at Bend give expected natural drought flows of 1403 and 940 cfs as the most probable and the once in 10-year values, respectively. This indicates that irrigation diverts about 90 percent of the natural flow under most probable drought and nearly 97 percent during the severe once in 10-year drought. Such excessive diversion drastically reduces waste assimilation capacity and precludes any extensive urban-industrial development.

Agricultural diversion is not limited to the arid regions. Supplemental irrigation is increasing in the humid regions to augment natural rainfall during the growing season. In most areas the practice is direct diversion from streams without compensation by upland storage. The maximum diversion occurs during the critical drought season—and the more severe the drought, the heavier the diversion. This practice is seriously detrimental to waste assimilation capacity during the critical season. In some states, legislation has been proposed to restrict such diversion except as compensated from storage.

Diversion from one basin into another basin for community water supply is not uncommon. New York City diverts water from the headwaters of the Delaware to metropolitan New York, but under certain limitations established by the U.S. Supreme Court managed by a river master. Diversion is permitted with compensation from storage to ensure a minimum basic flow of 1750 cfs in the Delaware River at Montague. On the basis of extremal theory, the natural drought flow of the Delaware as minimum monthly average and minimum daily average discharge expected on the average once in 10 years is 1013 and 734 cfs, respectively at Montague. The severity of the prolonged northeastern drought of 1964–1965 necessitated emergency steps to meet dwindling storage for New York City and releases to the Delaware did not meet the court's stipulation. With reduced flow in the Delaware River, seawater intrusion extended farther upstream, threatening the Philadelphia water supply intake.

Developments proposed for the Delaware River Commission in multipurpose works (including the New York City stipulations) have been reported as capable of ensuring a minimum flow at the Tocks Island dam of 3270 cfs. However, as initially planned, reregulation below Tocks Island hydrolectric dam was not included, and proposed rules of release would provide only a minimum on off-peak power periods no greater than that now required from New York City reservoirs alone. If such a plan is executed, the benefits of the extensive developments proposed would not be fully realized in terms of increased downstream flow.

A serious effect on waste-assimilation capacity occurred for a number of years on the Nashua River as a result of diversion of about one-third of the natural drainage to the metropolitan Boston water supply. Although storage was provided, no release was made to the river during the critical drought season; the heavily industrialized Nashua basin consequently suffered serious curtailment of waste assimilation capacity. It is recognized that community water supply is a prime water resource need, but this does not warrant extensive diversion without compensation from storage, particularly where proper planning can provide ensured streamflow in the critical season. Correction of this abuse has been made in the Nashua basin.

Another type of diversion, disastrous to waste assimilation capacity of estuaries, is rerouting freshwater from a highly developed harbor reach to prevent it from depositing silt in the navigable channel of the estuary. Such diversions have been seriously considered for the Savannah and Charleston harbors. One plan for Savannah proposed diversion of the entire freshwater flow to a nonnavigable undeveloped outlet to the ocean. This would completely negate the upstream multipurpose Clark Hill–Hartwell–Stevens Creek development discussed previously, which now ensures 7000 cfs of freshwater flow at Savannah Harbor. Diverting this freshwater flow would result in seawater intrusion to the diversion dam, and there would be no net seaward movement through the harbor. The Savannah area now is a highly developed urban-industrial complex; computations show that, if such diversion were undertaken, extensive areas of the harbor would be completely devoid of dissolved oxygen through the urban reach. Even with the best possible waste treatment, conditions would be intolerable and could be remedied by only drastic curtailment of industrial production.

A similar plan has been proposed for the tidal estuary of Charleston Harbor. Extensive developments have been constructed in upstream storage of freshwater runoff to supply the Charleston urban-industrial area with an ensured flow of some 14,000 cfs. It has been proposed to divert 90 percent of the freshwater runoff, bypassing the harbor. Again, computations show that such diversion of freshwater flow would result in drastic reduction in waste assimilation capacity in the estuary, and extensive reaches through this highly developed area would, even with the best treatment, become completely devoid of dissolved oxygen. A vigorous economic development campaign based on an ample water supply and streamflow has attracted many industries; the diversion would necessitate abandoning much of this development, since the situation with respect to satisfactory waste disposal would be intolerable.

These are dramatic illustrations of the impact of river developments on the waste assimilation capacity of streams and estuaries. Any river development can be a liability to self-purification; most can be assets under careful planning and management. Recognition of the potentialities becomes imperative as urban-industrial growth accelerates.

10

THE STREAM SURVEY

In this chapter the term "stream survey" is limited primarily to the collection of data essential to the rational method of stream analysis, with special attention to stream sampling. There are many types of stream surveys, and the kind of data collected depends on the purpose in mind.

TYPES OF STREAM SURVEY

Spot Check

For administrative purposes stream samples are taken occasionally to spot check water quality in relation to a limited local situation. Such data are usually of very limited value in stream analysis because there is seldom sufficient information concerning the waste loading that produced the sampled water quality or the prevailing hydrology before and during sampling. The spot checks usually involve grab samples taken under a wide range of conditions and are not adequate for statistical definition of any one stable condition of load and runoff. The extent and intensity of sampling are usually inadequate to define a profile along the receiving stream.

Routine Periodic

On critical rivers regulatory agencies, municipalities, and industries usually establish a system of *routine* river sampling on a regular schedule, such as a traverse of several stations along the course of the river. Some take the traverse runs daily except on weekends; others schedule runs once every week or once every month. In some instances daily waste loads are also determined and stream gaging stations are established. Such stream survey data are of value in providing a background of variation under the seasonal changes. Depending on the frequency of traverse runs, it may be possible to composite data to reflect quite a narrow range in critical stream runoff and water temperature and quite a stable level of waste loading.

Continuous Monitoring

Following the pioneering work of the Ohio River Valley Water Sanitation Commission [1–3], systems of continuous monitoring of streams are employed as tools in managing waste disposal and pollution control.

The system developed by Cleary [1–3] for the Ohio River employs 13 field stations, 10 equipped with transmitters for telemetering data to the central headquarters and 3 equipped with strip-chart recorders for on-site recording. Field stations provide multiple sensor units, which use electrodes or transducers for measuring seven water quality characteristics: pH, oxidation–reduction potential (ORP), chloride, dissolved oxygen, conductivity, temperature, and solar radiation. In addition, for subsequent more complete laboratory analysis, each station can automatically collect a sample of river water whenever an abnormal change in quality occurs.

Each transmitter field station is interrogated once every hour from the central headquarters and transmits readings of quality conditions as they are being measured at the moment. The data received, punched on paper tape and typed on a log chart, are processed by digital computer, edited, verified, and summarized daily. The daily printout shows such parameters as maximum, minimum, arithmetic average, and standard deviation.

ORSANCO staff correlate these data with streamflow and supplementary sampling station results, continuously evaluating and diagnosing river conditions to detect current significant deviations in water quality. The character of the water is measured against prescribed criteria and compared with the quality of river conditions in previous years. The monitoring is integrated throughout the river system and is particularly effective in detecting accidental spills of industrial waste, permitting warning to downstream users; in controlling releases of salt brine; and in detecting and isolating abnormalities of unknown origin.

An increasing number of large industries in isolated locations on rivers are resorting to continuous monitoring in the control of wastewater effluent release. In addition to monitoring quality at specific cross-sections, a longitudinal profile through the critical reach can be continuously available. From these and with the aid of computer adaptation, stream analysis studies can establish control curves that regulate allowable waste loading in accord with runoff and temperature to ensure at all times the prescribed water quality criteria in the critical reach.

Initial capital outlay and the cost of maintaining continuous monitoring equipment are high, but where control requires extensive data the unit cost of individual tests is greatly reduced. However, unless the data collection is routinely employed in the management of waste disposal, continuous monitoring is not justified. A single monitor located without the guidance supplied by detailed stream analysis studies is of questionable value. Simply to acquire great masses of data that are not effectively used in quality control is of dubious value.

Stream Classification

As a basis for classifying a stream for water quality for various uses, regulatory agencies usually undertake comprehensive surveys and river sampling programs. Here the stream sampling is intensive, providing the data with which to evaluate classification proposals.

Short-Term Intensive

The short-term intensive stream sampling program best serves the purposes of stream analysis studies. The objective is to obtain a series of samples under a stable hydrologic regime and known steady waste loading with sufficient number of samples at each station in the network to permit statistical treatment of the data. This warrants more detailed discussion.

STREAM SURVEY REQUIREMENTS FOR APPLICATION OF THE RATIONAL METHOD OF STREAM ANALYSIS

The Design of the Stream Survey

Preliminary Steps

It is important to define precisely the scope and content of stream analysis studies and the extent of the projections to be made. These criteria will narrow or expand the amount and kind of data necessary for the purpose.

An intensive library search will usually disclose a substantial body of useful data as background, including maps and reports on river developments of federal, state, and private agencies. Moreover, the state agencies and local industries often have unpublished data of previous surveys and river sampling.

Before completing the design of the proposed stream survey, a field reconnaissance provides an overview of the entire drainage basin. Traversing the basin by airplane or helicopter provides access to areas not approachable by road. An essential part of the reconnaissance is a trip on the river by boat. A good log supported by ample photographs keyed to location on a working map provides an intimacy that is not attainable by occasional views from road and bridge.

With these preliminaries, an efficient stream survey can be designed and executed. The stream survey logically divides into three parts: the hydrology, the sources of wastes, and the stream sampling.

Hydrology

The hydrologic data are usually available in published form or on tape. In some basins it may be necessary to supplement by field survey. The primary elements are basin physiography, climate, streamflow, and channel characteristics. Data for definition of these four elements and methods of analysis are discussed in detail in Chapter 3.

Sources of Wastes

Location and measurement of the existing waste loading discharged to the river usually entail a major survey in the field. The urban and industrial sources are best obtained from an onshore survey; the agricultural and natural sources, usually scattered, are reflected as residuals in the riverflow and tributaries, better obtained by watercourse sampling. The scope of the stream analysis study determines the type of waste product—organic, microbial, radioactive, inorganic, or thermal. In addition to data concerning the current waste loadings, generally it is necessary to obtain data that permit projecting future loadings. This extends the survey data to detailed population and economic factors of considerable magnitude, primarily available in published form from federal and state agencies. These phases are discussed in some detail in Chapter 2.

Stream Sampling

Since the object of stream sampling is to obtain a representative, reliable measure of water quality along a stream course (associated with a known waste loading), consideration must be given to the type of samples and to where, when, and how many of them should be taken; the staff, equipment, and laboratory facilities required to execute the program; and how the data are to be processed and analyzed. Sampling must reflect a certain degree of stability in conditions during the period in which the samples are taken; otherwise the results will be a heterogeneous mixture without meaning. Since a river is living and dynamic, subject to hydrologic and biologic variations and to the impact of man-made influences, stability during sampling is very difficult to achieve, and, at best, sampling results are subject to variation. Hence the observed sampled quality can be expressed only in statistical terms, a mean condition within a certain confidence range, depending on the number of samples taken at a given station and the variance of individual samples. Although we strive for stability during the sampling period, we must rely on the number of samples to ensure reasonable reliability. A number of factors that influence sampling results warrant further consideration.

The Design of the Watercourse Sampling Program

Types of Samples

The type of samples to be taken depends on the orientation of the proposed stream analysis study as to type of waste and self-purification (organic, microbial, radioactive, inorganic, or thermal). The primary classes of examination made on samples are physical, chemical, bacteriologic to determine sanitary quality, bioassay for evaluation of acute toxicity to fish, biologic (plankton and bottom fauna), and radioactivity. The analytical methods for the many specific constituents are standardized for water and wastewater as published in the current edition of *Standard Methods* [4].

Generally major concern centers on organic and bacterial wastes, and only a limited number of tests are made for stream analysis purposes. These may be augmented for special purposes as a unique condition demands. Since a great deal of laboratory work is involved, specific tests on samples should be kept to the minimum. For evaluation of organic and bacterial self-purification these are dissolved oxygen, BOD, pH, temperature, and coliform concentration for freshwater streams; for tidal estuaries concentration of chloride is added. If the abnormality of nitrification is anticipated, tests for nitrate and ammonia are included. Where fish life is of major concern, other chemical and biologic examinations serve as auxiliary data. Many refinements can be added, but if this is at the expense of reduction in the number of samples to ensure the statistical significance of the essential items, any additions should be carefully weighed.

Factors Governing Location of Sampling Stations

Ordinarily too many sampling stations are established with insufficient number of samples collected at any one. A few well selected locations with sufficient number of samples to define results in statistical terms provide a much more reliable profile of quality than many stations with only a few samples at each.

Major factors that influence the location of a sampling station are the relative position of waste effluent outfalls, the channel characteristics, river developments, and the self-purification characteristics of the stream.

Waste Effluent Outfalls

If a sampling station is established at or immediately below a source of waste discharge, it is very probable that sample results, influenced by streak pollution, will be highly erratic. The station should be some distance downstream, where dispersion throughout the section is essentially complete. If the stream is sluggish and wide, care should be taken to test for significant variability; if variability is detected, it may be necessary to sample at more than one position in the cross-section.

An initial upstream station should be located as a control above the zone of pollution to reflect the residual stream quality from sources of upstream waste and from land drainage and surface wash. There can be a significant organic BOD load and lowered dissolved oxygen as the river arrives at the head of the reach to be analyzed.

Channel Characteristics and River Developments

Configuration of the stream channel, irregularity of cross-section, depth, bends, and gradient should be considered in locating a sampling station. Wherever feasible, a single sampling position in the cross-section is desirable, but the position must be representative of the main body of flow. Sampling from shore is seldom satisfactory, particularly in an irregular section with shallow beach and deep channel. A good location is a reach of straight, regular cross-section without excessive turbulence downstream of bends and irregular reaches that have produced good mixing and uniformity throughout the section. Bridges may afford a convenient location, if other features are satisfactory. Generally, however, a boat is necessary for most stations; a boat offers the advantage of flexibility—it is possible to check transversely at a given section and longitudinally between sections for the occurrence of any abnormalities.

Physical characteristics of the channel of a wide, deep river may make it necessary to locate sampling points at several positions transversely in the section and at different depths. This greatly adds to the number of samples necessary to obtain a reliable mean for the section. Stations should not be located immediately below the junction of a major tributary. A better measure of the influence of the tributary is to locate the station on the main stream just above the junction and to establish a secondary station on the tributary above its mouth. Otherwise the sampling station on the main stream should be located sufficiently downstream to ensure adequate intermixing of the tributary.

Power dams, storage pools, and navigation works drastically alter the natural channel of a river and may introduce short-circuiting and stratification, with accompanying variation in water quality in the cross-section. One of the most troublesome variations in sampling introduced by man-made developments is the hydroelectric power plant. Radical diurnal variations in stream discharge are induced by heavy release to meet the peak power demands for a few hours of the day with curtailed flow at off-peak hours. This sets up a series of pulsations in discharge that persist down the course of the stream and make representative sampling exceedingly difficult. Arrangements should be made with the power company to stabilize the flow. Usually cooperation can be obtained if the survey is of short duration. Otherwise intensive sampling on a 1- or 2-hr schedule (or continuous monitoring) is necessary to reflect the cycle. As the pulsation of flow passes downstream, it smoothes somewhat, and the integration of the cycle will provide a mean condition.

A somewhat similar pulsation in discharge is induced by operation of locks for navigation. Usually, however, this is not as severe as that induced by hydro peaking practice.

Self-Purification Characteristics

In the evaluation of organic self-purification the major objective is to describe the shape of the entire dissolved oxygen profile along the river course, and particularly the critical low point in the sag. Hence more sampling stations are located through the zone of depletion than along the recovery reach. If the waste loading is concentrated in a single reach, the dissolved oxygen profile will normally be quite regular and a minimum of sampling stations will suffice. If waste loadings occur along extensive reaches, the dissolved oxygen profile may be composed of primary and secondary sags with intermediate recovery zones, and in such cases a considerable number of sampling stations are required.

It is therefore helpful to have foreknowledge of the shape of the expected dissolved oxygen profile in establishing sampling stations. Previous survey data may suffice, but, if they are not available, it is desirable to take a test run under a runoff regime approximating that contemplated for the sampling program.

Abnormalities in self-purification—such as immediate demand, sludge deposits, and biological extraction—can radically alter an otherwise normal dissolved oxygen profile. Such abnormalities usually can be anticipated from the reconnaissance survey, the onshore waste survey, and the channel cross sectioning; from these the sampling stations can be located to reflect any distortions.

Time and Intensity of Sampling

Correlation with Stable Hydrograph

The most important factor that affects sampling is the daily hydrograph. The ideal time for sampling is during a steady runoff pattern, but sampling should not commence until the hydrograph has attained stability for a period at least equal to the time of passage through the critical reach of the river to remove the influence of any preceding instability. In addition to stability, a level of runoff should be anticipated that will produce a significant sag in the dissolved oxygen profile but not deplete the dissolved oxygen at any location. Such a level can be decided on from previous surveys or from preliminary computations of dissolved oxygen profiles expected from the current BOD waste loading.

Since temperature affects self-purification, it is likewise necessary to select the sampling period when water temperature is stable. This is not as difficult to anticipate as is stability of the hydrograph.

From statistical analysis of long-term records of streamflow at a key gaging station it is possible to determine the month in which flow and temperature are most likely to achieve stability at the desired levels. General plans can then be made to sample during that month. Preceding this month, arrangements can be made with the U.S. Geological Survey to obtain daily discharge gage readings at the key station to follow the trend of the daily hydrograph. Arrangements can also be made with the local U.S. Weather Bureau staff to provide long-term weather forecasts (monthly or bimonthly) as well as the weekly predictions. For a gradually declining hydrograph usually a week of stability prior to commencement of sampling is adequate to establish an equilibrium condition along the river course. If the hydrograph is flashy, punctuated with sudden peaks and rapid declines, it is desirable to delay commencement of sampling for more than a week to dissipate the influence of the preceding freshet. The weather forecasts will aid in making the decision whether to delay for greater stability or to commence sampling before the next rainfall. A correlation of rainfall intensity with freshet discharge will also indicate the magnitude of storm required to influence stream discharge significantly.

These forecasts are only probabilities of occurrence, and staff and equipment should be in readiness to take advantage of conditions as they actually occur. It is a useless effort to proceed arbitrarily at a fixed time if stability has not been attained. Figure 10–1*a* shows the hydrograph of a flashy stream, with arrows indicating the days on which sampling took place. With such radical fluctuations in runoff, sampling

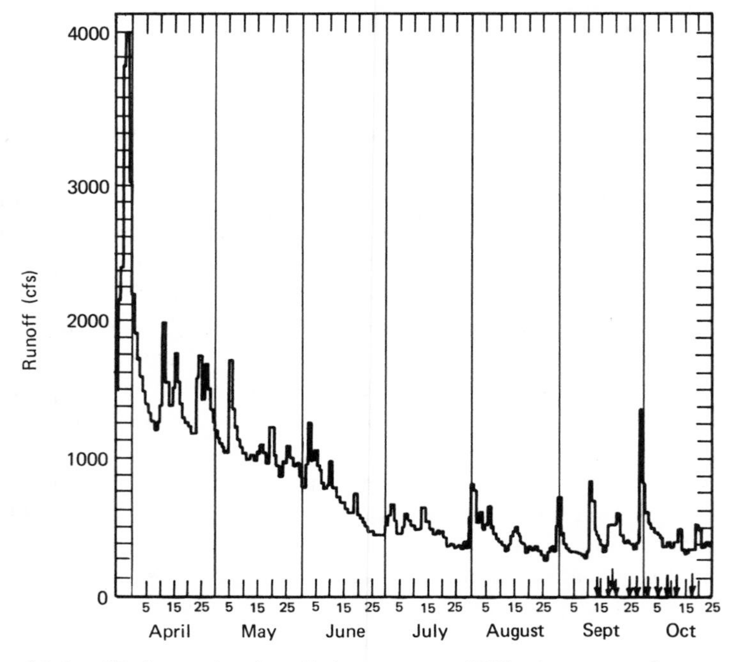

Figure 10–1*a* Hydrograph of a flashy stream. (Difficult to sample at a steady river regime. Samples are likely to be highly variable.) Arrows indicate sampling periods.

results are certain to be erratic. It cannot be expected that variation is normally ordered and that an average of the daily values would reflect a steady river regime equal to the average of the erratic daily discharges during these days. In contrast, the hydrograph of Figure 10–1*b*, also of a flashy stream, shows an ideal sampling period during the month of September when a period of steady flow prevailed, uninfluenced by previous flash flows.

Correlation with Industrial Production

Another factor that often leads to erratic sampling results is the operating pattern of major industries. Some operate 7 days a week, 24 hr a day, with a virtually continuous discharge of waste effluent; others shut down completely over weekends; still others operate only on an 8-hr shift. Monday morning samples after a weekend shutdown can hardly be expected to be representative. Similarly sampling during or immediately after holiday or vacation shutdowns is not representative, and it may take a week or more before production is normal.

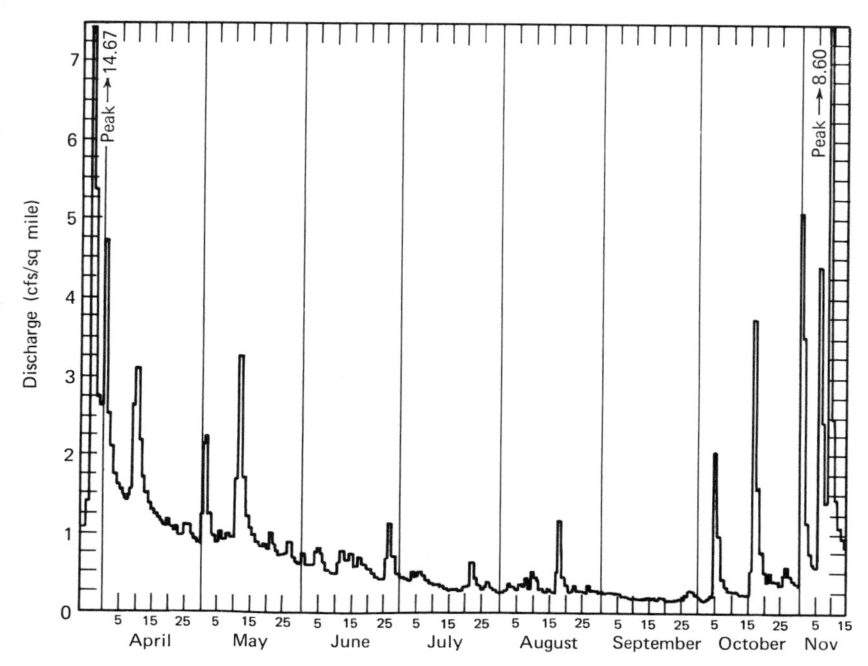

Figure 10–1*b* Hydrograph of a flashy stream, with ideal steady flow during September.

Correlation with Tide

What has been said of factors influencing sampling in freshwater streams applies equally to estuaries. As we have seen in Chapter 8, the tides and tidal currents induce periodic and nonperiodic variations that must also be taken into account in time and intensity of sampling. In addition to the moment-to-moment variation through the day it is necessary to correlate with the day-to-day variations in the cycles. The two most significant are the moon's phase and the moon's declination. During phases of new and full moon the range is considerably above the average, and during the first and third quarters it is substantially below average, as illustrated in Figure 10–2. The moon's declination induces diurnal inequality between succeeding tides and is more pronounced in certain geographical locations. There are other long-term factors, as well as local complicating influences, such as wind and physiographic features that can alter the normal tidal pattern.

The cycles of the moon's phase and the moon's declination are domi-

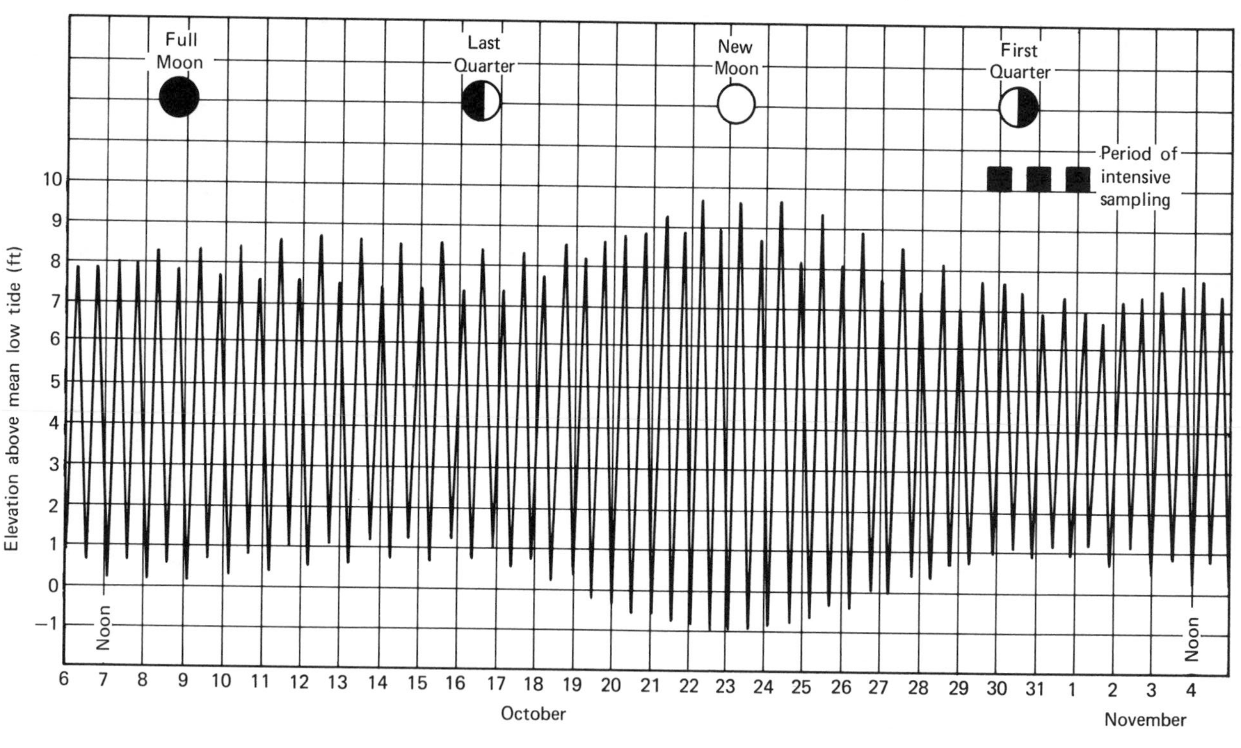

Figure 10–2 Estuary D. Predicted tide for October 6–November 4, showing tidal phases and diurnal inequalities.

nant and are readily reflected by plotting the predicted tides* for several months encompassing the anticipated sampling period. A sampling time is then selected when the phase produces a suitably stable range, and declination induces a minimum in diurnal inequality. Such a period is illustrated in Figure 10–2.

The level of runoff selected as desirable for sampling will usually be low flow, which will not appreciably affect tide stage and tidal translation but is very important in establishing the salinity gradient through the brackish reach (see Chapter 8). Stability in runoff before sampling therefore has a double significance. The coincidence of proper runoff and tidal conditions is not easily anticipated; hence alternative choices should be selected on the plot of predicted tides in the event the first choice in tide does not actually coincide with an adequate runoff level and stability. If ideal runoff conditions occur, tidal instability may be accepted and intensive sampling undertaken; but, if tidal conditions are ideal and runoff is erratic, it is virtually useless to conduct sampling. If the nature of the problem warrants, a second intensive sampling may be planned for the occurrence of a better coincidence of tide and runoff. However, the entire dry weather season may be punctuated by freshets, and then the sampling program might well be postponed.

Number and Intensity

With so many factors inducing variation in water quality of a natural stream (or estuary), it is obvious that random sampling or sampling over an extended period of time is almost certain to reflect a series of distorted values of heterogeneous conditions. It should be equally obvious that sampling must be intensive over a short period to ensure stability and thus permit relating the observed water quality profile to the set of stable conditions that produced it.

Freshwater Stream

For freshwater streams a minimum of six to eight samples taken through the day at all stations for 2 or 3 days usually provides a representative measure of stream quality profile with a minimum opportunity for significant change in hydrologic stability and normal waste loading. One such good set of data reliably defining oberved stream conditions serves as the basis for comparison with the independently determined stream quality profile computed for the same conditions. A close agreement between the computed and the oberved profiles under the one known stable regime then warrants application of the rational method

* Tide Tables, U.S. Coast and Geodetic Survey.

of stream analysis in defining expected stream conditions for a whole range of other waste loadings, hydrologic conditions, and man-made river developments.

Figure 10–3 represents an intensive, well planned survey carried out by the Michigan Water Resources Commission in which 18 samples were taken over a 2-day period preceded by and during a stable hydrograph and a normal waste loading. Normal distributions of dissolved oxygen developed for each sampling station, with the mean dissolved-oxygen profile as shown by curve *A*, and the 90-percent confidence range in individual dissolved oxygen values shown as the shaded band. Although there is variation in the results of individual samples, there is reasonable stability in the mean profile based on 18 samples. For example, at station 5 the summary statistics developed from the normal probability plot are as follows:

Mean: 30 percent of saturation
σ: 9 percent of saturation
n: 18 samples
Standard error of mean: $9/\sqrt{18} = 2.1$ percent
90-percent confidence range (individual samples): 16 to 44 percent
90-percent confidence range (other means): 26.6 to 33.4 percent

Thus with 18 samples the potomologist could be 90 percent confident that the means of other similar sets of 18 samples taken under the same conditions will fall within the range of 30 ± 3.4 percent of saturation. However, for individual samples the dissolved oxygen for a 90-percent confidence range would be within 30 ± 14 percent of saturation. This emphasizes the necessity for systematic sampling with a sufficient number of samples at each station to define a reasonably reliable observed mean dissolved oxygen profile for comparison with the independently computed profile. Curve *B* of Figure 10–3 shows the independently computed profile for the conditions that prevailed during the intensive sampling period, indicating quite a good agreement in shape and magnitude with the observed profile.

Since the standard error of the mean varies inversely as the square root of the number of samples, the range in means for any confidence level can be narrowed by increasing the number of samples. Such increase in reliability of the mean must be balanced against the extra effort and cost of field sampling and laboratory work. For example, lengthening the sampling period to 4 days with 36 samples would narrow the range in means at 90-percent confidence within 27.6 to 32.4 percent of saturation. This is a rather small gain for a doubling of effort. It is well to emphasize, however, that a minimum of approximately 18

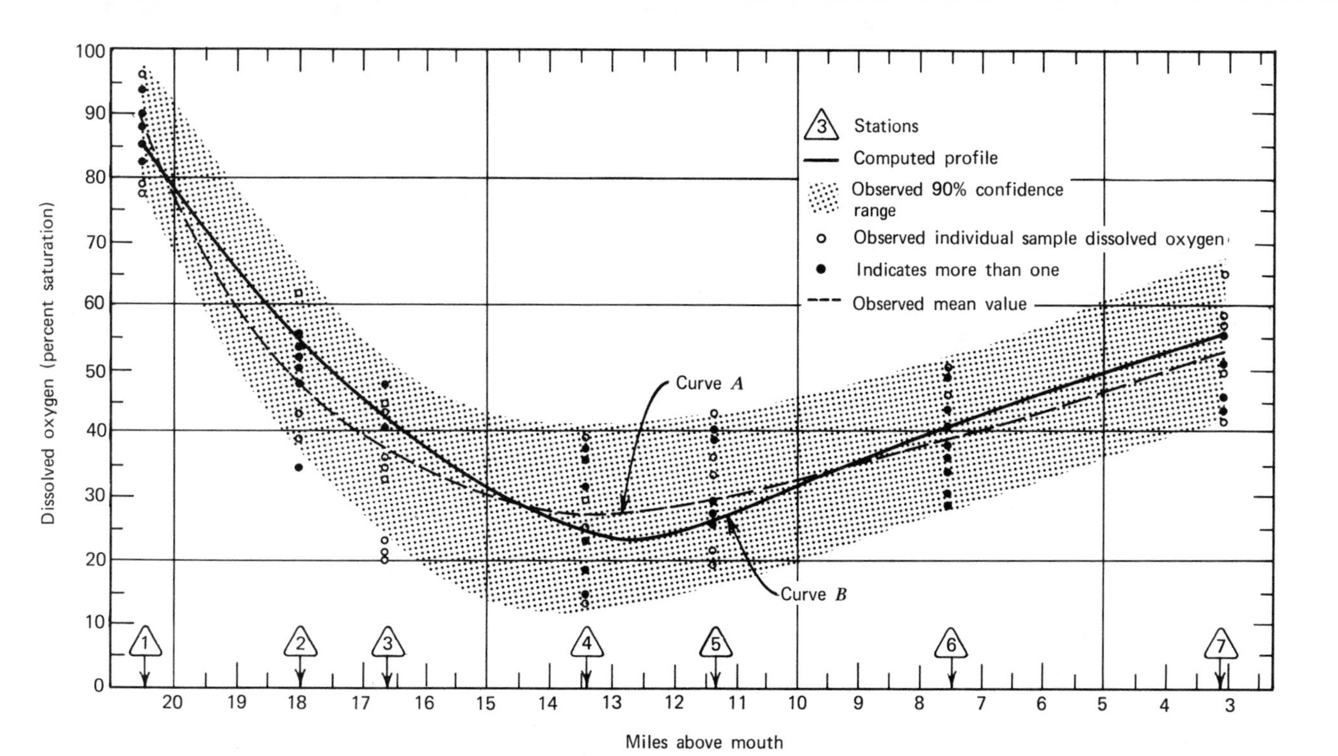

Figure 10–3 River H. Comparison of computed and observed dissolved oxygen profiles. August 11 and 12, 1954, temperature 17.5 to 20.5°C, average runoff 225 cfs.

samples is usually necessary to define a meaningful plot on probability paper from which a mean and standard deviation can be determined.

Estuary

In estuaries in which the quality gradients are steep the observed sample results are subject to a considerable range through the daily tidal cycle. On a declining water quality profile, ebb tide translates higher quality water to a fixed sampling location, and flood tide translates poorer quality water; conversely, on a rising quality profile better quality water passes on the flood and poorer quality water on the ebb. Thus samples systematically taken through the hours of a day synchronize with the tidal cycle, and water quality follows a definite cycle at each particular sampling station, depending on its position along the course of the estuary. The purpose of sampling is to define the cycle at each station in a manner that permits integration of a mean tide value and determination of values at the end of the ebb and of the flood. The best approach is to sample at frequent intervals throughout the daily cycle (every hour or every 2 hours) and thus describe the entire cycle.

A stable range in stage and minimum diurnal inequality usually can be achieved through a 3-day intensive sampling period. In some estuaries it is too dangerous to sample at night, and sampling is restricted to a 12-hr daylight period. This will include the range through a high and a low tide. If such a 12-hr sampling period is divided into six 2-hr periods of the mean tide cycle of the 3 days, with six samples taken in each 2-hr phase, in 3 days six sets of 18 samples are obtained, or a total of 108 for the cycle. Each set of 18 can be plotted on probability paper to obtain a mean and standard deviation from which the variation through a tidal cycle can then be plotted. Integrating such a cycle gives the value for mean tide, and the smoothed plot identifies values for the end of flood and the end of ebb phases. If the estuary is deep, it may be necessary to sample both the top and bottom waters, thus doubling the total samples to 216. It is apparent that, if reliability in a statistical sense is to be obtained, a great many samples are required.

If the precision of measurement is poor, as in determination of coliform bacteria MPN, it is essential to resort to intensive sampling. Unless MPNs are reduced to mean densities (see Appendix B), it will be difficult to define or even recognize a tidal cycle curve. For most other parameters, such as dissolved oxygen, precision of measurement is such that it may not be necessary to obtain 18 samples for each 2-hr phase of the cycle. One compensating influence of tidal phenomena with respect to sampling is that there are no sudden changes in the quality profile. Ebb and flood translation acts as a leveling device, reducing changes in quality to gradual transitions. Sampling stations can be spaced at considerable distance with some assurance that there will be no abrupt intervening changes. Also a few measurements by precise means establish a reliable position on the cycle without recourse to large numbers for statistical definition. Thus, except in the case of bacterial enumeration, various shortcuts are usually employed in sampling tidal estuaries to reduce the number of samples.

Since the lunar day is 24.84 hr, the moon passes a given meridian about 50 min later each day; thereby a fixed station sampled at the same time each day will reflect all phases of a complete cycle of rising and falling tide in approximately 15 days (Chapter 8). If stage and diurnal inequality do not vary greatly in a 15-day interval and runoff is stable, sampling a station at the same time each day for the 15-day period provides data from which the tidal cycle can be constructed. The sampling data at each station, depending on the date and time of sampling, are oriented to their equivalent positions in the tidal cycle. The daily sampling results are positioned with reference to a typical tidal cycle selected to represent the mean through the sampling period, using as a reference point low water or high water, depending on which is applicable for the particular date on which samples are taken.

Such an orientation of sampling in the period August 6 to September 7 is illustrated in Figure 10–4; curve *A* is the reference tidal cycle; curve *B*, the orientation of the chloride values; and curve *C*, the orientation of dissolved oxygen values. The sampling station is located on a declining dissolved oxygen profile; it will be noted that low and high dissolved oxygen levels are synchronized with low and high tide, respectively; chloride, of course, declines on the ebb and rises on the flood. Figure 10–4 also shows the results of hourly sampling on a single day (August 29) for a 12-hr period, describing the dissolved oxygen cycle (curve *D*). Considering the long period over which daily samples were taken, curve *C* is in close agreement with the short sampling results shown by curve *D*. The advantage of spreading the sampling over a longer period of 15 days or a month is that it can be executed with a small field and laboratory staff; the disadvantage, which in most instances overshadows the advantage, is the risk of sampling instability, reflecting too many different conditions.

Illustrated Example of Sampling an Estuary. A carefully designed intensive sampling program for an estuary, executed by an industrial group, illustrates the essential steps and some of the difficulties encountered. The purpose of the survey was to verify self-purification factors adopted in the computation of resultant dissolved oxygen pro-

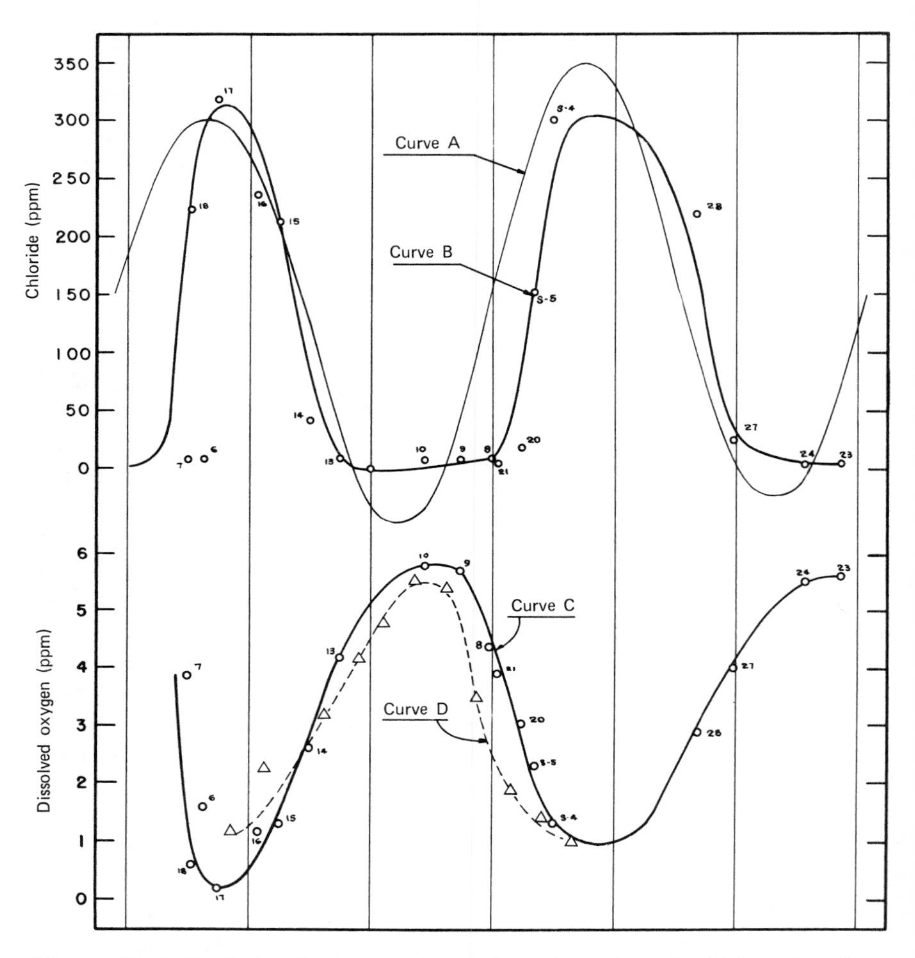

Figure 10–4 Typical orientation of sampling data through tidal cycle. Curve *A*: typical tidal cycle; curve *B*: salinity; curve *C*: orientation of daily sampling. August 6 to September 7 (○ 7, August; ○ S-4, September); curve *D*: hourly sampling through 12-hr cycle, August 29.

files and to project by the rational method the expected waste assimilation capacity under proposed regulated freshwater runoff from a system of upstream reservoirs.

The relation of the sampling period to the daily hydrograph reflected by an upstream gage is shown in Figure 10–5.

The sampling design was conceived to select either an early summer or fall period when a moderate dissolved oxygen profile was likely to

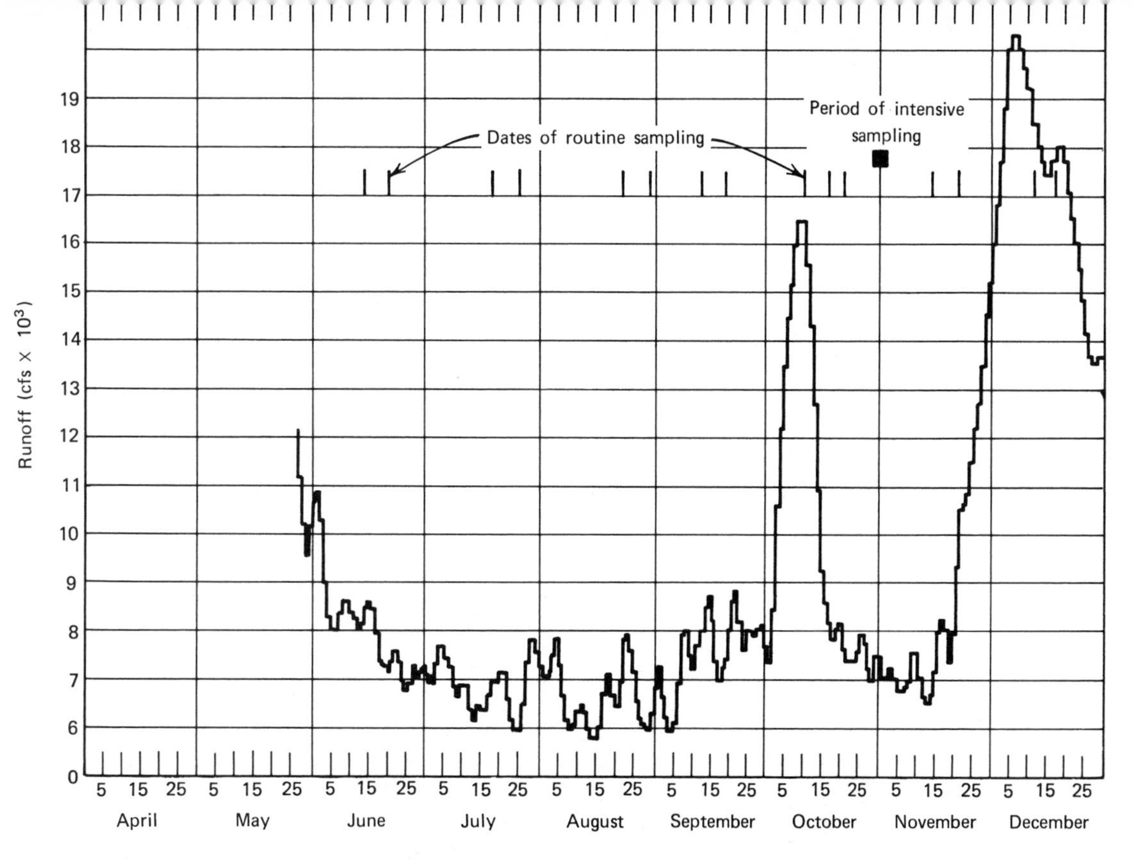

Figure 10–5 Estuary D. Daily hydrograph.

occur. Also, the design selected periods of minimum diurnal inequality of tides, protracted stable freshwater runoff, and normal industrial operation by the various industries. For several reasons the early summer period did not meet the sampling design criteria, and the survey was postponed until fall. Unfortunately a freshet of considerable magnitude occurred in early October, peaking at a daily average discharge of 16,500 cfs on October 11. This was followed by a rapid decline to 7000–8000 cfs for the end of October, and a decline in water temperature approaching 13 to 17°C. Predicted tides, as shown in Figure 10–2, indicated a minimum of diurnal inequality coincident with the first quarter of the moon at the end of October and early in November. A decision was made not to risk another freshet, the probability of which is great in early November. It was decided to undertake the intensive sampling on October 30, 31, and November 1, recognizing that equilibrium conditions in the harbor, principally the salinity gradient, would not be ideal.

Ten sampling sections through 27 miles of the brackish reach were selected. A preliminary survey with sampling at quarter points and main channel, with three depths at each, demonstrated that sampling in the main channel at 5 and 25 ft provided a representative value for the entire cross-section. As planned, hourly samples were taken during daylight through the tidal cycle for 3 days for dissolved oxygen, water temperature and chlorides, at 2-hr intervals for BOD, and at 4-hr intervals for pH.

During the sampling period water temperature in the upper reach averaged 13°C and in the lower reach to the mouth, 17°C. Mean freshwater runoff just prior to and during the survey at the control gage ranged between 6960 and 7930 cfs, averaging 7490 cfs, which, adjusted to the head of the brackish reach, was taken as 7800 cfs.

To facilitate comparison the individual sampling data at each station were oriented to their equivalent position in the tidal cycle, depending on the date and time of sampling, and smoothed tidal cycle curves developed, as illustrated in Figure 10–6. Such plots for each of the 10 sampling cross sections permitted obtaining an integrated mean through the tidal cycle and provided a method for approximating the condition at the end of the ebb and of the flood tides. Higher chlorides were observed in the samples at 25-ft depth than in the samples at 5-ft depth; this differential and (to some extent) the differentials in dissolved oxygen between top and bottom confirm that equilibrium conditions had not been completely restored after the freshet preceding the sampling.

Figure 10–7 is a profile plotting of the dissolved oxygen data, expressed as the average between pairs of sampling at 5- and 25-ft depths, unadjusted for tidal phase. As evolved from the tidal cycle plots superimposed

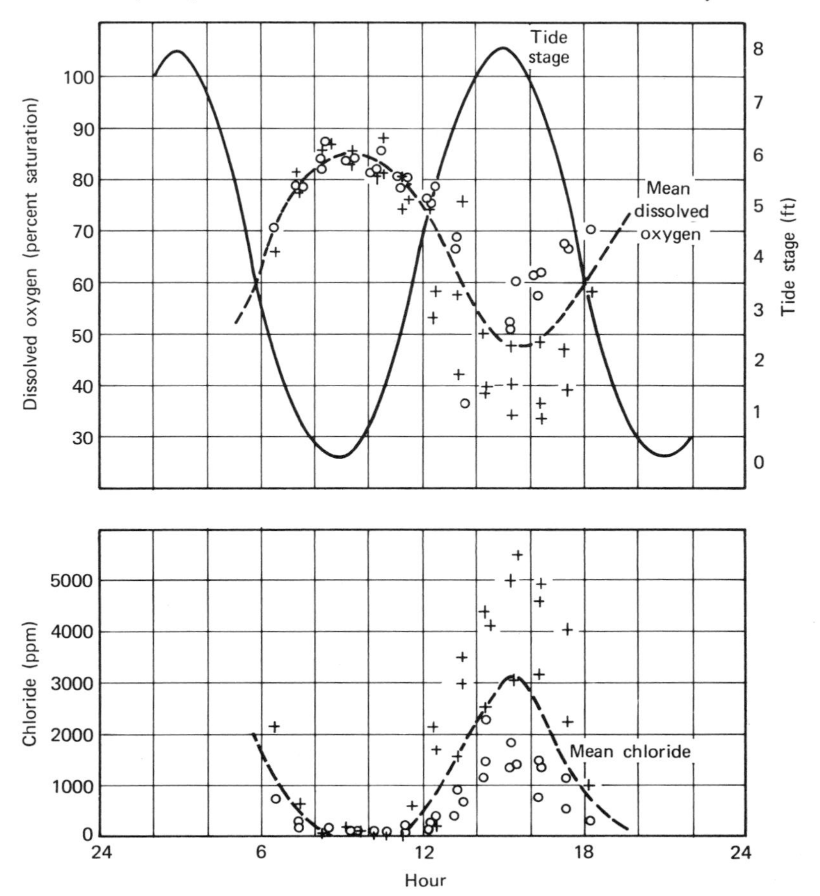

Figure 10–6 Estuary D. Typical orientation of sampling data through tidal cycle at station No. 3; + 25-ft depth; ○ 5-ft depth.

on Figure 10–7, curve *A* is the observed midtide mean dissolved oxygen profile through the tidal cycle, curve *B* is the observed mean profile at the end of flood tide, and curve *C* is the observed mean profile at the end of ebb tide.

The independently computed mean tide dissolved oxygen profile for the conditions that prevailed during the survey is shown as curve *D* superimposed on the observed survey data of Figure 10–7. It will be noted that the computed profile fits well through the midposition of the observed data and is in fairly close agreement with curve *A*, the observed mean midtide profile. The computed critical dissolved oxygen

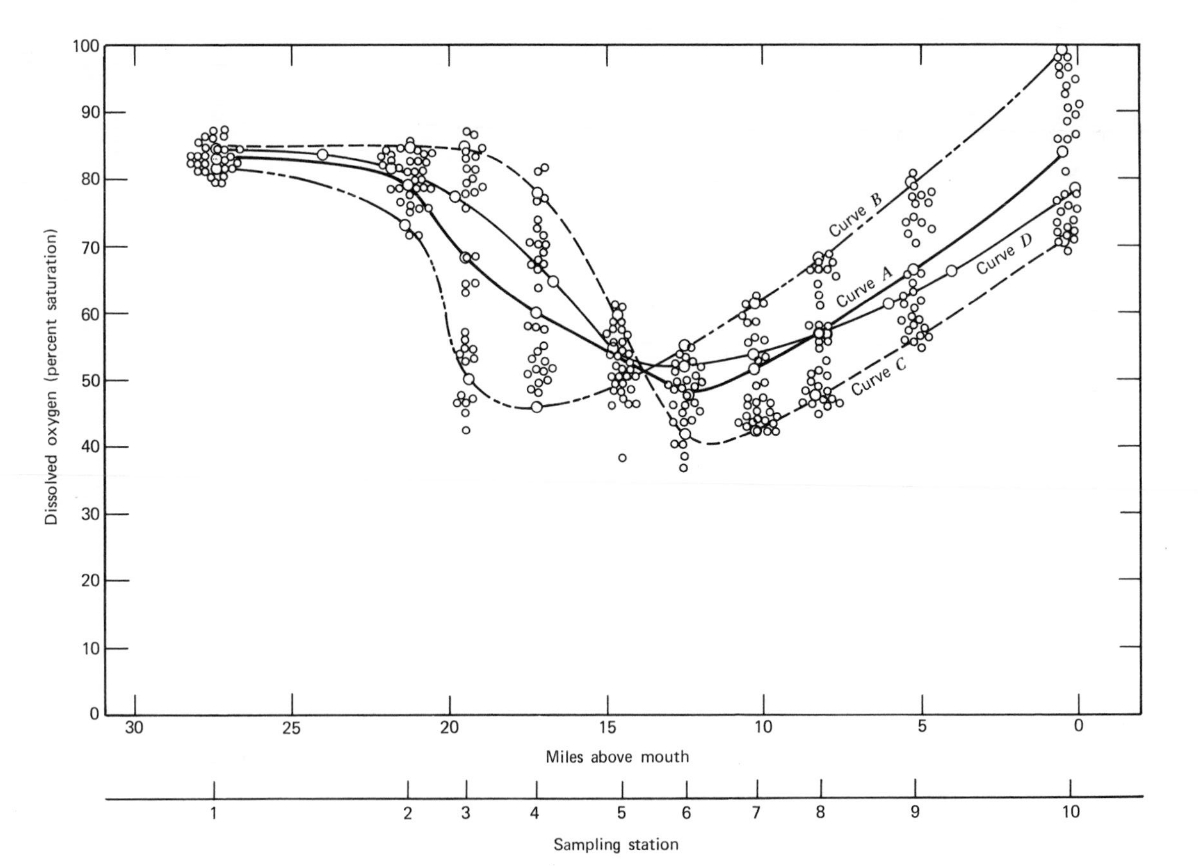

Figure 10–7 Tidal estuary D. Observed and computed dissolved oxygen profiles (October 30, 31 and November 1); average of 5- and 25-ft (or 10-ft) samples unadjusted for tidal phase. Curve A: mean through tidal cycle; curve B:

level declines to 52 percent of saturation in the vicinity of mile 13 as compared with 48 percent for the observed value; and the computed dissolved oxygen recovery is pronounced, to 78.5 percent of saturation at the mouth as compared with the observed value of 85 percent.

THE EXECUTION OF THE STREAM SURVEY

Critical-Path Control

The hydrologic data and the survey of waste sources should be undertaken prior to the stream sampling, since these assist in the design of the sampling program. However, the determination of the waste loading should not be so far removed from the sampling period that a significant change intervenes. Where major industries are involved, the current waste load can be related to the measured load by the level of production. This will also include current information on anticipated operating patterns during and before the stream sampling period to ensure as far as possible normal production uninterrupted by shutdowns.

Correlation of the three primary phases of the stream survey (hydrology, waste sources, and stream sampling) are most effectively controlled by development of a critical-path schedule, such as employed in large construction or logistic operations.

Laboratories

In many river basins a number of good laboratories are available in local industries, utility companies, and in public locations, such as water and wastewater treatment plants. In cooperative studies these resources can be coordinated so as to distribute the heavy laboratory work involved in the intensive stream sampling program. Although the laboratory load is heavy, it is for a short period; hence local participation usually can be obtained. In some situations a mobile laboratory may be necessary to supplement local resources, and usually extra glassware and consumable supplies are needed for the many samples to be handled.

Personnel

Again, local technical personnel may be available to perform the laboratory examinations and the field work of collecting and fixing samples. From these a staff can be assembled, coordinated, and instructed for specified tasks to ensure uniformity in practice. In addition to technical staff, personnel to operate boats and transport staff and samples are necessary. Reliable local employees, available for assignment on short notice for the short period involved, are preferable.

Equipment

Field equipment includes transportation vehicles, boats, and sampling apparatus. Insofar as possible, the sampling crew should be relieved of extraneous tasks and should be serviced by vehicle drivers and boat operators. Boats should be of a size and stability suitable to the watercourse exposure without risks from possible rough water and large enough to accommodate a workbench for fixing samples and storage of sampling equipment and bottles.

Sampling equipment depends on the type of samples to be collected and the depth of water to be sampled. Most equipment is standardized for minimum requirements but varies from the complex to the relatively simple. Descriptions of apparatus are found in *Standard Methods* [4], U.S. Geological Survey Water-Supply Papers [5], and in various technical journals. The intensive sampling program requires several sets of equipment, but usually these can be borrowed for the short period involved.

Preplanning

Since the stream sampling program is intensive and short, detailed preplanning is essential for successful execution. With the completion of the sampling design and location of sampling stations fixed, each laboratory is assigned to service a specific allocation of stations within its capacity to handle a definite schedule of samples. In turn sampling crews are assigned, distinguishing between stations serviced from bridges and those serviced by boat. Usually at least three stations can be serviced by one properly powered and manned boat. In wide, deep estuaries it may be necessary to assign a boat to each cross-section if samples are collected at more than one position in the section. Definite routes and schedules are established for the transportation personnel in moving sampling crews and in collecting samples and supplies.

Detailed written instructions and recording forms are prepared for each operation, and personnel are instructed as to specific duties. The entire crew and staff should assemble for an explanation of the purposes of the survey and the interrelation of each specific assignment. Special attention is given to coordination of laboratory procedures, equipment use, standard chemical solutions, and media. When the anticipated time for the survey appears imminent, a dry run is made of the entire operation to ensure smooth execution and full coordination of the entire technical staff and field crews. Interchange of samples among all laboratories verifies uniformity of results. No detail, including safety instructions, is too small to neglect; nothing of significance should be taken for granted. Failure of one segment of the operation, through ignorance or carelessness, may jeopardize the entire survey.

Records

Convenient detailed forms are prepared for identifying samples and reporting the results of examinations, including remarks concerning any observed abnormalities. Field crews identify each sample and report the exact time and date each sample is taken, with notation of weather (sunshine or degree of cloudiness), air temperature, wind direction, condition of the water (choppy or calm), and appearance of unusual color, slicks, or floating material. This assists in the interpretation of any abnormalities that may be reported by the laboratory examinations. Systematic and detailed reporting facilitates interpretation and statistical analysis of the raw data.

REFERENCES

[1] Cleary, E. J., *J. Am. Water Works Assn.,* **50,** 1219 (1958).
[2] Cleary, E. J., *Revue Suisse d'Hydrologie,* **22,** 420 (1960).
[3] Cleary, E. J., in *Proc. International Conf. on Water Pollution Research,* Pergamon Press, New York, 1964, p. 63.
[4] *Standard Methods for the Examination of Water and Wastewater,* 12th ed., American Public Health Association, Inc., New York, 1965.
[5] Rainwater, F. H., and Thatcher, L. L., Geological Survey Water-Supply Paper 1454, U.S. Government Printing Office, Washington, D.C., 1960.

11

EFFICIENT USE OF WASTE ASSIMILATION CAPACITY

Effective lines of defense and offense in waste disposal, pollution control, and water quality management depend on efficient use of the natural or enhanced waste assimilation capacity of the stream. Efficient use necessitates quantitative measurement. As we have seen, the application of the rational method of stream analysis is a reliable means of determining waste assimilation capacity in quantitative terms and thus affords the basis for evaluation of specific alternatives and combinations, such as water quality objectives and criteria, degree of wastewater treatment, low flow augmentation, direction of urban-industrial growth, plant site location, land and water use planning and zoning, multipurpose river development, and the effects of river developments and water uses.

In this chapter we discuss a number of practical applications of more efficient use of the total annual waste assimilation potential, such as drought control by streamflow equalization, storage of wastes and regulated effluent release, and tailored industrial production.

WASTE ASSIMILATION POTENTIAL UNDER NATURAL HYDROLOGIC VARIATIONS

Efficient use of waste assimilation capacity in any specific situation entails as a first step the quantitative definition of the potential under the full range of the expected natural hydrologic variation. This holds for any waste and especially for organic waste. Because of its demand on dissolved oxygen, the most sensitive indicator of water quality, and its complex relation to hydrology, organic waste is selected to best illustrate stream analysis methods.

Total Waste Assimilation Potential—A Case Study

As a specific illustration, river J is selected for quantitative definition of organic waste assimilation potential under natural hydrologic variations. This selection is made not only because of the wide variation in hydrologic conditions but also because it deals with a common difficulty, limited basic data and rational substitutions. The river J basin, a relatively small one, encompasses 4860 square miles, almost wholly within a single physiographic region. Formed by the junction of two major streams, tributaries A and B, it traverses 52 miles of lowland to its mouth at the upper end of a sound to the ocean. A subdrainage area, tributary C (1620 square miles) enters 12 miles below the junction. The relatively deep channel and flat low water profile result in a sluggish stream. During low runoff a slight tidal influence extends upstream from the sound but without seawater intrusion; though estuarial in its aspect, river J may be treated as a freshwater stream. Channel cross-section soundings along the upper reach taken by the Army Corps of Engineers and U.S. Coast and Geodetic Survey charts for the lower reach provide a reasonable basis for definition of channel parameters and computation of time of passage.

The two major natural hydrologic variables that determine waste assimilation potential are runoff and temperature. Runoff characteristics of river J were determined from analysis of discharge records of the tributaries and referred to runoff at its head at the junction of tributaries A and B. Based on statistical analysis of this record, an unusual seasonal pattern of runoff emerges (Figure 11–1), disclosing relatively high winter–spring runoff and low summer–fall drought flow, but subject to a wide range, with flow in some years being extremely high and in other years being extremely low. Analysis of the minimum monthly average flows of the summer–fall season indicates drought severities of 115 and 60 cfs once in 5 and 10 years, respectively. These are relative to an average annual runoff of 2140 cfs. It is to be expected, therefore, that waste assimilation capacity will also be highly variable; although on the average it is good, it is very poor during extreme drought.

The seasonal pattern of water temperature accentuates this with higher waste assimilation during the cool winter–spring and lower assimilation during the warm summer–fall season. Water temperature records are not available for river J; accordingly the seasonal pattern has been

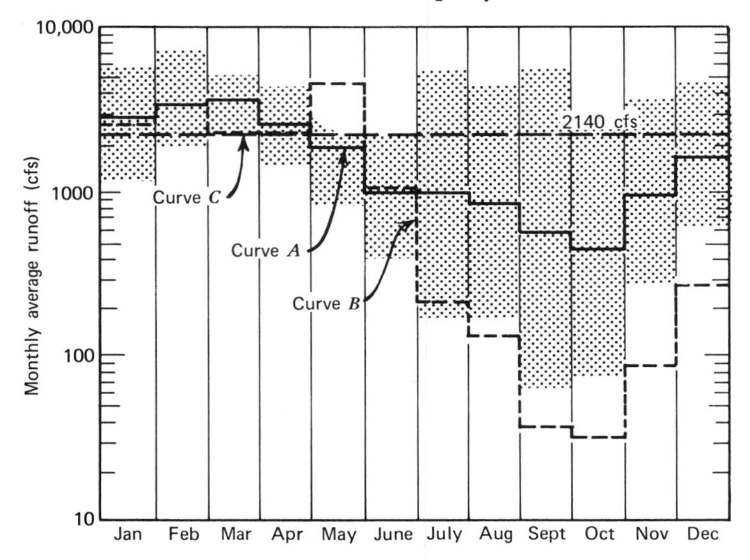

Figure 11–1 River J. Seasonal pattern of runoff. Junction of tributaries A and B. Curve A: most probable mean; curve B: monthly means for 1954; curve C: mean for period 1944–1956; shaded area: range within which monthly mean can be expected for 80 percent of years.

based on the monthly mean air temperature records and limited temperature data of adjacent streams.

The Framework of Computed Dissolved Oxygen Profiles

A series of dissolved oxygen profiles, computed by the rational method, constitutes the framework for determining the waste assimilation capacity for the recorded range of runoff and seasonal temperature. The specific conditions for the computation of the dissolved oxygen profiles assume organic waste loading free of settleable solids subject to deposit discharge at the head of the river J. Residual BOD_5 from tributaries is taken as 1.0 ppm, and dissolved oxygen at 85 percent of saturation. Amortization of organic waste loading is assumed normal at $k = 0.1$ at 20°C, and adjusted for other temperatures. Runoff is referred to the head of river J as the combined discharge of tributaries A and B; increments of runoff are assumed contributed along the course at the same yield as at the head of the river.

The framework of dissolved oxygen profiles was conceived for minimum computations to be executed by desk calculator and slide rule, without benefit of digital computer adaptation. For this purpose the computations were executed for four runoffs and three temperatures.

(With computer adaptation a much more extensive series can be provided readily for refinement.) The reference frame was achieved with 36 computed profiles for selected waste loadings to produce ranges in dissolved oxygen decline, but not exhaustion. Usually three profiles are required at each runoff–temperature combination to provide a basis for interpolation of allowable waste load for the prescribed critical dissolved oxygen level. Typical computed profiles are illustrated in Figure 11–2 for severe drought high temperature conditions, in Figure 11–3 for high runoff winter conditions, and in Figure 11–4 for the intermediate runoff and temperature of late fall and late spring.

From the series of these dissolved oxygen profiles the control curves, Figure 11–5, are developed (by methods described later), giving the relation between allowable BOD_5 waste load for any runoff at three specific temperatures, 15, 20, and 30°C, to ensure the critical dissolved oxygen content of 4 ppm prescribed by the regulatory agency. The control curves afford the basis for determining the total waste assimilation capacity under the expected natural hydrologic variations. In this case, to be conservative, seasonal temperatures were taken on the high side at only three levels, 15°C for the winter months December through April, 20°C for late fall (October and November) and for late spring (May and June), and 30°C for the warm summer months July, August, and Sep-

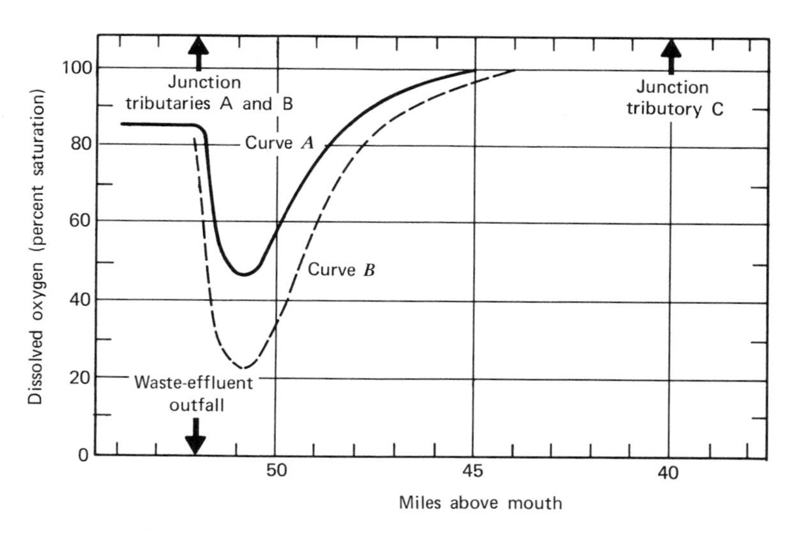

Figure 11–2 River J. Computed dissolved oxygen profiles for drought conditions—runoff 115 cfs water temperature 30°C. Curve *A*: 1250 lb BOD_5 per day; curve *B*: 2500 lb BOD_5 per day.

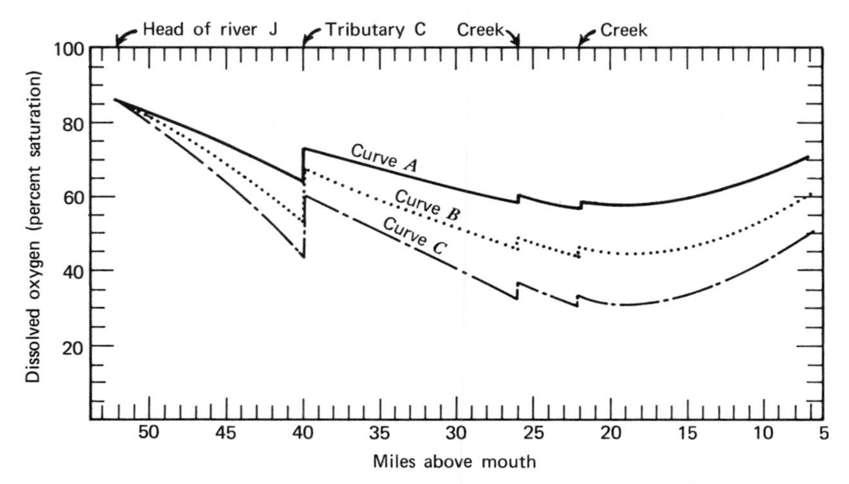

Figure 11–3 River J. Computed dissolved oxygen profiles. Runoff 5000 cfs, water temperature 15°C. Waste effluent loading (BOD_5): curve *A*: 200,000 lb/day; *B*: 300,000 lb/day; curve *C*: 400,000 lb/day.

tember. (A refinement in temperature increments is possible, as illustrated later.)

The Critical Reach and the Critical Dissolved Oxygen Content

The general shape of the dissolved oxygen profile determines the location of the reach of the stream in which the critical dissolved oxygen depletion occurs. Each stream, depending on its natural hydrologic setting and channel characteristics, develops unique dissolved oxygen profile patterns. The critical dissolved oxygen shifts in location and degree of depletion with each combination of organic loading, runoff, and temperature, as illustrated in Figures 11–2, 11–3, and 11–4. At high runoff and low temperature (Figure 11–3) the critical reach shifts downstream to the vicinity of mile 20. As runoff declines and temperature rises, the critical reach shifts upstream (Figure 11–4), at light loads to the vicinity of mile 35, at heavy loads to above the junction of tributary C. During extreme drought and high temperature the dissolved oxygen profile contracts radically, with a sharp decline and recovery, and the critical dissolved oxygen is confined above the junction of tributary C.

The critical dissolved oxygen may be expressed in concentration (ppm), but this has limited meaning unless temperature is also designated. It is preferable to express it in percentage of saturation, which incorporates the temperature factor. The location of the critical reach

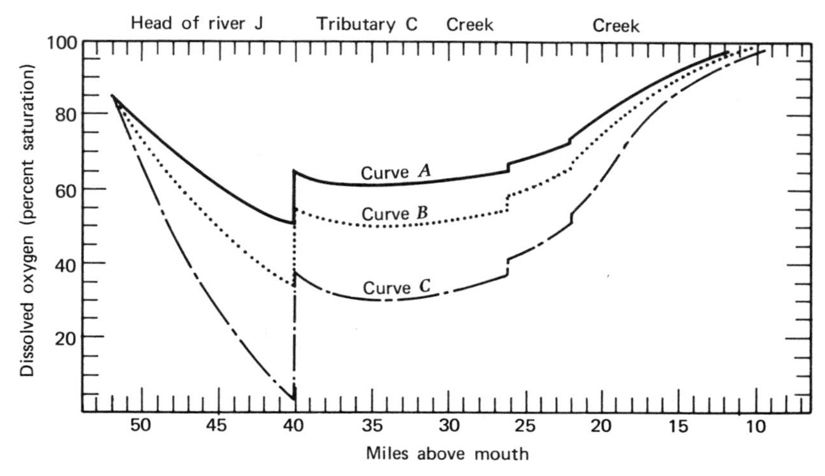

Figure 11–4 River J. Computed dissolved oxygen profiles. Runoff 1500 cfs, water temperature 20°C. Waste effluent loading (BOD_5): curve A: 33,000 lb/day; curve B: 49,000 lb/day; curve C: 80,000 lb/day.

is significant in relation to downstream uses, but with respect to compliance with regulatory criteria or standards it is the level of the critical dissolved oxygen that is of primary concern.

Annual Variation in Waste Assimilation Capacity

If full advantage is taken of the waste assimilation capacity month by month at the prevailing runoff and the seasonal water temperature, the total annual waste assimilation capacity can be determined readily from the control curves (Figure 11–5). For the available 12 years of runoff records the annual waste assimilation capacities were thus determined for river J under the prescribed condition that the critical dissolved oxygen will not decline below 4 ppm. In this instance the water year was taken as November through October of the following calendar year. As illustrative of the procedure, the results are tabulated by month for 2 years in Table 11–1, and summarized as annual totals in Table 11–2 (column 2). For example, for May 1945 the monthly average runoff is 2726 cfs and the seasonal temperature for May (assumed) is 20°C; from the control curve B, of Figure 11–5, the allowable waste load in BOD_5 per day at 2726 cfs is 91,000 lb, which for 31 days gives a total of 2.82 million lb for the month. The allowable waste loadings determined for each month of the year for the recorded runoff and seasonal temperature provide the total annual waste assimilation capacity.

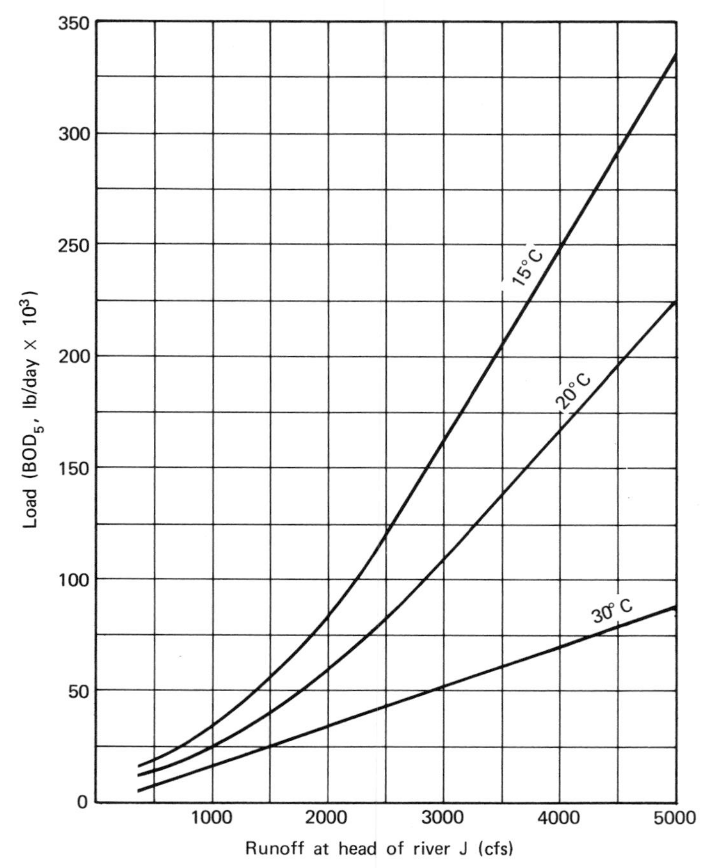

Figure 11–5 River J. Runoff-waste load relation control curves to maintain 4 ppm of dissolved oxygen in the critical reach.

As with all hydrologic phenomena, the annual waste assimilation capacity varies from year to year, ranging from 15.6 million lb of BOD_5 per year in 1954–1955 to 56.8 million lb in 1951–1952. Arranging the annual values in order of magnitude and plotting on normal probability paper develops a reasonably normal distribution, as shown by curve *A* of Figure 11–6, with the most probable value of 36 million lb and a standard deviation of 16.2 million lb, or a coefficient of variation of 45 percent. The expected variation from year to year is great and parallels the high degree of variation in hydrology. From this sample it can be expected that on the average once in 10 years the annual waste assimilation capacity will be as low as 15.5 million lb or less;

Table 11–1 River J—Computed Monthly Waste Assimilation Capacities Based on Recorded Runoff and Seasonal Temperature[a]

Year and Month (1)	Assumed Water Temperature (°C) (2)	Runoff at Head of River J (cfs) (3)	Allowable BOD_5 per Month (lb × 10^6) (4)
1944–1945:			
November	20	594	0.42
December	15	2664	4.20
January	15	2818	4.60
February	15	4298	8.31
March	15	3403	6.06
April	15	1772	2.08
May	20	2726	2.82
June	20	1873	1.60
July	30	8111	4.28
August	30	2857	1.53
September	30	3597	1.93
October	20	1207	0.88
			38.71
1945–1946:			
November	20	1123	0.80
December	15	3568	6.50
January	15	5763	12.03
February	15	6049	12.71
March	15	3003	5.08
April	15	2930	4.89
May	20	4010	5.33
June	20	2430	2.36
July	30	3201	1.71
August	30	1147	0.60
September	30	607	0.30
October	20	1000	0.69
			53.00

[a] Maintaining 4 ppm of dissolved oxygen in the critical reach.

and also once in 10 years as great as 56 million lb or more, ensuring at all times a critical dissolved oxygen level of 4 ppm.

Waste assimilation capacity not only varies with hydrology but depends on the prescribed critical level of dissolved oxygen. This is illustrated for river J in the annual waste assimilation capacity determined

Table 11–2 River J—Computed Annual Waste Assimilation Capacities Based on Recorded Runoff and Seasonal Temperatures

Year (1)	BOD_5 per Year (lb $\times 10^6$) Maintaining Critical Dissolved Oxygen of	
	4 ppm (2)	3 ppm (3)
1944–1945	38.7	50.1
1945–1946	53.0	66.0
1946–1947	24.5	30.6
1947–1948	51.7	63.6
1948–1949	51.2	64.4
1949–1950	24.4	31.6
1950–1951	19.9	25.0
1951–1952	56.8	71.2
1952–1953	31.2	38.7
1953–1954	26.0	32.2
1954–1955	15.6	20.7
1955–1956	31.8	39.8

in a similar manner but based on maintaining a critical dissolved oxygen level of 3 ppm (Table 11–2, column 3). These also form a reasonably normal distribution, as shown by curve B of Figure 11–6, with the most probable annual value of 44 million lb of BOD_5 and standard deviation of 20 million lb. This compares with 36 million lb with a standard deviation of 16.2 million lb for a critical dissolved oxygen level of 4 ppm, as shown by curve A.

The Seasonal Pattern of Waste Assimilation

An analysis of the monthly waste assimilation capacities of river J for the period of record for a critical dissolved oxygen level of 4 ppm, considering each month of the year separately, discloses distributions that are logarithmically normal. From these the seasonal pattern is developed, as shown in Figure 11–7. Curve A is the most probable monthly waste assimilation capacity; the shaded band about curve A is the range within which monthly capacity is expected for 80 percent of the years, 10 percent of the years above the upper limit and 10 percent of the years below the lower limit. Individual years may deviate from the

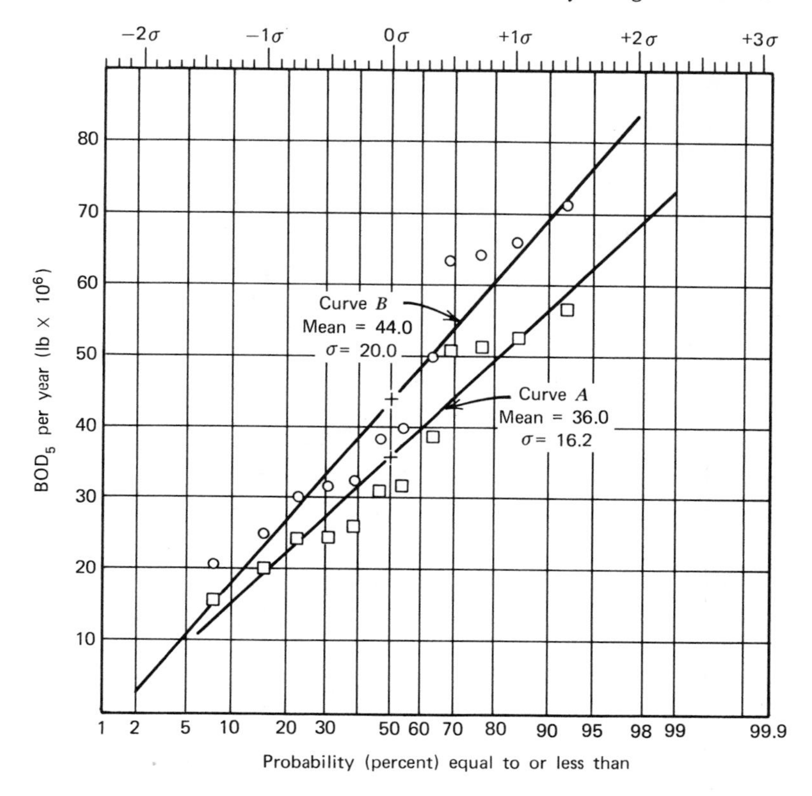

Figure 11–6 River J. Annual waste accumulation capacity. Critical dissolved oxygen level: curve *A*: 4 ppm; curve *B*: 3 ppm.

normal seasonal pattern (curve *A*) since not all months of the year will experience hydrologic conditions of the same probability of occurrence.

A similarity in seasonal pattern between the monthly hydrograph of runoff (Figure 11–1) and the monthly waste assimilation capacity (Figure 11–7) is noted, with capacity being high in the winter–spring season and low during the summer–fall season, but with wide variation and occasionally extremely low capacity during severe droughts.

The Severe Restriction of Drought Conditions

As a further illustration of the severe restriction of drought conditions on the waste assimilation capacity of river J, based on maintaining a critical dissolved oxygen level of 4 ppm, an analysis is made of the minimum monthly waste assimilation capacities selected each year from

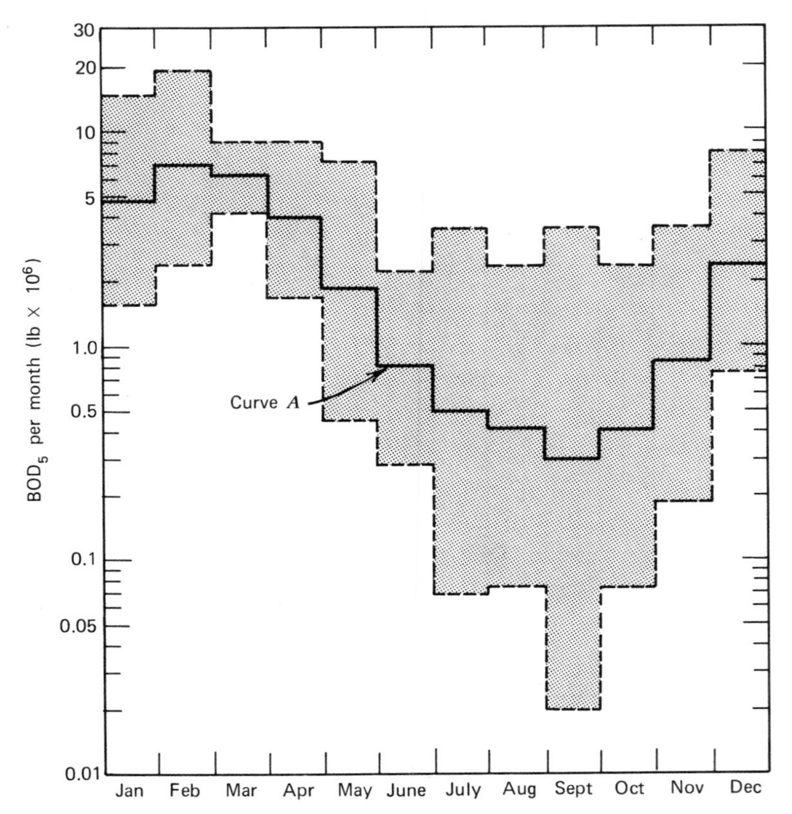

Figure 11–7 River J. Seasonal pattern of monthly waste assimilation capacity.

the warm weather, summer–fall drought season (June–October). The distribution is log-extremal, in reasonable agreement with the theory of extreme values, with a very high degree of variation, as reflected by a variability ratio of 0.046 (the once in 10-year value divided by the most probable). The most probable minimum monthly capacity is 290,000 lb, or 9650 lb/day; the once in 5-year minimum monthly capacity is 37,000 lb, or 1230 lb/day; the once in 10-year minimum monthly capacity is 13,500 lb, or 450 lb/day of BOD_5. These loadings maintain the critical dissolved oxygen level of 4 ppm.

The independently computed dissolved oxygen profile at the once in 5-year minimum monthly average drought flow of 115 cfs and a water temperature of 30°C for a waste load of 1250 lb/day of BOD_5 (Figure 11–2) produces a very sharp decline in dissolved oxygen to a critical value of 45 percent of saturation, or 3.4 ppm. Considering the steepness of the dissolved-oxygen profile, this is in close agreement with the once in 5-year allowable waste loading of 1230 lb/day of BOD_5 determined from the probability distribution of the minimum monthly waste assimilation capacities, for the critical dissolved oxygen level of 4 ppm.

It is apparent that, if waste effluent produced each day must be discharged to the stream daily, the restriction of the drought streamflow conditions is extremely severe indeed, limited to 450 lb/day of BOD_5 (or 2760 population equivalents) to ensure that the prescribed critical dissolved oxygen will not decline below 4 ppm more often than once in 10 years.

Even if waste treatment is extended beyond removal of settleable solids to an efficiency of 90 percent by a complete biological system, the allowable load (free of settleable solids) produced per day could not exceed 4500 lb/day of BOD_5 or 27,600 population equivalents. This approach to waste disposal represents a highly inefficient use of the natural waste assimilation potential of the river. For example, the once in 10-year minimum annual waste assimilation capacity, as determined previously, is 15.5 million lb of BOD_5 or 42,400 lb/day (260,000 population equivalents). This compares with the 450 lb/day (2760 population equivalents) restriction under the once in 10-year drought. Thus the once in 10-year minimum annual capacity is nearly 100 times that under the once in 10-year drought; and 10 times greater than that achievable with complete biological treatment. This also emphasizes that under natural drought severities satisfactory waste disposal cannot be provided by treatment alone.

Two obvious approaches to more efficient use of natural waste assimilation capacity are *drought control* through streamflow regulation and *regulated waste effluent release* tailored to variations in seasonal patterns of hydrology. These will be discussed in detail shortly.

Illustration of Construction of Control Curves

In order to illustrate the development of control curves a selection is taken from computations carried out for river K where channel parameters were available from detailed echo sounding cross sectioning along the course and river surveys were also available for verification of the adopted self-purification factors. An extensive series of dissolved oxygen profiles was computed by the rational method for ranges in BOD effluent loading discharged at mile 238 under various seasonal hydrologic conditions. In this instance the BOD loading is expressed in population equivalents and includes an upstream residual of 1.0 to 1.5 ppm of standard BOD_5, depending on runoff, with the waste effluent discharge at mile 238. An additional BOD load discharged at mile 206 is taken into account but is held constant while that at mile 238 is varied. BOD rate curves analyzed by Lee's grids (Chapter 4) approximate normal amortization at a k value of 0.1 at 20°C. Waste effluent is free of settleable solids and immediate demand. Runoff is referred to mile 238, but increments from tributaries are taken into account. Initial dissolved oxygen is based on river surveys, which indicated 85 percent of saturation during the warm weather season and 95 percent during the cool weather season.

Figures 11–8 and 11–9 are typical computed dissolved oxygen profiles at 32°C at 1000 and 5000 cfs, respectively; they illustrate the shift in the critical reach. Figure 11–10 is a typical computed cool weather profile at 12°C and 1000 cfs. From the extensive series of computed profiles at three temperatures, 12, 20, and 32°C, for ranges in runoff up to 6000 cfs, the control curves are developed in four steps.

Critical Dissolved Oxygen versus Load—Temperature and Runoff Specific

As a first step, the framework of dissolved oxygen profiles provides the basis for establishing the relation between critical dissolved oxygen and BOD load at specific temperatures and runoffs. For example, from

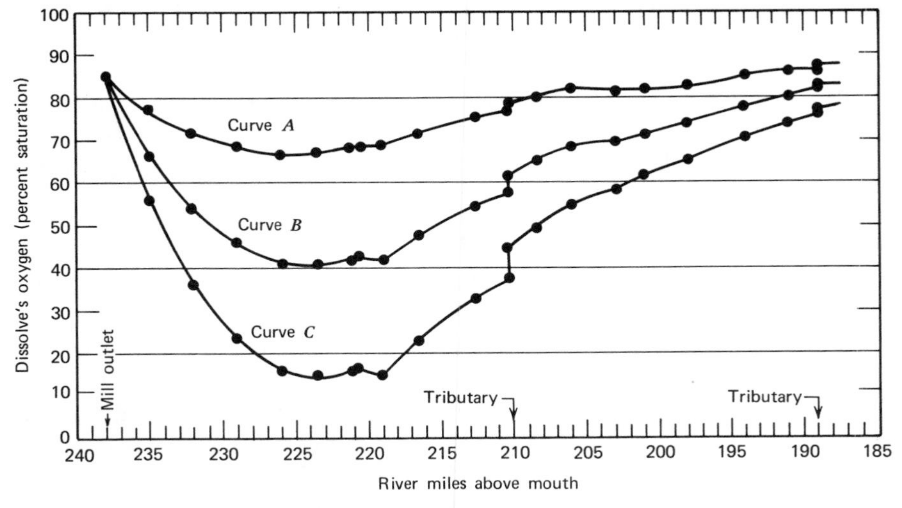

Figure 11–8 River K. Computed dissolved oxygen profiles at a runoff of 1000 cfs and a temperature of 32°C. BOD load (population equivalents): curve A: 100,000; curve B: 200,000; curve C, 300,000.

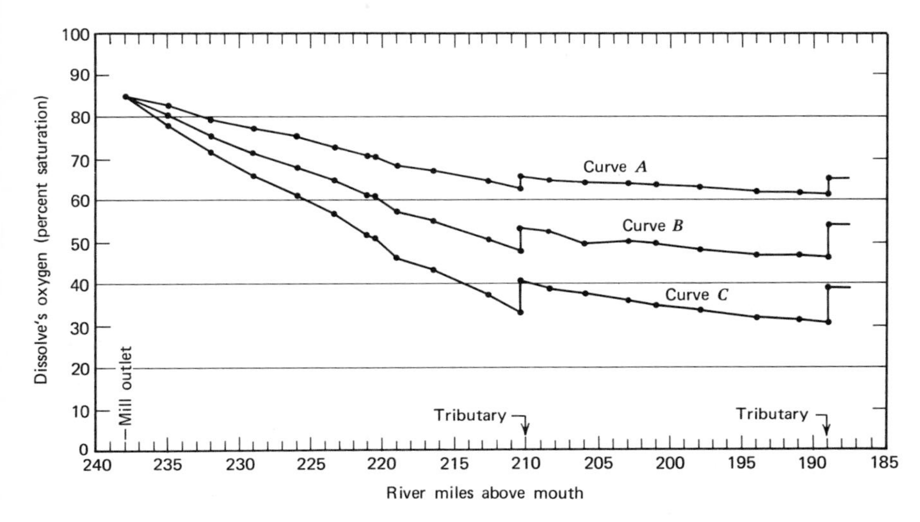

Figure 11–9 River K. Computed dissolved oxygen profiles at a runoff of 5000 cfs and a temperature of 32°C. BOD load (population equivalents): curve *A*: 554,000; curve *B*: 861,000; curve *C*: 1,168,000.

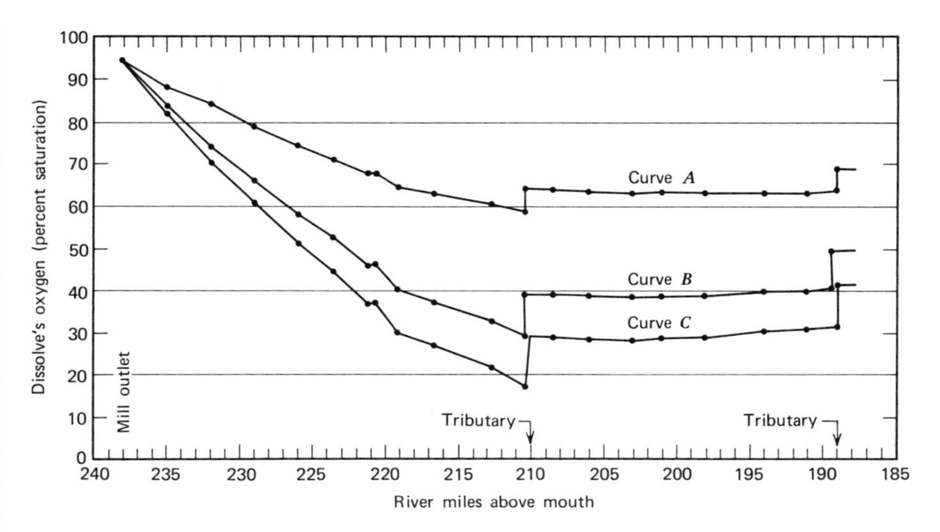

Figure 11–10 River K. Computed dissolved oxygen profiles at a runoff of 1000 cfs and a temperature of 12°C. BOD load (population equivalents): curve *A*: 203,000; curve *B*: 356,000; curve *C*: 418,500.

the three profiles critical dissolved oxygen values of 15, 40.5, and 66 percent of saturation, are obtained from Figure 11–8 at the three loads of 100,000, 200,000, and 300,000 population equivalents, respectively, specific for a temperature of 32°C and a runoff of 1000 cfs. These three sets of coordinates located on Figure 11–11*a* at (*a*) (*b*) and (*c*) establish curve *B*. Similarly curve *A* establishes the relation specific for 32°C and a runoff of 500 cfs, curve *C*, for 2000 cfs, and progressing at 1000-cfs increments to curve *H* at 6000 cfs.

Similar curves for critical dissolved oxygen versus BOD load can be developed from computed dissolved oxygen profiles at the specific temperatures of 20 and 12°C for specific runoffs. (For subsequent reference these are designated as Figures 11–11*b* for 20°C and 11–11*c* for 12°C but are not printed.)

For river K the relation between critical dissolved oxygen and waste load for each specific runoff is practically linear, a form convenient for interpolation and for moderate extrapolation above and below the computed points. These control curves thus provide a means for readily determining critical dissolved oxygen values for any BOD loading, specific for the three temperatures and specific for the runoffs as designated.

Critical Dissolved Oxygen versus Runoff—Temperature and BOD Load Specific

The second step is to establish control curves for the relation between critical dissolved oxygen and runoff. These are developed from Figures 11–11*a*, *b*, *c* at any desired increments of BOD loading, as shown in Figure 11–12*a*, specific for the temperature of 32°C. For example, from Figure 11–11*a*, curve *C* (2000 cfs and 32°C) the critical dissolved oxygen values at BOD loadings of 100,000, 150,000, and 200,000 population equivalents, progressing to 600,000 population equivalents, are 78, 71.5, 64, 50, 36.5, 22, and 7.5 percent of saturation, respectively. These coordinates locate points (*a*) to (*g*) on Figure 11–12*a*. Similarly other sets of coordinates are derived from Figure 11–11*a* and transferred to Figure 11–12*a* to establish the nest of control curves specific for 32°C and specific for the incremental BOD loads as designated. Similarly nests of control curves can be established specific for 20 and 12°C (Figures 11–12*b* and *c*, respectively, based on Figures 11–11*b* and *c*, respectively; figures not printed). Thus at any runoff up to 6000 cfs the critical dissolved oxygen is readily obtained at these three specific temperatures from any specific BOD load curve.

With critical dissolved oxygen versus runoff curves established, by the reverse process it is possible to develop additional curves for critical dissolved oxygen versus BOD load, specific for other runoffs. For exam-

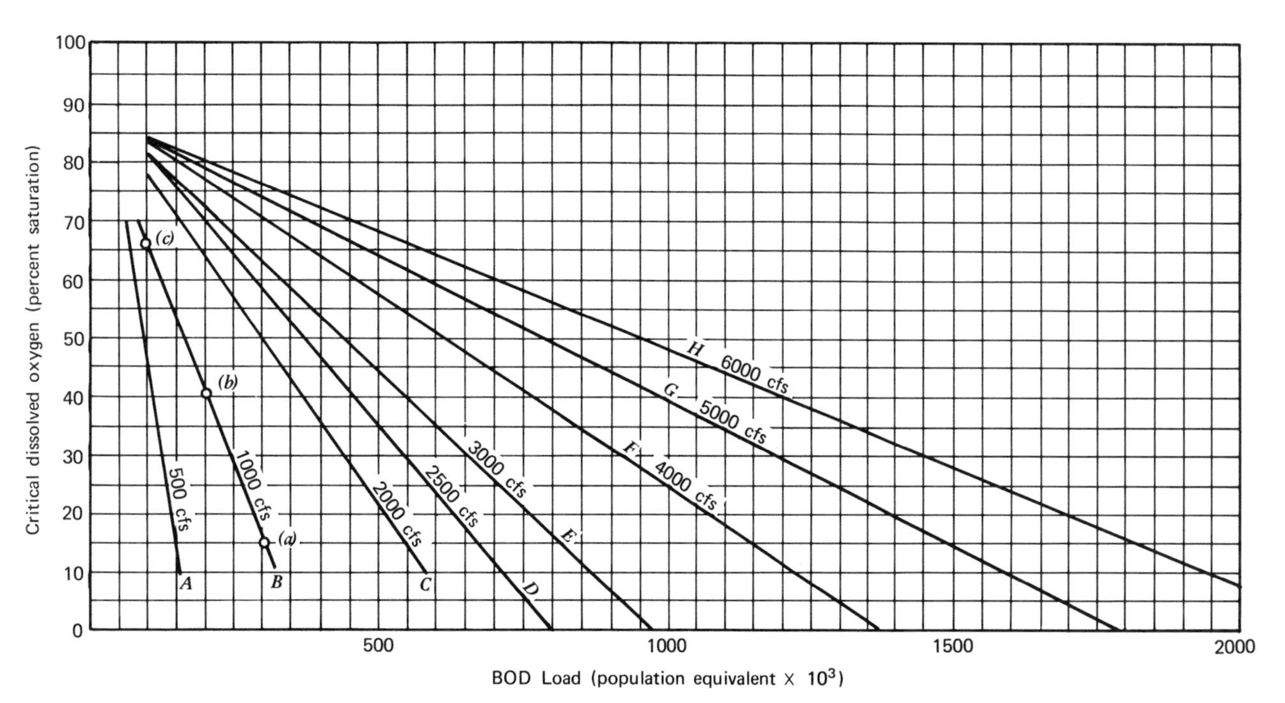

Figure 11–11*a* Critical dissolved oxygen versus BOD load, temperature (32°C) and runoff specific.

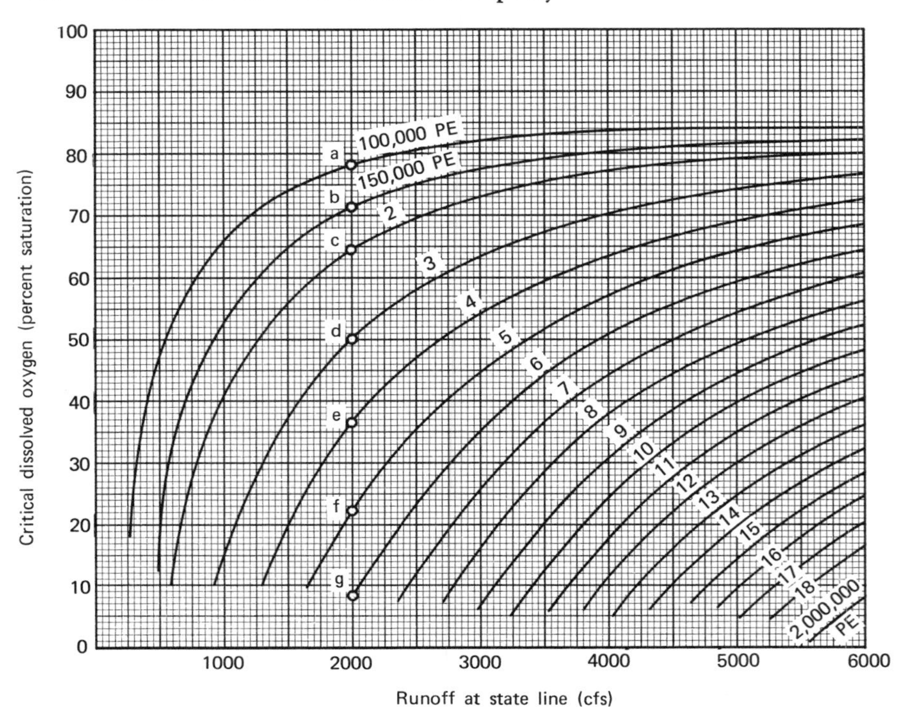

Figure 11–12*a* Critical dissolved oxygen versus runoff, temperature (32°C) and BOD load specific.

ple, curve *D* of Figure 11–11*a*, specific for 2500 cfs and 32°C, is constructed from Figure 11–12*a* by reading the critical dissolved oxygen values on the ordinate scale of 2500 cfs at intersections of the specific load curves and transferring to Figure 11–11*a*.

Critical Dissolved Oxygen versus Temperature—Runoff and BOD Load Specific

The third step is to provide the relation between critical dissolved oxygen and water temperature at desired specific runoffs for the range of specific BOD loads. Such nests of control curves are now developed from Figures 11–12*a*,*b*,*c*. Figure 11–13*a* is for the specific runoff of 3000 cfs. Smooth curves are formed for interpolation between 12 and 32°C, and extrapolation can be made to 10 and 35°C. For example, the specific BOD load curve of 600,000 population equivalents at the specific runoff of 3000 cfs, as shown in Figure 11–13*a*, is defined by point *x*, critical dissolved oxygen of 35 percent of saturation at 32°C

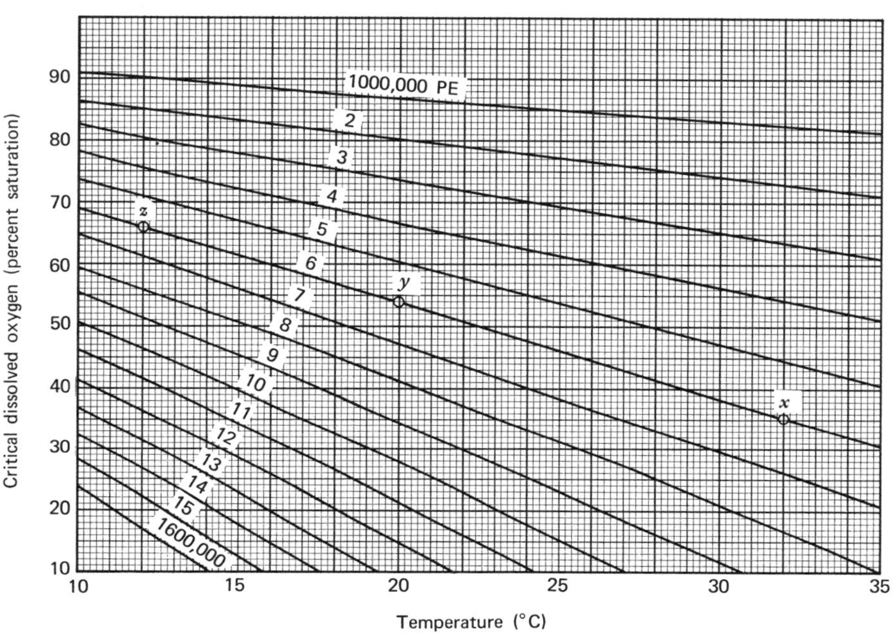

Figure 11–13*a* Critical dissolved oxygen versus temperature, runoff (3000 cfs) and BOD load specific.

is obtained from Figure 11–12*a;* the point *y* critical dissolved oxygen of 54 percent of saturation, from Figure 11–12*b;* and the point *z* critical dissolved oxygen of 66 percent of saturation, from Figure 11–12*c* (Figures 11–12*b* and *c* not printed).

Similar nests of control curves can be developed readily at any desired specific runoff. Thus these control curves afford a basis for determining critical dissolved oxygen at a specific runoff over the range of incremental BOD loads at any temperature from 10 to 35°C.

BOD Load versus Runoff—Critical Dissolved Oxygen and Temperature Specific

Finally, control curves are developed from which the allowable BOD load can be defined for any runoff and water temperature combination to ensure that critical dissolved oxygen will not decline below a precisely prescribed level. Figure 11–14 represents the nest of control curves specific for the critical dissolved oxygen level of 3 ppm. For example, coordinates for 32°C are readily obtained from Figure 11–11*a* where the horizontal line 3 ppm (40.5 percent of saturation) intersects curves *A*

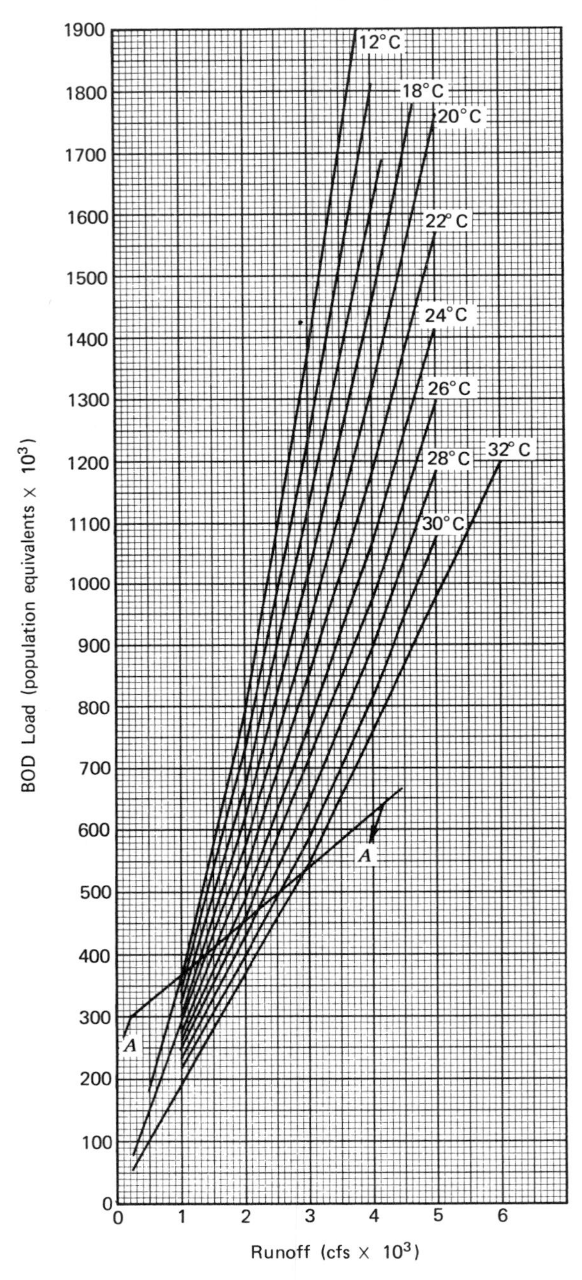

Figure 11–14 Runoff versus BOD load relation, specific for a critical dissolved oxygen level of 3 ppm and temperatures of 10 to 32°C.

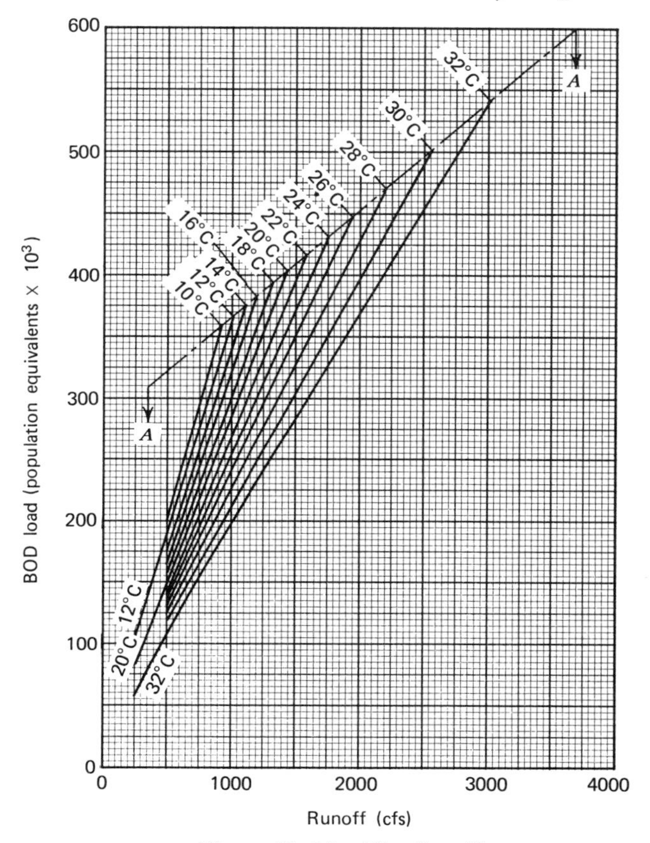

Figure 11–14 (Continued)

to H specific for the runoffs of 500, 1000, and up to 6000 cfs, giving allowable BOD loads of 103,000;200,000;370,000 population equivalents, and so on. These coordinates are transferred to Figure 11–14, through which the 32°C smoothed curve is drawn. Similarly the 20 and 12°C curves are developed from Figures 11–11*b* and *c* (not printed).

Other nests of control curves for BOD load versus runoff can be developed readily for other specific critical dissolved oxygen levels that are prescribed or under consideration. Figure 11–15 shows a portion of the nest specific to maintain the critical dissolved oxygen level of 5 ppm, developed in a manner similar to that for Figure 11–14. Comparing results of a decision for a critical level of 3 ppm (Figure 11–14) with that of 5 ppm (Figure 11–15), emphasizes the implications in terms of the allowable BOD loading. For example, at 2000 cfs and 30°C, by

Figure 11–14 the allowable BOD load to maintain a critical dissolved oxygen level of 3 ppm is 400,000 population equivalents; to maintain a critical dissolved oxygen level of 5 ppm, by Figure 11–15 the allowable load is 210,000 population equivalents—a reduction of almost one-half.

The importance of considering the implications of restriction on natural waste assimilation capacity in prescribing an unusually high critical dissolved oxygen requirement in streams with seasonally high temperatures is further emphasized by the control curves of Figure 11–16. These

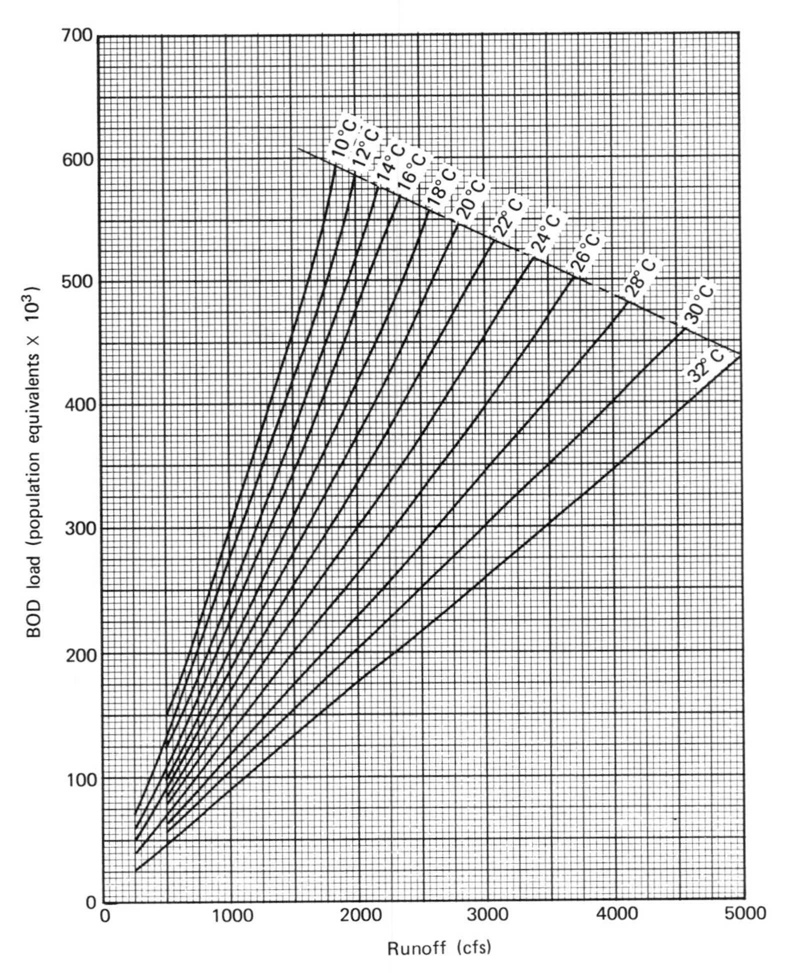

Figure 11–15 Runoff versus BOD, specific for a critical dissolved oxygen level of 5 ppm and temperatures of 10 to 32°C.

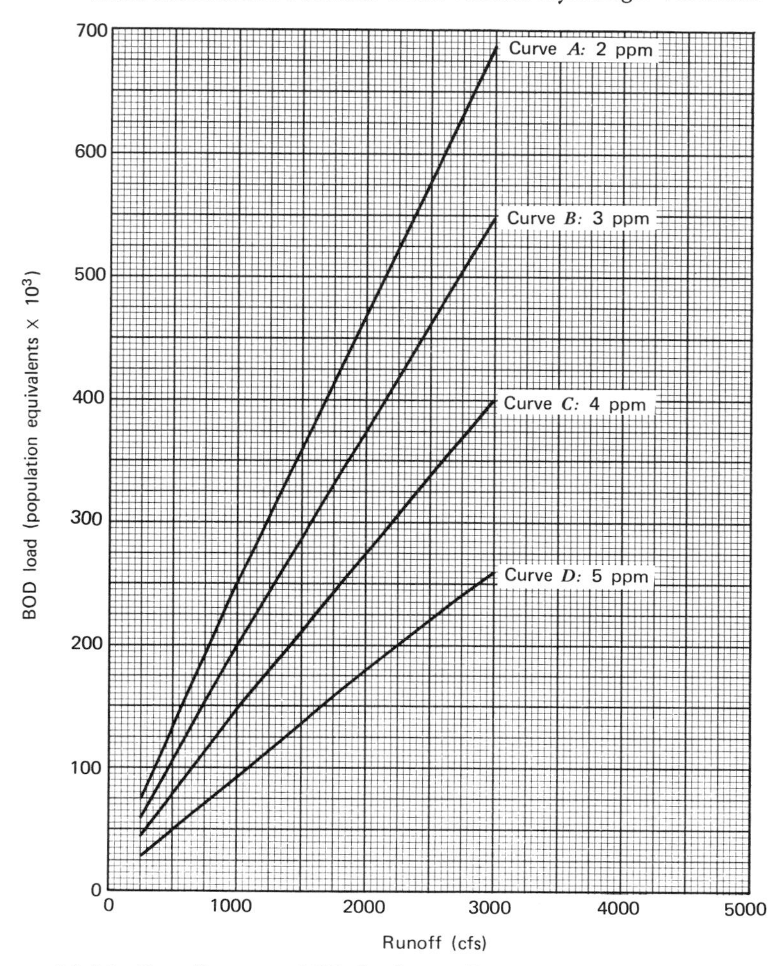

Figure 11–16 Runoff versus BOD load, specific for a temperature of 32°C and critical dissolved oxygen levels of 2, 3, 4, or 5 ppm.

curves show the BOD load versus runoff relation specific for a temperature of 32°C at four specific critical dissolved oxygen levels—curve *A*, 2 ppm; curve *B*, 3 ppm; curve *C*, 4 ppm; and curve *D*, 5 ppm. Dissolved oxygen saturation capacity at 32°C is low, 7.4 ppm; hence the corresponding percentage of saturation values for critical dissolved oxygen levels rise sharply from 2 ppm (27 percent), 3 ppm (40.5 percent), 4 ppm (54.1 percent), and 5 ppm (67.6 percent).

Correspondingly, the allowable BOD load declines sharply. For example, at a runoff of 2000 cfs the allowable load declines from 467,000 to 370,000 to 275,000 to 177,000 population equivalents for critical dissolved oxygen levels of 2, 3, 4, and 5 ppm, respectively.

The shape of the control curves is unique for each river and for the specific location or combination of locations of waste effluent outlets. It is obvious from the illustration of river K that a great range in allowable waste load is dictated by the prescribed critical level of dissolved oxygen. Control curves of natural waste assimilation capacity, such as illustrated, are valuable guides in reaching decisions as to the critical dissolved oxygen criteria to be established, particularly if accompanied by economic analyses and evaluations of rational water use quality requirements.

Although control curves have been illustrated for dissolved oxygen, the framework of waste assimilation capacity under natural hydrologic variations can and should be developed for other water quality parameters, depending on the type of waste.

DROUGHT CONTROL—EFFICIENT USE OF WASTE ASSIMILATION CAPACITY

The Principle of Drought Control

The principle of flood control has long been accepted as protection against the extremes of maximum stream runoff, but the principle of drought control as protection against the extremes of minimum streamflow has been, with few exceptions, only recently recognized. Drought control is the natural complement to flood control; both principles can and should be incorporated into the same reservoir development. The peak floodwaters are retained, at least in part, and released to fill in the deep deficiencies in streamflow during the drought season.

We have seen that under natural hydrologic variations the total waste assimilation capacity is relatively great, in some instances on the order of 100 times greater than that available under uncontrolled drought streamflow. Community and industrial development and the social and economic potentials of a river basin are drastically restricted if geared to the minimum waste assimilation capacity during uncontrolled drought. We cannot increase the total natural waste assimilation capacity, but we can, with moderate expenditure, remove the severe restrictions of uncontrolled drought and make much more efficient use of what is available.

Drought control, in the context considered here, is the assurance of availability of a larger river, the protection against the risks of shrinkage of streamflow during the natural variation of the dry weather season. So far man has not reached a stage that permits control of weather. Even if it were possible, perhaps it would not be entirely desirable to substitute a uniform pattern of rainfall. It is technically and economically feasible, however, to control to a degree the stream runoff associated with the irregular occurrence of rainfall. The ultimate attainable would be a completely balanced stream runoff between excess flow during floods and deficient flow during droughts, not only over the seasons of a year but also over the more extensive droughts of accumulated deficiency of several successive years.

Among promising approaches to more efficient use of the natural waste assimilation potential through drought control are headwaters storage, incremental storage along the river course, adaptation of large scale pumped storage, and multipurpose combinations. The approach to be adopted in any case is dictated by many factors peculiar to the river basin and by local resources and aspirations. The focus should extend beyond alleviation of existing acute pollution problems to consideration of long-range basin potentials, prevention of pollution, directional guidance of urban and industrial growth within the water resource frame, and rational allocation of other water uses and developments in relation to satisfactory waste disposal.

The Multiple Benefits of Drought Control

Drought control inherently provides benefits not only to waste assimilation capacity but to all existing and potential water uses along the course of the stream—municipal and industrial water supply, electric power, recreation, aquatic life, navigation, irrigation, and community environment enhancement and esthetics. Such multibeneficial endeavors usually require a cooperative effort by diverse interests and involve broad water policy—political, legal, financial, as well as technical considerations. Statesmanship and persistent effort in promotion are necessary to bring a development to fruition, but the ultimate benefits are great.

Although our focus here is on the efficient use of the natural waste assimilation capacity of the stream, waste disposal need not always be the primary beneficiary, nor should it always carry the primary financial assessment. Since benefits are multiple, all should share in the distribution of costs. Not all benefits are amenable to quantitative determination to the extent that is possible for waste assimilation capacity, but this does not free the others from responsibility in sharing in the cost distribution. Priorities may shift among the water use benefits of drought control, depending on social and economic objectives and the needs in different drainage basins. Drought control, in addition to its multibeneficial character, extends over a long time; hence the financial burden also should not be distributed only among current beneficiaries, but allocations of cost should be made against the future potentials created. The potential benefits of drought control are so great and important to the social and economic well-being of a river basin, state, region, and nation that there is also justification for broad public support, as explained in Chapter 12.

In any drought control development, regardless of the primary purpose, consideration should be given to all other water uses, and particularly to the key problem of ultimate waste disposal and water quality control. We have seen in Chapter 9 that the effect of other river developments and water uses (depending on how they are planned and operated) can be either detrimental or beneficial to the waste assimilation capacity of the stream. Efficient use of water resources demands multibeneficial drought control development for today's acute problems and for tomorrow's potentials.

Headwaters Storage and Regulated Release

The conventional method of augmenting low streamflow has been the construction of reservoirs in natural valleys at locations suitable for a dam and with sufficient tributary drainage area to provide the necessary natural runoff to supply the required storage capacity. These requirements usually restrict sites to the headwaters remote from urban-industrial developments, which are generally along the lower river valley. The value of drought control by augmenting low flow during the critical warm weather dry season is illustrated by excerpts from a detailed study of river L. Figure 11–17 represents the computed dissolved-oxygen profiles expected under once in 10-year natural drought runoff for two plans of wastewater treatment of the projected 1975 loads—curve *A* for intermediate treatment (65 percent reduction in BOD) and curve *B* for complete treatment (85 percent reduction in BOD).

Drought conditions on this river are extremely severe, with the once in 10-year drought of 133, 223, and 316 cfs at locations X, Y, and Z progressively downstream. Under these extremely low flows, even with a very high degree of treatment the prescribed dissolved oxygen level of 4 ppm in the critical reach Y cannot be met and the dissolved oxygen declines to 28 percent of saturation (2.1 ppm). This illustrates the limitation of lines of defense by treatment alone and emphasizes the necessity of considering lines of offense to enhance waste assimilation capacity.

Several natural reservoir sites are available above the highly developed urban-industrial reach for storage of runoff to augment dry weather flow. Figure 11–18, curve *A*, shows the computed dissolved oxygen profile

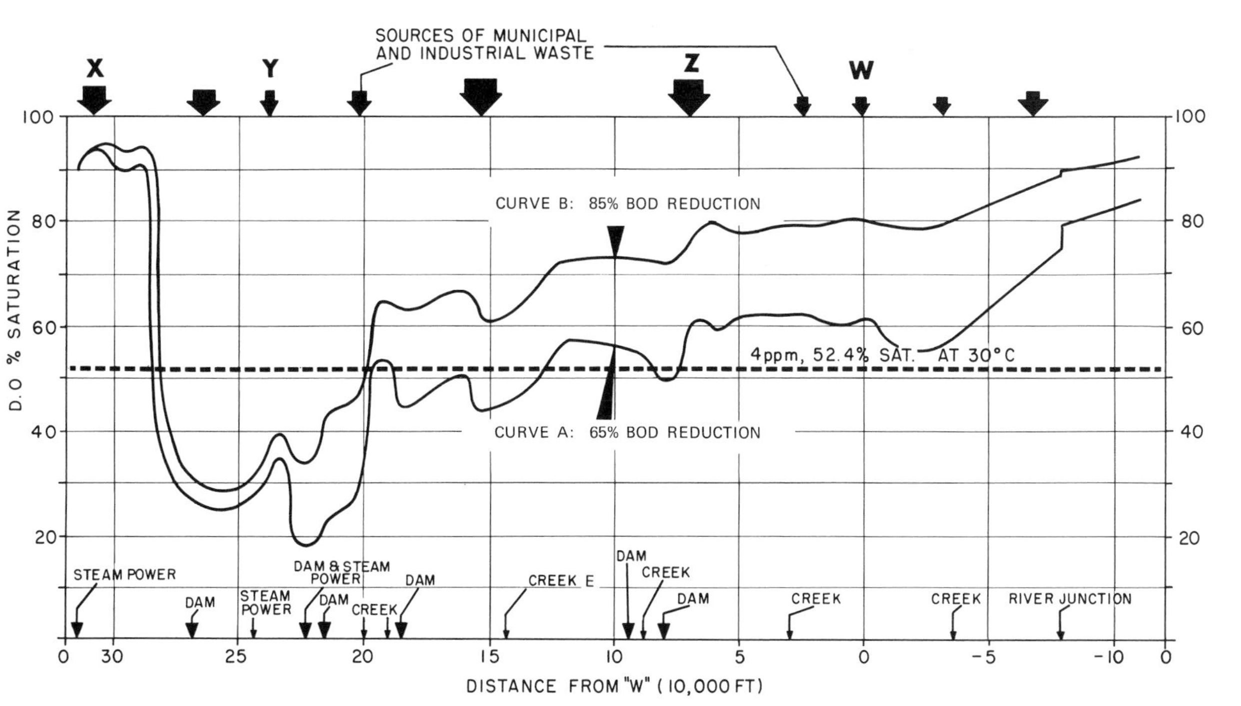

Figure 11–17 River L. Forecast dissolved oxygen profiles, 1975 loads. Reliance on treatment alone, no regulation of streamflow, once in 10-year drought runoff at Z of 316 cfs. Curve *A*: intermediate treatment; curve *B*: complete treatment.

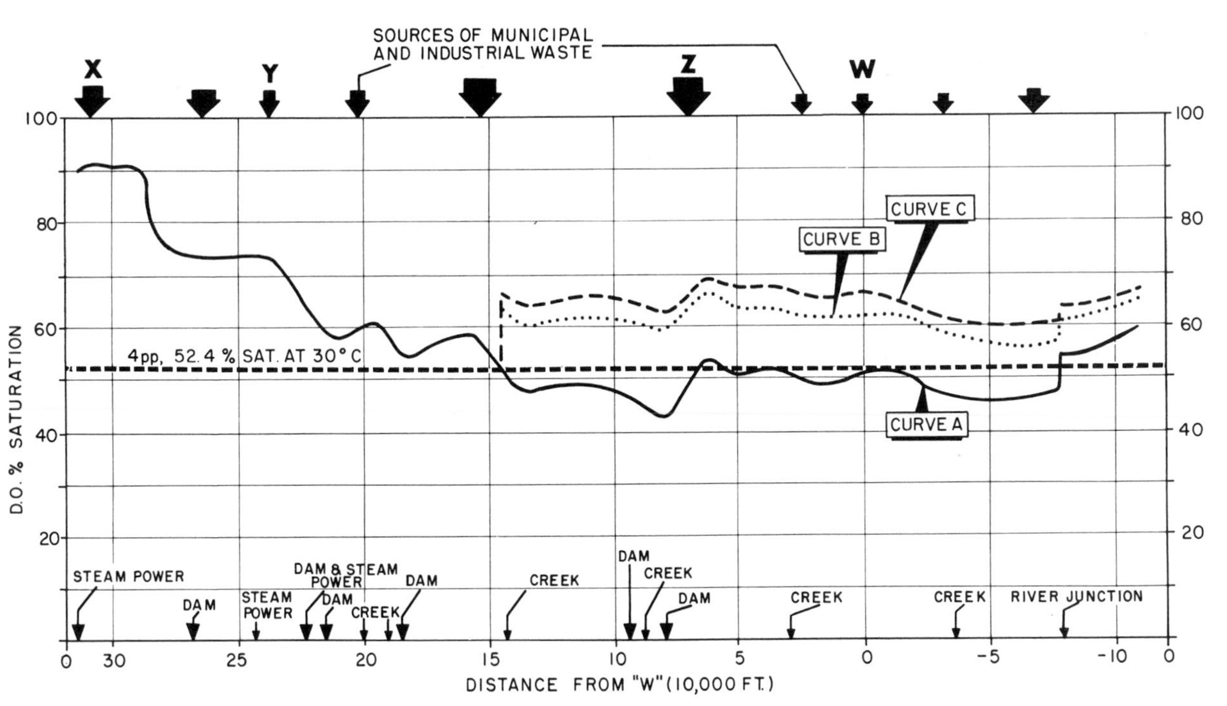

Figure 11–18 River L. Forecast dissolved oxygen profiles, 1975 loads. Primary treatment, with augmented streamflow. Curve *A*: once in 10-year drought flow, augmented by head water storage to 800 cfs at Z: curve *B*: additional increment from creek E storage to 1000 cfs at Z; curve *C*: additional increment from creek E storage to 1100 cfs at Z.

for moderate treatment, removing settleable solids subject to deposit, in combination with low flow regulation from a headwater reservoir, increasing the dry weather flow from 316 to 800 cfs at location Z. A substantial improvement, above 4 ppm, is shown at the critical location Y, but downstream the dissolved oxygen profile develops two secondary sags that fall somewhat below the prescribed level of 4 ppm. In part the secondary sags are associated with the shortened time of passage with increased streamflow from the headwaters, carrying a higher residual BOD downstream kaleidoscoping loading to induce the downstream dissolved oxygen sags. As headwaters augmentation is increased, more residual BOD is carried downsteam, but as the secondary sag moves farther downstream a point is reached at which flow augmentation brings about the desired improvement along the lower reaches. Such an unbalanced dissolved oxygen profile is inherent in augmentation from headwaters, particularly where the lower river receives a succession of BOD loads along the course.

Another illustration of the value of low flow augmentation from headwater reservoirs relates to a detailed study of estuary D. The once in 10-year minimum monthly average landwater discharge to the estuary is 3640 cfs; the seasonal warm weather water temperature is shown to peak with drought flow at 30°C. Under these conditions the allowable industrial waste BOD loading (exclusive of residual from upstream and from municipal sources) is approximately 500,000 population equivalents to ensure a dissolved oxygen level of 3 ppm in the critical reach (39.4 percent of saturation).

For the purpose of evaluating the benefits to increased waste assimilation expected from proposed low flow augmentation from headwater reservoir developments, a series of dissolved oxygen profiles was computed, and from this a set of control curves (critical level versus runoff, specific for 30°C and a range in industrial BOD loadings) was developed, as shown in Figure 11–19. These curves show that augmenting the drought flow from 3640 to 7500 cfs increases the allowable industrial load to maintain a dissolved oxygen level of 3 ppm (39.4 percent of saturation) from 500,000 to 1,050,000 population equivalents—a doubling of the industrial waste assimilation capacity.

Incremental Storage along the River Course

Referring again to river L, it was seen in Figure 11–18 that, although augmentation of flow from a headwater reservoir greatly improved water quality in the upper reach, secondary sags developed downstream. The ideal augmentation of low flow would be to contribute successive incre-

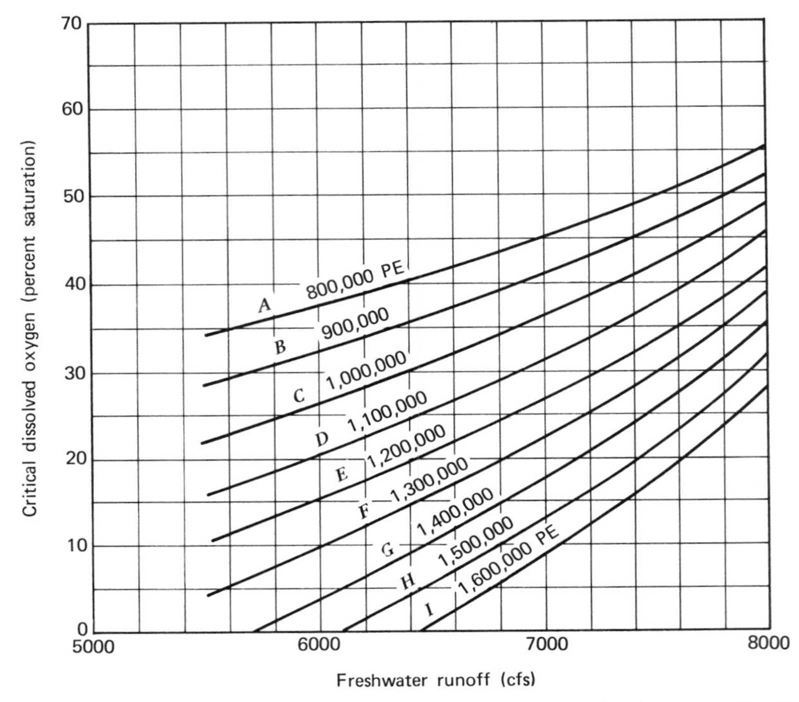

Figure 11–19 Estuary D. Relation between critical dissolved oxygen level and runoff at 30°C. Control curves *A–I* are for indicated industrial waste BOD loadings.

ments along the course of the river at the point of need and in amounts to ensure a balanced dissolved oxygen profile at the prescribed level.

Studies of river L provide an opportunity of illustrating this principle by curves *C* and *B* of Figure 11–18. A natural reservoir site (small relative to headwater sites) capable of developing 150- to 300-cfs augmentation is available on creek E, strategically located at the point where the dissolved oxygen profile commences to sag below the prescribed level. It will be recalled that curve *A* is based on augmentation entirely from a headwater reservoir, ensuring a streamflow of 800 cfs at Z. Increasing the flow at Z to 1000 cfs by an increment contributed from storage on creek E, it is noted on curve *B*, eliminates the sags of curve *A* and produces a marked elevation of the dissolved oxygen level, well above the prescribed 4 ppm. Increasing the increment to ensure a flow of 1100 cfs at Z produces further improvement, as shown by curve *C*. Computations indicate that, if augmented flow to 1000 cfs at Z is derived entirely from headwater reservoirs, the dissolved oxygen profile in the secondary sags would not quite attain the prescribed level of 4 ppm;

and at a flow augmented to 1100 cfs it would slightly exceed the 4-ppm level.

Thus the incremental augmentation from development on creek E is distinctly superior to dependence on headwaters alone. Incremental augmentation makes more efficient use of self-purification along the course and minimizes kaleidoscoping of residual BOD loadings which creates the downstream secondary sags. Unfortunately incremental augmentation from natural reservoir sites, with sufficient drainage area to fill by runoff, along the developed reaches of rivers at the points of greatest need is quite limited in most river basins. However, the principle of incremental augmentation of streamflow has great potential for the enhancement of waste assimilation capacity and water quality control, and with adaptation of pumped storage is quite feasible.

Adaptation of Pumped Storage

The Concept of Adaptation of Pumped Storage

The Dual Power–Water System. As conceived here, adaptation of large scale pumped storage is a dual power–water system. Two primary ingredients in our social and economic growth are simultaneously provided, electrical energy and water. The power aspect of the system is primarily oriented to meeting the growing demands for peak power and reliable reserve as protection against power failure. Pumped storage as adapted meets these two major demands in providing large blocks of electrical energy to meet the peak demands in the daily power cycle; inherently it has double reserve capability, instantly absorbing huge blocks of energy by pumping or instantly producing large blocks of energy by generating. The water aspect of the dual system is inherently a multipurpose water resource development, uniquely adapted to drought control by low flow augmentation, and in the flexibility of the system enhances waste assimilation capacity and meets other water resource needs as well.

Basically pumped storage functions as an energy accumulator that stores low value off-peak energy (usually generated by steam plants) by using it to pump water from a lower to a higher reservoir from which it can be returned to generate power during peak periods. In most instances both pumping and generation operations are accomplished by a single reversible machine. Figure 11–20 illustrates a schematic layout of a typical adaptation of pumped storage to a power–water system. The upper reservoir includes storage capacity requirements for both drought control augmentation water (and other purposes) and for power cycling; the lower pool, in addition to functioning in the conventional

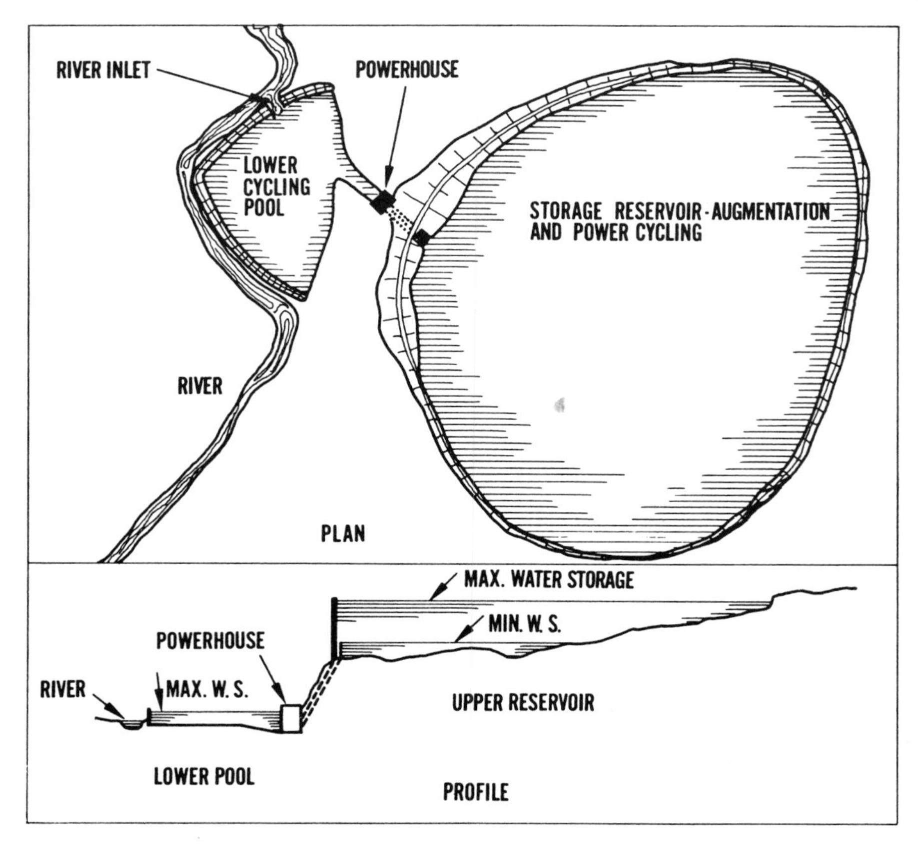

Figure 11–20 Schematic diagram of pumped storage hydro augmentation project.

manner for recycling water for energy storage and generation, serves as a reregulation reservoir for release of water to the river to augment flow.

Benefits to the Electric Utility Industry. From the standpoint of the electric utility industry, when a portion of the capacity is pumped storage hydro generation, the entire system can operate at a more efficient level [1]. Electric power demand is leveled and thermal capacity can be lower, with operation at a higher capacity factor and a lower unit operating cost. Energy used during pumping has a low incremental cost, and energy returned by pumped storage generation at times of the peak demand on the system has a high value. This favorable differential is a valuable economic asset.

The November 9, 1965, northeastern power blackout and subsequent failures in other areas emphasize the importance of fast acting reserve capacity inherent in the pumped storage developments. The inability of slow acting steam electric power plants to respond to emergency demand is a major contributing factor to failures. The pumped storage hydro has the unique advantage of being quickly available for full capacity use, often in less than a minute, to absorb an emergency electric load by pumping or by generating to meet sudden heavy demands. By virtue of the larger capacity of the upper reservoir, these favorable aspects of hydro operation can be sustained for much longer periods than is possible in conventional electric utility pumped storage installations.

As discussed in Chapter 7, large quantities of water are essential to the thermal power base for the condenser circulation systems and for the satisfactory dissipation of the tremendous waste heat loads. It is anticipated that by the early 1970s the bulk of electrical energy will be provided by nuclear fuel steam electric power generation. With the shift from fossil fuel and waste heat loads will be substantially increased (40 to 50 percent). This shift, coupled with the trend toward concentration of generating capacity at large plants, makes it imperative to locate the large nuclear fuel plants on large reservoirs, since few rivers are of adequate capacity to meet the demands. Large scale pumped storage power–water systems again can play a significant role in ensuring an adequate condenser water supply and extensive reservoir surface area for dissipation of the waste heat load, thus permitting location of large nuclear power stations in areas near power demand. As an illustration, in the studies of the Miami River Valley, Ohio, (discussed later) preliminary analyses of such a development at the Honey Creek site just above Dayton indicate a technical and economic feasibility of 240,000 acre-ft of usable storage for streamflow augmentation and other multipurpose water use, support for 250 MW of pumped hydro peaking capacity, and adequate condenser water and waste heat dissipation area for 1500 MW of nuclear base-load generating capacity.

The multipurpose water functions of dual power–water systems not only can meet utility industry needs but also, in supplying ample water, provide the base for community and industrial growth, thereby sustaining the growth in demand for electrical energy so essential to the prosperity of industry.

A remarkable feature of the dual power–water pumped storage system is the compatibility between the two functions of supplying power and water, without conflict or sacrifice in the efficiency of either; actually the two enhance each other. The symbioses should promote adoption of this promising system.

Multiple Water Resource Benefits

In dealing with the water phase in the dual power–water system the primary multiple water resource benefits are considered in five classifications:

1. Efficient use of water in stream quality control.
2. Reservoir water quality improvement.
3. Community and industrial water supply.
4. Water based recreation.
5. Flood protection.

These multiple benefits are illustrated in excerpts from detailed studies of feasibility in the complex Miami River basin [2]. Six promising dual power–water pumped storage sites were located, with a total storage capacity of 800,000 acre-ft and a total peaking generating capacity of 1345 MW.

Efficient Use of Water in Stream Quality Control

The efficiency of pumped storage is illustrated in the comparison of a conventional headwater reservoir located near Piqua above Dayton with a pumped storage development (Crains Run) on the lower Miami, located near an existing thermal electric power plant below Dayton at a critical reach of the river (Figures 11–20 and 11–21). The pumped storage peaking hydro power capacity at the Crains Run development ranges from 300 to 380 MW (average 325 MW).

To facilitate the application of the rational method of stream analysis in predicting the water quality parameters expected from various combinations a digital computer program was developed (Appendix C) and verified against observed stream sampling results. From extensive series of computed dissolved oxygen profiles control curves were constructed, establishing the relationship between the critical dissolved oxygen level and required flow augmentation, either from the conventional headwater reservoir in conjunction with the downstream pumped storage reservoir.

Figure 11–22 shows the relative advantage in the efficient use of waste assimilation by providing high quality water augmented by pumped storage. Curve *A* is the dissolved oxygen profile computed for natural drought flow conditions of 350 cfs with no augmentation; curve *B* shows the profile for 300 cfs of headwaters augmentation determined from the control curves as required to meet the prescribed dissolved oxygen level of 4 ppm; curve *C* is the profile showing that with downstream augmentation at the critical zone an increment of 150 cfs (one-half that required by headwaters alone) is adequate to maintain a dissolved

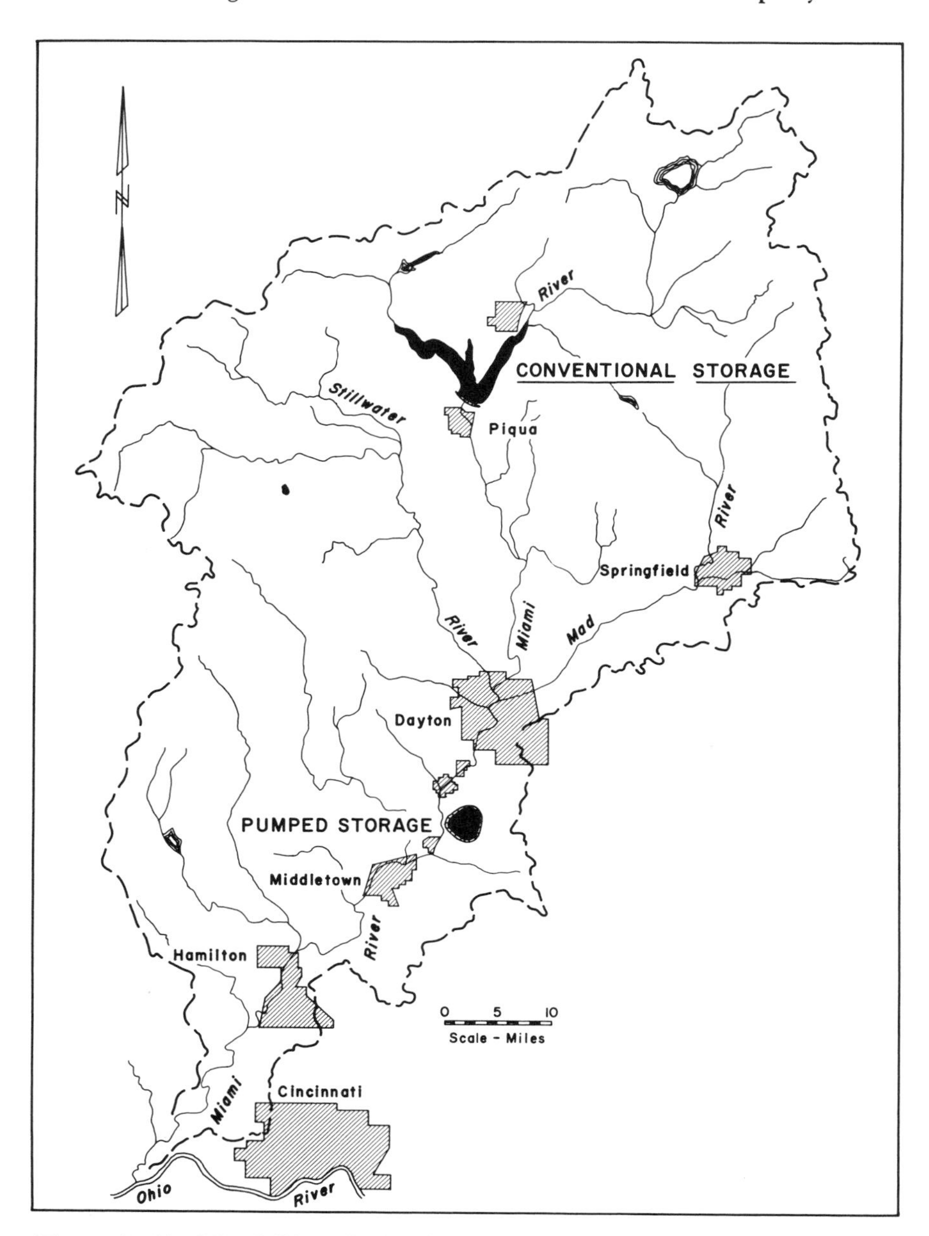

Figure 11–21 Miami River Basin. Combination conventional storage and downstream pumped storage.

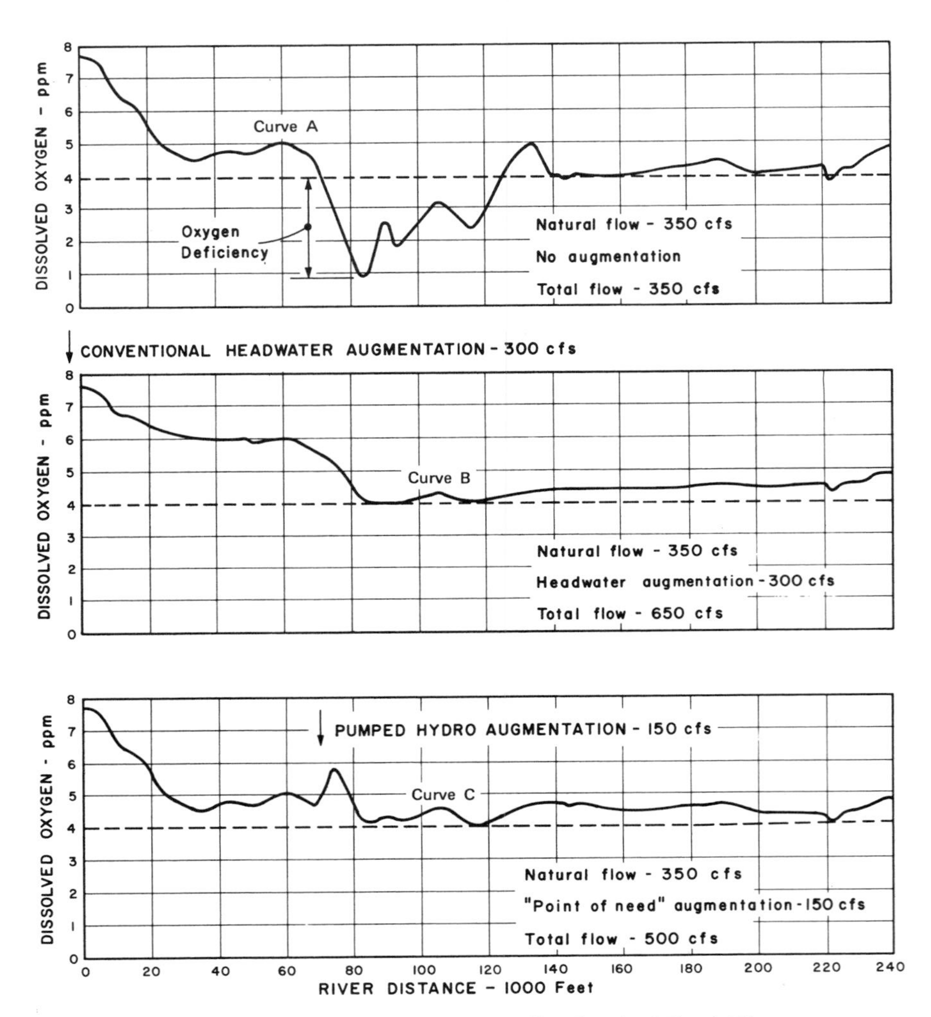

Figure 11–22 Dissolved oxygen profiles for the Miami River.

oxygen level of 4 ppm. Computations further show that this approximate relative efficiency of the downstream pumped storage is attained under all probable conditions for the various factors involved and for different critical dissolved oxygen levels above or below the prescribed 4 ppm.

In order to determine storage requirements, using the relationship developed between augmentation requirements and critical dissolved oxygen, the droughts over the period of record were simulated (in a computer program) for the two-reservoir system (Piqua and Crains Run). Plotted on extreme probability paper, these storage requirements form reasonably good straight line distributions in agreement with the theory of extreme values. From these distributions the storage required for the worst drought of record (1934–1935) is equivalent to a severity expected on the average once in approximately 30 years.

To meet the dissolved oxygen requirement of 4 ppm for this critical drought severity with streamflow augmentation from the conventional headwaters reservoir alone would require a storage volume of 262,000 acre-ft; for the two-reservoir combination storage requirements would be 108,000 acre-ft at the Piqua headwaters reservoir and 80,000 acre-ft at the downstream Crains Run pumped storage reservoir. The combination two-reservoir system incorporating pumped storage results in the substantial reduction is required storage of 74,000 acre-ft, a saving of approximately the capacity of the pumped storage reservoir. Droughts of other severities on the Miami River disclose similar advantages of downstream pumped storage over conventional storage alone.

As already discussed for incremental storage along the river course, the advantage of downstream pumped storage stems from augmentation at the point of need and more efficient use of intervening waste assimilation capacity, with less residual BOD carried downstream, avoiding overlapping of loads to form a secondary downstream critical reach. In addition, a substantial advantage of pumped storage water for augmentation is the marked improvement in quality induced by the daily pumping cycle, to be discussed more fully shortly.

There are two additional significant advantages inherent in pumped storage adaptation that should be emphasized. The availability of strategic sites with adequate tributary drainage areas for conventional reservoirs is limited in most river basins. Such sites are usually available only in the headwaters, remote from developed areas of need. Many sites nearer developments are already preempted for other uses or are encroached on by community and industrial expansion. In contrast, there are many more sites suitable for pumped storage because they do not depend on tributary drainage area to fill the reservoir. Hence sites can be strategically located along lower reaches as well as upstream. Advantage can be taken of adjacent small tributary drainage areas with valleys at suitable elevations above the main river; these can be augmented by excavation and berm fill. Since the tributary drainage area is relatively very small, the reservoir dam entails a simple design and in most instances with no costly flood spillways. Construction cost per unit of capacity is therefore lower than that for conventional reservoirs. The risk of adequate water to fill the reservoir is minimized by virtue of the large tributary drainage area of the main river from which water is pumped. Filling permits placing reservoirs in close proximity to needs for augmentation, for power generation, and for many other community water needs.

Flexibility of operation is the other unique advantage of pumped storage augmentation in controlling sudden changes in water quality and day-to-day variations in waste effluent loading, breakdowns in treatment processes, or accidental spills at industries. By providing augmentation near the location of normally critical water quality decline, a refined responsive means is available to correct deficiencies and ensure a much higher reliability in maintaining the desired water quality in the river. Also, large volumes of augmentation water can be readily available (in minutes) to reduce the concentration of contaminants or to provide large quantities of dissolved oxygen immediately and to correct an oxygen deficiency. Augmentation from a conventional headwaters reservoir that takes several days to reach the problem area cannot be used with effectiveness in meeting such situations. This flexibility of pumped storage also permits integration of a river monitoring system with operational control of releases to improve efficiency of river management not otherwise possible.

Likewise, there is flexibility in the seasonal refilling operations of pumped storage reservoirs not available to conventional reservoirs. The conventional impoundments, by nature, must receive any water that drains from the relatively large tributary catchment area. Such large drainage areas are exposed to risks of pollution from many sources on the watershed, which, even with careful surveillance and control, often reaches the reservoir. The relatively small tributary drainage area of pumped storage developments minimizes such risks. Also, in the control of pumping from the main river a degree of selectivity in withdrawal can be exercised to avoid periods of gross pollution or accidental heavy contamination.

If the pumped storage system is also to be employed as a river scrubber, pumping can selectively intercept heavily polluted water and the storage, if desired, can be divided for series operation into two or more reservoirs for more effective off-channel self-purification, reserving the final stage for maximum reaeration to ensure augmentation water high in dissolved oxygen.

Reservoir Water-Quality Improvement

It is well known that large conventional reservoirs undergo a seasonal cycle of stratification and turnover, and during the stratification phase the waters of the lower depths (hypolimnion) degenerate in quality, particularly in dissolved oxygen. As discussed in Chapter 9, drawdown of conventional reservoirs usually results in withdrawal of poor quality water, seriously deficient in dissolved oxygen during the critical warm weather season.

Reynolds [3] has shown by theoretical analysis confirmed by field observations that adaptation of large scale pumped storage power–water systems can effectively prevent stratification and ensure reservoir water of high quality throughout, with dissolved oxygen maintained at levels approaching saturation. In considering mixing a reservoir by pumping, Reynolds takes two approaches by which the effects of this water movement may be analyzed. One approach is considered that of simple displacement by which less dense water (warmer) pumped back from the lower reregulating pool moves through the upper reservoir, entraining some surrounding water, until it reaches a level of equal density or the surface. The other method takes into account the effects of momentum of the body of water. Applying these theoretical approaches to the Crains Run reservoir on the lower Miami River, he has shown by the first approach that on the critical day the entire volume of the reservoir at full capacity would be displaced in a day. Such daily reservoir turnover would ensure complete mixing and prevent stratification. Similarly, by the second approach, including moderate wind action, computations show that stratification can also be prevented. These computations and field observations indicate that the water movement associated with pumped storage developments causes very large eddy effects, preventing stratification and tending to circulate the entire body of water.

Reynolds [3] has also shown that with this induced turnover water by reaeration has oxygenation capacity available to satisfy a substantial BOD load, well above that usually encountered in conventional reservoirs, with dissolved oxygen throughout maintained virtually at saturation at all times.

Under these conditions there is a possibility of a pumped storage reservoir serving additionally in the role of a river scrubber, or treatment system. With a portion of riverflow recycled daily to the upper reservoir where retention time serves to reduce residual BOD, not only is augmentation water returned with its full quota of dissolved oxygen but the residual BOD is lowered as well.

With the prevention of stratification and daily turnover induced by pumped storage cycling, the temperature of the return augmentation water is lower than that encountered in the upper stratified layers (epilimnion) of conventional reservoirs. Since the saturation capacity of water increases substantially as temperature declines, augmentation would contain more oxygen than water withdrawn from the epilimnion of a conventional reservoir with an inlet tower permitting surface withdrawal.

With respect to other chemical parameters of water quality it is expected that the augmentation water of a pumped storage reservoir would be equal or superior to that of conventional reservoirs. The reaeration capabilities of pumped storage reservoirs designed for complete mixing will ensure maintaining the entire body of water near saturation with oxygen. Hence the usual chemical cycles involved in a conventional reservoir associated with stratification and oxygen depletion in the lower layers are not applicable. This should protect against reduced compounds, such as ammonia and soluble iron and manganese, found in conventional reservoirs. Continuous circulation throughout the year should disrupt the usual phosphorus cycle associated with seasonal biological productivity and may prevent unusual blooms induced by sudden large releases in the turnover of stratified conventional reservoirs; continuous circulation may also result in a net lowering of phosphorus in available form.

With respect to biological productivity, nutrients are generally considered most available to organisms in reduced form; hence the lack of stratification and the high dissolved oxygen content of a pumped storage reservoir would tend to retard productivity. Although circulation provides wide distribution of nutrients, it also tends to prevent zones of temporary buildup; continuous circulation at full depth in and out of the zone of light penetration also tends to reduce the average amount of light available to suspended phytoplankton. With respect to algal blooms, the eminent authority, Lackey [4], states that "when we note the conditions of blooms in the field, it is at once apparent that there is much variation of conditions under which they appear, and at best, our knowledge of what nutrient substances and amounts are related to algal densities is merely a guess." Unfortunately the net effect on biological productivity of all the factors arising out of circulation of a large impoundment are not precisely known; hence productivity cannot be predicted. However, it would appear that excessive blooms would not occur to any extent beyond that experienced in conventional reservoirs, and the likely trend would be toward retardation.

Community and Industrial Water Supply

The pumped storage dual power–water system also affords a multiple water resource benefit in providing an ample supply of raw water for community and industrial needs. Good raw water quality is obtainable by virtue of the substantial self-purification provided where there has been a long retention time in storage and by selective pumping as an aid in quality control in seasonal reservoir refilling. The location of pumped storage developments off the main channel in proximity to urban-industrial developments (Crains Run or Honey Creek in the Miami Valley) offers unusual opportunity for integrated municipal and industrial water supply systems serving large areas of need, with assurance of adequate water for orderly growth and development. There is also opportunity for integrated surface water and groundwater development. During years when all of the design flow augmentation is not required, excess water may be available for direct replenishment of groundwater or conservation of groundwater by utilizing the surface storage supply.

Water Based Recreation

Combinations of conventional reservoirs and pumped storage developments offer opportunities in operating schedules to ensure high stable reservoir levels for extensive water based recreation throughout the summer season. Referring to the Miami River basin studies, from the standpoint of use of the headwaters reservoir as a multipurpose development for recreation, in a severe drought such as 1934–1935, Piqua (without Crains Run) would have been completely drawn down by the end of August, the peak of the recreation demand. However, in combination with the pumped storage at Crains Run, at the end of September (toward the end of the recreation season) the drawdown at Piqua would be only about 7 ft and would average about 3 ft through the season. Operation of the Crains Run pumped storage so as to carry the downstream augmentation needs during the recreation season permits the upstream Piqua reservoir to function as a major water based recreation facility, even during the most severe drought of record; during the normal droughts only minor drawdown would be necessary. This can be achieved without interference with other multipurpose uses.

Combinations of two or more pumped storage dual power–water systems likewise provide flexibility in multipurpose operation to permit water based recreation. For example, in the Miami studies it was shown that the proposed Honey Creek pumped storage development just above Dayton in combination with the downstream Crains Run development could admirably serve the water based recreation needs of urban Dayton without interference with the other multipurpose water–power functions. Again, the downstream Crains Run installation would carry the major water quality augmentation needs through the recreation season, with only minor drawdown at Honey Creek. Although Honey Creek is only 15 miles from the center of metropolitan Dayton, water quality for recreation would be exceptionally good, since major waste effluents are downstream. With this proximity to a metropolitan center now served only by distant water based recreational areas, the studies indicate that within a 5-year period after closure of the Honey Creek dam the reservoir would attract a million or more visitors each year.

Water based recreation—bathing, sailing, boating, fishing, and picnicking—at reservoirs such as TVA and Corps of Engineers are increasing at the rate of 15 to 25 percent per year. A most pressing need is nearby facilities to provide ready access to the congested urban-industrial centers. An illustration of this need is demonstrated by the intensive use made of the reservoir (Lake Lanier) formed by Buford dam on the Chattahoochee River above Atlanta, Georgia (Chapter 9). It is reported that the annual visitor load for various water based recreational activities is close to 10 million, the largest attendance at any reservoir in the United States. Such intensive use of water based recreation areas reveals an instinctive craving of urban man for contact with the water environment. In serving such recreation demands the pumped storage dual power–water system offers a unique opportunity that is not ordinarily possible with conventional more distant headwaters reservoirs.

Quite apart from water based recreation on reservoirs, drought control by low flow augmentation provides reclamation benefits along the entire river course. Exposed stream channels during the drought season are unsightly and depress community amenity values and adversely affect property values. Drought control ensures riverside recreation through urban areas and enhances park areas. Most cities are astride rivers; properly managed low flow augmentation can convert the shorefront from a depressing liability to a beautifying civic asset providing relaxation and pleasure to thousands of citizens.

Flood Protection

Since drought control and flood control are complementary, a degree of flood protection is available in pumped storage dual power–water systems. The unusually high pumping capacity and quick response of pumped storage equipment offers opportunity to decapitate flood peaks, thus providing auxiliary flood protection. One Miami design provides pumping capacity on the order of 25,000 cfs, or about 50,000 acre-ft per day. This can be a substantial flood protection benefit, particularly if an extra increment of reservoir storage is reserved for this purpose. Such flood control operations would not appreciably interfere with the normal pattern of pumped storage, since the period of flood pumping would be short and infrequent. With location along the main river in the lower valley, pumped storage would be a critical aid in emergency protection of levees and low lying urban areas.

WASTE STORAGE AND REGULATED EFFLUENT RELEASE TAILORED TO VARIATION IN WASTE ASSIMILATION CAPACITY

Since drought control benefits all uses along the river valley, it involves a cooperative basinwide approach, usually as a public venture. Such ventures may involve years of effort to promote, plan, and bring to fruition. A second, less involved, approach to more efficient use of the waste assimilation potential is the storage of wastes and regulated release of effluent tailored to the natural variation in waste assimilation capacity. This approach is particularly applicable to industry located in areas where sufficient land is available. It has appeal to industrial management, since it provides a high degree of control of waste disposal, incorporates both storage and treatment aspects, is economical, and can be initiated by an individual industrial establishment. The system can be divided into two types—seasonal storage and release and storage with continuous release. In the former all waste is stored during the season of critical low flow and the lagoons are emptied during the high runoff season, when waste assimilation is great; in the latter a more refined practice involves continuous storage and continuous regulated release, taking fuller advantage of the annual waste assimilation capacity. The system is applicable to most types of waste, but is usually practiced for inorganic, organic, and thermal wastes as well as combinations of these.

Lagoon Waste Storage as a Treatment System and Regulator

Treatment Efficiency and Flexibility as a Regulator

Lagoon waste storage inherently serves a dual purpose, as treatment and as a regulation system to control the release of effluent in accord with the prevailing waste assimilation capacity to ensure prescribed water quality criteria. Lagoons are constructed by excavation and berm fill or by taking advantage of natural topography of adjacent areas with small tributary drainage areas. Multiple ponds provide segregation in treatment phases, with improved efficiency and flexibility over single pond arrangements.

For inorganic stable wastes the treatment aspect is minimal, but the lagoon system provides an opportunity to remove settleable solids, to neutralize acids and bases, and to an extent to bleach color.

For organic wastes efficiency varies with several parameters [5]. Multiple ponds in series provide for initial sedimentation; large deep ponds provide intermediate storage where such requirements dictate; and shallow final oxidation ponds provide efficient BOD reduction. Excessively high BOD concentration and temperature of waste inhibit efficiency. The BOD_5 concentration should preferably be below 300 ppm, and pH below 9.

Extensive experience in the southern kraft pulp and paper industry* indicates that non-odor-producing operation yielding BOD removals in excess of 90 percent can be attained with well designed lagoons that provide at least 20 days' detention capacity and have a BOD_5 surface loading not exceeding 50 lbs per acre per day. Nutrient supplementation is usually not necessary since the long detention period permits many cycles of reuse of nutrient elements by the microbial population. As BOD loading is increased, efficiency in removal declines and anaerobic conditions may prevail. With BOD_5 loadings in excess of 100 lb per acre per day, lagoons are capable of removal of more than 50 lb but are subject to odor generation. The performance of a number of systems operating for kraft mill wastes, as reported in Table 11–3, confirms these efficiency loading relationships. As one would expect, efficiency is best with long detention periods and shallow lagoons. However, depth may exceed 5 to 10 ft if the surface loading does not exceed the BOD_5 limit of 50 lb per acre per day.

The storage lagoons also serve as efficient cooling ponds for dissipation of waste heat. Usually where storage capacity is adequate for effective treatment of organic waste, pond surface area is also adequate to reduce the temperature of wastewater to levels approaching the equilibrium temperature of the prevailing meteorologic environment.

Lagoon storage capacity provides a flexibility in the management of

* Personal communication from I. Gellman, National Council for Stream Improvement, Inc.

Table 11–3 Typical Efficiency of Kraft Mill Lagoon BOD Removal

Mill	Influent BOD (ppm)	Loading and Removal (lb BOD/acre-day)		Removal (percent)	Depth (ft)	Retention (days)
Augusta (total)	280	52	48	93	5	75
Augusta (partial)	280	275	132	48	4	9
Catawba	155	120	40	33	12	40
Franklin (old)	100	53	43	80	10	50
Valdosta	200	30	27	90	3.5	80
Lufkin	160	106	60	55	5	20
Spring Hill	130	11	10	92	4	180
Naheola	130	298	68	23	4.5	6

waste disposal that is not possible in conventional treatment systems. The lagoon system readily absorbs shock loads, spills, and the usual rather large day-to-day variations in industrial waste loads. These are equalized in the long detention time, and the lagoon effluent is stabilized to a much lower coefficient of variation.

The primary function of the lagoon system is to serve as a regulator to release or to store wastes as the situation demands during critical drought periods, during fish spawning runs, and during the recreation season. The objective is to provide sufficient storage to serve such purposes and to escape the restrictions imposed on industrial development by natural drought risks. The level of industrial development that is permitted is dependent on the relation of lagoon storage capacity, water quality objectives, and the framework of the available total annual waste assimilation capacity.

Wastewater Reuse

Lagoon practice also affords an opportunity to reuse wastewater for certain industrial processes and particularly for cooling water. A good example of this practice is the wastewater lagoon near Shreveport, Louisiana, operated for the dual function of seasonal waste storage and as a cooling pond for reuse of wastewater for process and condenser circulation (discussed in Chapter 7).

Seasonal Storage and Release

In some situations it is desirable to design the lagoon system as a seasonal operation, storing all the waste during low runoff and releasing only during the high flood season. This is not the most efficient use of waste assimilation capacity, but it has the advantage of complete control of water quality during the critical season. The feasibility of seasonal operation depends on the characteristics of the seasonal pattern of runoff—that is, the duration of the drought flow period relative to the level and duration of the high runoff season. Analysis of the monthly average runoff record provides a basis for determining the required lagoon storage and the opportunity to empty annually all of the stored wastes.

The potential for such seasonal operation is illustrated in studies of river M. Figure 11–23, the seasonal pattern of runoff based on records at the control gage (17,192 square miles), shows a wide differential between the high runoff season (December–June) and the severe but relatively short drought season (August–October). The most probable and the once in 10-year minimum monthly average drought severity expected on the average are 590 and 245 cfs, respectively, as compared

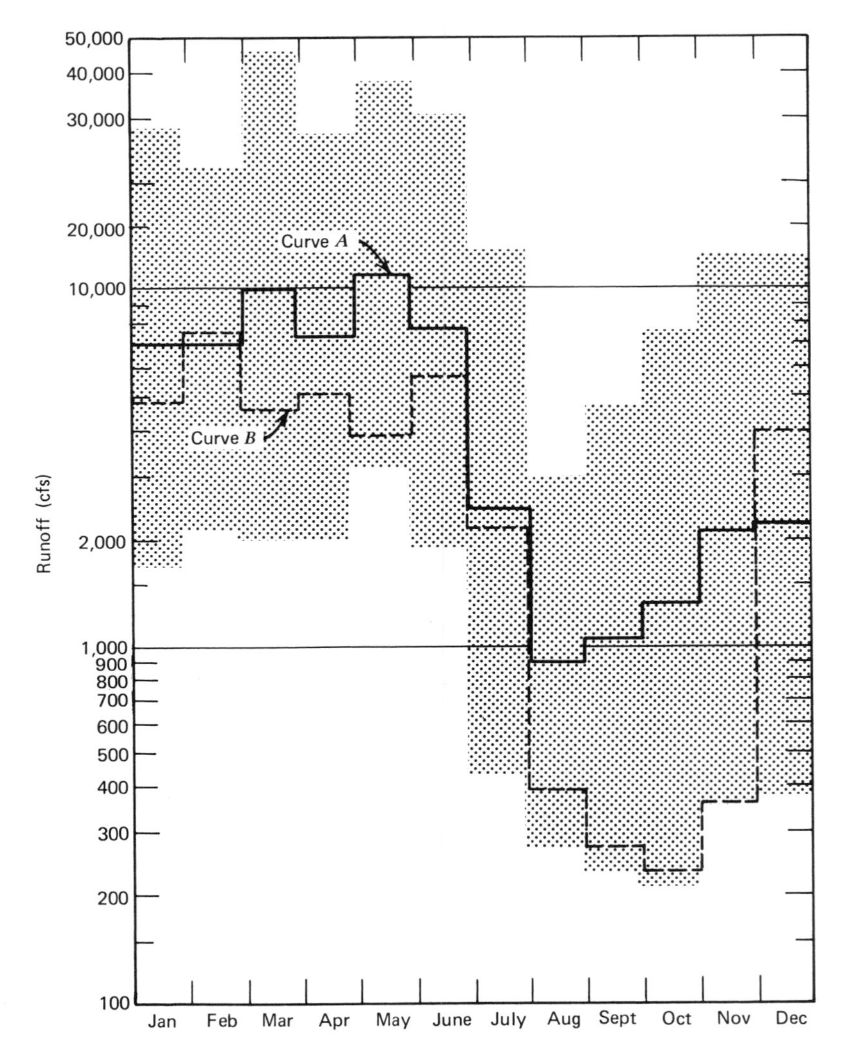

Figure 11–23 River M. Seasonal pattern of runoff. Curve A: most probable monthly mean for specific months; curve B: 1939 observed monthly means; shaded area: range within which monthly mean can be expected for 80 percent of years.

with the long term average flow of 7370 cfs and the normal seasonal high flow of about 8000 cfs.

Waste storage potential is approximated from a statistical analysis of the number of months each year that monthly average runoff is below certain selected levels for the low runoff season. Arbitrary runoff levels selected in this instance are 500, 800, 1000, and 1500 cfs. The distribution of the number of months each year of record when flow was below each of these four controlling flows arranged in order of magnitude and plotted on normal probability paper is reasonably normal, as illustrated by the distribution for 1000 cfs shown in Figure 11–24. Controlling flow is considered to be the monthly average below which all waste is diverted to lagoon storage, with no attempt to handle a fraction of it currently in the river.

From the four probability distributions Figure 11–25 is developed as an approximation of the number of months storage of waste is required at various probabilities at any flow from 500 to 1500 cfs. For example, if the controlling flow is taken as 1000 cfs, the most probable storage required by Figure 11–25 (or 11–24) is 2.4 months; or in 50 percent of the years, within the range 0.7 to 4.1 months; once in 10 years,

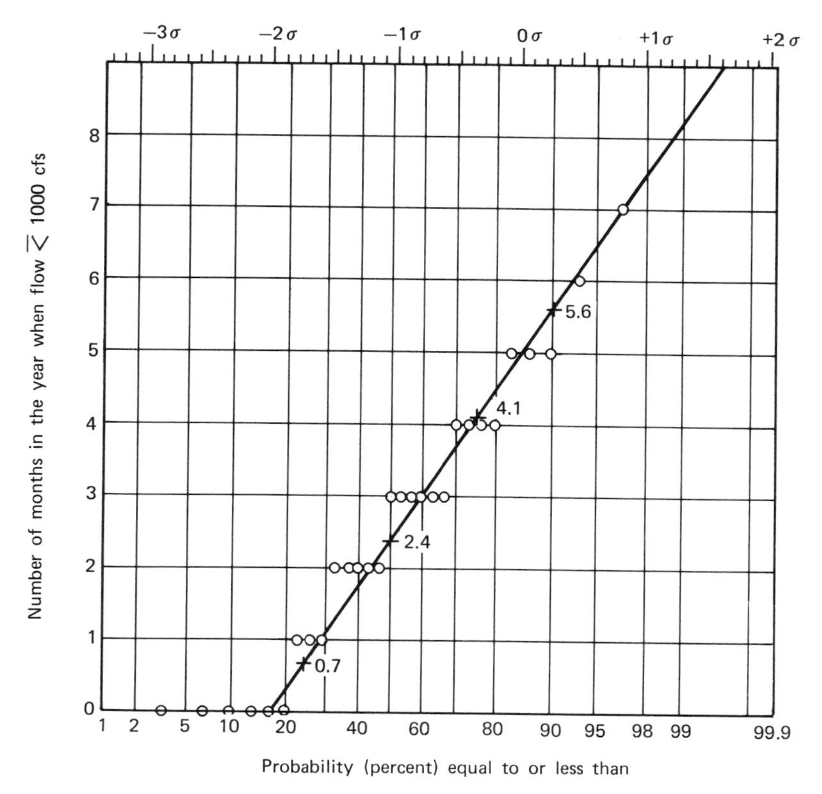

Figure 11–24 River M. Distribution of number of months flow is equal to or less than 1000 cfs.

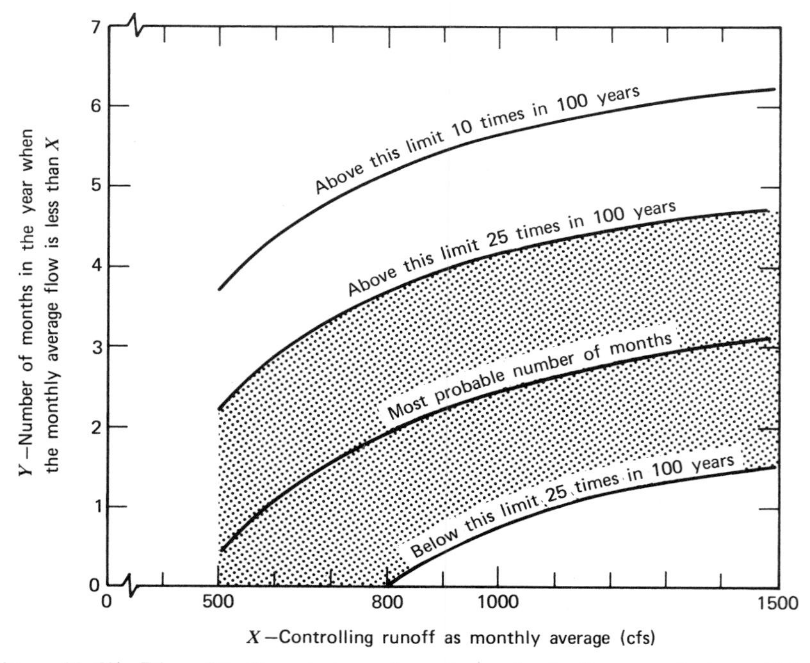

Figure 11–25 River M. Expected number of months in the year during which the monthly average flow is below specified controlling runoff. Shaded area: range expected for 50 percent of years.

5.6 months; once in 10 years no storage is needed. The high flows recorded during the wet season indicate ample opportunity to empty lagoons each year. A controlling flow of 1000 cfs, it is noted, is about four times the once in 10-year drought flow and nearly twice the most probable drought flow.

It is recognized that the storage requirements can be reduced if part of the waste is released to the stream at flows less than 1000 cfs under a controlled basis not to exceed the waste assimilation capacity of the prevailing runoff. Thus the practice of seasonal lagoon storage removes the restriction of severe drought conditions and permits a much higher industrial potential. The more efficient use of seasonal waste assimilation capacity by lagoon practice, providing both storage and treatment, is feasible in many river basins.

Waste Storage and Continuous Regulated Effluent Release

A more efficient use of the natural waste assimilation capacity is possible if waste storage is practiced with continuous regulated effluent release, taking fuller advantage of capacity at all runoff conditions. This is illustrated in continuing the analysis of river J for which the waste assimilation potential under natural hydrologic variations has previously been defined. There remains an important practical consideration—namely, the determination of the required size of lagoons in relation to the industrial production level within the framework of the total annual waste assimilation capacity under prescribed water quality criteria.

Since waste storage requirement is a function of inflow to outflow from the lagoons, the principle of the mass diagram is applicable for determination of lagoon capacity. A mass diagram, commencing with November 1944, of cumulative allowable lagoon waste effluent has been plotted as a basis for determining the waste storage capacity required for regulating lagoon effluent discharge. The maximum ordinate between the mass curve and a mill production rate curve, drawn tangential to the mass curve each year, provides a series of expected annual lagoon storage requirements. Three continuous mill production rates were investigated: 22.5, 15.5 and 10.0 million lb of BOD_5 per year. Applying these mill production rates to the mass diagram shows that in each year the lagoons could be emptied, with the exception of the critical period 1954–1956, when for mill production rates greater than 9 million lb of BOD_5 per year a carry-over of storage from 1954 through 1955 was indicated.

Figure 11–26 shows the section of the mass curve for this critical period. The irregular solid line represents the cumulative allowable lagoon waste effluent discharge, and curves *A*, *B*, and *C* are for the continuous mill production rates of 22.5, 15.5 and 10.0 million lb of BOD_5 per year, respectively.

The protracted period of low runoff and consequent low waste assimilation capacity, commencing in June 1954, extended through February 1955 and was followed by unusually low spring runoff. The low runoff period of 1955 was of short duration, and unusually high streamflow occurred during the fall and winter. It will be noted that for the mill production rate of 10.0 million lb of BOD_5 per year (curve *C*) the lagoon could almost be emptied in the spring of 1955, and the maximum waste storage capacity required to carry through the preceding 1954 dry period, given by the ordinate WX, is 5.3 million lb of BOD_5. However, for mill production rates greater than 10.0 million lb of BOD_5 per year the lagoon could not be emptied until the spring of 1956, requiring a carry-over of waste storage from 1954 into 1956. The maximum lagoon storage capacity required for the mill production rate of 15.5 million lb of BOD_5 per year (curve *B*) would have occurred at the end of

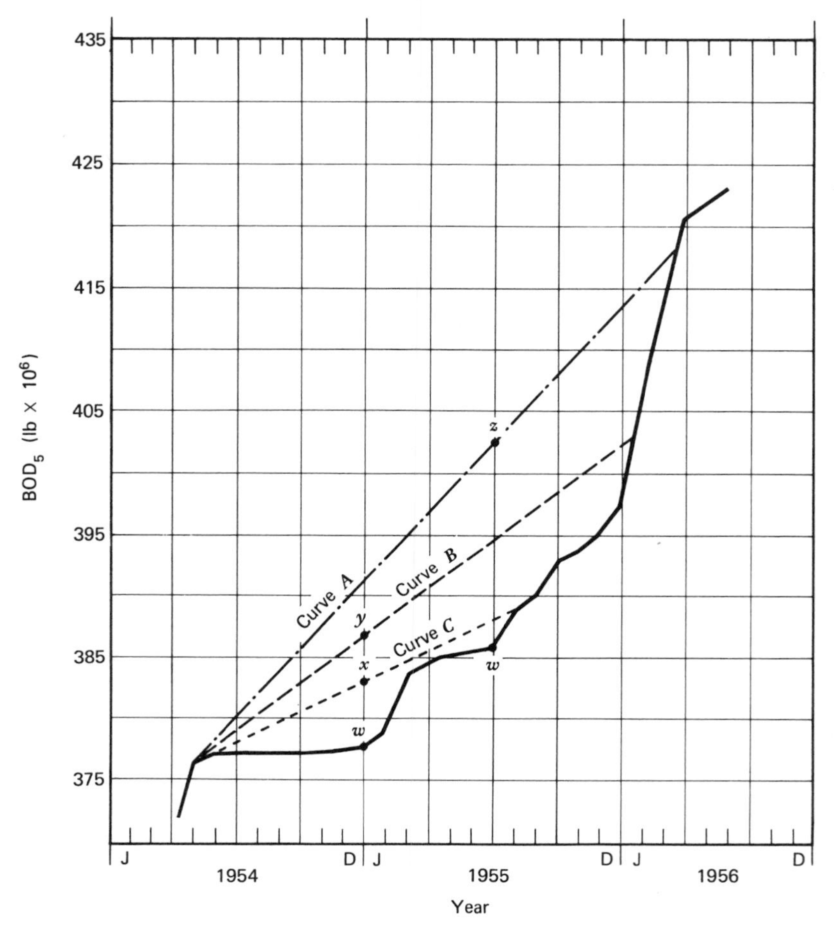

Figure 11–26 River J. Mass diagram for determination of lagoon storage requirements, based on cumulative allowable lagoon waste effluent discharge to maintain a dissolved oxygen level of 4 ppm in the critical reach (lagoon storage, allowable lagoon waste discharge, and mill production rate expressed as BOD of lagoon effluent). The curves represent the following continuous mill production rates (pounds of BOD_5 per year): curve *A*, 22.5×10^6; curve *B*, 15.5×10^6; curve *C*, 10.0×10^6.

1954, indicated by the ordinate *WY* at 8.9 million lb of BOD_5. With increase in mill production rates, such as curve *A*, the maximum ordinate shifts to mid-1955, with lagoon storage indicated by *WZ* at 16.5 million lb of BOD_5.

Table 11–4 lists the lagoon storage capacity required each year from

Table 11–4 River J—Annual Lagoon Waste Storage Requirements for Regulation of Effluent Discharge to Maintain a Critical Dissolved Oxygen Level of 4 ppm[a]

Year	Lagoon Waste Storage Required (lb of $BOD_5 \times 10^6$) for a Mill-Production Rate of		
	10.0×10^6 lb of BOD_5 per Year	15.5×10^6 lb of BOD_5 per Year	22.5×10^6 lb of BOD_5 per Year
1945	0.0	0.9	2.1
1946	1.0	2.8	5.7
1947	2.6	5.4	9.0
1948	1.9	3.8	6.5
1949	0.5	1.0	3.8
1950	0.3	1.6	4.9
1951	3.2	6.3	10.2
1952	3.1	5.5	8.7
1953	4.3	7.5	12.3
1954	5.3	8.9	14.0
1955	2.3	8.6	16.5
1956	1.8	3.6	8.5

[a] Based on mass diagram of monthly waste assimilation capacity from November 1944 to October 1956.

1944 to 1956, taken from the mass diagram, for three mill-production rates. As shown in Figure 11–27, these annual lagoon storage requirements form reasonably normal distributions. Curve *A*, *B*, and *C* represent continuous mill production rates of 22.5, 15.5, and 10.0 million lb of BOD_5 per year. From these distribution curves it is possible to determine for any probability the storage requirement for the three mill production levels. For example, lagoon storage requirements expected to be exceeded on the average 10 times in 100 years would be obtained where the 90-percent probability extension vertically intersects curves *A*, *B*, and *C*, giving storage requirements of 15.2, 9.3, and 4.8 million lb of BOD_5 per year, respectively. Similarly for any other probability corresponding storage requirements can be obtained from Figure 11–27.

From these data the relationship between lagoon waste storage requirement and the annual mill production rate is developed, as shown in Figure 11–28, from which the expected range in lagoon storage require-

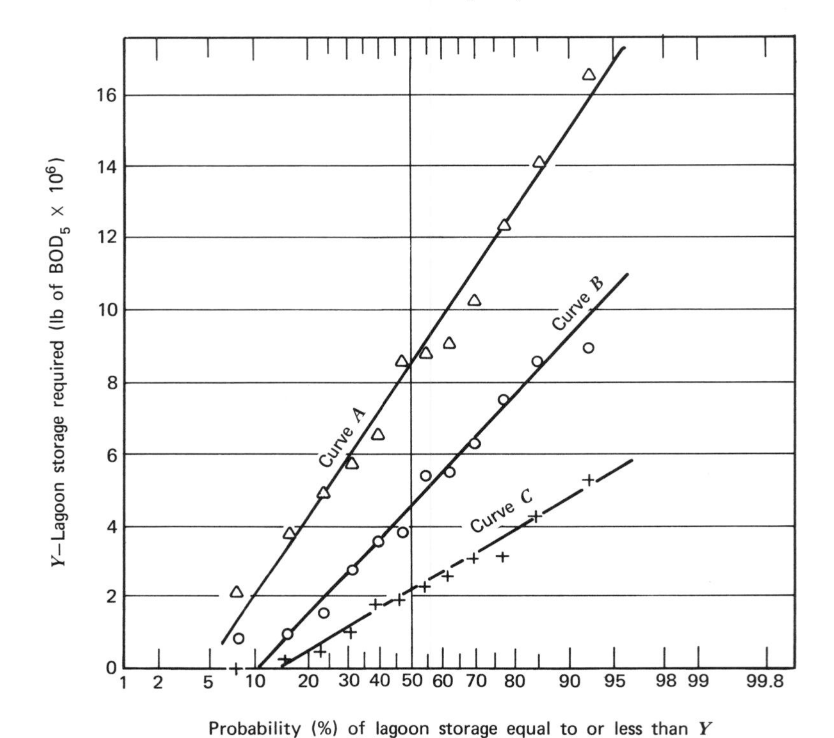

Figure 11–27 River J. Expected lagoon storage required to maintain a dissolved oxygen level of 4 ppm in the critical reach (lagoon storage and mill production expressed as BOD of lagoon effluent). The curves represent the following continuous mill production rates (pounds of BOD_5 per year): curve *A*, 22.5×10^6; curve *B*, 15.5×10^6; curve *C*, 10.0×10^6.

ment can be readily obtained for any mill production rate. The shaded central band reflects the range in storage requirement expected for 50 percent of the years, with the central solid line reflecting the most probable storage requirement any year. The outer curves reflect the range in storage requirement expected for 80 percent of the years, the upper curve being the requirement above this limit 10 times in 100 years, and the lower curve being the requirement below this limit 10 times in 100 years. For example, at a mill production rate of 16 million lb of BOD_5 per year the most probable storage requirement would be 4.8 million lb of BOD_5; for 50 percent of the years the range in storage requirement would fall between 2.3 and 7.3 million lb; 10 times in 100 years, once in 10 years, a storage of at least 9.8 million lb of BOD_5

would be required; and 10 times in 100 years, once in 10 years, no storage would be required. Many years runoff and waste assimilation capacity are such that relatively small lagoon storage would be adequate; however, variation from year to year is great, and it would be desirable to provide storage of a magnitude that would not be exceeded on the average more often than once in 10 years.

It should be pointed out that lagoon storage, allowable lagoon waste effluent discharge, and mill production rate as employed are all expressed in terms of million pounds of BOD_5 of lagoon effluent. To convert lagoon storage requirement to a volume basis would require determination of an efficiency of BOD reduction through the lagoons and in turn determining the BOD and volume of waste produced per ton of mill production. A further consideration that affects lagoon capacity is the net balance between rainfall, evaporation, and seepage.

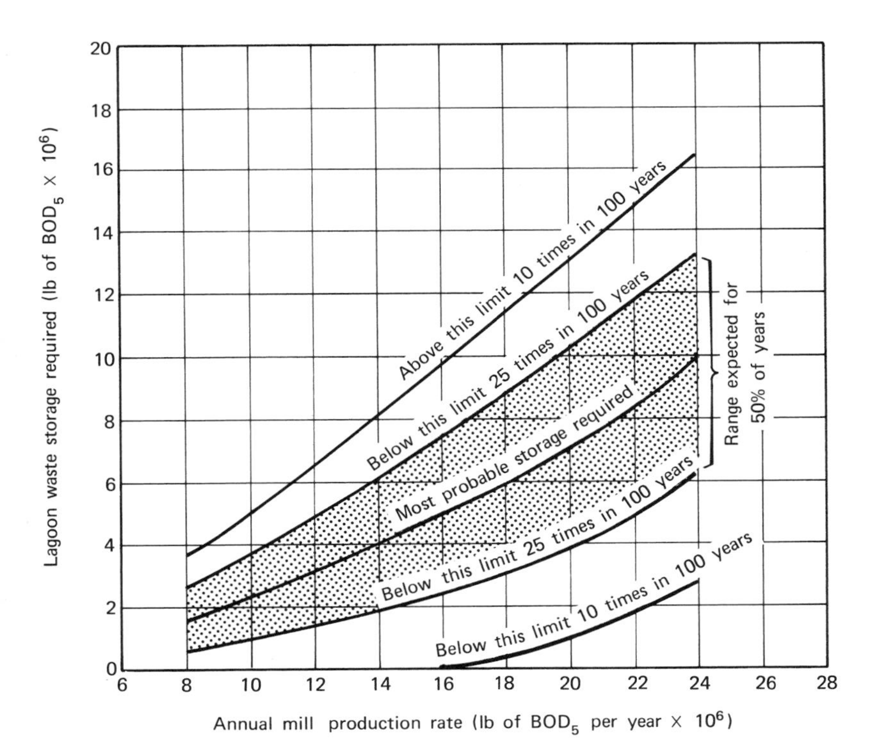

Figure 11–28 River J. Relation between lagoon waste storage required and annual mill production rate under regulated lagoon waste effluent discharge correlated with river runoff to maintain a dissolved oxygen level of 4 ppm in the critical reach (lagoon storage and mill production expressed as BOD of lagoon effluent).

The following example illustrates the great increase in industrial potential under such lagoon storage–treatment system as compared with restriction under natural drought conditions.

Conditions:	
Raw waste production BOD_5	80 lb/ton of product
Raw waste volume	36,000 gal/ton of product
Efficiency of lagoon treatment system:	
BOD_5 reduction (conservative estimate)	50%
Desired mill production capacity	1200 tons/day
Prescribed critical dissolved oxygen level	4 ppm
Lagoon capacity required:	
Annual mill waste lagoon effluent BOD_5 ($80 \times 0.5 \times 1200 \times 365$)	17.52×10^6 lb/yr
From Figure 11–28, most probable BOD_5 storage required to maintain 4-ppm critical dissolved oxygen level	5.8×10^6 lb
or waste accumulation of $(5.8/17.52) \times 365$	121 days
Most probable volume of storage: ($121 \times 36{,}000 \times 1200$)	5227 MG or 16,100 acre-ft
Once in 10-year required storage volume:	
From Figure 11–28, BOD_5	11.2×10^6 lb
or waste accumulation of $(11.2/17.52) \times 365$	233 days
$233 \times 36{,}000 \times 1200$	10,060 MG or 30,800 acre-ft

Thus, with storage of 30,800 acre-ft, the natural waste assimilation capacity can support a mill producing 1200 tons per day and ensure a 4-ppm critical dissolved oxygen level for 90 percent of the years. We have already seen that under once in 10-years monthly average drought flow the waste load to the stream may not exceed 450 lb of BOD_5 per day; allowing for a 90-percent reduction by biological treatment, the daily BOD_5 waste production is limited to 4500 lb, which at 80 lb per ton limits production capacity to 56 tons per day. The lagoon storage system permits a production capacity of 1200 tons per day—more than 20 times that possible under the once in 10 year drought with complete biological treatment. A 50-percent efficiency in BOD reduction for well designed lagoon systems is considered conservative; if increased to 75 percent, mill production capacity could be doubled to 2400 tons per day.

Operational Control

The foregoing analyses are based on complete theoretical utilization of waste assimilation capacity under steady monthly average flow and under certain other assumed conditions. In the day-to-day operational control a number of modifying factors must be taken into consideration, such as deviations in residual dissolved oxygen and BOD_5 of the tributaries and the differences in runoff yield. Runoff is not steady during the month; consequently the most difficult problem is to anticipate the daily flow, particularly in a stream with a flashy hydrograph, such as river J. The risk of anticipating daily runoff on a rising hydrograph is not serious, but an overload of waste released on a falling hydrograph can result in stream degradation below prescribed quality criteria.

The waste effluent released for a given runoff must pass beyond the critical reach in the stream before runoff declines to a lower level. If runoff suddenly declines when waste effluent has been released on the basis of the high flow, a high residual BOD remains to be assimilated by the reduced flow at a longer time of passage. This results in an overburden on the dissolved oxygen. Thus it is essential to define the nature of the falling hydrograph and to determine waste effluent release on an anticipated reduced runoff several days in advance. Detailed analyses of daily hydrographs over the years of record indicate definable hydrograph declines to be expected seasonally, and this provides a rational basis for anticipating runoff. In addition, daily and long-term weather forecasts for the drainage basin are indispensable. Thus, under conditions of declining hydrograph, all runoff that occurs in excess of the projected minimum value at which waste effluent is released is, in effect, wasted.

An important adjunct to efficient operational control is a monitoring system strategically located on tributaries and in the critical reach of the main stream to provide continuous central recording of runoff, temperature, stream quality, and waste effluent characteristics. A period of conservative operating experience serves as a practical guide for improved control. Where the problem warrants, more sophisticated control is possible by adaption of high speed computers to execute direct computation of dissolved oxygen profiles based on monitored data and a range of rational anticipated runoff conditions. Such a program for computer adaptation is outlined in Appendix C. Even with the most sophisticated day-to-day control practice, operation inherently falls somewhat short of attaining the full theoretical waste assimilation potential.

On the basis of the studies on river J, the mill expansion and the wastewater lagoon storage–treatment system was constructed, and actual operation has consistently met regulatory requirements with a margin of safety.

With continued economic pressures for greater industrial production within limited water resources, lagoon storage–treatment systems will increase in importance as a practical means of more efficient and fuller use of seasonal range in the waste assimilation capacity of streams and assurance of desired water quality.

INDUSTRIAL PRODUCTION TAILORED TO WASTE ASSIMILATION CAPACITY OR ALTERNATIVES—A CASE STUDY IN A DEVELOPING RIVER BASIN

As urban-industrial concentration intensifies, problems of location of industry and difficulties of satisfactory waste disposal become more complex. Alternatives and new approaches deserve consideration. A relatively new approach not to be overlooked is tailoring industrial production operations to variations in the waste assimilation capacity. An example based on a detailed study of river N illustrates the possibilities in regard to alternatives.

Because of markets, raw materials, and other reasons, the industry selected a location on a stream near an existing urban-industrial complex. Location above the city required meeting high stream quality criteria; downstream location allowed lower criteria but resulted in a substantial residual BOD loading, lowered dissolved oxygen, and temperature superelevation from the urban-industrial area. The decision was made by the industry to locate downstream, where the critical dissolved oxygen level was prescribed at 30 percent of saturation.

Drought Restrictions and Other Impacts

A statistical analysis of drought flows disclosed the minimum daily average and minimum monthly average severities expected once in 10 years to be 390 and 600 cfs, respectively. An analysis of the water temperature profile from the waste heat load of a large steam electric power plant upstream disclosed a river temperature at 35°C at the industry site during drought conditions. Arrangements for upstream reservoir release were made to ensure a minimum streamflow of 400 cfs.

It was desired to install industrial capacity of 1200 tons of product per day; with a raw waste load on a BOD basis of 500 population equivalents per ton, or 600,000 population equivalents, reduced by partial treatment (42.5-percent efficiency) to 345,000 population equivalents, or 287 per ton. The residual upstream BOD load from the urban-industrial area was determined to be 174,000 population equivalents.

Waste Assimilation-Capacity Control Curves

With reasonably good channel cross sections available, an extensive series of dissolved oxygen profiles was computed for ranges of runoff water temperature, and waste effluent loading from the industry. From these, by methods that we have already described the set of control curves for temperatures of 20 to 35°C was developed, as shown in Figure 11-29; from these curves the allowable industrial waste load to maintain a critical dissolved oxygen level of 30 percent of saturation is readily obtainable for any runoff from 400 to 900 cfs.

Alternatives

Industrial Potential under Drought Flow

Under the proposed regulated minimum assured flow of 400 cfs (which is nearly equivalent to the once in 10-years drought severity) and elevated water temperature of 35°C, from Figure 11–29 the allowable industrial BOD load is calculated to be 130,000 population equivalents, which on the partially treated basis is equivalent to an industrial production

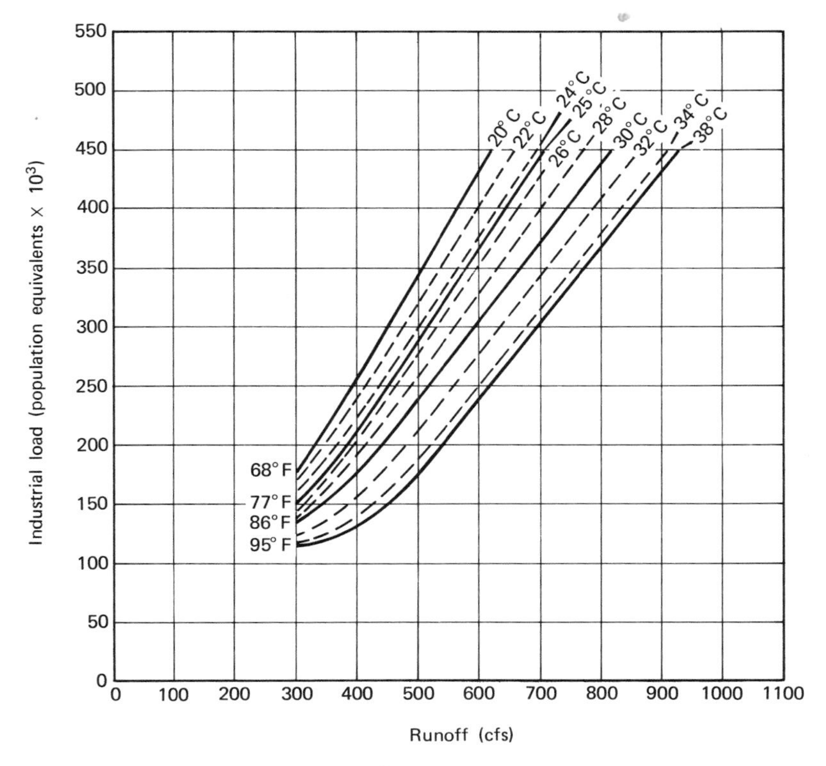

Figure 11–29 Limiting industrial production to ensure a dissolved oxygen level not declining below 30 percent of saturation in the critical reach at various runoffs and water temperatures.

level of 453 tons per day, or slightly more than one-third of the desired 1200 tons per day. Thus with dependence on the regulated minimum of 400 cfs the desired industrial potential cannot be met. Alternatives are considered.

Reduced Heat Load

If waste heat is reduced to the 30°C equilibrium temperature of the warm season meteorological environment, by Figure 11-29, the allowable industrial load at a minimum flow of 400 cfs can be increased to 175,000 population equivalents, or a production capacity of about 610 tons per day—a substantial increase, but only about one-half that desired.

Low Flow Augmentation

If the heat load remains unchanged but low flow augmentation is provided to reduce the temperature to 32°C, the augmented flow required for the effluent load of 345,000 population equivalents by Figure 11–29 is 700 cfs. This is possible at some sacrifice of hydroelectric practice at the upstream reservoir. Under these conditions the industrial BOD allowed to maintain a critical dissolved oxygen level of 30 percent of saturation is 345,000 population equivalents, or just equivalent to the desired production of 1200 tons per day. Negotiations, however, failed to produce agreement to raise the assured minimum flow from 400 to 700 cfs.

Increased Waste Treatment

Another alternative is further treatment of industrial waste to reduce the effluent BOD load. As determined previously from Figure 11–29, the allowable BOD load at a runoff of 400 cfs and a temperature of 35°C to maintain a critical dissolved oxygen level of 30 percent of saturation is 130,000 population equivalents. On the basis of the raw waste load of 600,000 population equivalents for the desired production of 1200 tons per day, this would require an increase in the efficiency of treatment from 42.5 to 78.3 percent, which implies some form of biological system, an expensive process.

Lagoon Storage–Treatment System

A lagoon storage–treatment system would be highly effective but is not feasible at this site because land is not available.

Tailored Industrial Production

Rejection of the alternatives outlined above leads to consideration of reducing industrial production during low-runoff periods in accord with the available waste assimilation capacity. As a basis for estimating the probable extent of curtailment of industrial production, a statistical analysis was made of the number of days each year of record that runoff occurred within three ranges, 400 to 499, 500 to 599, and 600 to 699 cfs in the warm summer–fall season (May through October). The number of days in each range described reasonably normal distributions, from which the expected number of days at three probabilities of occurrence was determined, as shown in Table 11–5.

Since the largest number of days occurred during September and October, when temperature was beyond the seasonal peak, a water temperature of 29°C was taken as an average, including the influence of upstream waste heat load. The allowable BOD waste loading at 29°C at the runoff ranges taken at midvalue is obtained directly from Figure 11–29 as 215,000, 285,000, and 350,000 population equivalents at runoffs at 450, 550, and 650 cfs, respectively, or allowable production levels of 749, 993, and 1200 tons per day (full capacity), respectively. The net losses in production then are 451 and 207 tons per day at 450 and 550 cfs, respectively, with no loss at 650 cfs. Applying these daily net production losses to the number of days of runoff expected in the ranges 400 to 499 and 500 to 599 cfs, as shown in Table 11–5, gives the total seasonal production loss.

Thus to ensure that the critical dissolved oxygen level will not decline below 30 percent of saturation, it can be expected on the average that once in 3 years the seasonal net production loss would not exceed 4900 tons, equivalent to only 4.1 days of full capacity operation, involving, however, curtailment on 19 days; once in 5 years the seasonal net production loss would not exceed 8860 tons, equivalent to 7.4 days of full capacity operation, involving curtailment on 31 days; and once in 10 years the net production loss would not exceed 12,800 tons, equivalent

Table 11–5 Total Number of Days Flow Expected between Stated Intervals for the Summer-Fall Season May through October

	Total Number of Days between Stated Intervals		
Probability	400–499 cfs	500–599 cfs	600–699 cfs
Once in 3 years	4	15	22
Once in 5 years	10	21	27
Once in 10 years	16	27	34

to 10.7 days of full capacity operation, involving curtailment below design capacity on 43 days. The total loss of tonnage below capacity production expected once in 3, 5, and 10 years is only about 1, 2, and 3 percent, respectively, of the total for the year under full capacity production. However, these relatively minor losses expected once in 3, 5, and 10 years are spread as planned curtailment of production on 19, 31, and 43 days, respectively. It is this spread of planned curtailment that is usually objected to by management.

Tailoring industrial production to variations in the waste assimilation capacity during the summer–fall drought season implies a system of operational control discussed in the section on lagoon waste storage, and in addition involves many in-plant adjustments along the entire line of production. Industries usually require shutdown periods for cleanup and repairs; these periods can be scheduled to coincide with drought periods. Careful planning can minimize disruption in routine production. The alternatives are costly and restrictive; to meet water quality requirement during a drought severity expected on the average once in 10 years by normal production practice limits the industrial production design capacity to 450 tons per day. This compares to the design capacity of 1200 tons per day for production tailored to waste assimilation capacity with a sacrifice in production not exceeding 3 percent more often than once in 10 years. As the cost and consequences of alternatives become greater, tailored industrial production offers an opportunity for industrial expansion with efficient use of waste assimilation capacity.

REFERENCES

[1] Calvert, J. D., *Power Engineering,* **70,** No. 5, 46 (1966).

[2] Velz, C. J., Calvert, J. D., Deininger, R. A., Heilman, W. L., and Reynolds, J. Z., in *Proceedings of the National Symposium on Quality Standards for Natural Waters,* University of Michigan School of Public Health, July 1966, p. 123.

[3] Reynolds, J. Z., *Some Water Quality Considerations of Pumped Storage Reservoirs,* Ph.D. Thesis, University of Michigan, 1966; *ASCE Proceedings,* **93,** No. PO 2, p. 15–35, (October 1967).

[4] Lackey, J. B., U.S. Department of Health, Education, and Welfare, Public Health Service, Robert A. Taft Sanitary Engineering Center, Technical Report 61-3, 1961.

[5] Gehm, H. W., *Water Poll. Control Fed. J.,* **35,** 1174 (1963).

12

MANAGEMENT OF WASTE DISPOSAL AND STREAMFLOW QUANTITY AND QUALITY

PERSPECTIVE OF A BROADER VIEW

The foregoing chapters have dealt with the technical details of applied stream sanitation. In closing we return to the broader view of water resources and waste disposal. This chapter, therefore, is limited to consideration of principles of management that are generally applicable; details will vary from region to region, state to state, and basin to basin.

Management, like the technical aspects, can easily bog down in administrative detail, stifling imaginative effort. Management then becomes a narrow preoccupation with meeting *standards* rather than a progressive, innovative action to achieve *objectives*. It was the late William Thompson Sedgwick who pointed out that standards are devices to keep the lazy mind from thinking, and Phelps the father of modern stream sanitation has cautioned [1]

> It will be well for the potomologist in the interest of his own clear thinking, to examine the data and the reasoning which underlie our present day standards, codes, and rules, as they apply to commodities (including water) and to the physical environment in general. One who is to enforce these provisions should have a much better understanding of their true merit than is implied in the printed documents.

Satisfactory management of waste disposal cannot be isolated from management of the total water resource and the interrelation of land and water use and development. The interactions are inextricably interwoven. Similarly all wastes are involved—the solid, the gaseous, and the liquid—and final disposal of inevitable residuals is dependent on the waste assimilation capacity of the land, the atmosphere, or the watercourse. Although management of water quality is the primary focus, it is pursued in the context of the broader perspective of the total water resource and the total water problem.

THE DUAL OBJECTIVES OF MANAGEMENT

As outlined in Chapter 1, the two primary objectives in the practice of stream sanitation are the satisfactory disposal of the water-carriage waste products of man's activities and the enhancement of satisfactory stream quantity and quality for intended multiple uses. The plan of action to achieve these objectives rests on the design of the combination of lines of defense and lines of offense that is specific for each particular river system.

By their nature, waste disposal and water quantity and quality control are complex and extensive in scope, usually not easily comprehended by the public nor fully appreciated by the multiple interests involved. Effective action necessitates cooperative effort and a rational balance among diverse interests. It is one thing to arrive at a sound plan of action from a scientific and technical standpoint; it is quite another matter to bring such a plan to mutual acceptance for public and private action. It is here that the social sciences play an important role.

Public relations and public administration skills must be effectively employed in gaining public understanding and support. Often sound plans founder on the rocks of misunderstanding and controversy among special interests. The social science skills are particularly needed in promoting offensive lines of action—for example, drought control by low flow augmentation.

Inevitably management of offensive and defensive lines cuts across political boundaries; many separate political units may be involved—in some cases two or more states. Each level of government—federal, state, county, and municipal—plays a role, and they do not always work harmoniously. The skills of political science are effective in integrating political units and in devising new institutional arrangements appropriate for unified action and financing.

INSTITUTIONAL ARRANGEMENTS

Sophisticated combinations of offensive and defensive strategy require long-term financing and a complex day-to-day management. New institutional arrangements are necessary to cope with the new complexities.

The Role of Government

Defense in the form of waste treatment alone is not adequate for today's and tomorrow's needs. In the past primary responsibility fell on individual municipalities and industries for financing and for day-to-day management, with general regulatory control vested in the states, usually centered in the health department. There has been a gradual evolution toward broader concepts, a broadening of financial support in the form of federal grants to municipalities for construction of treatment works. Regulation has remained vested primarily in the states, but it has broadened beyond health departments to new agencies, such as water resource commissions to include conservation, agriculture, recreation, and other interests. Recently financial aid for treatment works and broad water resource planning has been provided by some states.

Before 1960 water resource development at the federal level was primarily restricted to flood control, navigation, hydroelectric power, and irrigation. But in that year federal agencies were authorized and required to consider low flow augmentation for stream quality enhancement. This innovation was followed with an increase in regulatory power at the federal level in the Water Quality Act of 1965 and the transfer of authority from the Public Health Service to the Federal Water Pollution Control Administration in the Department of the Interior. Although regulatory power will probably remain predominantly at the state level, greater federal involvement is likely to continue—with potentialities for division of responsibilities and conflict of policies. In a country as large as the United States, the diverse regional needs and philosophies make it imperative that management of waste disposal and related water resource development remain at the local, state, or regional level. However, the complexities of the problems preclude the day-to-day administration exclusively by individual local communities. More sophisticated management is imperative, and the required staff and financing can be provided only by a more encompassing unit of government.

Interstate Compacts or Drainage Basin Commissions

On the larger interstate streams, compacts among states afford a mechanism by which a basin can be encompassed and managed as a unit. One of the most successful efforts is that of the Ohio River Valley Water Sanitation Commission (ORSANCO), organized as an interstate compact among the eight bordering states for a regional assault on common problems of stream pollution. It pioneered in development and managerial use of an extensive continuous monitoring system (Chapter 10), gearing waste disposal to variations in the assimilation capacity of streams. It also has effectively employed public relations skills in promoting abatement proposals and in development and acceptance of rational criteria and stream standards.

The interstate compact mechanism is, however, not without inherent constraints on performance and achievement. The compact has limited authority in matters of administration and must work through the states in most matters of significance. It is financially dependent on budget contributions from the signatory states. Limited and divided responsibility between the compact organization and states can be a serious impediment in implementation of progressive activity. And, of course, an association of states in so complex a venture as a regional water authority can scarcely be stronger than its most reluctant member state legislature. Although much can be achieved by persuasion and publicity, efficient management requires the backing of independent, adequate authority for enforcement.

Although interstate compacts are essentially a mechanism for cooperation on interstate rivers by the states, independently of federal agencies, the interstate commission, such as that for the Delaware River, include the federal government as a partner. This inclusion recognizes the federal interest in interstate rivers and takes advantage of the experience of such federal agencies as the Army Corps of Engineers in large-scale construction and of the financial capacity of the large unit of government. The autonomy of the commission approach provides a greater degree of freedom of action than that inherent in interstate compacts. There is, however, a natural reluctance on the part of some states to employ this form for fear that a direct partnership with the federal government will ultimately mean federal domination.

Drainage Basin Authorities or Conservancy Districts

In the promotion of federal approaches a great deal of emphasis has been given to the interstate river systems. Although the great main-stem rivers are important, the bulk of the waste disposal problems and the water resource development needs are on intrastate streams. Many of these intrastate streams serve highly developed urban-industrial corridors involving several political units. Increasingly the states recognize that the municipal-state relationship in the management of water resource and pollution problems is inadequate. State legislatures are beginning to authorize the formation of drainage basin authorities or conservancy districts with broad powers to finance, plan, construct, and operate water resource developments and provide overall day-to-day management much in the nature of a corporate structure. The states, in delegating these extensive powers over water resources to the specifically defined basin corporation, retains only minimal administrative functions, principally a restraint to ensure that the public corporation operates in the interest of the public. This public corporate structure is not without risks, because he who controls water controls the lifeblood of community and industrial growth. Adequate safeguards can be retained in an appropriate state agency as an overseer. A corporate structure facilitates development of sophisticated management of lines of defense and offense. Financial independence with power to raise funds, assess for benefits, and charge for services (with adequate state supervision and possibly backing) ensures a businesslike, but nonprofit public operation with a minimum of state or federal constraints.

The pioneering action of the State of Ohio in authorizing the creation of conservancy districts after the disastrous 1913 Miami Valley flood, set an early example for independent action for flood control without the assistance of federal funds. The organic Conservance Act which has been so successful in flood control can equally apply to drought control and, with broader applications, to water resource management. This is exemplified in the progressive activities of the Miami Conservancy District.

Direct State Action

Two serious constraints are usually inherent in any voluntary organization of special institutional arrangements: years of delay in gaining acceptance among the many political units encompassing a logical drainage basin, and delays and disagreements over means of financing and the high interest rates charged the small units of government. As a means of circumventing these constraints some states, following the early example of the Province of Ontario, Canada, are taking direct action in providing facilities and services in logical drainage areas, collecting assessments or service charges of local governmental units and industries. It is claimed that such direct action expedites needed works, facilitates planning, permits development of multiple lines of defense and offense, derives economies of scale, reduces local bonded debt burdens, and provides substantial saving in lower interest rates. Concern, however, is expressed over loss of local initiative and control, possibilities of political influence, and conflict of interest in the state serving in the dual capacity of an action agency and of a regulatory agency.

Public–Private Cooperatives

The new need to give attention to the offensive strategy of water management and to ensure adequate streamflow for both public and private users may well in the future give rise to joint public–private institutional arrangements. A natural cooperative exists in the dual development of electric power and water resources in the adaptation of pumped storage (Chapter 11) to streamflow augmentation. Institutional arrangements can be made to the mutual private and public benefit in cooperative ventures. These may be in the nature of a public–private partnership, a public project for water and power with contractual arrangement for the sale to private utilities of hydroelectric energy during peak demand and purchase of steam generated energy from the utilities during off-peak periods. Conversely, the development can be primarily a private venture, with contractual arrangements with a public drainage basin authority for such water benefits as flow augmentation, water supply, and recreation. In the latter instance the public benefits can be paid as a service charge, a tax relief, a depreciation acceleration, or some other incentive. It is argued that more economical design of facilities and more efficient management will be obtained by incorporating the private business sector.

ECONOMIC CONSIDERATIONS

Cost of Wastewater Treatment

Municipal

The high cost of wastewater treatment is a significant constraint on economic growth. In order to obtain an objective measure of demonstrated practice an extensive investigation has been made of actual costs of construction of municipal treatment works based on contractors' bid prices or final contract payments. A sample of 371 plants, including conventional primary, chemical coagulation, trickling filter, and activated sludge types, was selected from the urban-industrial regions of northeastern and central United States. The majority of the data were obtained from the files of state regulatory agencies where price and basis of design of plants could be confirmed. These were supplemented by an earlier survey made in cooperation with *Engineering News-Record* [2] and by costs reported in the literature. The number of treatment works, the period covered, and the sizes and types included approach a regional random sample, thereby providing a reflection of demonstrated experience.

Cost is limited to construction of treatment works exclusive of sewerage systems, booster pumping stations, intercepting sewers, and outfalls. Also excluded are the costs of preliminary investigations, land, engineering and legal fees, and interest during construction. Only plants submitted to public bidding and constructed under contract are included, and no attempt is made to minimize variations in cost by deleting high or low extremes.

For comparability each plant cost as of the year bid is referred to 1926 as the base year of construction, adjusted by means of the *Engineering News-Record* Construction Cost Index. Also, all types are referred to 100-percent efficiency base by considering primary, chemical coagulation, trickling filter, and activated sludge systems achieving 35, 65, 85, and 90 percent BOD removal, respectively.

It is generally recognized that unit cost of construction declines with increasing size of treatment works. This is reflected in Figure 12–1, which shows the relation between unit cost per million gallons per day capacity (100-percent efficiency basis) and the size of the works in million gallons per day design capacity. The line of best fit through the data indicates a linear relation between the logarithms of unit cost and size of works. Curve *A* represents the average or normal demonstrated practice, from which the most probable unit cost for any size plant can be obtained. Lines of scatter based on the standard error of estimate are established about the most probable line, enclosing a range within which unit costs for 50 percent of treatment works can be expected. The upper boundary, designated curve *B*, represents the limit for the least economical quarter of demonstrated practice; curve *C*, the lower boundary, represents the most economical, or best, quarter of demonstrated practice. A marked decline in unit cost is reflected as the size of treatment works increases; unit cost declines by one-half in the size range from 0.1 to 2.5 mgd and again is halved in the size range from 2.5 to 60 mgd. In addition to economy in design, it is recognized that cost of plants deviating above or below the 50-percent range depends on such local factors as unusual foundation, rock excavation, or high groundwater. In some instances high first costs may be justified by lower operating costs.

Of interest is the close agreement of the most probable line *D*, based on the earlier sample of 185 plants constructed prior to 1948 [2], with the most probable line *A*, based on the 371 plants including the more recent construction up to 1958. The *Engineering News-Record* Cost Index is widely accepted as a reliable indicator of the trend in construction costs. The close agreement in the two samples indicates no significant change in treatment works cost other than that accounted for by the general rise in construction costs. Inflation of construction costs has been dramatic and continues to rise; taking 1926 as the base, costs have increased 3.5 times by 1957, and 5.3 times by 1968; and cost in 1926 doubled (2.08 times) that of 1913.

It is emphasized that the cost curves of Figure 12–1 are a reflection of urban-industrial regions where wages and material costs are high

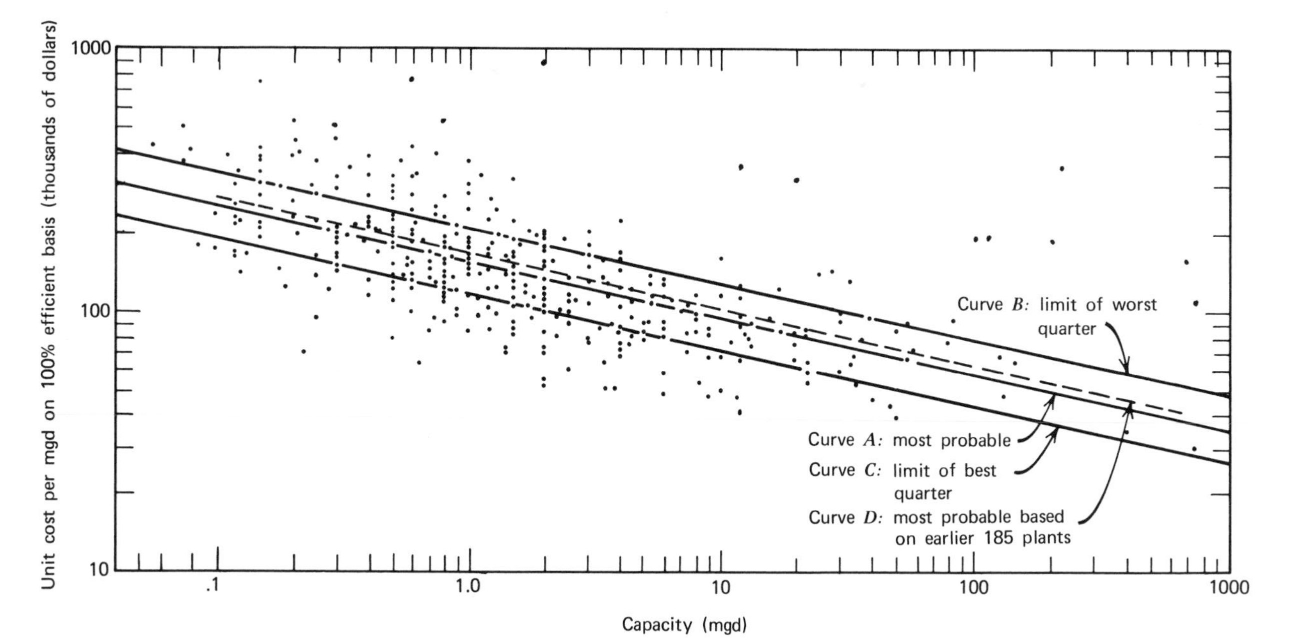

Figure 12–1 Construction cost of sewage treatment works. Demonstrated practice in northeastern and central United States. All types adjusted to 100-percent treatment efficiency and *Engineering News-Record Cost* Index Base 1926.

relative to rural regions and therefore may not be generally applicable without an adjustment for regional cost differentials. Differences among cost curves developed by others workers [3–5] in part stem from different definitions of regional boundaries, items included or excluded in cost, and the basis of plant design capacity. Any cost curve should be intended to serve as a preliminary approximation, not a substitute for engineering estimates based on contract plans and specifications for a specific facility. However, the cost curves of Figure 12–1, based on actual demonstrated practice, serve as an indicator of what a conventional wastewater treatment plant can be built for and should cost; deviation beyond these ranges should elicit explanation. An important economy factor not reflected in the curves is the basis of the design capacity; some designs are conservatively projected for population growth 10 or 20 years hence, others are overdesigned for future growth that may never be attained. Conservative projections with provisions for plant expansion as needed is the more economical approach (Chapter 2).

The following example illustrates the use of the cost curves. Assume a community of 100,000 population (projected 10 years hence) requiring a wastewater treatment works of 12.5-mgd capacity and 90-percent reduction in BOD is to be bid in early 1968 when the *Engineering News-Record* Cost Index is 530 (1926 base). From curve *A* of Figure 12–1, the most probable unit cost for a 12.5-mgd works is $91,000 per mgd referred to 100-percent efficiency and 1926 cost base. The 50-percent confidence range extends from $69,000 (curve *C*) to $121,000 (curve *B*). The most probable cost for January 1968 bidding for the 90-percent efficiency plant would be $5,426,000 ($91{,}000 \times 5.3 \times 0.90 \times 12.5$), and the best and worst quarters are $4,114,000 and $7,215,000, respectively. These estimates cover only the cost of construction, which usually represents 80 to 85 percent of the total bonded project cost, the other costs being preliminary expense, land, legal, engineering, and interest during construction.

In addition to the treatment works, usually the appurtenances, intercepting sewers, pumping stations, and outfall sewers are also necessary. The costs of these vary radically from 50 to 200 percent of the cost of the treatment works, depending on topography and location of the treatment works in relation to the community and the receiving stream. Assuming appurtenances equal the cost of treatment works and the project cost represents 80 percent of the bonded cost, the most probable total investment would be $13,565,000, with the 50-percent range from $10,285,000 to $18,038,000. On the design population basis this represents an investment of $135.65 per capita, with a 50-percent range of $102.85 to $180.38 per capita.

Like the capital investment, the operating expenses of treatment works are also high, and increasing. Included are salaries of technical supervisor and experienced operating personnel, supplies, maintenance, insurance, and power. For primary plants current (1969) annual operating costs range from 5 to 15 percent (average 10 percent) of the cost of the treatment works, and for complete treatment range from 6 to 10 percent (average 8 percent). In the above example the most probable annual operating cost would be \$4.35 per capita, with a 50-percent range of \$3.30 to \$5.30 per capita.

Thus for a city of 100,000 the total annual cost for amortization and interest on the treatment works and appurtenances (at 7 percent), and the operating cost would most probably be \$13.85 per capita, with a 50-percent confidence range of \$10.50 to \$17.90 per capita. This constitutes a substantial annual per capita expense.

Industrial

In addition to the normal municipal waste treatment cost, there is a substantial economic effect on industries that are not served by the municipal systems. The cost of industrial waste treatment works is somewhat lower than that of municipal works, in part because industrial works are usually designed for open weather conditions (unhoused) and storage, administrative, and laboratory services are absorbed by the industrial establishment. Greatest divergence is generally in primary treatment; for the more extensive works needed to provide intermediate and complete biological treatment the agreement between municipal and industrial costs is quite close. The technical complexities of treating the wide range of industrial wastes offset to a large extent any differential in cost; in the absence of specific industrial cost data, the curves of Figure 12–1 serve as a reasonable guide. Since the drain on capital is substantial and operating costs are high, the economic burden of high degree of treatment is a significant constraint on industrial growth.

Economic Constraints and Evaluation

It has been stressed throughout the text that the level of community and industrial growth to a large extent is determined by the available water resources and the way in which these resources are developed, used, and managed. Many factors besides water resources are involved, but ultimately it is the availability of adequate water, and more particularly the waste assimilation capacity of the stream, that will limit growth in any river basin. This undeniable constraint is only too evident in the arid regions. As long as growth remains well below the optimum use of the water resource, the economic importance of water is often ignored. However, since water is a dynamic resource subject to the laws of chance occurrence, a dependable supply of quality water and the cost of maintaining the quality of the resource become paramount considerations as economic growth approaches that optimum.

In the more sophisticated institutional arrangements for management of water resources and waste disposal the economic factor becomes an increasingly relevant consideration in the design of lines of defense and offense. The economics of water resource development and waste disposal are not well defined, and there is disagreement among economists and engineers as to methods of evaluation. Costs and benefits, direct and indirect, are difficult to define and evaluate in monetary terms. The choice of alternatives in an efficient design of offense and defense depends on more precise definitions and a clearer understanding of the importance of water resources in the social and economic structure of regions. These limitations, however, do not justify ignoring the economic factor and proceeding on an arbitrary, unreasonable course without regard to cost.

Economists unfamiliar with the role of water resources and the waste assimilation capacity of streams have overlooked the significance of water in deciding where to locate industry. Economic base studies usually center on the interregional (general area) level, and ignore the constraints of the intraregional (actual site) level. At the general area level primary concern is with such factors as distance from markets and raw materials, and labor availability and productivity. These items make up a large proportion of the total production cost, and water has traditionally appeared to be relatively unimportant by comparison. However, after a general area of location has been selected, constraints of water, particularly the high cost of waste disposal due to the limited assimilation capacity of the stream, can become major factors in local industrial growth. Within the general interregional area there are available choices of location on a number of streams. Rivers with good natural or enhanced waste assimilation capacity attract industrial growth; those with poor waste assimilation capacity or those adversely affected by other river developments are a serious restraint on growth, as industry must contemplate high capital and operating cost for waste treatment. Industry is free to locate, relocate, or expand in any one of a number of sites in the interregional basins and still maintain proximity to markets or to take advantage of other favorable regional factors without bearing the high cost associated with intraregional differentials. Many industries now devote intensive study to favorable location to ensure satisfactory waste disposal (with a margin for future expansion) at reasonable cost. The economic effect on intraregional basins, therefore, can result in either a boon or a devastating depression.

The decision of an industry not to locate or not to expand in a river basin strikes at the very economic base of the community. It is reflected directly in personal income foregone by loss of employment. In turn this direct personal income loss affects the income of others in and outside the basin by a multiplier effect of two to three times.

As urban-industrial expansion continues, many river basins increasingly will face the economic constraint of limited natural waste assimilation capacity within the restrictions imposed by regulatory agencies for satisfactory stream water quality.

It is a fallacy to assume that all river basins are equally competitive. Actually for some years site selection on the basis of adequate waste assimilation capacity has been standard practice of many industrial organizations, and this practice will continue to affect the distribution of personal income in many communities. Some communities are aware of the economic impact and are proposing various means to improve their competitive position; others are unaware or unconcerned. However, no community or river basin is likely to allow its industry and its population to be moved in order to meet a water quality objective. Remedial action of one form or another would most certainly be demanded.

It therefore becomes increasingly necessary to consider a broader economic base than is usually employed in evaluation of the efficiency and economic justification of investment in various lines of defense and offense for waste disposal and for enhancement of quantity and quality of streamflow. Viewed from the standpoint of community economic growth, substantial benefits accrue to local residents as personal income returns to justify a substantial community investment. An efficient combination of offensive lines and defensive lines offers opportunity for broad economic and social benefits that are shared not only by direct beneficiaries but also by the entire population of the river basin, even extending to interregional areas. These specific and yet broad benefits pose a challenge to economists in devising equitable bases for sharing costs.

QUALITY CONSIDERATIONS

Objectives, Criteria, and Standards

The terminology of water quality has traditionally been rather loosely used. Weston [6] suggests differentiation among the terms "objectives," "criteria," and "standards." "Objective" is the general term, defined as a goal toward which a program of management is aimed. It denotes a desirable end or ideal to be achieved eventually, but not necessarily immediately. "Standard," in contrast, denotes a specific legal requirement for immediate compliance. "Criterion" denotes a method of measurement (physical, chemical, biologic, and hygienic), a means of forming a judgment. Where, when, how, and under what conditions measurements are made and how the data are interpreted are the criteria by which compliance with a standard or movement toward an objective are judged.

Interrelation of Quantity and Quality

A specific standard of stream quality is expressed as a concentration, in parts per million or milligrams per liter. Since standards and criteria so frequently ignore it, at the risk of appearing overly elementary it should be emphasized that a concentration consists of two parts, an absolute amount of the constituent material and an absolute amount of water in which the constituent is mixed. Since we deal with a dynamic condition in the stream, the amount of water is expressed as amount in a unit of time or the streamflow in cubic feet per second or million gallons per day. If a given amount of waste matter per day is discharged into the stream at a steady flow rate, assuming complete mixing, a definite *concentration* results. If the streamflow is reduced by one-half, the concentration of the waste doubles. Thus any stream standard that specifies a concentration of constituent matter and does not also specify a *streamflow probability* is meaningless. For example, frequently a dissolved oxygen standard of 5 ppm is specified without any designation under what streamflow condition this standard is to apply. Obviously when streamflow is great in a large river, the standard may be met easily, even if the organic waste load is great; but in a small stream, and especially during a severe drought, it may be very difficult to meet the standard even with a very small organic load. We then must ask, at what probability of drought severity is it expected to maintain 5-ppm level of dissolved oxygen—the severity expected once in 100 years, the most severe recorded, a severity in the dry season expected to occur on the average year, or a severity expected once in 5 or once in 10 years? The drought flow to protect against is as important as the stipulation of a concentration; the two must go together. Water quality levels must be associated with a streamflow.

Stream versus Effluent Standards

Since interest is in the resultant quality of the receiving stream, it is logical that standards should relate to the stream and only incidentally to the effluent discharged to the stream. There has been a tendency for simplicity of administration to specify effluent standards without consideration of the quality of the receiving stream. Even if detailed analyses are made of resultant stream conditions expected under the range of waste effluent loading and streamflow, it is highly desirable to retain stream standards and criteria rather than effluent standards. Effluent criteria should be reserved as administrative tools of management, not stipulated as the primary standard. It is important to keep the focus on the quality of the stream, lest the overall objectives of pollution control be dimmed or lost sight of. There is no substitute for good objectives, criteria, and standards related to the quality and quantity of the streamflow.

General Considerations

In the development of stream objectives, criteria for judgments, and specific standards a number of general considerations are relevant. To gain acceptance and responsible action, above all, the developments of these stream requirements must be rational—that is, it is expected that they will reflect careful thought, extensive analysis and research, and responsible action. The relation between water quality criteria and specific stream standards for a particular situation is readily understood and accepted only if they are reasonable, measurable, and attainable.

The principle of *reasonable use* is at the root of wise conservation and development of water resources. As we have already stated, nonuse is total waste, and appropriation for a single use to the detriment of other uses is abuse.

Quality standards must be measurable in quantitative rather than qualitative terms to be administratively enforceable. If standards are set beyond available analytical methods of measurement or at levels beyond the accuracy or precision of such methods, it is not likely that they will be respected or enforced.

If quality requirements can be attainable by proven processes, at feasible costs, and with reasonable applied research and development, there should be no delay in acceptance. However, progress should not be delayed because of lack of technical perfection. What is reasonably known and understood should be practically applied.

Quality in Relation to Intended Use

Water quantity and quality requirements of a stream take on different meanings, depending on the intended use at a particular site along its course. We have identified seven basic uses of the stream waters:

1. Community and industrial water supply.
2. Electric power generation (hydro, fossil fuel, and atomic fuel).
3. Recreation (bathing, boating, and water sports).

4. Irrigation.
5. Navigation.
6. Fish, shellfish, and wildlife.
7. Ultimate disposal of wastewater effluents.

Within any one of these there are further distinctions. Requirements for community potable water supply are concerned primarily with hygienic quality and limitations on harmful or esthetically objectionable substances. Supply for some industrial purposes may require chemical limitations that are more restrictive than those for potable purposes but may tolerate bacterial pollution. Hydroelectric power has primary concern with quantity and little concern with quality; requirements for boiler feed may be high with respect to corrosion and scale, but condenser cooling requirements may be less stringent. For fish dissolved oxygen is more important than bacterial contamination; conversely, for shellfish the bacterial factor is highest, even exceeding that for bathing and water contact sports. And so it is that each special use has its specific quality objectives and standards. Frequently it is necessary to accept less than the ideal objective.

Since water is a multiple use resource, location of the site or point of use in relation to other uses of the stream is as important as the quality requirements for a specific intended use. Rivers as well as reaches of a given stream vary as to natural suitability for different purposes. Man, to some extent, can modify and enhance suitability for some purposes, but it is naive ever to expect that all reaches and all rivers can be made suitable for all uses. Yet in the development of criteria and standards such irrational concepts are too frequently apparent.

Classification and Zoning

Before objectives are set and criteria and standards are adopted, decisions concerning intended uses and allocation of these uses along the course of the stream must be made. Then, and only then, can rational requirements be prescribed. This process may be referred to as classification and zoning.

It has been repeatedly stressed that waste assimilation capacity is a dynamic natural resource and varies in time and place from river to river and reach to reach. Thus some rivers and some locations are more suitable for ultimate disposal of residual waste burdens than others. Furthermore it has been repeatedly shown that, although the disposal of residual wastes may adversely affect other water uses, other water uses may also have a detrimental or beneficial impact on the waste disposal potential.

The objective of stream classification and area zoning is to take advantage of the natural potentials along the course of the river system and to avoid misplacing uses unsuited to certain reaches or incompatible with other uses. For example, headwater streams are naturally superior as source of potable water supply and sport fishing. Since urban areas have uncontrollable sources of waste, the reaches immediately below cities are certainly not ideal for location of water supply intakes or for bathing areas. Areas downstream are more appropriate for location of the waste effluent outfalls and for large industrial developments. This does not imply relegating the downstream reaches to classification barren of aquatic life or to open sewers. In the decentralization of industry in the urban-industrial complex it is equally unsuitable to allow such expansion along the small headwater tributaries where the waste treatment requirement is excessively high and where the waste assimilation capacity cannot handle even highly treated effluents. Large industries belong on large rivers, where streamflow and the associated assimilation capacity are at their maximum.

One of the constraints in stream classification and area zoning is the inheritance of existing mislocations resulting from lack of forethought and effective direction of growth. The existing uses must be recognized, but if recognized as mislocations they can be dealt with more effectively than if ignored. At least new growth can be better controlled until obsolescence removes some of the earlier blemishes.

Classification of zones of the river system and related land use zoning warrant careful analysis in evaluating the consequences of any plan. As part of this process the rational method of stream analysis as developed in this text serves as a powerful tool. Although classification and zoning provide patterns for growth, the processes are not frozen and should be continuing in character, with opportunity to evolve changes as new situations demand. The system, however, should be more than a weather vane subject to every minor demand for change when pressure from special interests is exerted.

In a democratic society all interests should have an opportunity to be heard. There are many voices, some with fixed prejudices, others entirely self-motivated, but some with the vision and truly concerned for the public long-term interest. Rivers belong to the people; given the facts and intelligently informed as to the consequences and costs of alternatives, the public usually makes the right decision. In this process the public is entitled to impartial guidance of the best qualified and experienced, scientific and engineering intelligence available.

When rational balance among uses is crystallized into zones of classification, it is then feasible to decide on objectives for each and to establish criteria and specific administrative standards for quantity and quality of the stream as it passes through the different zones.

Specifications and Tolerances

In the field of applied engineering *specifications* usually set up a desired result and require compliance within certain limits, known as *tolerance*. Specifications for a part of a highly delicate mechanism may call for an exact dimension, plus or minus a very small variation. In a less precise mechanism the same dimension may accept a greater tolerance, because it is more difficult and expensive to secure a more precise fit. The tolerance is determined by the requirements of the job weighed against the cost of additional refinement. The shaft of a wheelbarrow does not need the refined tolerance of the shaft of a delicate scientific apparatus.

The concept of engineering specifications and tolerance is applicable to standards in general and has particular relevance to water quality. It suggests physiological and esthetic limitations, the amount of impurity or the extent of departure from the optimum that will be tolerated by the individual without impairment of health or offense to the senses. It also suggests the amount of impurity or the extent of departure from the optimum that will be tolerated by various forms of aquatic life of the stream or tolerated in association with other beneficial uses of water.

The Principle of Expediency

This concept also recognizes an important principle that tolerances are a concession to expediency, that the optimal or ideal condition seldom is obtainable in practice, and that it is wasteful and therefore inexpedient to require nearer approach to it than is readily obtainable under current engineering practices and at justifiable costs. The history of the evolution of drinking water standards is a good illustration.

The specification *no evidence of sewage pollution,* an ideal condition, can seldom be met by employment of even the best current practice. In 1900 the installation of slow sand filters was followed by an immediate and spectacular drop in deaths from typhoid fever. On the basis of filter performance and epidemiological evidence, the standard of acceptable performance adopted at that time was *not over 50 percent of 1-ml samples examined to be positive for* B. coli, *the characteristic sewage organism.* This was equivalent to a permissible concentration of 69 organisms per 100 ml.

In 1910, when the U.S. Public Health Service adopted its first standard, it was entirely feasible to require not more than 2 bacteria of the coliform type per 100 ml of water. This was feasible primarily because in the interim chlorination had been introduced, a process so simple, so economical, and so effective that without operational or financial hardship the more stringent specification could readily be met.

Revisions were made in the Public Health Service Standard in 1925, 1942, 1946, and 1962, upgrading, adding, and clarifying requirements. The current standard sections 3.21 and 3.22 [7] specify the following:

> 3.21 When 10-ml standard portions are examined, not more than 10 percent in any month shall show the presence of the coliform group. The presence of the coliform group in three or more 10-ml portions of a standard sample shall not be allowable if this occurs:
>
> (a) In two consecutive samples;
> (b) In more than one sample per month when less than 20 are examined per month; or
> (c) In more than 5 percent of the samples when 20 or more are examined per month.
>
> When organisms of the coliform group occur in 3 or more of the 10-ml portions of a single standard sample, daily samples from the same sampling point shall be collected promptly and examined until the results obtained from at least two consecutive samples show the water to be of satisfactory quality.
>
> 3.22 When 100-ml standard portions are examined, not more than 60 percent in any month shall show the presence of the coliform group. The presence of the coliform group in all 5 of the 100-ml portions of a standard sample shall not be allowable if this occurs:
>
> (a) In two consecutive samples;
> (b) In more than one sample per month when less than five are examined per month; or
> (c) In more than 20 percent of the samples when five or more are examined per month.
>
> When organisms of the coliform group occur in all five of the 100-ml portions of a single standard sample, daily samples from the same sampling point shall be collected promptly and examined until the results obtained from at least two consecutive samples show the water to be of satisfactory quality.

It will be noted that two statistical criteria are applied: a *mean* concentration for the month based on the percentage of portions positive in the month, or an MPN of one per 100 ml (Appendix B); and a *tolerance* in deviation of a limited number of samples taken during the month as set forth in (a), (b), and (c). For a sample of five portions of 10 ml with three positive and two negative, the tolerance is an MPN of 9.2 per 100 ml; for four positive and one negative, an MPN of 16.1 per 100 ml; with all positive we do not know, since theoretically the sample could contain an infinite number, but a very restricted number of such samples are tolerated.

As an alternative in testing procedure section 3.22 of the standard prescribes use of 100-ml (instead of 10-ml) portions of each sample. In terms of this procedure 60 percent of the 100-ml portions in a month positive are also equivalent to a density of one coliform organism per 100 ml, but with a greater statistical reliability. Again a tolerance is given as in (a), (b), and (c).

Thus since 1900 the drinking water standard evolved from 69 to 2, to 1 coliform organism per 100 ml. There is little, if any, evidence that the use of water within the range of two to one coliform organisms per 100 ml has ever caused disease. The adoption of the more stringent standard was justified on two principles, the principle of *expediency* and the principle of *relative safety*. The second principle warrants further clarification.

The Principle of Relative Safety

A principal aim of standards is to provide protection, to establish a safe environment. Yet safety itself is merely relative, because complete safety is an ideal that is beyond practical attainment. Protective devices yield always a percentage improvement; successive applications merely multiply percentages of overall benefits, with always some residual risk. Referring again to drinking water standards, appreciation of the basic fact that safety is relative and that water purification yields only relative improvement justifies adoption of a more stringent standard that has become feasible, and thus expedient, with the practice of chlorination. By the same reasoning, we recognize that the higher standard provides a *safer* water but not necessarily a *safe* water.

Also by the same reasoning, there is justification for measures to eliminate hazards that, judged on the basis of general knowledge and experience, may, under certain contingencies, produce an epidemic or other injury.

Both situations, the one dealing with a definite numerical reduction of an existing and measurable danger and the other dealing with a hazard (a danger from a situation not yet developed), yield a probability as a measure of relative safety.

Thus in establishing standards it is attempted to reduce the numerical measure of probable harm, or the logical measure of existing hazard, to the lowest level that is practicable and feasible within the limitations of financial resources and engineering skill. Progressive improvement is increasingly more expensive, and a level is reached in every case where it is either impracticable from an engineering point of view or excessively expensive to proceed.

MANAGEMENT CONSIDERATIONS

Administrative Standards

From the discussion at the beginning of this chapter it would be logical to conclude that objectives and criteria are the important elements and that standards, as defined, are not necessary. This is true from a technical standpoint, because the criteria are the basis for formulating a judgment and as such should be the last step applied in each specific case as to quality for intended use. As more sophisticated management of water resources and waste disposal comes into common use, rigid standards may give way to the more rational criteria for administrative control. In the meantime it is well to keep in mind that standards represent, in general, imaginary lines between good and bad; employed for the enactment and enforcement of rules and regulations, but nonetheless inadequate expressions of real facts. For these reasons it is preferable to refer to them as *administrative standards.*

In many fields of regulation and enforcement the simplicity of administrative standards is found essential by virtue of the magnitude of the task for the limited staff available. Administrative standards must be sufficiently precise for legal enforcement. However, in establishing and administering such arbitrary lines it is essential that objectives and criteria be carefully considered, because in the final ruling the courts invariably adjudicate case by case on the basis of what is reasonable.

Legal Definition of Pollution

Gradually over the years laws have been enacted by the states sharpening the definition of what constitutes pollution in order to strengthen administrative control and, if necessary, legal enforcement through the courts. Most early laws were based primarily on protection of public health, aimed principally at municipal sewage, with the legal interpretation that injury in fact has occurred. Later, in many states protection of livestock and fish was included, aimed at both sewage and industrial wastes, but again legal interpretation placed the burden of proof on demonstrated injury. More recently the laws have been based on *public health, safety,* and *welfare,* and include protection for all beneficial uses of water. Furthermore most legislatures have eliminated the necessity to show that injury occurred in fact, by stipulating that it is unlawful to discharge any substance that is or may become injurious or detrimental. This greatly increases the power of regulatory agencies and opens the door to prevention.

Such progression in legal power is reflected in the State of Michigan Act 245 (1929), as amended by Act 117 (1963) and Act 405 (1965), which stipulates the following [8]:

> It shall be unlawful for any person directly or indirectly to discharge into the waters of the state any substance which is or may become injurious to the public health, safety or welfare; or which is or may become injurious to domestic, commercial, industrial, agricultural, recreational or other uses which are being or may be made of such waters; or which is or may become injurious to the value or utility of riparian lands; or which is or may become injurious to livestock, wild animals, birds, fish, aquatic life or plants or the growth or propagation thereof be prevented or injuriously affected; or whereby the value of fish and game is or may be destroyed or impaired.

With this growth in regulatory authority there also occurred dispersion among various state agencies which had a tendency to confuse and divide responsibility. More recently states in administrative reorganizations are correcting this defect and collecting the dispersed functions dealing with water resources and waste disposal into a single agency or at least into fewer agencies.

General Desiderata for Regulatory Agencies

Regulation of waste disposal is predominantly vested in state agencies, with a direct relation between the regulatory agency and the individual municipality or industrial establishment. As more sophisticated management becomes necessary for offensive lines as well as the defensive lines of control, new institutional arrangements will develop, whereby more of the detailed management responsibility now carried out at the state level will be delegated to river basin authorities to deal with problems on a unified drainage basin basis. Hence consideration is given to some of the more general requirements of regulatory practice.

Rules and Regulations

All regulatory agencies operate under rules and regulations in administering the authority and powers granted by law. Objectives of the lines of offense and defense and the criteria and standards are stipulated and explained. Rules of procedure must be reasonable, equitable, and consistent. What applies to one must equally apply to all. There may be some differences in approach between that for municipalities as public bodies (creatures of the state) and that for private industries. Both must receive equitable treatment under the rules.

This does not imply that all portions of the state be attacked at the same time. Problems vary in degree of severity, magnitude, and need. The most effective approach is to develop a program that takes drainage basins as units, giving priority to the basins with more pressing problems. Unless satisfactory waste disposal is provided for an entire drainage basin, there is danger that one area along a stream may make a heavy investment only to have it negated by an inadequate intervening reach. The unit basin approach also affords opportunity to undertake the necessary comprehensive stream analysis studies for sound solutions. In promotion of the unit basin approach an order of priority also affords an orderly method for communities and industries to make preparation for their turn in the program.

It is also essential that regulation be persistent over time. A steady pressure for enforcement is respected and brings about action much more effectively than a wavering policy or a sudden "clean-up campaign." The purpose is not to persecute, but to bring about continuous responsible action on the part of those regulated.

Plans and specifications supported by an engineering report are required for sewerage systems and treatment works before a permit is granted for construction and operation. The permit may stipulate conditions to be maintained and a time schedule for execution.

An effective administrative measure to bring about necessary construction of treatment works is to deny permits for any further extensions to the sewerage system until adequate treatment is provided.

When persuasion fails to induce action, it is necessary to serve notice to abate. This is followed by formal hearings at which all interested parties can present their arguments and the regulatory agency staff can present basic facts. A time schedule for action is then prescribed by the regulatory agency. If action is not taken, the agency has authority to institute legal action in which the court rules and prescribes a course of action backed by penalties for noncompliance. The individual communities and industries also have the right of court action if they feel the hearings were improperly held or any action of the regulatory agency was capricious or unreasonable.

In evaluating compliance, criteria and standards are established, preferably relative to the quality of the receiving waters rather than the effluent. It is often submitted as an argument against stream classification and zoning and stream standards that the regulatory agency does not have the budget or staff to make the comprehensive studies necessary to arrive at a rational basis. It is also argued that the time involved in making such studies and classifications unduly delays action. The temptation is then open to set arbitrary effluent standards, requiring wholesale a prescribed treatment of all wastes regardless of the need, of the magnitude of the waste load, or of the size and character of the receiving stream. The specious argument is often added that this regulation treats all sources alike and is therefore most equitable. Nothing is farther from the fact. It penalizes those who by forethought and study have located where their waste effluent, following reasonable treatment, can readily be handled by the waste assimilation capacity of the stream. At the same time it exercises favoritism to those who arbitrarily located without regard to suitability of the site, and by providing the wholesale treatment are relieved of any further responsibility, even if after the prescribed treatment the effluent load degrades the stream below its waste assimilation capacity. In the long run there is no substitute for criteria and standards tailored to the specific requirements of a stream.

Municipal Constraints and Incentives

Two primary constraints, financial and political, hinder progress in adoption of lines of offense and defense by municipalities. The burden of local taxation is heavy, and the urban areas face many demands on the tax dollar. Sewage treatment is not exactly a popular program, people usually prefer a school or an expenditure from which they see a more direct benefit. In the layman's mind the heavy capital and operating cost of wastewater treatment benefits downstream communities at his expense. It takes a great deal of selling to obtain voter support for a large bond issue that results in higher tax rates.

In most states municipalities face bonded debt limits and many municipalities are at or near that ceiling. In some states bonds for sewage treatment are exempt from the limit if the community is under order from the state regulatory agency to proceed. This is to an extent a relief, but it also tempts communities to delay until an order is issued.

These constraints are quite successfully circumvented by regarding wastewater treatment as a utility service to be financed by revenue bonds, which are wholly exempt from the debt limit. The maturity of the revenue bonds and operating cost are met by a service charge against users, generally as a percentage of the water bill. This is not entirely equitable, as the entire burden is on the connected user, and many large land holdings pay nothing, although their property values are enhanced by the potential access to service. In part, attempts are made to meet this objection by placing a portion of the initial cost on general obligation bonds or on special assessment bonds. On the whole the revenue bond approach is quite successful and is a relief to the tax roll.

Political factors are often constraints on rational action, particularly where a unified approach to a whole drainage basin involves several separate political units. The separate political units are reluctant to give up their autonomy even if it means a saving to the taxpayer or the user. Occasionally joint action among political units is prevented by fear of loss of patronage. These constraints are overcome only by public education in the merits of unified action, and in this the social and political sciences can make an effective contribution.

As incentives to municipalities to build wastewater treatment works grants in aid and low interest loans of various kinds are offered by the federal government, and increasingly by the states. The larger units of government have a legitimate stake in the enhancement of the quantity and quality of stream water and in satisfactory waste disposal; the entire financial burden should not fall on the local community.

These incentives are not without some drawbacks and inequalities. The uncertainty of continuity of grant funds may act as a delaying force on communities that are ready to proceed but reluctant in anticipation of a grant. Communities that have, in good faith, proceeded on their own financing are penalized, as the grant mechanism is not retroactive and supports only new construction.

The grants are restricted to public bodies and are not available to private industry. Various proposals are under consideration from time to time to offer industry incentives in the form of a tax writeoff or accelerated depreciation on investments for treatment facilities. Some industries located within reach have integrated with the municipal system under contractual arrangements and thus received an indirect benefit from the public grant. Where technically feasible, such integrated systems are usually superior to proliferation of many smaller independent facilities.

Industrial Constraints

The two primary constraints that inhibit voluntary industrial waste treatment action are *technical* and *economic*. Municipal wastewater in volume and characteristics is relatively uniform among cities and can be reasonably projected on the basis of community composition—residential, commercial, and industrial. This does not generally hold for industrial wastewaters, as volume and composition vary greatly, depending on the products that are manufactured and the processes that are employed. Accordingly the task of establishing stream quantity and quality requirements, and the degree and type of treatment necessitate extensive investigation. These technical uncertainties are a constraint on both regulatory agency and industry action.

In an effort to meet these technical constraints realistically, a two-step approach to industrial waste disposal (such as developed by the Ohio River Valley Water Sanitation Commission [9]) can be applied, a minimum basic initial requirement, followed by a supplementary more detailed specification. The intent of the basic minimum requirements is protection of the stream from gross and obvious pollution, such as settleable matter subject to sludge deposit, unsightly scum and floating matter, excessive color and odor, and freedom from concentrations of substances toxic to man, animals, and aquatic life. These minimum requirements can generally be applied without detailed study or delay; the supplementary requirements then can be developed by research and detailed analysis of each specific problem and applied as practical solutions become available. However, whenever technically feasible, the initial approach should be based on a rational design of lines of offense and defense since this avoids the risks of misunderstanding and delay in fulfillment of the full requirement.

The major industrial constraint is economic. Treatment of waste, as we have seen, is costly and for industry constitutes either a drain on profit or an addition to price. If the cost is excessive, it cannot be passed on to the consumer in increased price without affecting the industry's competitive position. As costs of treatment increase the economics of waste disposal constitute a significant constraint on industrial and community growth.

Surveillance

Good management implies good surveillance. Investment in the best of facilities can be lost by poor operation. This applies equally to offensive and defensive lines. There are many aspects of good management surveillance, but three primary phases warrant emphasis—namely, operation and maintenance of facilities, stream monitoring and survey, and continuous analysis and promotion.

Operation and Maintenance

The operation of wastewater treatment works requires the skilled supervision of competent technical staff. Most states require treatment works to be operated by licensed personnel. Licenses are granted on the basis of education, experience in operation, and a written examination, with gradations based on the size and complexity of the treatment facilities. Some states have voluntary licensing through technical organizations, but licensing under law is believed to be more effective. In addition to ensuring skilled operation, the licensing procedure is a protection to supervisors and operators in appointment and tenure of employment.

Wastewater is subject to variation in volume and characteristics, and to ensure consistent efficiency treatment requires constant supervision throughout the 24 hours of the day. In addition to mechanical manipulation (manual or automatic) the basic control is directed by laboratory tests appropriate for the various principles of treatment involved, necessitating a well equipped laboratory and competent staff. The state regulatory agencies in surveillance require routine operating reports of key laboratory test results for various stages in processes as well as the final effluent and stream sampling. The regulatory agency also provides staff advisory service to local operators and carries out periodic inspection and testing.

An excellent test of normal operation is afforded in probability plots of test results (Appendix A). This provides a ready visual identification of abnormal results and offers a clue to changes to improved operation. Some abnormal deviations are associated with weather conditions, principally rainfall. If the community is served by combined sewers, stormwash may induce radical fluctuations in operation; if it is served by a separate sanitary sewerage system, the impact of rainstorms is less severe, but, since complete separation is impractical, some upset in operation is inevitable.

Where the community system includes industrial wastes, the possibility of operating difficulties is increased. Exclusion of industries from access to the community system is no solution, since usually the wastes can be more efficiently handled at the central system than at numerous small industrial waste treatment works. For some wastes, such as products toxic to biological treatment, it may be necessary to pretreat before discharge to the community system.

In the more sophisticated management involving offensive lines to enhance streamflow and stream quality by low flow augmentation, an elaborate system of surveillance of operation of reservoirs, flow retardation, and release works requires constant supervision. The streamflow regulation entails a net of hydrologic and hydraulic recorders to guide and coordinate operation.

Stream Monitoring and Survey

In the management of water quality of streams a system of watercourse monitoring may involve a net of automatic continuous sampling stations, such as discussed in Chapter 10 or may depend on routine periodic stream sampling performed by supervisors of treatment works and by industries. As abnormalities are detected or as proposals are under consideration for new industries or urban expansion, intensive stream surveys are undertaken to permit reevaluations by stream analysis methods.

Continuous Analysis and Promotion

Stream monitoring and watercourse sampling require analysis and interpretation in evaluation of effectiveness of current lines of defense and offense, and in determination of alterations for improvement of efficiency. Surveillance also implies alertness to proposals for changes in river developments under consideration by other agencies, industries, utilities, and water users as to their effects on the quantity, quality, and waste assimilation capacity of the stream. Good management also takes the initiative in promotion of programs to remedy and prevent pollution and to enhance the quantity and quality of the streamflow. A positive approach to regulation and management in the form of guidance to industry and communities in location and direction of growth is a highly effective area of effort; efficient use of land and water resources is wholly dependent on ability to direct growth.

In these regulatory and managerial efforts the science and art of applied stream sanitation play significant roles, and the rational method of stream analysis provides a powerful tool. The degree to which we can predict the consequences of any form of human activity is a measure of our maturity. Much is known that can be applied in the wise use, development, and management of our water resources. We meet the future with prudence and forethought, and in our affluence it is within our reach to provide not merely an environment in which we can survive but a quality environment in which we can thrive.

REFERENCES

[1] Phelps, E. B., *Public Health Engineering,* Vol. 1, Wiley, New York, 1948.
[2] Velz, C. J., *Engineering News-Record,* October 14, 1948, p. 85.
[3] Logan, J. A., Hatfield, W. D., Russel, G. S., and Lynn, W. R., *J. Water Poll. Control Fed.,* **34,** 9, 860 (1962).
[4] USPHS Publication No. 1229, U.S. Government Printing Office, Washington, D.C., 1964.
[5] Smith, R., *J. Water Poll. Control Fed.,* **40,** 9, 1546 (1968).
[6] Weston, R. F., in *Proceedings of the National Symposium on Quality Standards for Natural Waters,* School of Public Health, University of Michigan, July 1966, p. 297.
[7] *Public Health Service Drinking Water Standards, 1962,* Public Health Service Publication No. 956, U.S. Government Printing Office, Washington, D.C., 1962, p. 5.
[8] State of Michigan Water Resources Commission, Act 245 (1929), as amended by Act 117 (1963) and Act 405 (1965).
[9] Cleary, E. J., *The ORSANCO Story,* Johns Hopkins Press, Baltimore, 1966, pp. 133–141.

Appendix A

STATISTICAL TOOLS

GRAPHICAL APPROACH TO STATISTICS

Statistical methods provide efficient means of quantitatively defining variations, which, as we have seen throughout this text, are constantly encountered in all phases of stream sanitation. In dealing with these methods emphasis is given to graphical approaches that are not ordinarily found in conventional texts. It is well to remember that any statistical method, regardless of how sophisticated, is not an end in itself but only a tool. Some of the more elaborate analytical methods, to the professional mind, can easily appear to be complicating frills rather than aids to judgment. The neat simplicity of the graphical approach avoids this complication and affords maximum opportunity to exercise professional judgment in the interpretation of data. The graphical methods are especially appropriate for complex applied fields, such as stream sanitation.

A brief review of elementary statistical concepts and probability provides a basis for development of the graphical methods.

STATISTICAL CONCEPTS

The principal concept of statistics is that of variation. If there were no variation in data, there would be no need for statistical methods. This concept of variation is in conflict with most of the mathematics we have been taught. In mathematics we usually start with a preconceived theory, which, if followed, automatically must lead to a single exact answer without any deviation. Such inflexible finality is not consistent with the facts of life as we observe them. Even with the most precise methods of measurement there is always variation.

In statistical reasoning we start with observations and with full acceptance of their inevitable variation, which leads not to a single answer, but rather a range of answers. For example, suppose we desire to know the weight of an object. A single measurement may be 10.1 mg, a second made under the same circumstances may be 9.8 mg, and, similarly, repeated measurements will show other variations. The question arises, Which of these measurements is correct? or, What is the true weight? The true weight remains unknown and at best can be only estimated. The best estimate is the average of the repeated measurements, in this case, say 10.0 mg. But confidence in any estimate can be had only in defining not a single value but a *range* within which we believe the true weight lies.

From a statistical analysis of the variation of the measurements confidence and range can be related. We are confident of enclosing the true value 68 times in 100 within the range 10.0 ± 0.1; or we are confident of enclosing the true value 997 times in 1000 within the range 10.0 ± 0.3. In the second instance our confidence is high; only 3 times out of 1000 would we be wrong in believing that the true value was within this wider range. But we have gained this added confidence only at the expense of precision; we are less definite as to the true weight in the sense of a single value.

In fact under this new concept we can never determine with finality the true weight, because, as we narrow the range working toward a single value, we lose confidence in enclosing it, with total loss of confidence if an attempt is made at expression as a single value. Thus in expressing measurements we always state a *range;* the mean plus or minus some increment of standard deviation, the increment depending on the degree of confidence desired.

The Nature of Variation

If a series of measurements, say weight, made under the same conditions on the same object are sorted in order of magnitude, it will be found that the results vary in a quite systematic manner. Many of the results will cluster near the midvalue, and a few will be more or less evenly distributed, with greater spread above and below the central majority. Another type of quantitative data consisting of a sample from a large population, universe, or supply—such as a single weight measurement of a number of such objects selected at random—will likewise show this tendency toward central cluster and symmetrical spread.

If a large number of such measurements are made and plotted on the scale of measurements, each individually represented by a circle, a symmetrical bell-shaped pile develops, as shown in Figure A–1. The

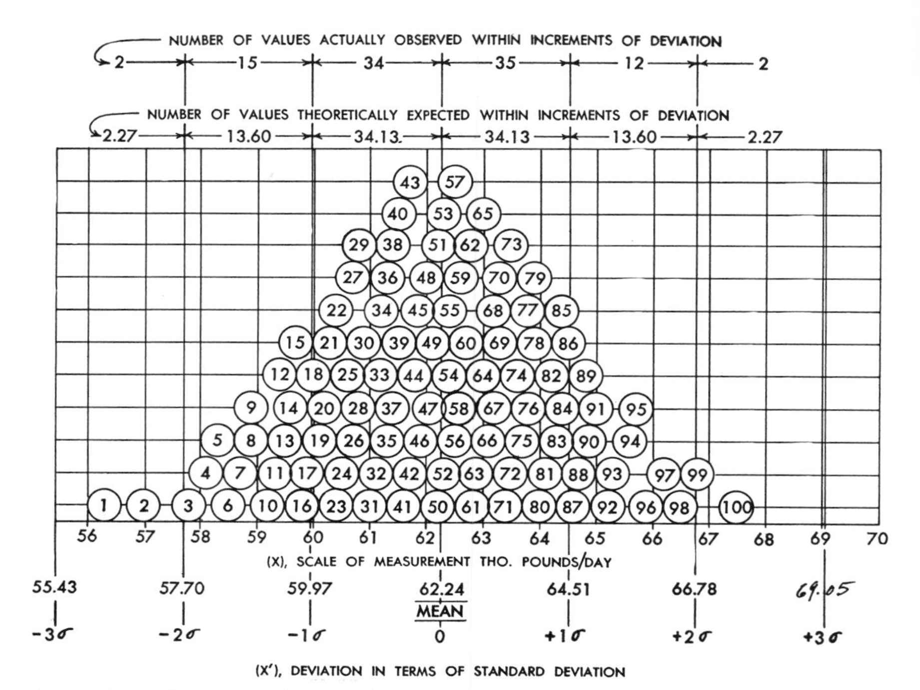

Figure A–1 Demonstration of the normal probability curve. Distribution of 100 suspended solids determinations; mean 62.24×10^3 lb/day; standard deviation 2.27×10^3 lb/day.

center of each circle is vertically above the magnitude of its measurement on the X scale; the number in each circle is the rank of each measurement arranged in ascending order of magnitude. The measurements cluster so closely about the mean that it is necessary to pile up the circles in order to locate them opposite their position on the X-scale, thus forming the bell-shaped curve of distribution. This frequency curve of distribution of measurements is referred to as the *normal probability curve.* The nature of this normal distribution is indicated by the proportion of the measurements found in equal increments of scale shown across the top of the pile. A more exact definition of the normal probability curve will be developed later.

Suffice it to observe that the distribution is nicely symmetrical about a centering value; and deviations above and below the centering value spread in a very orderly fashion. Other random samples taken from the same universe under the same conditions will reproduce quite similar distributions both as to centering position on the scale and as to deviation, or spread. Measurements of different phenomena will develop similar bell-shaped distributions, but it will be observed that each has its own centering value and some vary in the degree of deviation, or spread about the central value.

Centering Values

Three centering values are commonly employed: the *median*, the *mode*, and the *mean*. The median value is the value exceeded by one-half of the values and of which one-half of the values fall short. It represents the series, if at all, by virtue of its position of dividing the series in half. The mode is the value in the series most frequently repeated and around which other values cluster most densely. It is the high point in the distribution, irrespective of the number of terms above and below it. The mean is the arithmetic average of the series of data (the sum of the individual measurements divided by the number of measurements). Where the distribution is *normal* (symmetrical), the median, the mode, and the mean are identical and coincide. The mean is the most useful centering value and is most frequently employed.

However, the average by itself is not adequate. Too frequently series of data are summarized merely by an "average" (the mean), without designating the number of measurements or the nature of variation or spread in the data. For example, consider the following three series of data, X, Y, and Z, which have the same average:

1. X composed of two values, 1 and 99; average = 50.
2. Y composed of 19 values, ranging from 10 to 90; average = 50.
3. Z composed of 19 values, ranging from 47 to 53; average = 50.

Although all produce the same average (50), it is not difficult to see that the three averages come from quite different data. In the first instance two measurements hardly give meaning to an average, particularly when they have a great spread; in the second and third instances, although both have the same number of measurements (19) and the same average (50), the second has a wide spread, ranging from 10 to 90, whereas in the third the data are compact, with a spread ranging from 47 to 53.

Deviation from the Central Value

In addition to the mean and the number of measurements it is necessary to know something of the nature of deviation of measurements from the central value.

The *standard deviation*, σ (sigma) is the fundamental measure of variation and constitutes a major "building block" in many statistical procedures. It is the root-mean-square deviation obtained by squaring each individual deviation from the mean, summing, dividing by the number of measurements, and extracting the square root of this mean square.

$$\sigma = \sqrt{\left[\sum_{1}^{n} (X - \bar{X})^2/n\right]}. \tag{A-1}$$

(There are various other transformations of this equation that are more useful if computations are made by machine; also a simple graphical method of obtaining the standard deviation is presented later.)

The standard deviation is important because through it the fundamental characteristics of the normal probability curve are defined and precision can be related to probability or a confidence level.

PROBABILITY

There are two concepts of probability, one based on what *can* occur and the other based on what *has* occurred. If an event can occur or has occurred in a ways and can fail or has failed in b ways, then the probability of a success is $a/(a+b)$, which is a ratio of the number of ways in which an event can occur or has occurred to the total number of ways of occurrence, each being equally likely to occur. Probability is expressed as a ratio and ranges from zero to unity, zero designating impossibility and unity designating certainty.

The first concept, based on what *can* occur, is referred to as *a priori* probability. In tossing a coin there are two ways in which an event can occur, either a head or a tail. The probability of a specified event, head, is therefore .5. There are six ways in which a die can fall. The probability of the event, an ace, is therefore $\frac{1}{6}$. These are probabilities before trial, based purely on what can occur.

The second concept, based on what *has* occurred, is referred to as *a posteriori* probability—probability after the trial. A coin is tossed 1000 times, and the event, a head, has been observed to occur 506 times. The *a posteriori* probability of a head, based on what has occurred, is 506/1000, or .506, as compared with *a priori* probability based on what can occur of .5.

The *a priori* or true probability of a universe can exist, but it cannot be measured. Repeated random samples drawn from a universe show results varying about the true value, results from the small samples showing greater variations than those from the larger samples; or, as the number of trials approaches infinity, the *a posteriori* probability determined from observations approaches the true *a priori* probability.

It is customary to express probability in numeric terms, from which it follows:

1. The sum of the probabilities of occurrence of a complete and mutually exclusive set of events is unity. Impossibility has a probability of exactly zero. The entire range of probabilities extends from impossibility to certainty, or numerically from 0 to 1.

2. If the probability of success is p and of failure q, then $p + q = 1$; and $1 - p = q$; and $1 - q = p$.

3. The probability of either one or another of mutually exclusive events occurring is the sum of their individual probabilities. The probability of drawing a spade or a heart is

$$\frac{13}{52} + \frac{13}{52} = .5;$$

of drawing a spade, a heart, or a club:

$$\frac{13}{52} + \frac{13}{52} + \frac{13}{52} = .75.$$

4. The probability of occurrence of a compound event composed of two or more independent events is the product of the individual probabilities. What is the probability of two heads in two throws of a coin? (One head on the first throw and one head on the second throw.)

Probability of head on the first throw = .5.

Probability of head on the second throw = .5.

Probability of two heads on two consecutive throws $.5 \times .5 = .25$.

The Normal Probability Curve

The normal probability curve is a standardized form of distribution where the scale of measurement is in terms of the standard deviation measured from the center of the distribution as the origin. For example, referring to Figure A–1, the X-scale is weight in thousand pounds per day, and the corresponding X'-scale is in terms of the standard deviation. The zero of the X'-scale is opposite the center of the distribution, the mean 62.24×10^3 lb/day. Unit distance on the X'-scale is equivalent to a standard deviation, in this instance 2.27×10^3 lb/day, measured along the X-scale but labeled $+1\sigma$, $+\sigma 2$, $+\sigma 3$ and $-\sigma 1$, $-\sigma 2$, $-\sigma 3$, on the X'-scale. To convert a series of scaled measurements X to the standard probability scale X' it is necessary to know the mean $\bar{X}$ and the standard deviation σ. The mean $\bar{X}$ is the arithmetic average, and the standard deviation can be computed from the data or, as will be illustrated later, both can be obtained readily graphically.

The significance of the standard deviation σ as a measure of variation is apparent from the following analysis of the distribution of the 100 measurements on Figure A–1. Extending two vertical lines at $+1\sigma$ and -1σ on either side of the mean, we count 69 of the total 100 measurements falling within this range; 31 fall outside it, 14 above and 17 below. The probability of measurements falling within this range based on what has occurred in 69/100, or 69 percent; and of falling outside it, 31/100, or 31 percent. Similarly we count 96 measurements within the range of the mean $\pm 2\sigma$, and 4 measurements outside it, 2 above and 2 below. The probability of measurements falling within is 96/100, or 96 percent, and of falling outside, 4/100, or 4 percent.

The Integration Graph

The results discussed above are based on a sample of 100 measurements. As the number of measurements is increased, the *a posteriori* distribution approaches the *a priori* distribution defined by the normal probability curve. The *a priori* probability of measurements falling within any range is given by the area under the curve between the limits of the range defined in terms of standard deviations from the mean. These areas have been computed for any increment of deviation and are available in tables.

One useful form is the integration graph shown in Figure A–2. The areas are given between the central value (the mean) and x/σ (deviation in terms of standard deviation) measured in only one direction from the mean.

Characteristics of the Normal Probability Curve

The normal probability curve has the following primary characteristics:

1. All events are distributed symmetrically about the mean or the most probable (MP) value.

2. The probability integral for the complete distribution of events is the area between $\pm\infty$, which is unity. The probability of all events is 1.0.

3. The probability integral for all events between the limits of $\pm 1\sigma$ ($\text{dev}/\sigma = 1$) from the most probable value is the area *abcd,* which from the integration graph is $2 \times .3413 = .6826$. Thus 68.26 percent of all events lie within a range of $\pm 1\sigma$ from the most probable value, and 31.74 percent of the events lie outside this range.

4. The probability integral for all events between the limits of $\pm 2\sigma$ ($\text{dev}/\sigma = 2$) from the most probable value in the area *efgh,* which from

x/σ	Area	Scale
0.50	1915	1900
.49	1879	
.48	1844	1850
.47	1808	1800
.46	1772	
.45	1736	1750
.44	1700	1700
.43	1664	1650
.42	1628	
.41	1591	1600
0.40	1554	1550
.39	1517	1500
.38	1480	
.37	1443	1450
.36	1406	1400
.35	1368	1350
.34	1331	1300
.33	1293	
.32	1255	1250
.31	1217	1200
0.30	1179	1150
.29	1141	
.28	1103	1100
.27	1064	1050
.26	1026	1000
.25	0987	
.24	0948	0950
.23	0910	0900
.22	0871	0850
.21	0832	0800
0.20	0793	
.19	0753	0750
.18	0714	0700
.17	0675	0650
.16	0636	
.15	0596	0600
.14	0557	0550
.13	0517	0500
.12	0478	0450
.11	0438	
0.10	0398	0400
.09	0359	0350
.08	0319	0300
.07	0279	
.06	0239	0250
.05	0199	0200
.04	0160	0150
.03	0120	0100
.02	0080	
.01	0040	0050
0.00	0000	0000

x/σ	Area	Scale
1.00	3413	3400
.99	3389	
.98	3365	3350
.97	3340	
.96	3315	3300
.95	3289	
.94	3264	3250
.93	3238	
.92	3212	3200
.91	3186	
0.90	3159	3150
.89	3133	
.88	3106	3100
.87	3078	
.86	3051	3050
.85	3023	
.84	2995	3000
.83	2967	
.82	2939	2950
.81	2910	2900
0.80	2881	
.79	2852	2850
.78	2823	
.77	2794	2800
.76	2764	
.75	2734	2750
.74	2704	2700
.73	2673	
.72	2642	2650
.71	2612	2600
0.70	2580	
.69	2549	2550
.68	2518	
.67	2486	2500
.66	2454	2450
.65	2422	
.64	2389	2400
.63	2357	2350
.62	2324	
.61	2291	2300
0.60	2257	2250
.59	2224	
.58	2190	2200
.57	2157	2150
.56	2123	
.55	2088	2100
.54	2054	2050
.53	2019	
.52	1985	2000
.51	1950	1950
0.50	1915	

x/σ	Area	Scale
1.50	4332	
.49	4319	
.48	4306	4300
.47	4292	
.46	4279	
.45	4265	
.44	4251	4250
.43	4236	
.42	4222	
.41	4207	4200
1.40	4192	
.39	4177	
.38	4162	
.37	4147	4150
.36	4131	
.35	4115	
.34	4099	4100
.33	4082	
.32	4066	
.31	4049	4050
1.30	4032	
.29	4015	
.28	3997	4000
.27	3980	
.26	3962	3950
.25	3944	
.24	3925	
.23	3907	3900
.22	3888	
.21	3869	
1.20	3849	3850
.19	3830	
.18	3810	3800
.17	3790	
.16	3770	
.15	3749	3750
.14	3729	
.13	3708	3700
.12	3686	
.11	3665	3650
1.10	3643	
.09	3621	
.08	3599	3600
.07	3577	
.06	3554	3550
.05	3531	
.04	3508	3500
.03	3485	
.02	3461	3450
.01	3438	
1.00	3413	

x/σ	Area	Scale
2.00	4773	
.99		
.98	4762	
.97		
.96	4750	4750
.95		
.94	4738	
.93		
.92	4726	
.91		
1.90	4713	
.89		
.88	4699	4700
.87		
.86	4686	
.85		
.84	4671	
.83		
.82	4656	
.81		4650
1.80	4641	
.79		
.78	4625	
.77		
.76	4608	
.75		4600
.74	4591	
.73		
.72	4573	
.71		
1.70	4554	4550
.69		
.68	4535	
.67		
.66	4515	
.65		
.64	4495	4500
.63		
.62	4474	
.61		
1.60	4452	4450
.59		
.58	4430	
.57		
.56	4406	4400
.55		
.54	4382	
.53		
.52	4357	
.51		4350
1.50	4332	

x/σ	Area	Scale
3.00	4986.5	
.98		
.96		4985
	4984	
.94		
.92		
2.90	4981	
.88		
.86	4979	
.84	4977	
.82	4976	4975
2.80	4974	
.78	4973	
.76	4971	
.74	4969	
.72	4967	
2.70	4965	
.68	4963	
.66	4961	
.64	4959	
.62	4956	
2.60	4953	
.58	4951	4950
.56	4948	
.54	4945	
.52	4941	
2.50	4938	
.48	4934	
.46	4931	
.44	4927	4925
.42	4922	
2.40	4918	
.38	4913	
.36	4909	
.34	4904	
.32	4898	4900
2.30	4893	
.28	4887	
.26	4881	
.24	4875	4875
.22	4868	
2.20	4861	
.18	4854	
.16	4846	4850
.14	4838	
.12	4830	
2.10	4821	4825
.08	4812	
.06	4803	4800
.04	4793	
.02	4783	
2.00	4773	

FIGURE A – 2

INTEGRATION GRAPH FOR NORMAL PROBABILITY CURVE

AREA BETWEEN MEAN AND x/σ

(TOTAL AREA UNDER CURVE 10,000 UNITS)

x/σ	Area
5.00	4999.99713
.50	4999.966
4.00	4999.683
.80	4999.277
.60	4998.409
.50	4997.674
.40	4996.631
.20	4993.129
3.00	4986.5

Figure A–2 Integration graph for normal probability curve. Area between mean and x/σ (total area under curve 10,000 units).

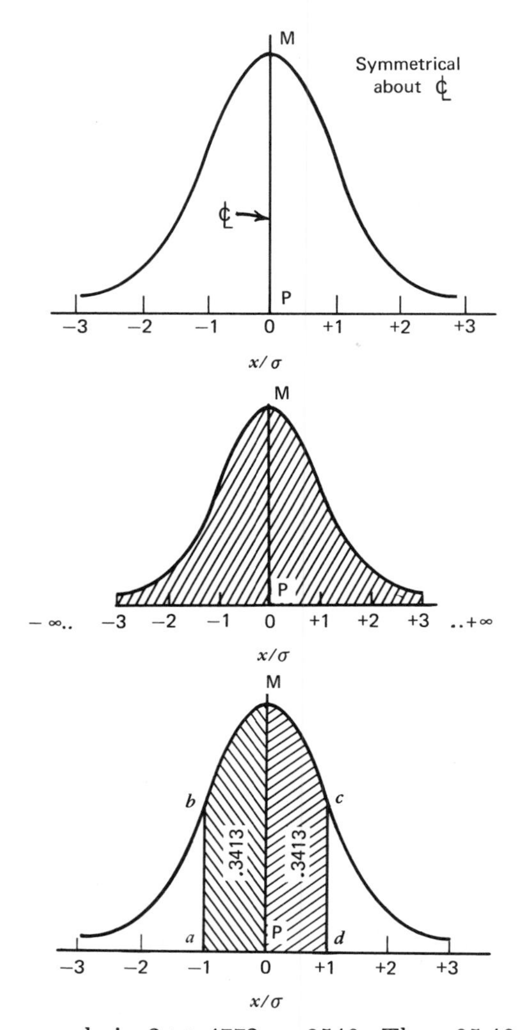

the integration graph is $2 \times .4773 = .9546$. Thus 95.46 percent of all events lie within a range of $\pm 2\sigma$ from the most probable, and 4.54 percent of events lie outside this range.

5. The probability integral for all events between the limits of $\pm 3\sigma$ ($\text{dev}/\sigma = 3$) from the most probable is the area *klmn,* which from the integration graph is $2 \times .49865 = .9973$. Thus 99.73 percent of all events lie within the range of $\pm 3\sigma$ from the most probable, and 0.27 percent of the events lie outside this range.

6. The probability integral for all events between the limits of $\pm 0.6745\sigma$ ($\text{dev}/\sigma = 0.6745$) from the most probable is the area *pqrs,*

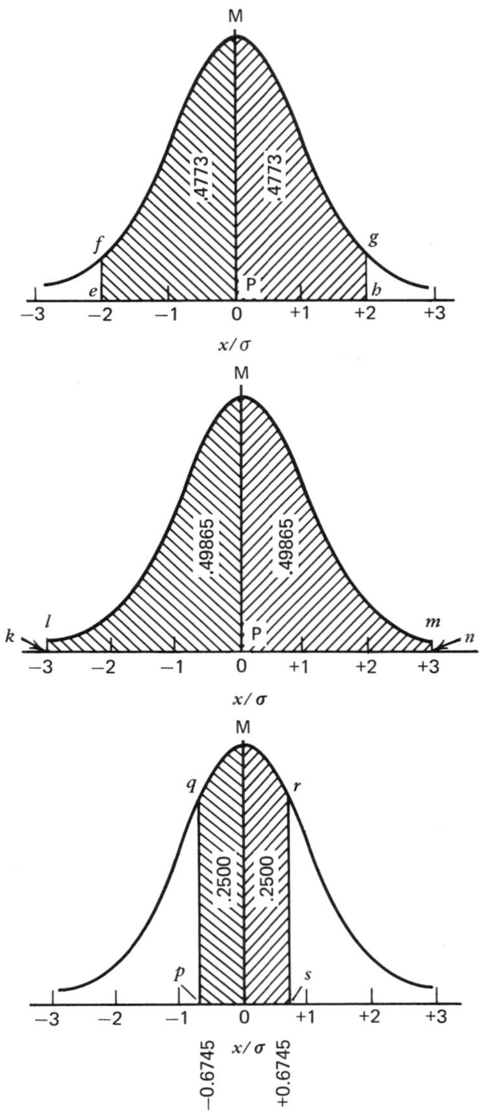

which from the integration graph is $2 \times .2500 = .5000$. Thus 50 percent of all events lie within the range of $\pm 0.6745\sigma$ from the most probable, and 50 percent of events lie outside this range. This is sometimes referred to as the range of probable error.

Suppose we wish to define the quality of effluent of a sedimentation tank in terms of suspended solids. The mean of 30 determinations is

102.0 ppm, but values vary about the mean with a standard deviation from the mean (σ) of 5.5 ppm. From what has been said heretofore, we must define quality of the effluent by a range, and we wish to know the confidence of enclosing the quality in that range.

For example, suppose we make the statement that the quality of effluent is within the range of 102.0 ± 10 ppm; what is our confidence that this range encloses the quality? or what risk do we take of having the quality measurement fall outside this range? This we determine from the integration graph:

Deviation from the mean $= x = 10$ ppm.
Deviation from the mean in terms of standard deviation,

$$\frac{x}{\sigma} = \frac{10}{5.5} = 1.82.$$

Opposite x/σ of 1.82 on the integration graph, Figure A–2, read area 46.56 percent or probability for the range, mean $\pm 1.82\sigma$, is 2×46.56 percent, or 93.12 percent.

We are about 93 percent confident of enclosing the quality of the effluent within the range of 102.0 ± 10 ppm. In accepting this range as defining the quality we run the risk of being wrong about 7 times in 100, as 7 times in 100, deviations greater than ± 10 ppm are expected.

Again suppose we do not wish to be wrong in our range of quality more often than once in 100 times; then in 99 times in 100 we must be right. This requires an area in one direction from the mean of 99/2, or 49.5 percent. Opposite area of 49.5 percent on the integration graph, Figure A–2, read x/σ of 2.567. The range then is defined by the mean $\pm 2.567\sigma$, or $102.0 \pm 2.567 \times 5.5$, 102.0 ± 14.1 ppm. In other words, in saying that the effluent quality is 102.0 ± 14.1 we are 99 percent confident of being right and only once in 100 times would we expect to be wrong by having quality fall outside of this range.

If we are willing to accept an even chance of being right or wrong, the range can be narrowed. This requires an area in one direction from the mean of 50/2, or 25 percent. Opposite area of 25 percent on the integration graph, Figure A–2, read x/σ of 0.6745. The range then is the mean $\pm 0.6745\sigma$, or $102.0 \pm 0.6745 \times 5.5$ ppm $= 102.0 \pm 3.7$ ppm. Now we have a narrower range, but we do not have great confidence in it because half the time the quality of the effluent is expected to fall outside the limits of this narrower range.

Although the integration graph is a useful device, a more convenient means of obtaining answers such as these (as well as other advantages) is provided by normal probability paper, as illustrated later.

Abnormalities and Asymmetrical Distributions

The normal distribution presumes homogeneous unchanging conditions without dominance of any one of many influences involved. In the accumulation of the sample series of data, events are independent and selections or occurrences are random, unbiased in favor of any particular outcome. These conditions are not always completely met, and, although most distributions tend to be normal, actually there may be an infinite variety of forms, some skewed, lopsided; others symmetrical but more peaked or more flat and stubby than the normal form.

If such abnormalities in form are inherent in the phenomenon, then increasing the number of measurements does not bring it back to that approaching a normal shape. These inherent abnormalities of peakedness and particularly those of skewness change the areas between the increments of standard deviation from those expected in the standard normal distribution. It is therefore essential to investigate the *form* of the distribution before using the mean and the standard deviation as summary statistics.

There are analytical methods for quantitative measures of skewness and peakedness (kurtosis) that stem from the concept of moments in engineering mechanics. These complex measures add little of value in most practical situations unless the series of measurements is large (250 or more) and is quite distinct in form. Since the third and fourth moments define skewness and peakedness, respectively, it is readily understandable that in a small series a few individual measurements that deviate from the mean beyond that normally expected when raised to the third and fourth power can contribute unduly to the moments and produce a false measure of skewness or peakedness, when actually the general form may be predominantly normal.

Since most series of data with which we deal in applied stream sanitation involve less than 250 measurements, the tedious computations for the analytical coefficients of skew and peakedness are not usually undertaken. A quick and reliable guide as to significant distortion from normal can readily be gained visually by plotting the data on normal probability paper. If there is reason to believe that the distribution is logarithmically normal, recourse is had to plotting on log-normal probability paper. Similarly, if the data are deliberately selected with a distinct bias, such as a series of extreme values (floods or droughts), recourse is had to plotting on extremal probability paper. (Probability papers are available commercially.)

Generally skewness is the most significant distortion. A quick visual approximation of the degree of distortion can be obtained by comparing the curvature of distributions of known coefficient of skew with that developed by a plot of the data on normal probability paper. Comparison is facilitated by converting the data to deviations from the mean expressed in terms of standard deviations.

Normal Probability Paper

The normal bell-shaped distribution curve can be transformed into a more useful form, as probability paper. Its property and special usefulness are that the summation of a normally distributed array will plot on this paper as a straight line, a form that facilitates the graphical solution of many statistical problems.

Construction of the Normal Probability Paper

The first step in the construction of normal probability paper in the transformation of the bell-shaped distribution curve to the probability summation, or the ogee curve. The bell-shaped curve and its integration graph (Figure A–2) give the areas of the central portion about the mean working outward toward the tails of the curve. The probability summation, or ogee curve, starts in the extreme left tail at minus infinity and sums the area by successive intervals toward the right. These two forms are shown in Figure A–3*a* and *b*, which illustrate the bell-shaped distribution curve and the summation, or ogee curve, respectively, both referred to an X'-axis scale in deviations from the mean expressed in terms of standard deviation.

The ogee curve can be constructed readily from the increments of area obtained from the integration graph, Figure A–2. The percentage of the area up to any deviation below the mean is simply obtained by subtracting from 50 percent, the area obtained from the integration graph for that deviation; and for deviations above the mean, by adding 50 percent to the areas of the integration graph. For example, the summation of area up to a deviation of -1σ below the mean is $(50 - 34.13)$, or 15.87 percent, whereas the summation of area up to $+1\sigma$ above the mean is $(50 + 34.13)$, or 84.13 percent. These two points are located on Figure A–3*b*, and similarly any number of other transformations can be made, forming a complete smooth ogee curve.

The next step is to reduce the ogee probability summation curve to a straight line. This is done simply by projecting the points on the ogee curve vertically to the linear X-axis scale and writing opposite each its corresponding percentage. For example, point a is projected to a', and its probability summation is written as 15.87 percent; b is projected to b' and is written as 84.13 percent.

Figure A–4 is such a scale arranged as the X-axis to form the normal probability grid. The zero of this scale is at minus infinity and 100 percent is at plus infinity, and the divisions are symmetrical about 50

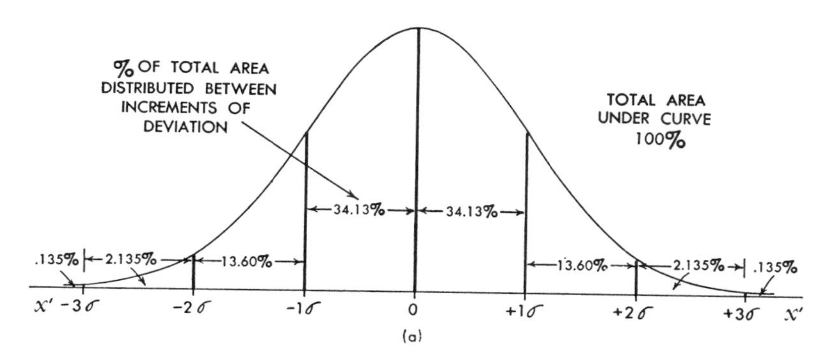

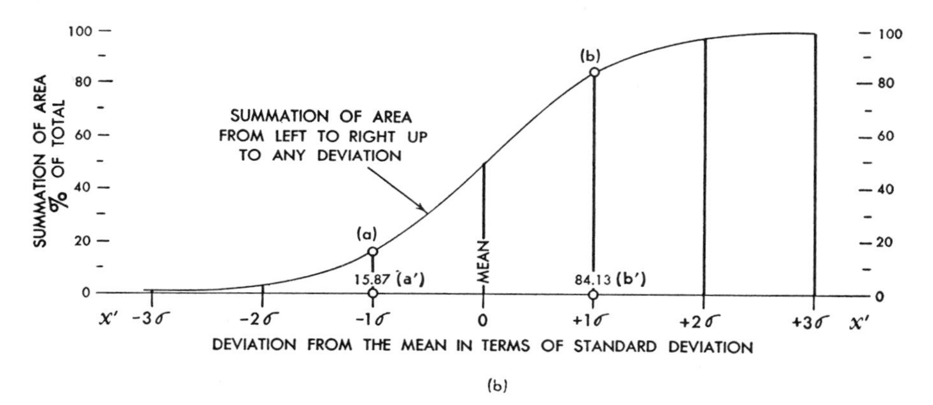

Figure A–3 The two forms of the normal probability curve: (*a*) the bell-shaped distribution curve; (*b*) the ogee summation curve.

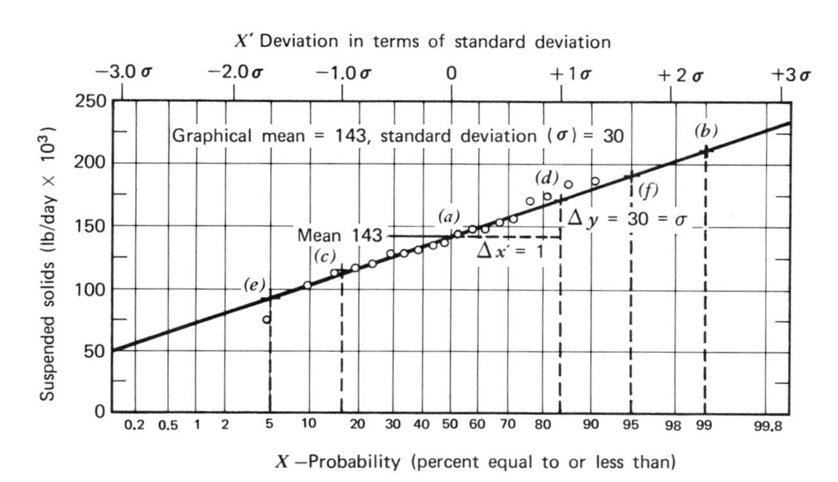

Figure A–4 Suspended solids—raw sewage.

percent, the midpoint. The Y-axis is linear and is assigned to the scale of the measurements. The X'-scale in terms of standard deviation is linear in relation to the Y-axis and is affixed as a convenience in fitting a straight line and in determining graphically the standard deviation σ.

Since the probability summation is the cumulative area from left to right, it represents probability *equal to or less than,* or, conversely, probability *of not exceeding.* Thus the simple technique of arranging measurements in ascending order of magnitude automatically places them in the order of their position on the probability summation scale of normal probability paper. There remains the problem of determining the plotting position of a series of measurements on the probability summation scale.

The Plotting Position and Procedure

Let us refer for a moment to the definition of probability, the ratio of the number of occurrences to the number of trials. Applying this to the series of 20 determinations of suspended solids in sewage shown in Table A–1 leads to the following:

Table A–1 Suspended Solids in Daily Raw Sewage Flow

Suspended Solids ($lb/day \times 10^3$) Arranged in Order of Magnitude	Rank Serial Number (m)	Plotting Position, Probability—Equal to or Less than [$m/(n + 1)$ as percent]
72.6	1	4.8
103.4	2	9.5
112.8	3	14.3
117.1	4	19.0
120.7	5	23.8
127.9	6	28.6
128.9	7	33.3
131.3	8	38.1
135.4	9	42.9
137.5	10	47.6
143.9	11	52.4
148.4	12	57.1
148.8	13	61.9
153.9	14	66.7
155.2	15	71.4
170.9	16	76.2
176.0	17	81.0
185.0	18	85.7
188.6	19	90.5
194.2	$20 = n$	95.2

Case I. The probability of a value equal to or less than the smallest value, the first, would be 1/20, or 5 percent; two values were observed to be equal to or less than 103.4×10^3 lb, probability 2/20, or 10 percent; and finally 20 values are equal to or less than the largest value (20/20, or 100 percent). Immediately we face a dilemma: if we plot the first point at 5 percent on the probability summation, we will not be able to plot the last point because it is positioned at 100 percent, which is located at plus infinity, completely beyond our scale. This is equivalent to saying that the largest value that was actually observed to occur once in 20 times would occur only once in an infinite number of observations.

Case II. Similarly, if we reverse the scale and start at the large end and number the values in descending order of magnitude and express probability as equal to or greater than, the largest value would be plotted at 1/20, or 5 percent; however, now we would be unable to plot the smallest value since its probability would be 20/20, or 100 percent, which would be located at minus infinity, completely off the scale.

Case III. Hazen [1] suggested what appeared to be a solution: plot the first value as the midpoint of equal increments on the scale of 100 percent. For example, the increment for the 20 points would be 100/20, or 5 percent. The first point being plotted at $\frac{1}{2}$ of 5, or 2.5 percent; the second point at $2.5 + 5$ percent, or 7.5 percent; etc., successively to the 20th point at 97.5 percent. This permits plotting all of the values on the scale and at first thought appears to solve the problem; but this does violence to the probability of the extreme values, the first and last points. Placing the smallest value of 20 values in a position on the probability scale at 2.5 percent is ascribing to it a probability as though it were the smallest value of a series of 40 measurements, 100/2.5; and likewise placing the largest value at 97.5 percent is as though it were the largest value of 40 measurements. Obviously such positions of the probability of the end points cannot be accepted.

Case IV. Theoretically the plotting position for the mean of any series regardless of the number of values is at 50 percent, and one-half of the values should be above and one-half below. If the mean is considered as part of the series, the number of plotting positions becomes $n + 1$. On this basis the mean plotting positions of the observed values are obtained from the relation $m/(n + 1)$ in which m is the serial number of the measurements arranged in ascending order of magnitude and n is the total number of measured values to be plotted. Thus the plotting position for the first value of a series of 20 values would be 1/21, or 4.77 percent; for the second, 2/21, or 9.53 percent, etc.; and for the 20th value, 20/21, or 95.23 percent.

Under these conditions it will be noted that the first and last values in the series of 20 will be plotted in positions as though they were the smallest and largest values in a series of 21 values (100/4.77), with 10 values above 50 percent and 10 values below 50 percent. These plotting positions are most consistent with theory and with the facts as observed, and case IV is therefore recommended.

In summary, the procedure for plotting observed or experimental data on normal probability paper is as follows:

1. Arrange the data in order of ascending magnitude.
2. Assign a serial number m to each of the n values 1, 2, 3, . . . , n.
3. Compute the plotting position of each serial value, giving the probability equal to or less than for each value by the expression $m/(n+1)$ expressing the ratio as a percentage.
4. The scale by which the observed or experimental data were measured is then laid off on the Y-axis, and the plotting positions are the probability scale on the X-axis. The points are plotted by using these X, Y coordinates.

If a straight line develops in the plotting, it indicates that the data have a normal distribution; that is, in accordance with the theory of probability we expect results distributed in this manner. Such a plot and straight line are the best possible representation of all the data in concise form.

If a straight line does not develop in the plotting, we suspect that something is changed in the conditions affecting the observed measurement beyond what is normally expected. This might mean carelessness in measurements or it may mean that we have not measured the same characteristics under the same conditions or that we have been measuring at the same time two or more distinctly different things. These very useful interpretations are also immediately visually apparent in the graphical presentation on probability paper.

Graphical Determination of the Mean and the Standard Deviation

Following the above procedure, the 20 determinations shown in Table A–1 are plotted on normal probability paper (Figure A–4). The first significant contribution immediately visually apparent is that the limited small sample of 20 measurements defines a good straight line through the bulk of the data; from this we can conclude that the form of the distribution is normal. With this established, we are in a position to determine from the straight line the two summary statistics, the mean $\bar{Y}$ and the standard deviation σ.

The mean ($\bar{Y}_g$) is determined graphically where the straight line of the distribution intersects the vertical line extending from 50 percent on the X-axis at a. The magnitude of the mean is read directly on the Y-axis as 143×10^3 lb/day. The subscript g is affixed to distinguish the graphical mean from the arithmetic mean.

The standard deviation is the slope ($\Delta Y/\Delta X'$) of the distribution line on probability paper. If there were no variation in the individual measurements, all values would be exactly alike and would plot exactly as a horizontal straight line. The angle from the horizontal position is therefore a measure of the degree of variability of the data; the steeper the slope, the greater the variability.

Since the probability summation scale is constructed linearly in terms of standard deviation from the mean, the slope of the distribution line measured on the standard deviation scale provides the standard deviation. Extending vertically from $+1\sigma$ (one standard deviation from the mean) ΔY can be read directly from the graph on the Y-scale as 30×10^3 lb/day. (The slope σ can be determined from any corresponding $\Delta Y/\Delta X'$ along the line.)

In addition to visually summarizing the data and graphically supplying the mean and the standard deviation from the mean, plotting on normal probability paper affords an opportunity to extend the series beyond the number of observations and determine the expected magnitude of a value at any probability equal to or less than. For example, referring to Figure A–4, if we wish to know the expected magnitude of a value that not more than 1 percent of the values are equal to or greater than (that is, 99 percent equal to or less than), we extend vertically from 99 percent on the X-axis to intersect the distribution line at b reading on the Y-axis 212×10^3 lb/day.

We also can make interpretations concerning confidence and precision in any range of deviation about the mean. For example, we are 68 percent confident of enclosing the quality within the range of 113×10^3 to 173×10^3 lb/day, located directly on the distribution line at the points c and d, vertical extensions from 16 and 84 percent (that is, 34 percent on either side of the mean). Similarly we are 90 percent confident of enclosing the quality within the range of 93×10^3 to 193×10^3 lb/day, located by e and f, vertical extensions from 5 and 95 percent on the X-axis (45 percent on either side of the mean).

Thus from the plot on probability paper we can read directly the probability of any range of deviation or the probability of not exceeding any magnitude of measurement without recourse to the integration graph (Figure A–2) for the bell-shaped distribution curve.

Plotting a Large Series

The example of Figure A–4 deals with a small number of measurements, and no difficulty is encountered in plotting each individual measurement. If the series of measurements consists of a large number of values it becomes confusing to plot each individual value, and it is preferable to group the data. For example, suppose we have taken 500 measurements of the hardness of a softened water supply and we wish to summarize the data and define the variation by plotting on probability paper. The scale of measurement in parts per million has been divided into 5-ppm intervals, and the hardness results have been sorted into each group as shown in Table A–2.

Table A–2 Hardness of Softened Water Supply

Group Hardness (ppm) (Y) (1)	Frequency (2)	Summation of Frequency (m) (3)	Probability Equal to or Less than [$m/(n+1)$ as percent] (X) (4)
30–34	2	2	0.4
35–39	6	8	1.6
40–44	16	24	4.8
45–49	32	56	11.2
50–54	53	109	21.8
55–59	74	183	36.5
60–64	87	270	53.8
65–69	85	355	70.9
70–74	66	421	84.0
75–79	42	463	92.4
70–84	22	485	96.8
85–89	10	495	98.8
90–94	4	499	99.6
95–99	1	500	99.8
	$n = 500$		

Summing the number of values found in each group gives the number less than the divisions of the group intervals, which correspond to the serial number m as shown in column 3. The plotting position (column 4) on the X-axis of the probability paper is $m/(n+1)$, expressed as percent, where n is 500, the total number of values. The coordinates

Y (column 1) and X (column 4) are then plotted on probability paper as shown in Figure A–5. The straight line formed summarizing the entire 500 values indicates a normal distribution, with a mean located at a of 63.6 ppm and a slope or standard deviation from the mean of 11.4 ppm. The probability of any range or of not exceeding any magnitude of measurement can be read directly from the graph. For example, we expect not more than 1 percent to exceed 90 ppm located at b; 1 percent less than 37.3 ppm located at c; and 50 percent within the range 56.0 to 71.3 ppm about the mean between d and e.

Reproducing a Series of Data

There are many occasions when we may wish to reproduce graphically on probability paper a series of data from a published summary that does not tabulate each individual value. For example, if we are given the mean, the number of measurements and the standard deviation from the mean, it is a very simple operation to reconstruct on probability paper the expected normal distribution of the individual values of the series. This is just the reverse process of plotting a series of given data. For example, we are given the summary of a series of BOD determinations, the number of measurements n is 24, the mean $\overline{Y}$ is 10.0 ppm,

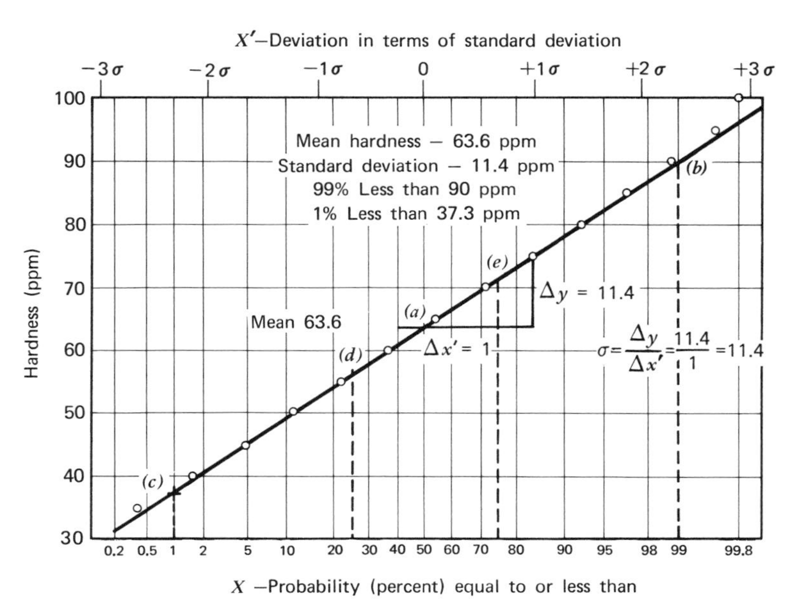

Figure A–5 Variation in hardness in softened water supply based on 500 determinations.

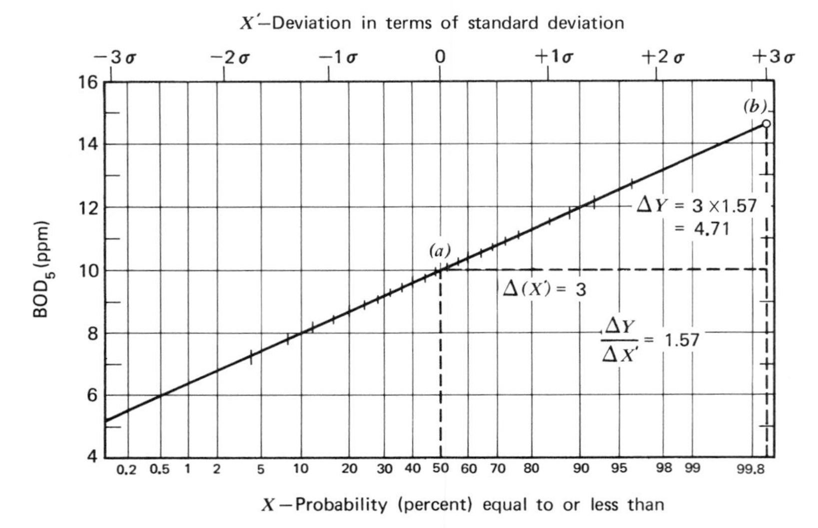

Figure A–6 Reproduction of distribution of a series of BOD determinations from number, mean, and standard deviation.

and the standard deviation σ from the mean is 1.57 ppm. The series is reproduced in Figure A–6 as follows:

1. Locate the mean on the Y-axis and draw a horizontal line to intersect with the vertical line, extended from 50 percent on the X-axis at a. This is the midpoint of the distribution.
2. From this point lay off a slope $\Delta Y/\Delta X'$ equal to the standard deviation of 1.57, locating point b.
3. Draw a straight line from point b through the mean a. This line represents the most probable distribution of the series of 24 individual measurements.

If the original distribution were truly normal, the most probable values of each measurement would be located where $m/(n + 1)$ or $m/(24 + 1)$ expressed in percent extended vertically cut the line.

With this reconstruction we now can readily make any interpretations concerning confidence and range or probability of not exceeding any magnitude of measurement just as though we had been given the original data.

Log-Normal Probability Paper

Certain types of data, particularly those derived from biological phenomena, are logarithmically normal. Such distributions will plot as a straight line on normal probability paper providing the logarithms of the measurements are substituted for the original data. The use of log-probability paper permits plotting directly the original data without converting to logarithms.

Figure A–7 illustrates a logarithmically normal distribution plotted on log-normal probability paper. The X- and X'-scales are identical with those previously described for normal probability paper: the X-scale at the bottom is probability (in percent) equal to or less than or probability of not exceeding; the X'-scale at the top is deviation from the mean in terms of standard deviation. The left ordinate scale Y is a logarithmic grid assigned to the scale of measurement of the original data; the Y'-scale at the right is a linear grid on which logarithms of the original data are located. The procedure in plotting on log-probability paper is identical with that employed for normal probability paper outlined above.

In interpreting a plot on log-probability paper it is cautioned that while the original data are plotted on the Y scale, actually it is the corresponding Y' values, the logarithms of the original data, that form the normal distribution. Hence the mean and the standard deviation are expressed in terms of logarithms. The subscript "log" is employed to indicate that a logarithmically normal distribution is dealt with. The procedure for determining graphically the $\text{mean}_{\log}$ and $\sigma_{\log}$ is identical

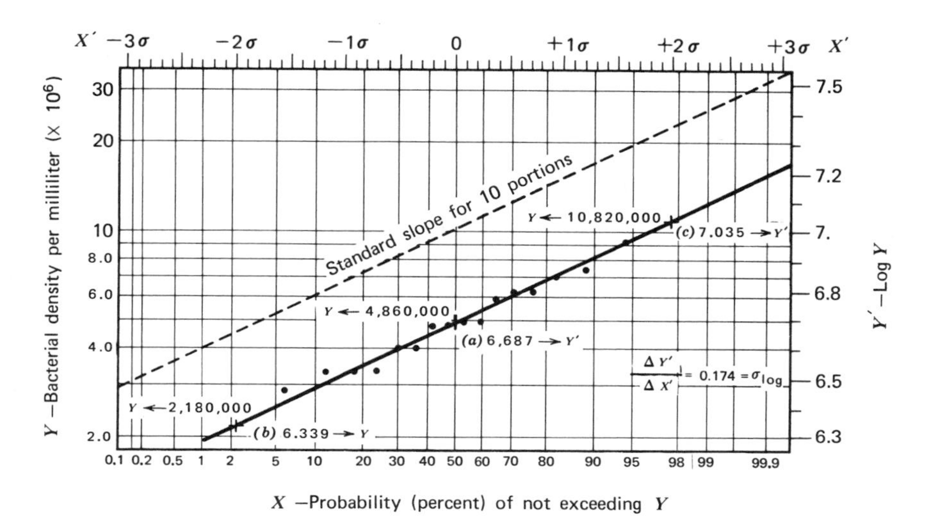

Figure A–7 Log-normal distribution of bacterial density by the dilution method (10 portions in each of three-decimal dilutions).

with that described for normal probability paper except that the Y'-scale is employed. Referring to Figure A–7,

$$\text{mean}_{\log} = 6.687,$$

located on the Y'-scale opposite a where 50 percent cuts the distribution line. This is equivalent to the arithmetic mean of the logarithms of the original values.

The slope of the distribution line is the measure of variation of the data and provides the standard deviation in terms of logarithms.

$$\sigma_{\log} = \frac{\Delta Y'}{\Delta X'} = 0.174.$$

$\Delta Y'$ is measured on the Y'-scale and $\Delta X'$ is measured on the X'-scale.

Similarly in computing analytically ranges for any confidence it is essential first to deal in terms of logarithms. After the logarithms of the limits of any range are determined their antilogarithms can be obtained and the results expressed in terms of the original units of measurements. The advantage of log-probability paper is that these conversions are accomplished automatically. Again referring to Figure A–7, the $\text{mean}_{\log}$, 6.687, is converted to units of the original measurement simply by reading directly on the Y-scale, giving a value 4,860,000 per milliliter. Similarly the range for $\text{mean}_{\log} \pm 2\sigma_{\log}$ in terms of logarithms is $6.687 \pm 2 \times 0.174$, or (6.339 to 7.035), located by b and c where -2σ and $+2\sigma$ intersect the distribution line, reading on the Y'-scale. The limits of this range converted to units of the original measurements are directly obtained by reading on the Y-scale (2,180,000 per milliliter to 10,820,000 per milliliter).

Distribution of Extreme Values

As discussed in Chapter 3, the extremes of hydrologic phenomena such as flood and drought streamflow do not follow a normal symmetrical distribution, but rather are skewed (the more severe values deviate beyond the mean to a much greater extent than the less severe values deviate below it). Gumbel [2] developed a standard skewed distribution based on the theory of largest values, which, when reduced to straight-line form on special probability paper, as suggested by Powell [3], affords a convenient graphical method for dealing with such data.

A comparison of the normal probability curve and Gumbel's standard skewed distribution of largest values is shown in Figure A–8. Figure A–8a is the normal probability curve, which is perfectly symmetrical, with the mean and mode coinciding at the midpoint. Figure A–8b is Gumbel's skewed distribution, which is asymmetrical, rising sharply to

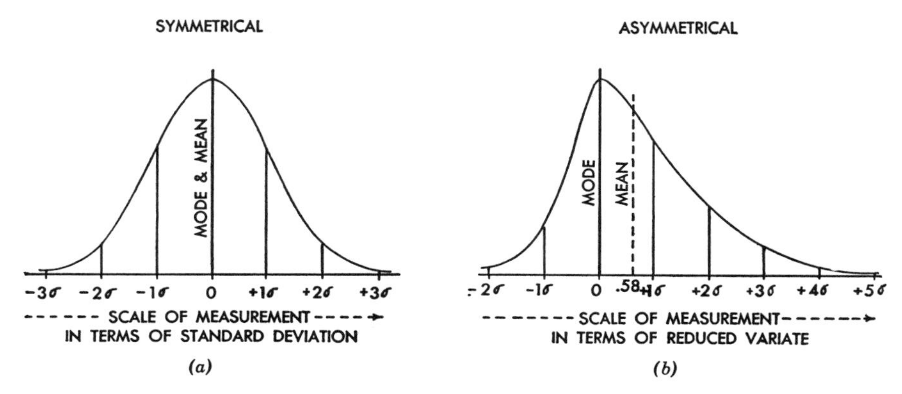

Figure A–8 Comparison of (*a*) the normal probability curve and (*b*) Gumbel's standard skewed distribution of largest values.

the peak and then falling off gradually in a long tail to the right, with the mode to the left of the mean. As with the normal curve, the probability of events is represented by the area under segments of the curve, total area being unity, or 100 percent. Gumbel has given these areas for increments of the curve, and by summing from left to right a probability summation scale is obtained for the skewed distribution similar to that developed for normal probability paper.

Basically Gumbel proposed for floods:

$$W(x) = e^{-e^{-y}}, \tag{A–2}$$

where $W(x)$ is the probability of a value's being equal to or less than x, e is the Napierian base (2.718), and y is the function of steamflow.

Expressing equation A–2 in terms of return period gives

$$T_x = \frac{1}{1 - e^{-e^{-y}}}, \tag{A–3}$$

where T_x, return period, is defined as the mean chronological interval between two values, equal to or greater than x. If the observations are equally spaced in time and the unit of time is taken as the interval between two successive observations, the return period T_x is defined as the mean number of observations, or the average time necessary to obtain once a value equal to or greater than a certain value x.

Solving the return period equation for y gives

$$y = -\log_e \left[-\log_e \left(1 - \frac{1}{T_x} \right) \right] \tag{A–4}$$

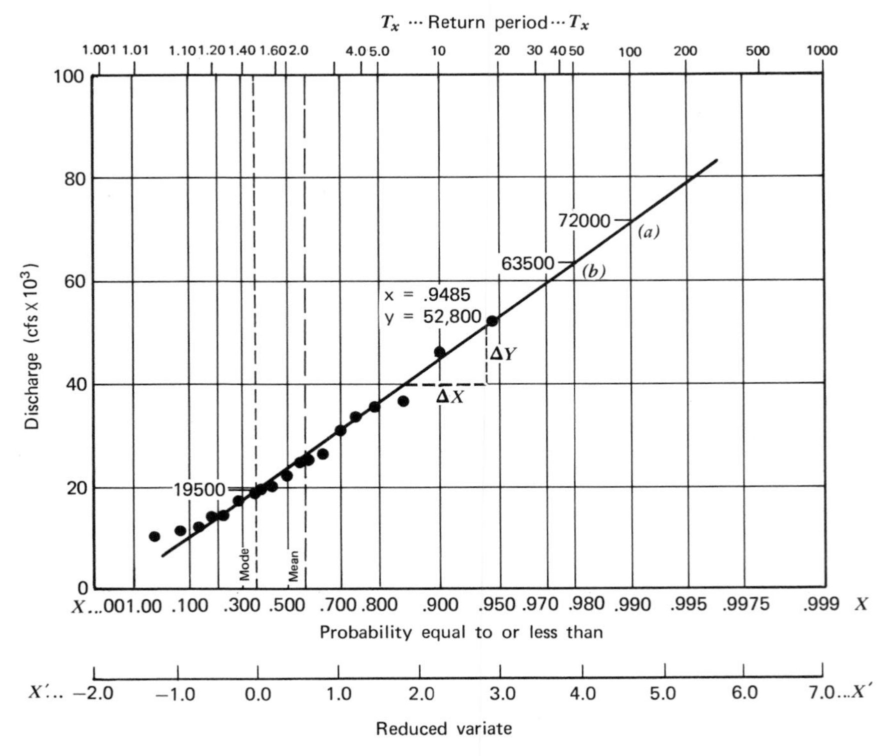

Figure A–9 Probability of annual flood discharge of the Clark Fork River below Missoula, Montana.

Linear Extremal Probability Paper

Powell [3] used these basic equations in the development of linear extremal probability grid where equation A–4 was solved for y for various given values of T_x. Figure A–9 is such a sheet of linear extremal probability paper on which are plotted the series of maximum annual instantaneous flood peak discharges of the Clark Fork River. The ordinate Y is a linear scale assigned to the observed extreme values. Scale X is the probability summation giving the probability of a value equal to or less than a certain value Y. The X'-scale is the reduced variate, introduced for linearity; the T_x scale at the top is the return period. From the basic definitions and equations it follows that the return period is equal to the reciprocal of 1 minus the probability summation equal to or less than:

$$T_x = \frac{1}{1 - x}.$$

Maximum and Minimum Extreme Values

Although while Gumbel's standard distribution of extreme values is based on the theory of *largest* values, it is not thus limited. By the simple device of arranging series of extreme values in order of *severity* rather than in order of magnitude, the standard extremal distribution is equally applicable to series of *minimum* values. In dealing with such maximum series as floods, the least severe value would be the smallest observed flood and the most severe value would be the largest observed flood. However, in dealing with such minimum data as droughts expressed in runoff (cfs) the least severe drought would be the largest runoff among the observed droughts and the most severe drought would be the smallest runoff among the observed values, just the reverse of floods. This places the most severe flood and the most severe drought in the same position in terms of probability of severity. Thus by the simple technique of arranging maximum or minimum data in order of *severity* we place them in order of their position on the probability summation scale of the skewed probability paper.

Series of data that are truly extreme values will plot as a straight line on the skewed probability paper, which facilitates interpolation and extrapolation. From such a straight line the probability of any severity can readily be determined graphically.

The Plotting Position

In discussing the plotting position on the probability summation scale we have recommended using $m/(n + 1)$. This simple relation may also be employed for series of extreme values where m is now the serial number of data arranged in order of severity and n is the total number of values to be plotted.

Gumbel [2], however, has suggested a refinement that is recommended when dealing with a small series of less than 20 observations. He develops the probability of the most probable least severe and the most probable most severe value as the plotting position of the two extreme observations m_1 and m_n, and calculates the remaining $(n - 2)$ plotting positions by dividing the intervening intervals into $(n - 1)$ equal units.

The most probable plotting position of the most severe value is found

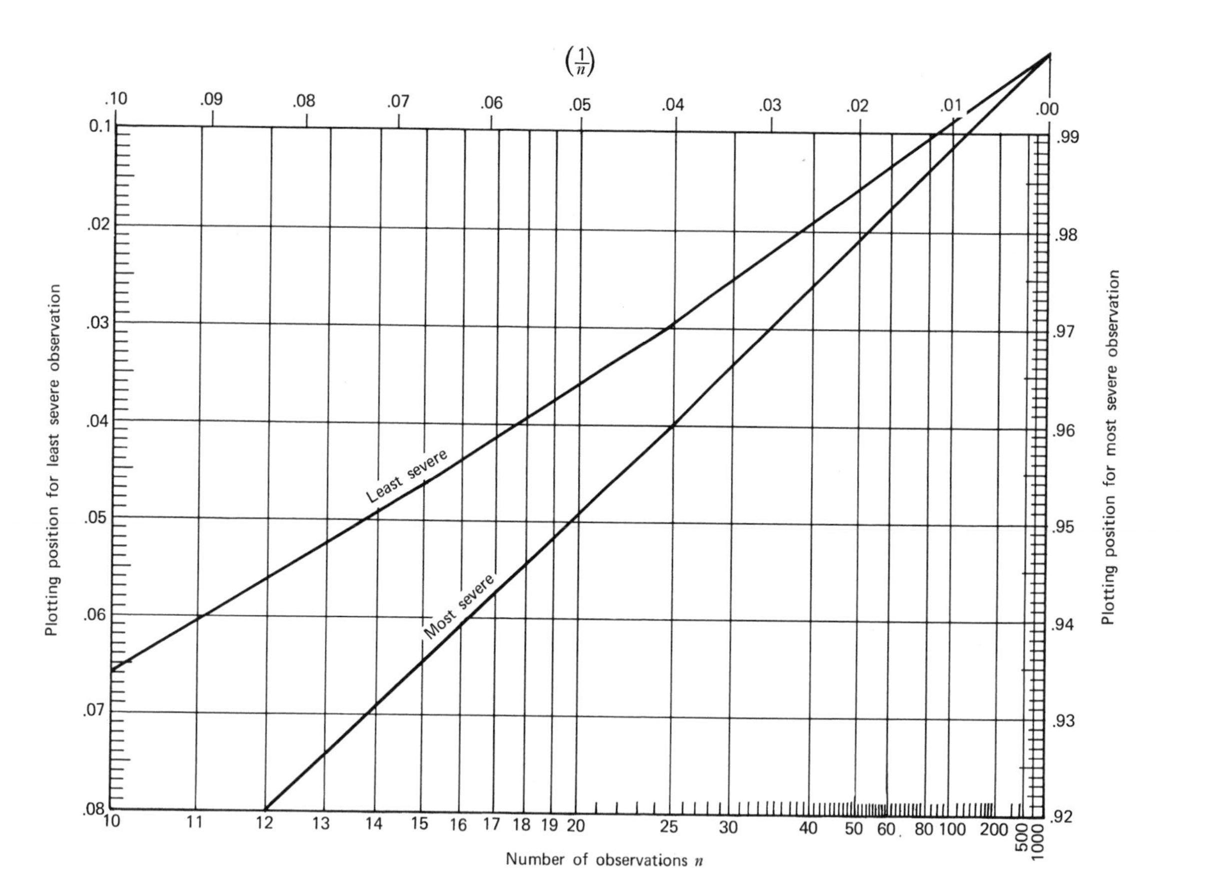

Figure A–10 Plotting positions for the least and the most severe of any number of observations for a series of either maximum or minimum data (according to Gumbel).

to be

$$m_n = e^{-1/n}$$

and that of the least severe value (m_1) is related to the period of record, n, by

$$\frac{1}{n} = \frac{m_1(-\log_e m_1)}{-\log_e m_1 - 1 + m_1}.$$

The plotting positions of these two end values of the series are given in the diagrammatic form in Figure A–10 as functions of n.

Table A–3 illustrates the procedure for plotting series of extreme maximum values. With a short record of 19 years, the plotting positions of column 5 are in accordance with Gumbel's refinement obtained as

Table A–3 Maximum Annual Flood—Highest Peak Discharge in Each Water Year, Clark Fork Below Missoula, Montana

Year (1)	Discharge (cfs) (2)	Floods Arranged in Order of Severity (Y) (3)	Serial Number (4)	Plotting Position (X) (5)
1930	17,500	10,400	1	.0373
1931	12,200	11,700	2	.0879
1932	25,000	12,200	3	.1385
1933	36,800	14,200	4	.1892
1934	25,700	14,400	5	.2398
1935	20,200	17,500	6	.2904
1936	26,700	19,100	7	.3410
1937	11,700	19,700	8	.3917
1938	35,700	20,200	9	.4423
1939	22,000	22,000	10	.4929
1940	14,200	25,000	11	.5435
1941	10,400	25,700	12	.5941
1942	30,500	26,700	13	.6448
1943	33,200	30,500	14	.6954
1944	14,400	33,200	15	.7460
1945	19,100	35,700	16	.7966
1946	19,700	36,800	17	.8473
1947	45,900	45,900	18	.8979
1948	52,800	52,800	19	.9485

follows:

Position of the most severe flood obtained from most severe curve, Figure A–10	.9485
Position for the least severe flood read from least severe curve, Figure A–10	.0373
Interval between least severe and most severe	.9112
Increment for successive points .9112/(19 − 1)	.050622

Columns 3 and 5 of Table A–3 provide the coordinates for plotting the data on linear extremal probability paper, as shown in Figure A–9. A straight line is indicated as represented by the line fitted through the data. Thus the 19-year record, although short, is in accord with the theory of extreme values. If we are willing to accept this small sample as representative of flood expectancy, extrapolation beyond the period of record affords an estimate of more severe occurrences. Since the mean return period interval between observations is a year, it is customary to speak of flood severity as equaled or exceeded on the average once in 100 years as the 100-year flood. For the Clark Fork the 100-year flood is located by extending vertically from 100 on the T_x scale to intersect the distribution line at a, reading on the Y-scale 72,000 cfs. Similarly on the average once in 50 years a flood of 63,500 cfs is located at b.

Linear extremal probability paper may be employed for any series of maximum values selected from other phenomena such as temperature or rainfall intensity. Certain series of minimum values that do not approach zero as a lower limit may likewise be plotted on linear extremal probability paper.

Log-Extremal Probability Paper

As has been discussed and illustrated in Chapter 3, minimum drought streamflow series are plotted on log-extremal probability paper. The extremal probability grid is identical with that developed for linear extremal probability paper, except that the ordinate scale is a logarithmic grid. The series of minimum streamflows arranged in order of severity plotted on the logarithmic grid automatically are located as the logarithms of the original data. The logarithmic scale fulfills the requirement of an unlimited distribution, since a zero value is approached but never reached. The procedure for obtaining the plotting position on the extremal probability summation scale is identical with that employed for linear extremal probability paper as illustrated above.

TESTS FOR STATISTICAL SIGNIFICANCE

Comparison is a common method of appraisal. However, comparisons should be approached cautiously as inherently all measurement or sampling, no matter how precise, is always subject to chance variation. Hence observed differences or similarities may not be significant as they may arise solely by chance. It should be emphasized at the outset that tests for statistical significance cannot prove a cause and effect relationship. The contribution of statistics is to define the expected variations due to chance. Actually any difference in results, no matter how great, can be ascribed to chance, but a point is reached where the probability that the difference is due to chance becomes so small that the investigator prefers to ascribe the difference to other causes.

The decision that the difference is not due to chance and the subsequent search for the underlying cause of the difference is completely outside the realm of statistical technique and is entirely a matter of professional judgment of the specialist in the field to which the investigation relates. The ideal is to combine in one person both the professional judgment of the specialist and the skill of the statistician. The statistical method is only a tool, not a substitute for reasoning.

The graphical procedure of plotting data on probability paper not only affords an excellent summary of series of measurements but also provides a quick visual test of statistical significance that may be adequate for many purposes. Although analytical procedures may provide a more refined test, they are more complicated and may not materially assist the investigator in making the basic decision of accepting or ruling out chance. Some situations are so complex that only by graphical portrait can the professional specialist rule out chance and grasp the significance of other relationships. The first step therefore should be a graphical approach, followed where warranted by analytical refinement.

Although the illustrations of graphical tests for statistical significance employ data pertaining to municipal wastewater characteristics, the methods are not in any way thus limited and can find wide application in any field involving evaluation of quantitative measurements.

Testing a Single Series

Normality

The most important test applied to a single series is to determine whether the distribution of individual measurements is normal. As we have seen, this is accomplished graphically by plotting the data on probability paper. If a straight line is formed, the distribution is normal and symmetrical; if there is a curve upward or downward, the distribution is skewed and asymmetrical. This is at once visually apparent. If a distribution plots as a straight line (is normal), then, and only then, the raw data may be condensed, without loss of significance, to three summary statistics: the mean $\bar{Y}$, the standard deviation σ, and the number of measurements n. Such a normal distribution was illustrated in Figure A–7, summarized as:

Mean ($\bar{Y}$ graphical)	143×10^3 lb/day
Standard deviation (σ graphical)	30×10^3 lb/day
Number of observations n	20

A graphical probability plot also affords visually an appraisal of such questions as:

Does the arithmetic mean approach the midpoint of the distribution at 50 percent, or plotted at 50 percent, is it in line with the bulk of the data?

Are any individual measurements "out of line" in relation to the other values? Do they deviate beyond what can reasonably be expected by chance variation, or is one led to suspect their initial reliability? Should they be accepted or rejected in defining the distribution, the mean, and the sigma?

Are the maximum and minimum values in line with the normally expected in a sample series of this size?

How good is sigma? Visually, how well do the points define the slope (σ) of the distribution? Is there wide scatter, or are the points closely knit about the line? What variation in slope results from reasonable graphical approximations of fit?

Abnormal Breaks in Data

Figure A–11 illustrates the effectiveness of a graphical test applied to a single series of 73 measurements of BOD, particularly as a method of detecting abnormal breaks in the data. It is at once visually apparent that the four largest values in the total distribution of 73 distinctly break away from an otherwise reasonably normal distribution. It is also apparent that this single series of 73 measurements contains data from two distinctly different universes: 69 values are consistent, but the four largest valves are out of line and very likely from a different level. Such breaks would not be so readily recognized in tabulations nor detected by the usual analytical procedures. Treating the 73 values as a random sample from one and the same universe or supply by the usual analytical procedure distorts the mean and the standard deviation.

The advantages of the graphical test procedure are apparent from

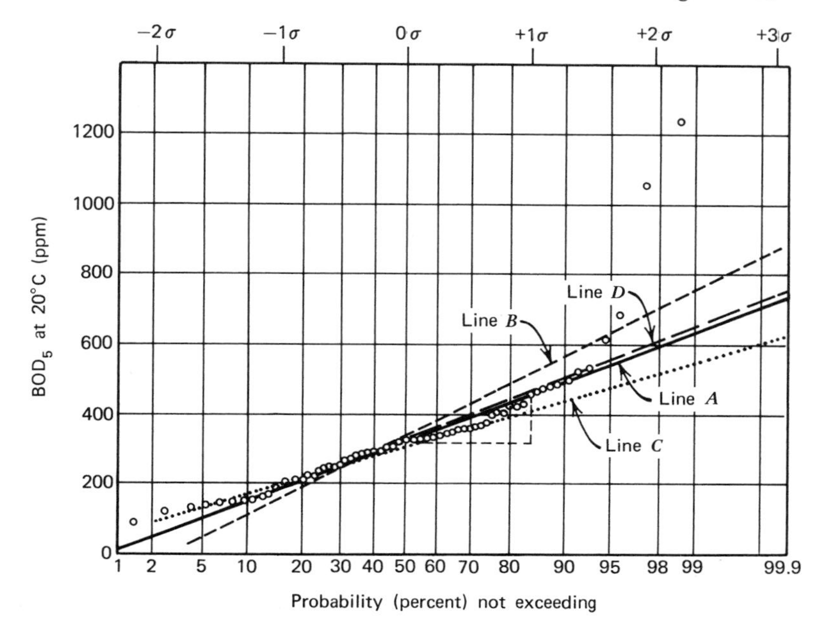

Figure A–11 Graphical test of single series. Detection of abnormal breaks in data—comparison of graphical and analytical methods. Line *A*: graphical distribution; line *B*: analytical distribution using all data; line *C*: analytical distribution without four largest values; line *D*: second random sample from dominant universe.

a comparison of the following results. Graphically plotting the 73 values, a reasonably well defined distribution line *A* is formed by 69 of the 73 values which cuts 50 percent to establish a graphical mean of 324 ppm and to define a slope (σ) of 133 ppm.

Thus the graphical method, in addition to permitting detection of a break in data, makes possible differentiation between the two universes or supplies and determination of a representative mean and sigma from the bulk of the data of the dominant universe. The plotting positions, however, are assigned as of a random normal series of 73 points, but the distribution line is fitted to the 69 points, ignoring the 4 points that are obviously from another universe.

In contrast, the analytical method applied to the series of 73 values (including the 4 values from another universe) produces a computed arithmetic mean of 341.8 and a sigma of 179.9, and defines a distribution shown as line *B* in Figure A–11. This distribution line is too high and too steep, and is obviously inconsistent with the bulk of the data. In fact in attempting to define both universes it defines neither.

In the analytical approach the four largest values cannot be deleted and the remaining 69 dealt with as a random sample. Obviously the deletion of the 4 largest values of a series of 73 values, even if they were in line with a normal distribution, would unduly lower the mean and sigma. Such treatment for this example would reduce the mean to 309.3 and sigma to 105.6. These statistics would now define a distribution line C falling below the bulk of the data and at a slope that is too flat. Again the analytical method failed to properly represent either universe.

A confirmation of the superiority of the graphical test procedure for detecting and differentiating breaks in data and defining the characteristics of the dominant universe is demonstrated by a second random sample drawn exclusively from the dominant universe. For this purpose a random sample of 40 values was taken and plotted on probability paper, producing a graphical mean of 335 and a sigma of 138, indicated by line D. The random sample of 40 values (line D) practically coincides with the original sample of 73 (line A). These statistics confirm the original graphical values, as defined initially by line A.

As further emphasis of the value of the graphical approach, Table A–4 contrasts the graphical with the analytical summary statistics of the operating data of the wastewater treatment works for 7 months, July through January.

It will be noted that the normal suspended solids load to the treatment works from month to month, uninfluenced by rainfall flushes, is approximately constant, as reflected by the graphical means shown in column 5. This is to be expected from a given population and normal industrial activity not subject to seasonal patterns. The day-to-day variability, as reflected by the graphical standard deviation, column 6, is also practically constant.

In contrast to this, there is wide variation reflected in the analytical summary of computed monthly averages and standard deviations as shown in columns 2 and 3. November and January are of special significance because the complete distributions were normal without any breaks. Rainfall during these months was light and probably occurred as snow; hence flushing of the combined sewers with distortion of suspended solids did not take place. For the graphical summaries, it is noteworthy that the means and standard deviations of the other months tie in closely with these two months. In contrast the results of the analytical summaries show wide variations in means and standard deviations for the other months when compared with November and January. Note, also, that only when the complete monthly distribution is normal (November and January) do the analytical results approach the graphical.

Table A-4 Comparison of Analytical and Graphical Statistical Summaries of Daily Suspended Solids Wastewater Treatment Works Influent, July–January

	Analytical Summaries			Graphical Summaries			
Month (1)	Computed Mean (lb/day × 10^3) (2)	Computed σ (lb/day × 10^3) (3)	n (4)	Graphical Mean (Normal Segment) (lb/day × 10^3) (5)	Graphical σ (Normal Segment) (lb/day × 10^3) (6)	n (7)	Remarks (8)
July	182.9	139.3	31	153	39.0	31	85% normal, followed by sharp break, with 15% distinctly abnormal
August	197.6	137.3	31	146	37.3	31	65% normal, followed by sharp break, with 35% abnormal
September	206.1	173.5	30	166	48.0	30	75% normal, followed by sharp break, with 25% abnormal
October	174.0	123.1	31	139	40.0	31	80% normal, followed by sharp break, with 20% abnormal
November	134.9	30.8	30	135	34.0	30	Complete distribution normal
December	167.2	104.6	31	140	42.7	31	80% normal, followed by sharp break, with 20% abnormal
January	141.4	39.4	31	142	41.8	31	Complete distribution normal
July–January	172.0	118.1	215	147	42.7	215	85% normal, followed by sharp break with 15% abnormal

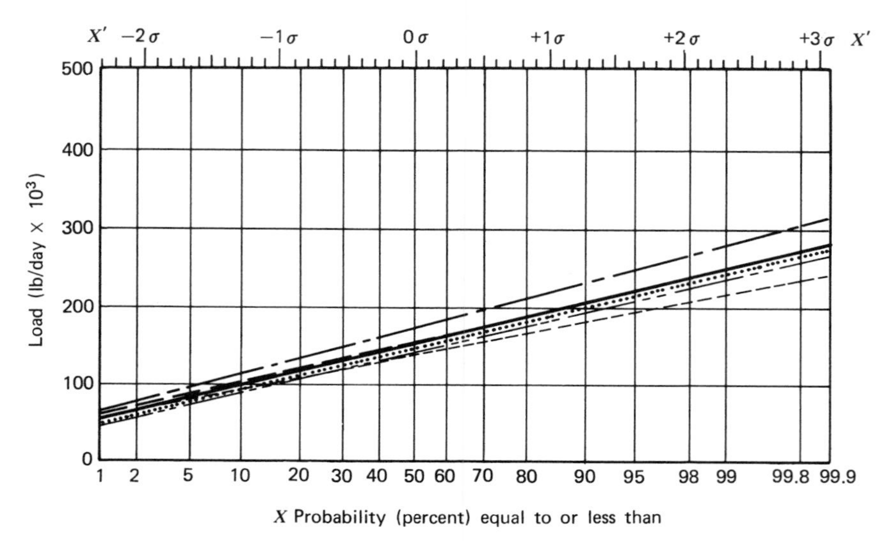

Figure A–12 Illustration of graphical test for similarity among several distributions.

In appraising treatment works loading and performance (or any other kind of data) it is obvious that the all too frequently employed "monthly average" may be a very misleading figure. It is essential to differentiate between the normal and the abnormal before drawing conclusions or making comparisons. There is no statistical substitute for the neat sophistication of a simple plot on probability paper in making such distinctions.

Testing Two or More Series

There are various graphical methods for comparing two or more series when testing for significant differences. Graphical procedures afford a quick visual appraisal of such questions as:

Are two series of measurements independent? Are two separate lines formed on probability paper, or do they approximately coincide?

Assuming there is no difference between the two series, in combining the data do the points fit as one line in the larger sample or are there segregations of values?

How do the deviations vary among different series? Are the slopes (σ's) the same or radically different?

What range in *individual measurements* can be reasonably expected in different series?

What range in *means* can be reasonably expected in different series?

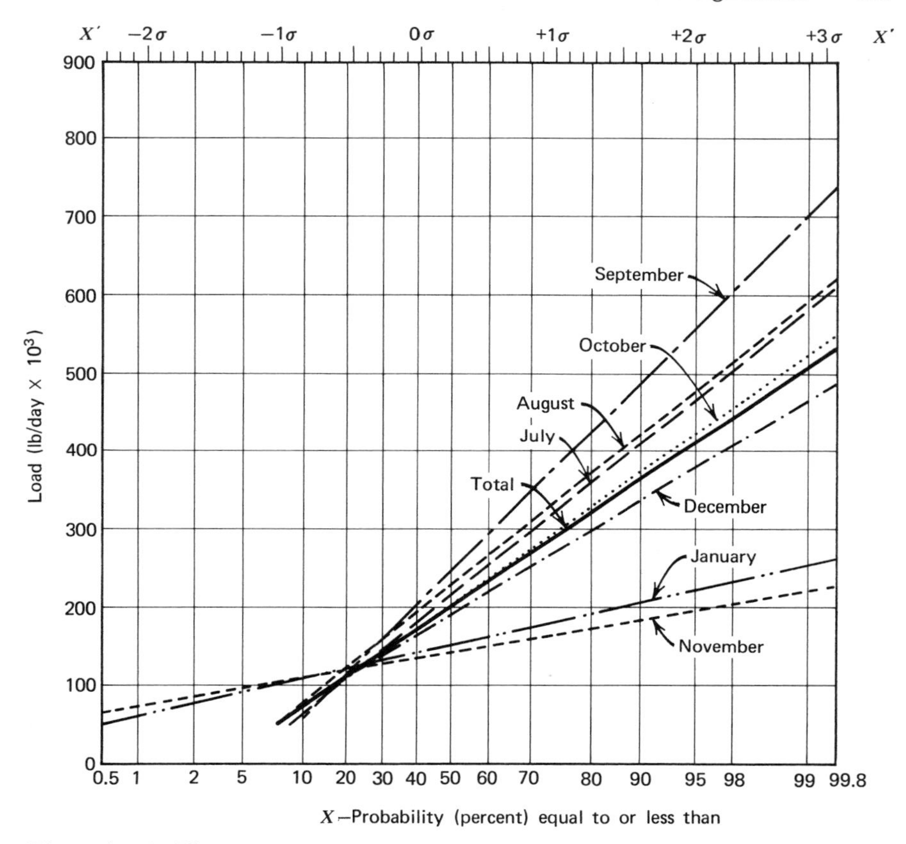

Figure A–13 Illustration of graphical test for difference among several distributions.

Is there a statistically significant difference between means?

Differentiating among Series

Plotting two or more series of data on the same sheet of probability paper affords a quick visual appraisal of similarities or differences. For example, assume that the distributions summarized in Table A–4 are all normal distributions as represented by the means and the standard deviations. Figure A–12 represents the reconstruction as normal distributions of the graphical summaries (columns 5 and 6 in Table A–4). The eight distributions cluster about the common mean and have similar slopes. The visual pattern strongly suggests similarity among the distributions.

In contrast, Figure A–13 shows the reconstruction of distributions based on the analytical means and standard deviations (columns 2 and

3 in Table A–4). A marked difference among these distributions is evident, not only in the level of means but also in their slopes. Such graphical differentiation affords a quick appraisal and identifies areas where further testing between distributions may be warranted.

Comparing Variability among Series in Terms of Ratios to Mean

A convenient method for graphically comparing variability among a number of series (or the relative magnitude at any probability) is to express each series of data as *ratio to its mean* and plot the ratios on probability paper. Regardless of the scale of measurement and size of mean, this makes each series directly comparable with a mean value of unity and a common ordinate scale as ratio to mean. The slopes of each distribution, which are the measures of variability, are then also directly comparable. Since data are plotted as ratios to the mean, the slope gives directly the *coefficient of variation:*

$$\text{C.V.} = \frac{\sigma}{\text{mean}}.$$

Overlapping Test

The statistical significance of the difference between two series of data can readily be evaluated by the overlapping test. In comparing two series, the first operation is to differentiate by plotting both on the same probability paper. However, it is not sufficient to demonstrate two distinct distribution lines to be confident that there is a statistically significant difference between the two series. We expect that by chance alone there will be shifts in the distribution lines upward and downward in plotting similar repeated sample series. Hence to demonstrate a statistically significant difference two distributions must remain beyond such shifts in position as chance alone can reasonably be expected to produce.

The Standard Error of the Mean. The standard error (S.E.) of the mean is one measure of such shifts in a distribution line.

$$\text{S.E.} = \frac{\sigma}{\sqrt{n}}.$$

The stability or reliability of the mean, as defined by its standard error, varies directly as the standard deviation σ of the individual measurements and inversely as the square root of the number of measurements in the series. The variability of the individual measurements is inherent in the phenomenon being observed, and not much can be done about this. However, it is possible to control the reliability of a mean simply by increasing the number of individual measurements. Increasing the number of measurements fourfold reduces the standard error of the mean by one-half, increasing the number of measurements sixteenfold reduces it by one-quarter, etc.; or more generally, from the transformation of the equation, the number of individual measurements required for a specified standard error of the mean is given by

$$n = \frac{\sigma^2}{(\text{S.E.})^2}.$$

For example, the mean $\bar{Y}$ and standard deviation σ as determined graphically are 135,000 and 34,000 lb/day, respectively, for the series involving 30 individual measurements. From this single series it is then possible to evaluate the reliability of the mean:

$$\text{S.E.} = \frac{34}{\sqrt{30}} = 6.2.$$

The variation of the mean, like variation of the individual measurements, follows the normal probability distribution, and from this it can be expected for repeated series of 30 measurements that

68 percent of the means would fall within the range of 135 ± 1 S.E., or 135 ± 6.2;

95 percent of the means would fall within the range of 135 ± 2 S.E., or 135 ± 12.4;

99.73 percent of the means would fall within the range of 135 ± 3 S.E., or 135 ± 18.6.

This defines the range in shift upward or downward that can be expected by chance alone from plotting repeated series of only 30 individual measurements made under the same conditions.

With the range in shifts of the distribution thus defined by the standard error of the mean, it is now possible to apply the overlapping test for statistical significance of the difference between two distributions. Marking off these ranges about the mean of each series graphically discloses the variation expected by chance alone at any confidence level.

If no overlap develops between the ranges, chance may be ruled out, and the difference between the two series being compared is accepted as being statistically significant.

If an overlap develops between the two ranges or if one range encloses the other, chance cannot be ruled out, and the difference between the two series is not accepted as being statistically significant.

It will be noted that the range of the mean is dependent on the choice of a confidence level. The acceptance of a 95-percent confidence range within which the mean is expected to vary by chance defines a specific range ±2 S.E. However, ruling out chance at this range implies a risk of being wrong about 5 times in 100, as chance alone can be expected to show a mean outside the range of ±2 S.E. about 5 times in 100. Setting a range of ±3 S.E. is generally interpreted as practically certain to account for variation associated with chance, but even here there is a risk of being wrong 27 times in 10,000, as chance can be expected to show a mean outside the range of ±3 S.E. about 3 times in 1000. The decision as to a confidence level always rests with the investigator and will vary with the nature of the problem and the consequences of being wrong.

The other factor to be considered is that the reliability of a mean can be changed by the number of individual measurements. Hence comparison between two means, each based on a small number of observations, may develop an overlap, and for these sized series the difference would be considered not statistically significant. Increasing the number of measurements would reduce the range and may conceivably not develop an overlap; for the larger series the difference then would be considered statistically significant.

Figure A–14 illustrates an application of the overlapping test in comparing the suspended solids content of the influent for the month of November with that for the month of January. Line A represents the distribution of the 30 individual measurements for November, giving a mean of 135,000 and a standard deviation of 34,000 lb/day. Line B represents the distribution of the 31 individual measurements for January, giving a mean of 142,000 and a standard deviation of 41,800 lb/day. Two distributions are formed with an apparent difference between the means of (142,000 — 135,000) or 7000 lb/day. The standard errors of the means are $34{,}000/\sqrt{30}$, or 6200 for November; and $41{,}800/\sqrt{31}$, or 7500 for January. The question arises, Is this difference statistically significant, or can it reasonably be ascribed to chance variations associated with such small number of observations? In making a decision in this instance about chance variation in the means we do not wish to be wrong more often than 5 times in 100. This requires defining the range in means by ±2 S.E.; 135,000 ± (2 × 6200), or 122,600 to 147,400 for $\bar{A}$; and 142,000 ± (2 × 7500), or 127,000 to 157,000 for $\bar{B}$. Laying off ±2 S.E. about each mean produces a large overlap; and therefore there is *no* statistically significant difference between the two distributions. Chance alone can easily account for the apparent difference.

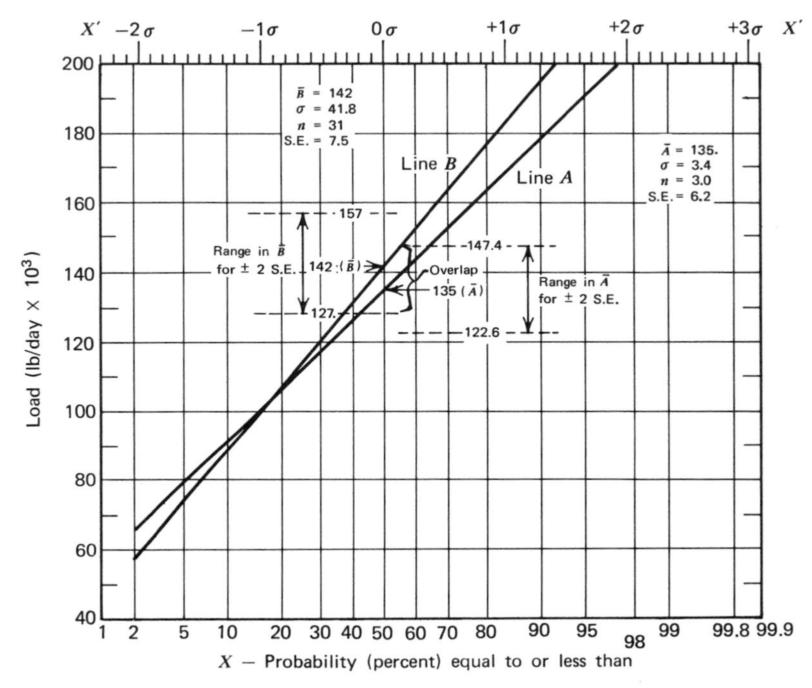

Figure A–14 Illustration of overlapping test between the November and January distributions of suspended solids.

Figure A–15 is a second illustration of the application of the overlapping test. Line A represents again the distribution of the influent suspended solids for November, whereas line B represents the suspended solids of the effluent for November. The mean of the influent is 135,000, with an S.E. of 6200 lb/day, whereas the mean of the effluent is 64,000, with an S.E. of 4700 lb/day. As is to be expected, there is a substantial difference between the influent and the effluent distribution lines (135,000 — 64,000), or 71,000 lb/day. In this instance we wish to be *practically certain* (99.73 percent) that the difference is beyond chance occurrence. This defines a range about the means of ±3 S.E. Laying off ±3 S.E. about each of the means does *not* produce an overlapping—in fact a considerable gap still separates the two distributions; hence we conclude that there *is* a statistically significant difference.

Since in this illustration we are comparing influent and effluent, an opportunity is afforded to appraise the monthly efficiency of the treatment works. The most probable efficiency would be that reflected by the difference between the two means, a removal of 71,000 lb/day, or

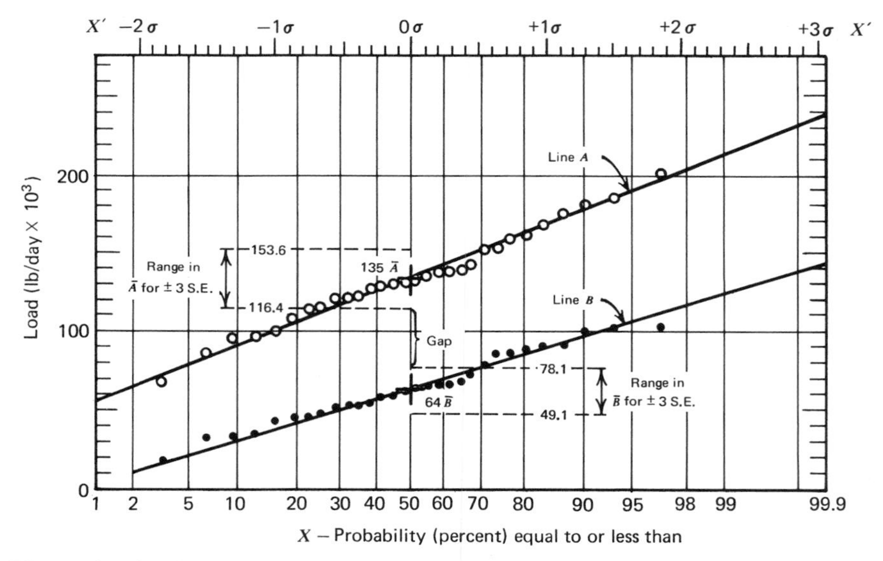

Figure A–15 Illustration of using the overlapping test for differences between the suspended solids content of the influent and of the effluent.

an efficiency of 71/135, or 52.5 percent. A very severe criterion by which monthly efficiency could be judged would be to compare the lowest range of the influent with the highest range of the effluent. This would give a removal of (116,400 — 78,100), or 38.3, an efficiency of 38.3/116.4, or 33 percent.

Another approach to efficiency would be to compare each daily influent and effluent and then plot the daily efficiencies on probability paper to define the nature of the daily variation, as shown in Figure A–16. Here the graphical mean daily efficiency for the month is 53 percent, which is in close agreement with the 52.5 percent previously determined from monthly mean results. It will be observed from Figure A–16 that there is considerable variation in day-to-day efficiency, with a standard deviation of 12.7 percent.

Number of Measurements Controlling Reliability of Mean

Frequently one is asked the question, How many measurements are necessary? The answer to this depends on the reliability that will be accepted and also foreknowledge of the nature of the inherent variability of the individual measurements. The first is a matter of decision; the second may be estimated from previous experience or from a trial run. Take, for example, the monthly mean influent suspended solids for No-

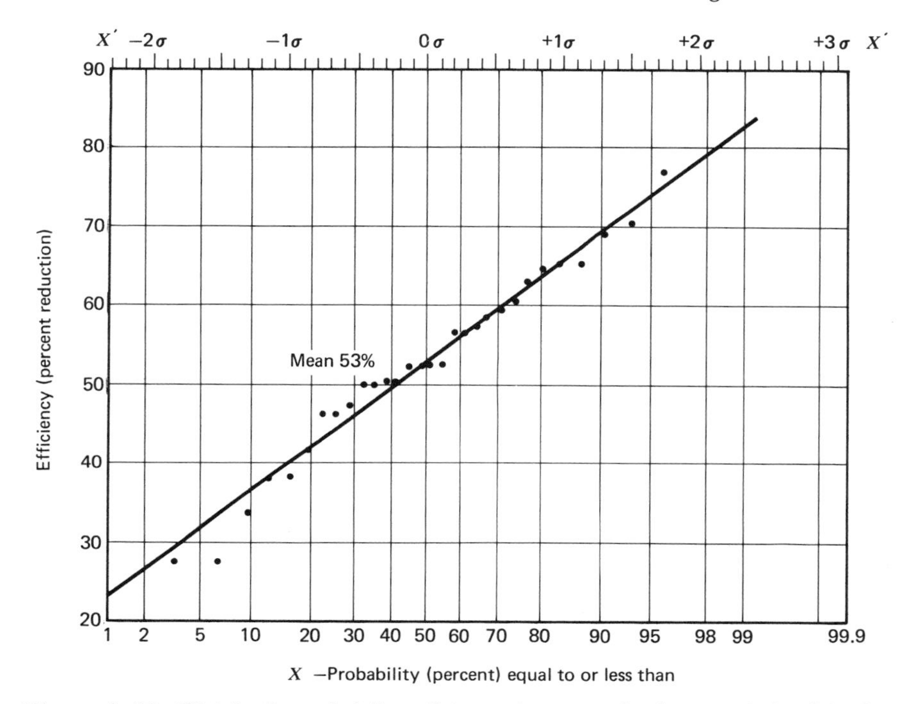

Figure A–16 Distribution of daily efficiency in removal of suspended solids, by primary treatment.

vember, 135,000 lb/day, with a standard deviation of 34,000. If it is decided to control the variation of the mean to a standard error of 2000, how many individual measurements will be required? Accepting the trial σ of 34,000 and the specified standard error of 2000, the required number of individual measurements is then readily determined from the following:

$$n = \frac{\sigma^2}{(\text{S.E.})^2}, \quad \text{or} \quad \frac{34^2}{2^2}, \quad \text{or} \quad 289.$$

Or let us suppose it is decided to be *practically certain* that the mean would not vary more than ± 3000 lb/day. Practically certain is equivalent to ± 3 S.E., hence 3 S.E. = 3000, or the required S.E. = 1000. For this specification

$$n = \frac{34^2}{1^2} = 1156.$$

The Standard Error of the Difference between Two Means

The simple overlapping test is a severe criterion for determining the statistical significance of a difference between two distributions. The assumption is made that the distributions approach each other; that is, as the smaller increases, the larger decreases. This requires a relative shift of 180 degrees, and hence the distance required to keep the distributions from overlapping is some increment of the sum of the standard errors of the two distributions. For example, if the standard error of the mean of series A is four units and that of series B is three units, 68 percent certainty of a statistically significant difference would require the two distributions to be at least $3 + 4$ units apart; 95 percent certainty, $2(3 + 4)$ apart; and practical certainty (99.73 percent), $3(3 + 4)$ apart. The overlapping test therefore defines the standard error of a *difference* between two means as the sum of standard errors of the two means.

If two series are free to shift independently of each other, they need not necessarily always vary by 180 degrees. Actually the independent shifting of two distributions has an average variation in relation to each other at 90 degrees. The relative difference between two means then would be equivalent to the hypotenuse of the right triangle whose legs are the two standard errors of the means.

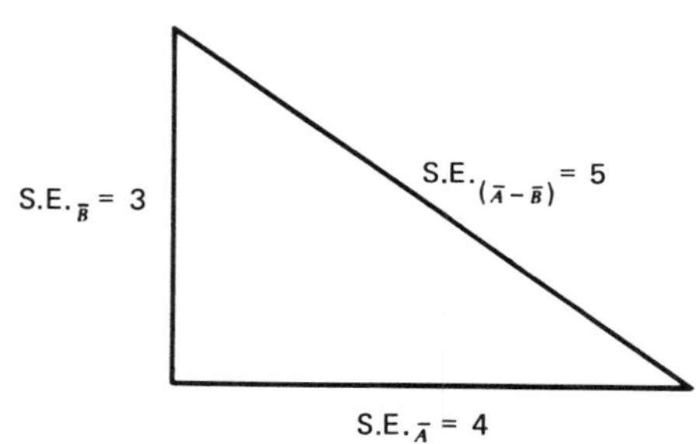

The standard error of the difference between the two means is the square root of the sum of the squares of the two standard errors of the means, which is the hypotenuse distance.

$$S.E._{(\bar{A}-\bar{B})} = \sqrt{(S.E._{\bar{A}})^2 + (S.E._{\bar{B}})^2}.$$

It will be noted that, although the standard error of the difference between the two means by the overlapping test is $3 + 4$, or 7, by assuming complete independence the standard error of the difference is

$$\sqrt{3^2 + 4^2}, \quad \text{or} \quad 5.$$

Thus the overlapping test is unduly severe.

Assuming the independence between two means and employing the hypotenuse as the unit of standard error of the difference, the test for statistical significance between two means is tested at, say, the 95-percent confidence level by subtracting from the difference twice the standard error of the difference. If a positive difference remains, it can be concluded that the difference is statistically significant at this level of confidence. If, after subtracting, no positive difference remains or if a negative value occurs, then the difference was completely washed out by chance alone, and it can be concluded that the apparent difference between the two means is *not* statistically significant.

Let us apply this criterion to the two illustrations employed in the overlapping test: from Figure A–14,

series A (November)	series B (January)
mean = 135	mean = 142
$\text{S.E.}_{\bar{A}} = 6.2$	$\text{S.E.}_{\bar{B}} = 7.5$

difference = 142 .– 135 = 7

$$\text{S.E.}_{(\text{difference})} = \sqrt{6.2^2 + 7.5^2} = 9.7$$

test at 95-percent confidence = $7 - (2 \times 9.7) = -12.4$

The difference is completely washed out by chance alone; therefore the apparent difference is *not* statistically significant. From Figure A–15,

series A influent	series B effluent
mean = 135	mean = 64
$\text{S.E.}_{\bar{A}} = 6.2$	$\text{S.E.}_{\bar{B}} = 4.7$

difference = 135 – 64 = 71

$$\text{S.E.}_{(\text{difference})} = \sqrt{6.2^2 + 4.7^2} = 7.8$$

test at (99.73 percent) practical certainty = $71 - (3 \times 7.8) = +47.6$

A substantial positive difference remains after deducting practically all that can be accounted for by chance; therefore there *is* a statistically significant difference.

Trend in Means

Where a trend is involved, comparisons among groups of means by adjacent pairs may indicate lack of statistical significance, but, if the means are arranged in chronological order, the *trend* may disclose a significant difference over its range. The quickest method to test the signifiance of a trend is by a graphical plot, such as illustrated in Figure A–17. The tabulation in Figure A–17 gives for each year a yearly *mean* elevation of the groundwater (column 2) based on 52 weekly measurements. The standard deviation σ of the individual weekly measure-

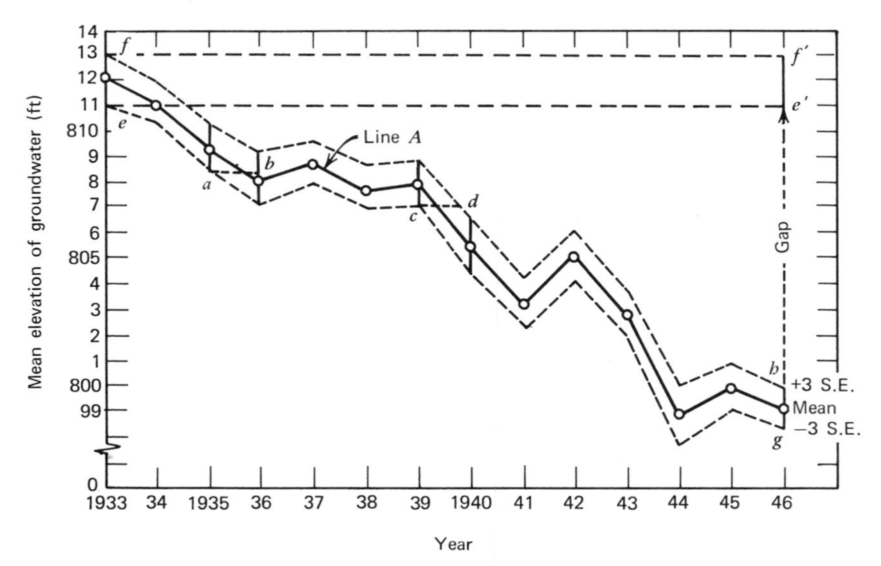

Figure A–17 Trend in groundwater level. Mean annual elevation based on weekly readings. (Tabulation not printed)

ments is shown for each year in column 3. This defines how individual weekly measurements varied each year. Column 4 shows the standard errors of each year's mean, determined from $\sigma/\sqrt{52}$. This defines how we may expect other similar yearly mean values to vary and is used to define ranges of means expected at any desired confidence level.

Graphically the mean value for each year is plotted on the ordinate scale against the year on the abscissa scale, as shown by the solid line *A*. If we decide on a confidence level of 99.73 percent, ± 3 S.E., as a limiting range for the means, we may plot two boundary lines +3 S.E. and —3 S.E. on either side of the yearly mean trend.

A simple test for overlapping for adjacent pairs of means can then be made by drawing a line horizontally from the lower boundary of one year to the next year to note whether the end of this line falls within or outside the range for that year. Referring to 1935 and 1936, the line *a* to *b* is drawn and *b* falls *within* the expected range for mean of 1936. Hence there is overlap and *no* statistical significance between mean elevations for these two years. For the years 1939 and 1940, drawing the line *c* to *d*, *d* falls *outside* the expected range for the mean of 1940. Hence there is no overlap, and we say that the difference between these two years *is* statistically different.

It is also immediately apparent that in some years there was actually a rise rather than a decline—1936 to 1937; 1938 to 1939; 1941 to 1942; and 1944 to 1945. Thus we have a confusing picture if we limit comparisons to adjacent years, with some pairs showing statistical significance, some not, and still others actually reversing and showing a rise. It is, however, visually apparent that there has been a *trend* downward. Comparison of the beginning and end of the trend, 1933 with 1946, is facilitated by projecting the range for 1933 (e to f) horizontally to (e' to f') above the year 1946. There is no overlap; in fact a wide gap exists between the range e' to f' for 1933 and the range g to h for 1946. Since there is no overlap and with such a wide gap between, we may rule out chance with considerable confidence and accept the downward trend as being statistically highly significant.

Linear Regression, the Straight Line Fit

Graphical Test for Linearity

As has been illustrated throughout this text, we seek to define relevant relationships between two variables in the conveniently usable form of the simple straight line. As a first step in this process the x and y coordinates are plotted on an arithmetic grid as a direct visual test for linearity. If the data show a straight line trend, frequently a good balanced line through the bulk of the plotted points can be fitted by eye; where there is considerable scatter the linear trend is not so readily apparent; a mathematical fit is usually obtained by the least squares method. However, as we have already seen in the text, if there are abnormal breaks in the data a mathematical fit may be misleading and a visual approach is preferable.

If the initial plot on an arithmetic grid develops a curvilinear relation, in many instances this can be reduced to linear form by replotting on semilog paper or on log-log paper. If a linear trend is produced on such logarithmic grids, it is then possible to proceed with fitting the straight line by working with the logarithms of the appropriate original data. Similarly, fitting the straight line on other special grids constructed to produce linearity may proceed by employing the appropriate reduced variates for the original data.

The Least Squares Line of Best Fit

The line of best fit through the data by the least squares method is usually termed the *regression line*. It is customary to designate the x-coordinate as the *independent* and the y-coordinate as the *dependent* variate. The line is the best expression of the dependent relation of one variate on the other in the same sense that the mean is the best single value representation of one variate. In each case the sum of the squared deviations about the line of best fit is a minimum.

In this two-dimensional series there are two directions in which the deviations may be measured, one parallel to each axis. There are, correspondingly, two possible regression lines, the regression of y on x and the regression of x on y. It is for this reason that it is desirable to designate as independent that variate (x) whose influence on the dependent variate (y) is to be evaluated. This distinction may be wholly arbitrary and does not preclude a reversal of the designations under appropriate conditions. In most situations with which we deal the distinction between variates is inherent from our basic concept of the phenomenon under investigation, although not always necessarily so.

Three simple characteristics, it will be recalled, define a straight line through scattered data, as illustrated in Figuure A–18.

1. The center of gravity of the data.
2. The slope of the line.
3. The elevation of the line.

The slope is the pitch of the line from the horizontal, which may be positive or negative; the elevation is the intercept on the Y-axis where x equals zero, which may also be positive or negative. The center of gravity is located at the coordinates of the means of the x and y data $\bar{X}$, $\bar{Y}$).

These characteristics are reflected in the equation of the straight line

$$y = C_0 + C_1 x$$

and the coefficients are best defined from the least squares relations as follows:

$$C_1 = \frac{\Sigma(dx\,dy)}{\Sigma(dx)^2} = \frac{\Sigma(dx\,dy)}{N(\sigma_x{}^2)}, \qquad C_0 = \bar{Y} - C_1\bar{X},$$

where dx and dy are the deviations of the respective values of x (independent variable) and y dependent variable) from their means $\bar{X}$ and $\bar{Y}$, N is the number of observations, and σ_x is the standard deviation of the x data.

It will be noted that, if the origin of the x- and y-axes is translated to the center of gravity ($\bar{X}$, $\bar{Y}$), C_0 is zero and the equation of the straight line is

$$y = C_1 x.$$

Thus the most significant regression parameter is C_1, referred to as the *slope coefficient.* It evaluates the relation between two variates in the form of *an average rate of change of* y *with change of* x.

From the regression equation for a known value of x an estimated value of y can be obtained. It must be recognized that this is only an estimate of y (the best estimate under the circumstances) and the true value of y may deviate from this estimated value. The magnitude of this deviation will depend on how well the original data approximated a straight line.

Standard Error of Estimate

The reliability of any estimated value y_e is defined by the *standard error of estimate* σ_e from

$$\sigma_e = \left[\frac{\Sigma(y_o - y')^2}{N}\right]^{1/2},$$

where y' is the estimated value of y computed from the equation $y = C_0 + C_1x$ for a given value of x and y_o is the *observed* value of y for the particular x value under consideration. This is not a convenient equation and a more practical transformation in application is

$$\sigma_e = \sigma_y \sqrt{1 - r_{xy}^2},$$

where σ_y is the standard deviation of the y data and r_{xy} is the correlation coefficient (subsequently explained).

In drawing the regression line through the plotted data the coordinates of the center of gravity $\bar{X}$, $\bar{Y}$ are located, and from this point the line is laid off at the slope C_1, and intercepting the Y-axis at C_0. The standard error of estimate σ_e provides a very important addition to the line of best fit. A band commonly referred to as *scatter lines* can now be constructed about the line for any particular confidence level, say 95 percent, $2\sigma_e$, above and below, parallel to the line of best fit, as shown in Figure A–18. It is assumed that the errors in estimates of y are normally distributed about the line of best fit, which should be true in most cases.

It is convenient to resort to the graph rather than the equations to obtain the most probable estimate of y from the line of best fit for any particular given value of x within the range of the basic data and likewise to read directly from the scatter lines the upper and lower limits of the range within which we expect estimates of y to fall for the particular confidence level prescribed. For example, in the illustration (Figure A–18) for the given value of $x = 10$ the most probable estimate of y is read as 13.0, and the lower and upper limits of the 95-percent confidence range are read from the scatter lines as 11.2 and 14.8.

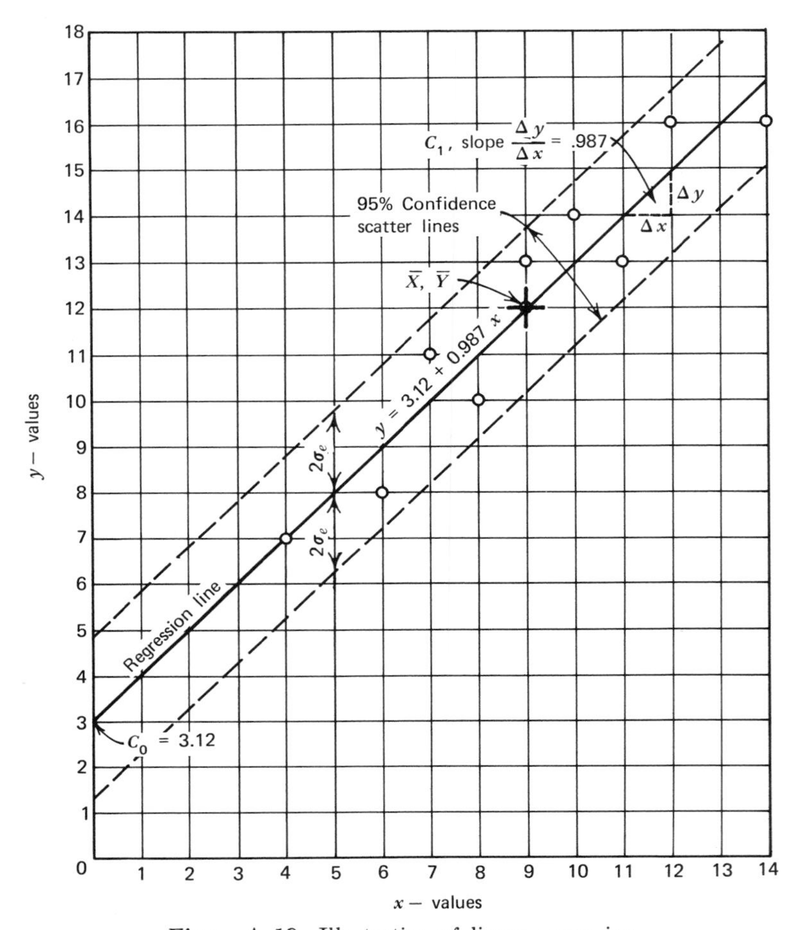

Figure A–18 Illustration of linear regression.

A word of caution. An unfortunate connotation of the term "regression line" is the idea that it may be extended indefinitely beyond the range of the data from which it is derived. A reasonable *extrapolation* may be made based on understanding of the phenomenon under investigation; otherwise unreasonable extrapolation may yield absurd results. However convincing and useful, extrapolation beyond the range of the immediate observations is the responsibility of the investigator. On the other hand, *interpolation,* the evaluation of specific relations within the range of the experimental observations, even though not actually encountered in any one experimental trial, is a major purpose of the regression line and its associated confidence lines of scatter.

CORRELATION

Correlation indicates how closely two variates are related and their degree of interdependence, whereas the regression line indicates the extent to which one follows the other even though in a very scattered form. The concept of correlation is one in which covariation is viewed as a mutual relation, whereas the regression line is essentially a unilateral relation. The slope coefficient in regression provides the average rate of change but no measure of degree of relation. The *coefficient of correlation* provides a measure of degree of interdependence but no measure of rate of change. Thus regression and correlation are complementary.

The Correlation Coefficient

There are various methods of measuring and expressing the degree of association among variables; all are arbitrary. The best known and most useful of these is the Pearson product-moment coefficient. It is usually designated by r with appropriate subscripts and is given by

$$r_{xy} = \frac{\Sigma(dx\,dy)}{\sqrt{\Sigma(dx^2)\,\Sigma(dy^2)}} = \frac{\Sigma(dx\,dy)}{N\sigma_x\sigma_y},$$

where dx is the deviation of x values from their mean $\bar{X}$ and dy is the deviation of the y values from the mean $\bar{Y}$.

In numerical value r may range from unity, indicating perfect correlation of a positive sort, through zero (no correlation) to minus unity—that is, perfect negative (inverse) correlation.

Reliability of the Correlation Coefficient

As with the means, correlation coefficients, based on random sampling from a large universe, are distributed about the true value, the coefficient of the universe. Unfortunately the distribution of r is not normal where true correlation exists in the universe nor does it lend itself to simple treatment. However, if *no* basic relationship exists between the two variables (parent universe $r = 0$), deviations from zero can be expected by chance alone in random samples. This distribution about zero fortunately *is normal.* Thus in testing of a sample coefficient of correlation for statistical significance two cases are considered:

Case I: Parent Universe $r = 0$. In the case where no true correlation exists in the universe the distribution of sample coefficients of correlation is normal and deviation from zero is measured by the standard deviation σ_r.

$$\sigma_r = \frac{1}{\sqrt{N - 1}}.$$

Thus for practical certainty (99.73-percent confidence) of a statistically significant relation a sample coefficient r_{xy} must fall outside the range of $0 \pm 3\sigma_r$; for 95-percent confidence outside the range of $0 \pm 2\sigma_r$; or other increments of σ_r for other levels of confidence in accord wtih the normal distribution.

Case II: Parent Universe True Correlation. Where true correlation exists between two variables of a universe the sample coefficient r_{xy} and its standard deviation σ_r do not apply. Recourse is had to an approximation through the z-function of Fisher [4]. Fisher proposed a transformation derived from r by the relation

$$z = \tfrac{1}{2} \log_e \frac{1 + r}{1 - r} = 1.15 \log \frac{1 + r}{1 - r},$$

and a standard deviation of z by the relation

$$\sigma_z = \frac{1}{\sqrt{N - 3}}.$$

The distribution of z is not exactly normal nor is its standard deviation constant under all conditions, but the function approaches these conditions with a sufficiently close approximation to make it acceptable as normal for most practical applications.

It is to be noted that in both cases reliability is predominantly related to the size of the sample, inversely proportional to the square root of N. In either case, therefore, small samples can be expected to portray large standard deviations (σ_r, σ_z). The finding of no statistically significant relation on the basis of a small sample may conceivably convert to significance if the sample is enlarged.

Illustrated Example

Figure A–18 is a plot of a small sample of 10 points with coordinates x, y as listed in Table A–5; it illustrates the principles of linear regression and correlation discussed above.

The regression line of best fit passes through the coordinates of the means ($\bar{X}$, $\bar{Y}$) at the slope defined by C_1. From this line is obtained the best estimate of y corresponding to a given x. For example, for $x = 10$, y_e is read as 13 on the graph or from the regression equation is computed as

$$y_e = 3.12 + (0.987)(10) = 12.99.$$

Table A–5 Illustration of Regression and Correlation Computations

x	dx	dx^2	y	dy	dy^2	$dx\,dy$
4	−5	25	7	−5	25	25
6	−3	9	8	−4	16	12
7	−2	4	11	−1	1	2
8	−1	1	10	−2	4	2
9	0	0	12	0	0	0
9	0	0	13	+1	1	0
10	+1	1	14	+2	4	2
11	+2	4	13	+1	1	2
12	+3	9	16	+4	16	12
14	+5	25	16	+4	16	20
Σ 90		78	120		84	77

$$\bar{X} = \frac{\Sigma x}{N} = \frac{90}{10} = 9; \qquad \bar{Y} = \frac{\Sigma y}{N} = \frac{120}{10} = 12;$$

$$C_1 = \frac{\Sigma(dx\,dy)}{\Sigma(dx^2)} = \frac{77}{78} = 0.987; \qquad C_0 = \bar{Y} - C_1\bar{X} = 12 - (0.987)(9) = 3.12;$$

$$r_{xy} = \frac{\Sigma(dx\,dy)}{\sqrt{\Sigma(dx^2)\,\Sigma(dy^2)}} = \frac{77}{(78)(84)} = 0.951;$$

$$\sigma_y = \sqrt{\Sigma(dy^2)/N} = \sqrt{84/10} = 2.9;$$

$$\sigma_e = \sigma_y\sqrt{1 - r_{xy}^2} = 2.9\sqrt{1 - 0.951^2} = 0.894;$$

$$\sigma_r = \frac{1}{\sqrt{N-1}} = \frac{1}{\sqrt{9}} = 0.333.$$

Regression equation: $y = C_0 + C_1x = 3.12 + 0.987x.$

However, in repeated samples of this small size it can be expected that y_e will vary considerably as measured by σ_e, (0.894); for 95-percent confidence between $12.99 \pm 2\sigma_e$, or from 11.20 to 14.78. Five times in 100 we expect y_e to fall outside this range. If we do not wish to accept this risk in the estimate of y exceeding this range, higher confidence scatter lines can be drawn, which then necessarily widens the range.

In evaluating the degree of correlation between x and y as a first step it is assumed that no true correlation exists in the parent universe from which this small sample was taken; hence r_{xy} and σ_r are applicable,

The computed coefficient of correlation is

$$r_{xy} = \frac{77}{\sqrt{(78)(84)}} = 0.951.$$

This appears to be of a high order approaching unity, but the standard deviation σ_r is large.

$$\sigma_r = \frac{1}{\sqrt{10 - 1}} = 0.333.$$

If there is no correlation, we expect r_{xy} in repeated samples of 10 sets of coordinates to vary about zero as follows:

68 percent confidence	0 ± 0.333
95 percent confidence	0 ± 0.667
99.73 percent confidence	0 ± 0.999

Thus, if we wish to be practically certain of a significant relation, the sample coefficient of correlation must exceed 0.999. Since the computed r_{xy} is below this (0.951), we *cannot* say with practical certainty that there is a significant relationship for this number of measurements.

Before accepting this conclusion, knowing that σ_r varies inversely as the square root of N, a large sample is obviously in order. For example, if we increase the sample size N to 101, σ_r would reduce to

$$\sigma_r = \frac{1}{\sqrt{101 - 1}} = 0.1.$$

If the correlation coefficient remained the same ($r_{xy} = 0.951$), we now would be practically certain that by chance alone, if no true relation existed, variation would not exceed $0 \pm 3\sigma_r$, or 0 ± 0.3. With the larger sample we are quite certain that a significant relation exists.

If it is desired to measure further the degree of this demonstrated relationship indicated in the analysis of the larger sample, the Fisher z-function is then employed.

Machine Equations

The equations we have presented are not especially adapted to the ordinary calculating machines. The following transformations proposed by Dwyer [5] facilitate solutions of linear regression and simple correlation problems:

$$L_{xx} = N\Sigma x^2 - (\Sigma x)^2,$$
$$L_{yy} = N\Sigma y^2 - (\Sigma y)^2,$$
$$L_{xy} = N\Sigma(xy) - \Sigma x\Sigma y.$$

These equations then provide

$$C_1 = \frac{L_{xy}}{L_{xx}},$$

$$C_0 = \frac{\Sigma y - C_1 \Sigma x}{N};$$

$$r_{xy} = \frac{L_{xy}}{\sqrt{L_{xx}L_{yy}}},$$

$$\sigma_e = \frac{1}{N}\left[\frac{L_{yy}L_{xx} - (L_{xy})^2}{L_{xx}}\right]^{1/2}.$$

Using Dwyer's machine equations, our illustrated problem would be solved as follows:

$$\Sigma x^2 = 888, \qquad \Sigma y^2 = 1524,$$

$$(\Sigma x)^2 = 8100, \qquad (\Sigma y)^2 = 14{,}400,$$

$$\Sigma(xy) = 1157, \qquad \Sigma x \Sigma y = 10{,}800,$$

$$L_{xx} = (10)(888) - 8100 = 780,$$

$$L_{yy} = (10)(1524) - 14{,}400 = 840,$$

$$L_{xy} = (10)(1157) - 10{,}800 = 770,$$

$$C_1 = \frac{770}{780} = 0.987,$$

$$C_0 = \frac{120 - (0.987)(90)}{10} = 3.12,$$

$$r_{xy} = \frac{770}{\sqrt{(780)(840)}} = 0.951,$$

$$\sigma_e = \frac{1}{10}\left[\frac{(840)(780) - 770^2}{780}\right]^{1/2} = 0.894.$$

LOGISTIC POPULATION PAPER

In dealing with human population growth, biological multiplication, progression of epidemics, and the like the logistic growth curve is transformed into a linear form, which facilitates solution by graphical methods.

Mathematical Formulation

Mathematical expression of the thesis of limited population increase is represented by the logistic, or S, curve (Figure A–19). Population growth, whether of plants, animals, or people, appears to follow a similar pattern. The rate of increase is expressed by the slope of the curve. It is positive throughout, increasing to a maximum at some midpoint and then decreasing toward zero at an upper asymptotic limit.

This rate has the differential form

$$\frac{dy}{dt} = Ky(A - y),$$

y being the change up to any time t and A being the total ultimate growth or population saturation.

Differentiating and equating to zero, the maximum rate is located by the relation

$$\frac{d}{dt}\left(\frac{dy}{dt}\right) = K(A - y)\,dy - Ky\,dy = 0,$$

$$y = A - y = \frac{A}{2}.$$

The rate is symmetrical about this midpoint.

An integral, but not the most general one, is

$$y = \frac{A}{1 + e^{-AKt}},$$

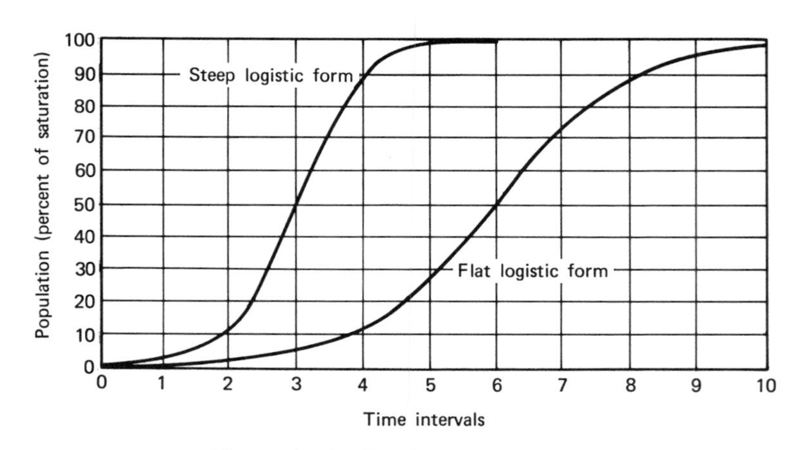

Figure A–19 The logistic curve.

whence

$$\frac{y}{A - y} = e^{AKt}$$

and

$$\log_e \left(\frac{y}{A - y}\right) = AKt.$$

If population y is expressed in percent of saturation (y'), the saturation value A is common to all cases at 100 percent and

$$\log_e \left(\frac{y'}{100 - y'}\right) = 100\ Kt.$$

To change to common logarithms merely changes the constant K. For convenience we write

$$\log \left(\frac{y'}{100 - y'}\right) = 100\ kt.$$

In this form of the integral the time axis is so located that, at $t = 0$, $y' = 50$ percent, the midpoint of the curve, which is the point of maximum rate. Substituting z for the left-hand term reduces to

$$z = \log \left(\frac{y'}{100 - y'}\right) = kt,$$

in which the function z is linear with t. At $t = 0$, $z = 0$ and $y' = (100 - y') = 50$ percent.

The Logistic Grid

A logistic grid representing this linear transformation is obtained by laying off x units as abscissa and z units as ordinates to the same scale. The ordinates are marked in terms of y', percent of saturation, but their dimensions are z, $\log\,[y'/(100 - y')]$.

Population data, logistic in character, plot as straight lines on this grid. The time intervals x to attain stated percentages of population saturation y' are listed in Table A–6, and the construction of the logistic grid is illustrated in Figure A–20. It will be observed that selecting a few of these coordinates from Table A–6 and plotting them on a Cartesian grid develops a perfect S-curve (Figure A–20a); plotting the same coordinates on the logistic grid develops a perfect straight line (Figure A–20b).

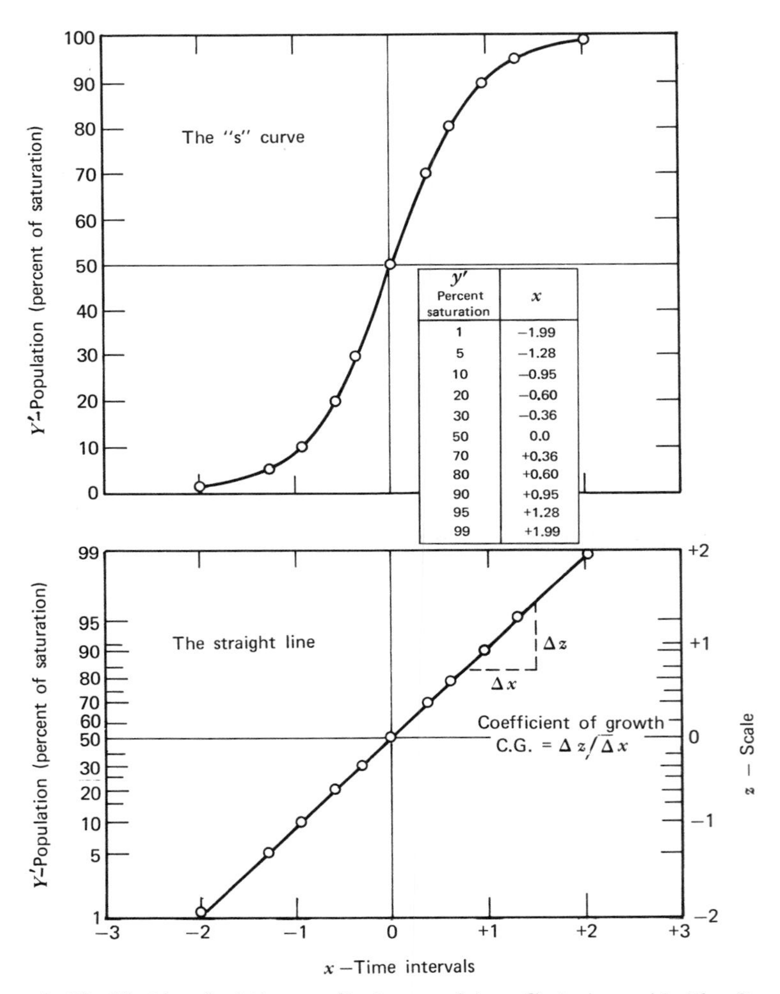

y' Percent saturation	x
1	−1.99
5	−1.28
10	−0.95
20	−0.60
30	−0.36
50	0.0
70	+0.36
80	+0.60
90	+0.95
95	+1.28
99	+1.99

Figure A–20 Plotting logistic coordinates on (*a*) a Cartesian grid (the S-curve) and (*b*) on a logistic grid (the straight line).

Application to Community Population Growth

In addition to the discussion of population projection presented in Chapter 2, the following illustrates the application of the logistic method to community population growth.

If population is expanding freely in a society of unlimited economic opportunity, the rate of increase is constant; in an area of limited economic opportunity the rate of increase must become less and less as

Table A–6 The Logistic Grid

y' (percent saturation)	$\frac{y'}{100 - y'}$	$\log \frac{y'}{100 - y'} = x = z$ Positive for $y' > 50$ Negative for $y' < 50$	y' (percent saturation)
50	1.0	0.0	50
55	1.22222	0.0871	45
60	1.50000	0.1761	40
65	1.85714	0.2688	35
70	2.33333	0.3680	30
75	3.00000	0.4771	25
80	4.00000	0.6021	20
81	4.26315	0.6297	19
82	4.55556	0.6585	18
83	4.88235	0.6886	17
84	5.25000	0.7202	16
85	5.66667	0.7534	15
86	6.14285	0.7884	14
87	6.69230	0.8256	13
88	7.33333	0.8653	12
89	8.09090	0.9080	11
90	9.00000	0.9542	10
91	10.11111	1.0048	9
92	11.50000	1.0607	8
93	13.28571	1.1224	7
94	15.66667	1.1950	6
95	19.00000	1.2787	5
96	24.00000	1.3802	4
97	32.33333	1.5096	3
98	49.00000	1.6902	2
99	99.00000	1.9956	1
99.5	199.00000	2.2988	0.5
99.9	999.00000	2.9996	0.1

the population grows and is limited to a saturation value. This basis is essentially Verhulst's theory of limited human increase applied to a technological, rather than an agrarian, society. In a complex technological order economic opportunity is a more important determinant of population saturation than land area, the Verhulst's theory is reconstructed accordingly. As the community grows in population it also grows in area; the boundaries are defined in terms of the sphere of economic influence extending outward from the central core.

Application of the logistic grid to human population growth simplifies projection and provides two important measures of the character of the community growth, its age and its coefficient of growth [6].

Age or Stage of Community Growth

Expressing population in percent of saturation changes the concept from numbers of people to that of the age of the entire unit or colony, expressed as a percentage of its ultimate growth. Furthermore on this percentage basis all colonies are directly comparable, irrespective of the number of individuals within each. If observed census data converted to percentages of saturation are plotted on the logistic grid, they will approach a straight line if growth is logistic. If not strictly logistic, the variation from a straight line is an indication of the skew from true logistic form. Furthermore in the *slope* of this line lies the answer to the rate of maturity of the community.

Coefficient of Growth

The rate of maturity of a logistic growth is measured by the overall pitch of the logistic curve rather than by the slope of the curve at any point or through any particular decade, a slowly maturing community describing an easy, flat S-curve and a rapidly maturing community describing a sharp, steep S-curve. Since the logistic grid reduces the S-curve to a straight-line form, this pitch is measured as the slope of the straight line $\Delta z/\Delta x$ on logistic paper and is termed the *coefficient of growth* (C.G.).* The coefficient of growth is an aid in the comparative study of the patterns of growth among communities.

Population Saturation

The application of this technique of population analysis is predicated on a determination of population saturation. No one simple method for predicting the ultimate saturation value of population is available. Mathematical techniques may be employed but are no more reliable than the assumptions on which they are based and often lead to irrational saturation values. Since the ultimate limit to growth is determined by economic opportunity, long range economic base projections are an aid in estimating population saturation. According to the thesis propounded, as the rate of increase approaches zero the population approaches its limit, saturation. This limit is approximated graphically if ratios of increments of increase (for intervals of time) to population are plotted

* In the mathematical formulation the slope ($\Delta z/\Delta x$) is k in

$$\log\left(\frac{y_1'}{100 - y_1'}\right) - \log\left(\frac{y_2'}{100 - y_2'}\right) = k(x_1 - x_2).$$

against population at the ends of the periods. Such plots, although zig-zag in character, usually demonstrate a trend toward decreasing rate of increase; the extension of the trend to the population intercept at zero rate of increase determines the saturation value. A graphical extrapolation is preferable to a mathematical fit since more weight can be given to the recent than to the early rates. Usually three estimates are made: a high saturation based on assumption of a combination of factors favoring growth, a low saturation based on a combination of factors unfavorable to growth, and a medium saturation between these extremes.

The concept of population saturation may be difficult to accept in a young society accustomed to think in terms of ever expanding economy. In the short range this view is understandable, but in the long range it is irrational to assume population growth without limit. Larsen [7] suggests the term "plateau," rather than "saturation," as less absolute with recognition of attainment in stages of intermediate levels. As applied in the logistic method, population saturation is envisaged in the long range as population stability (or plateau) with no implication of degeneration. In fact to support and maintain a high level of population in a competitive society necessitates continuous economic growth and improvement in the quality of the environment. Within a stable population there is unlimited opportunity to absorb an expanding economy in terms of quality of living.

Illustration of the Logistic Method

Figures A–21*a* and *b* illustrate a simple application of the graphical logistic projection to a self-contained community based on census data 1850 to 1960. Table A–7 lists the census population from which the increase per decade and the percentage of increase are determined. Figure A–21*a* is the plot of percentage of increase (column 4) against population (column 2), through which a low, medium, and high saturation are estimated as 60,000, 80,000, and 100,000, respectively; columns 5, 6, and 7 are the conversions of the census population to percentage of saturation for the respective saturation levels.

These population values expressed as percentage of saturation plotted on the logistic paper (Figure A–21*b*) approximate straight lines, indicating that the community growth follows the logistic theory; from these lines the characteristics of community growth are defined. Line *A*, the low saturation, provides the best fit, with an average deviation* from the census values for 1880–1960 of 1.8 percent; the community age repre-

* The average deviation is merely a measure of the degree to which the census values deviate from the logistic curve; it is not a measure of reliability of the projection beyond the last census value.

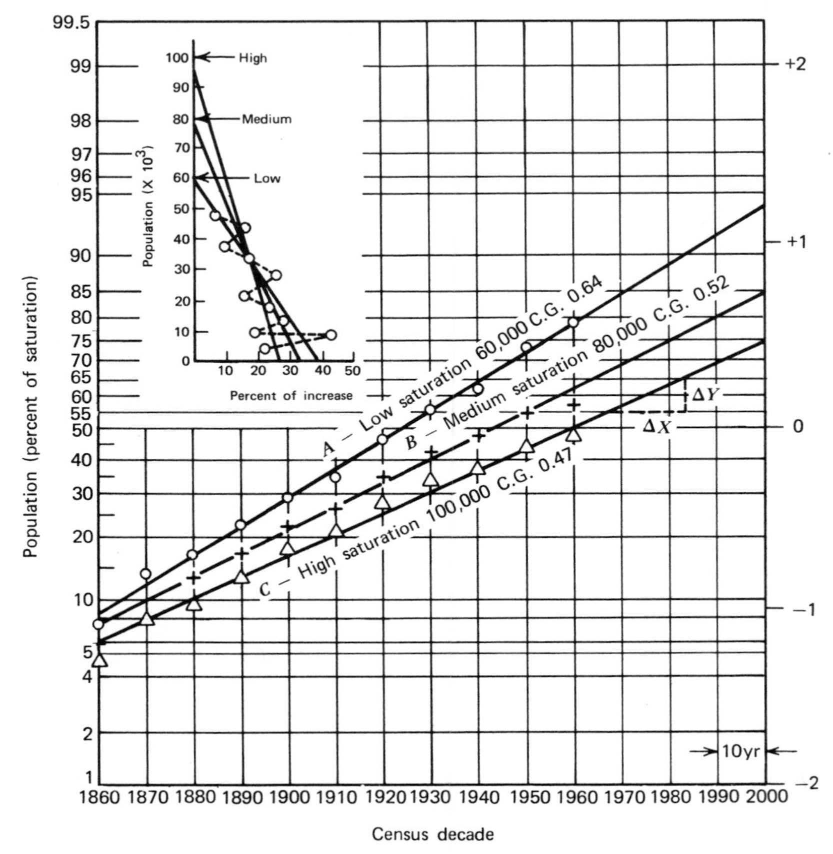

Figure A–21 Application of graphical logistic projection of community population growth.

sented by the percentage of saturation as of 1960 is 79 percent; the coefficient of growth represented by the slope $\Delta Y/\Delta X$ (measured in any linear scale) is 0.64. Similarly line B, the logistic based on the medium saturation, provides a logistic fit, with an average deviation of 3.1 percent, a stage of growth, as of 1960, of 63 percent of saturation, and a growth coefficient of 0.52. Line C, based on the high saturation population, it is noted, tends to curve downward slightly, indicating that this saturation value may possibly be too high. (If a logistic plot tends to bend upward, it indicates that possibly the saturation value is too low.) The average deviation from the logistic line C is 4.1 percent,

Table A–7

Year (1)	Census Population ($\times 10^3$) (2)	Increase (3)	Percent of Increase (4)	Percent of Saturation: Low Saturation (60) (5)	Medium Saturation (80) (6)	High Saturation (100) (7)
1850	3.6			6.0	4.5	3.6
1860	4.6	1.0	21.7	7.7	5.7	4.6
1870	8.0	3.4	42.5	13.3	10.0	8.0
1880	9.8	1.8	18.4	16.3	12.3	9.8
1890	13.5	3.7	27.4	22.5	16.9	13.5
1900	17.6	4.1	23.3	29.3	22.0	17.6
1910	20.8	3.2	15.4	34.7	26.0	20.8
1920	27.8	7.0	25.2	46.3	34.8	27.8
1930	33.5	5.7	17.0	55.8	41.9	33.5
1940	37.2	3.7	9.9	62.0	46.4	37.2
1950	44.2	7.0	15.8	73.7	55.3	44.2
1960	47.3	3.1	6.6	79.0	59.1	47.3

age as of 1960 is 50 percent, or midstage on the growth curve, and growth coefficient is 0.47.

The population projection of future growth along the logistic curve is readily obtained by extending the straight lines *A*, *B*, and *C*, reading percentage of saturation at each decade and converting to population by applying the respective saturation values. These are summarized in Table A–8 from 1850 to the year 2000, and plotted on a Cartesian grid as a projection belt (high, medium, and low), as shown in Figure A–22. Since the saturation values are theoretically attained only at an infinite time, the projection belt widens gradually beyond 1960, and hence, although the saturation values selected may be quite wide, the differences in population estimates 10 or 20 years beyond 1960 are not so divergent.

Two-Stage Logistic Growth

The impact of socioeconomic change may be of sufficient magnitude to induce a two-stage logistic curve, each stage with its distinct coefficient of growth. For example, the introduction of mass production industry has resulted in marked two-stage growth for such areas as Detroit. Before 1910 the standard metropolitan area (Wayne, Macomb, and Oakland counties) shows a growth coefficient of 0.5, whereas after that date and coincident with the introduction of the automobile industry the

Table A–8 Community Logistic Projection

Year	Census ($\times 10^3$)	Low Saturation (60)		Medium Saturation (80)		High Saturation (100)	
		Percent of Saturation	Population ($\times 10^3$)	Percent of Saturation	Population ($\times 10^3$)	Percent of Saturation	Population ($\times 10^3$)
1850	3.6	6.0	3.6	5.3	4.2	5.0	5.0
1860	4.6	8.6	5.2	7.4	5.9	6.5	6.5
1870	8.0	12.0	7.2	9.8	7.8	8.0	8.0
1880	9.8	16.3	9.8	13.0	10.4	10.5	10.5
1890	13.5	22.0	13.2	16.9	13.5	13.0	13.0
1900	17.6	29.3	17.6	21.5	17.2	17.0	17.0
1910	20.8	37.0	22.2	27.0	21.6	20.8	20.8
1920	27.8	46.3	27.8	33.5	26.8	25.0	25.0
1930	33.5	56.0	33.6	40.0	32.0	31.0	31.0
1940	37.2	65.0	39.0	47.0	37.6	37.0	37.0
1950	44.2	73.0	43.3	55.0	44.0	44.2	44.2
1960	47.3	79.0	47.3	63.0	50.4	50.0	50.0
1970		85.0	51.0	69.0	55.2	57.0	57.0
1980		89.0	53.4	75.0	60.0	64.0	64.0
1990		92.1	55.3	80.0	64.0	70.0	70.0
2000		94.3	56.6	84.5	67.5	75.0	75.0

growth shifted to a steeper form, with a growth coefficient of 1.0. Such changes in characteristics of growth are not so readily recognized by other methods of analysis. Similarly great social changes induce two-stage growth—transition from an agrarian to an urban-industrial economy.

Changes in coefficient of growth from decade to decade should not be taken as multistage growth, but should be considered as variations inherent in the complexities of community growth. Two-stage characteristics are only identifiable over several decades; a substantial, sustained social or economic change is necessary to establish a new coefficient of growth pattern. A high coefficient of growth indicates an explosive type of community, with a high middle-age growth, but is a danger sign of a short community lifespan with rapid transition toward population saturation in old age. Whether automation and the atomic age will be a change of sufficient magnitude to raise saturation levels and induce new patterns reflected in measurable changes in coefficients of growth remains to be seen.

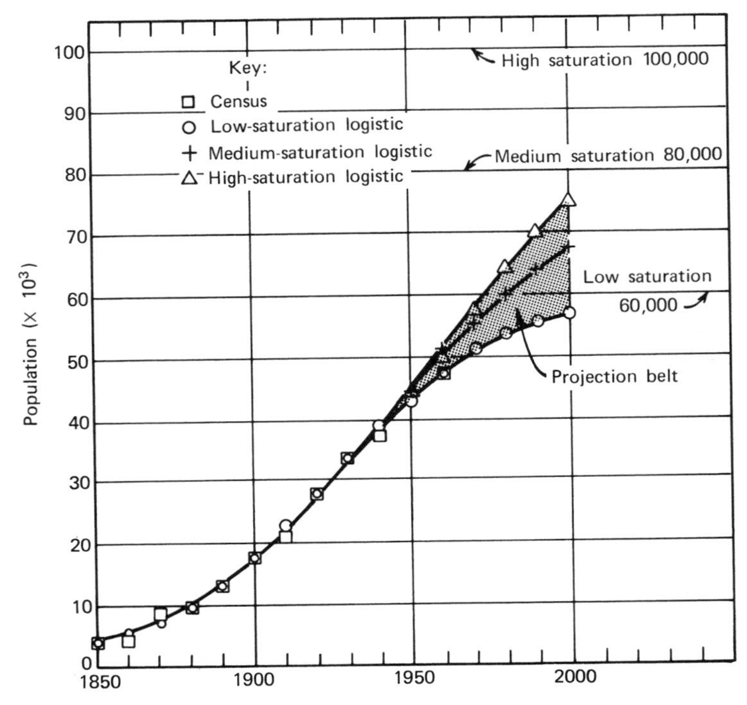

Figure A–22 Community logistic growth projection belt.

REFERENCES

[1] Hazen, A., *Flood Flows,* Wiley, New York, 1930.

[2] Gumbel, E. J., *Ann. Math. Statistics,* **12,** 163 (1941).

[3] Powell, R. W., *Civil Engineering,* **13,** 105 (1943).

[4] Fisher, R. A., *Biometrika,* **10,** 507 (1915).

[5] Dwyer, P. S., *Linear Computations,* Wiley, New York, 1951, p. 302.

[6] Velz, C. J., and Eich, H., *Civil Engineering,* **10,** 619 (1940).

[7] Larsen, H. T., *J. San. Engr.* (December 1964); *Proc. ASCE, 90,* SA 6, 140, (1964).

Appendix B

BACTERIAL ENUMERATION

MEAN DENSITY AND THE LAW OF MIXING

The nature of distribution of bacterial cells in a larger volume may be referred to *mean density* (MD). If it were possible to count the entire volume, the mean density would be the total count divided by the total volume expressed in some unit such as milliliters. If less than the total volume were employed in counting, it would be found that the densities in the smaller samples would vary from the mean density. Perfect admixture as represented by the mean density would require exactly the same number of bacterial cells in each unit volume. Complete segregation, on the other hand, would require all cells to be located in a single unit volume. Such extreme distributions do not prevail as the *law of mixing* comes into play. Whether approached from either initial distribution of perfect admixture or complete segregation, statistical study of results of many repetitions of mixing experiments show that the end result is the same. Any mixing tends to redistribute the cells so that a partial admixture results, leading to a statistical mean state, departures about which in either direction, become increasingly unstable and improbable.

THE DILUTION METHOD

The Binomial and the Poisson Series

The dilution procedure is fundamentally a demonstration of mere presence or absence of organisms. Consider the case of 1000 bacterial cells in a 1000-ml volume and consider one specific milliliter as a unit volume and number the 1000 bacterial cells serially (1, 2, 3, . . . , 1000). The probability that cell 1 will be located in the unit volume is 1/1000 and the probability that cell 1 will *not* be located in this unit volume will be (1 — .001), or .999. Likewise, for cell 2 probability of a positive result in the unit volume is .001 and the probability of a negative result is .999. Similarly for the other cells. Thus there may be 1001 results, 0, 1, 2, 3, . . . , 1000 bacterial cells in the unit volume, the probability of each of which is given by the expansion of $(p + q)^n$, which in this case is

$$(.001 + .999)^{1000}.$$

The binomial expansion of $(p + q)^n$, describes the normal distribution best when p equals q and in the central portion. In the dilution method of evaluating bacterial density p is very small in relation to q. This type of distribution is best represented by the Poisson series.

$$e^{-x}\left(1 + x + \frac{x^2}{2} + \frac{x^3}{2 \times 3} + \frac{x^4}{2 \times 3 \times 4} + \cdots\right), \quad \text{(B–1)}$$

where e is the base of natural logarithms, 2.718, and x is the true mean density of the bacteria. The sum of the series is unity, and the terms represent the respective probabilities of there being 0, 1, 2, 3, 4, etc., bacteria within any random unit volume. The probability of *no* bacterial cells being within a unit test volume is given by the first term of the Poisson series, e^{-x} and the probability of a positive result (one or more organisms present) is $(1 - e^{-x})$. Since the probability of a *compound event* composed of two independent events is the product of their individual probabilities, the probability of one unit test volume's being positive simultaneously with another unit test volume's being negative is given by

$$P = e^{-x}(1 - e^{-x}) \quad \text{(B–2)}$$

or for test volumes of different unit sizes

$$P = e^{-v_1x}(1 - e^{-v_2x}), \quad \text{(B–3)}$$

where v_1 is the volume in milliliters of the negative portion and v_2 is the volume in milliliters of the positive portion.

Extending this reasoning and remembering that the probability of a compound event is the product of the probabilities of the individual events involved, we may write more generally

$$P = [(e^{-v_1x})q_1(1 - e^{-v_1x})p_1][(e^{-v_2x})q_2(1 - e^{-v_2x})p_2] \\ [(e^{-v_3x})q_3(1 - e^{-v_3x})p_3]\ [\cdot\ \cdot\ \cdot], \quad \text{(B–4)}$$

where $v_1, v_2, v_3, \ldots$ are the sizes in milliliters of portions of the sample tested; $q_1, q_2, q_3, \ldots$ are the number of portions of these respective sizes giving *negative* test results; and p_1, p_2, p_3, . . . are the number of portions of those respective sizes giving positive test results.

Most Probable Number

It is observed from the preceding equations that the probability P of a test result is a function of the size and number of portions and also of the density of bacteria. Obviously a great many events can occur. If these equations of probability are solved by successive trials by substituting values of x, it will be observed that P will approach a maximum as x approaches some definite value, above or below which P declines. This value of x is therefore defined as the *most probable number* (MPN). It means simply that *among all possible values, this particular value would produce the observed test result with greater frequency than any other.* However, many densities other than the most probable number will produce the observed test result at almost the same frequency.

The most probable number in itself has little significance without also a measure of its reliability—that is, its probability in relation to the probability of the other densities that also can produce the observed positive and negative test results. This is considered separately in the two types of samples, the single dilution and the multiple dilution.

Multiple Portions of a Single Dilution

In considering a sample of multiple portions of a single dilution from the previous development the proportion of negative results is represented by

$$\frac{q}{n} = e^{-vx}, \tag{B-5}$$

where v is the volume in milliliters of the portions, n is the number of portions constituting a sample, and q is the number of portions showing negative test results. Taking the log of both sides and transposing, setting x equal to the most probable number gives

$$\text{MPN} = \frac{1}{v} \log_e \frac{n}{q}$$

or, in common logarithms,

$$\text{MPN} = \frac{2.303}{v} \log \frac{n}{q}. \tag{B-6}$$

Table B–1 lists in column 3 the computed MPN densities per milliliter for the most commonly employed water samples of 5 and 10 portions each.

Table B–1 MPN Density of Bacteria for Combinations of Positive and Negative Test Results from Single Dilution—Multiple-Portion Sample[a]

Number of Portions			
Positive (1)	Negative (2)	MPN per Millimeter, 1-ml Portions (3)	Usual Expression of MPN per 100 ml, 10-ml Portions (4)
Five-Portion Sample			
0	5	0	< 2.2
1	4	0.223	2.2
2	3	0.511	5.1
3	2	0.916	9.2
4	1	1.609	16.1
5	0	∞	—
Ten-Portion Sample			
0	10	0	< 1.0
1	9	0.105	1.0
2	8	0.223	2.2
3	7	0.357	3.6
4	6	0.511	5.1
5	5	0.693	6.9
6	4	0.916	9.2
7	3	1.204	12.0
8	2	1.609	16.1
9	1	2.303	23.0
10	0	∞	—

[a] Although our concern here is with a single sample, the MPN can likewise be applied to a series of samples. For example, a commonly accepted standard for potable water specifies, that of all the 10-ml portions examined per month, not more than 10 percent shall show the presence of organisms of the coliform group. The MPN of this limiting value is likewise obtained from the proportion of negative results.

$$\text{MPN} = \frac{1}{10} \log_e \frac{1}{0.9} = 0.0105 \text{ per milliliter,}$$

or 1.05 per 100 ml, or commonly stated as 1 per 100 ml.

It will be noted from the equation (and Table B–1) that the most probable number is a function of the *proportion* of negative test results in the sample; hence all samples, regardless of the number of portions employed, that have the same proportion of negative results have the same most probable number. It does not follow, however, that the reliability is the same.

Evaluation of Single Sample of Multiple Portions

Since the probability of a test result is a function of the size and number of portions and also the density of bacteria, it is desirable to develop the distribution of bacterial densities expected for each possible test result. This will define the range in bacterial density expected for each test result for any prescribed confidence level.

If p is the true probability of a positive test result showing, then the probability that out of n tubes innoculated exactly r will be positive is the $(r + 1)$th term of the binomial expansion of $(p + q)^n$. Also the probability that out of n tubes innoculated r or less will be positive is the summation of the probabilities from $r = 0$, $r = 1$, $r = 2$, . . . , up to r', or

$$\Sigma p = \sum_{r=0}^{r=r'} \frac{n!}{r!(n - r)!} p^r(1 - p)^{n-r}. \qquad \text{(B–7)}$$

Tables [1] of this summation have been computed for $n = 5$, 10, 15, 20, and 30 for the full range of assigned true values of p from .025 to .975.

As in the case of the Poisson series, the true probability p of a positive test result as employed in these summation tables can be expressed in terms of bacterial density from

$$p = 1 - e^{-x}, \qquad \text{(B–8)}$$

where x is the true mean density. From such summation tables and this conversion it is possible to construct probability distribution curves of bacterial density x expected for each test result of portions that are positive.

The probability distributions for single samples of 5 and 10 portions are shown on log-normal probability paper as Figure B–1a and b, respectively. (Similar plots are readily constructed for 15-, 20-, and 30-portion samples based on the summation tables.) The MPN densities are also located on the respective distribution curves. For the usual samples of 5 or 10 portions the distributions are curvilinear; the range in expected bacterial density is wide; and the MPN value has a probability of a low order. As the number of portions in a sample is increased, the distributions approach a linear form; the slope is less steep ($\sigma_{\log}$ decreases); and the MPN approaches the probability of 50 percent. For example, for a 5-portion sample (Figure B–1a) the highest probability of MPN is that for the test result (4/5) at 34 percent for density of 1.6 per milliliter. However, other (4/5) test results can be expected

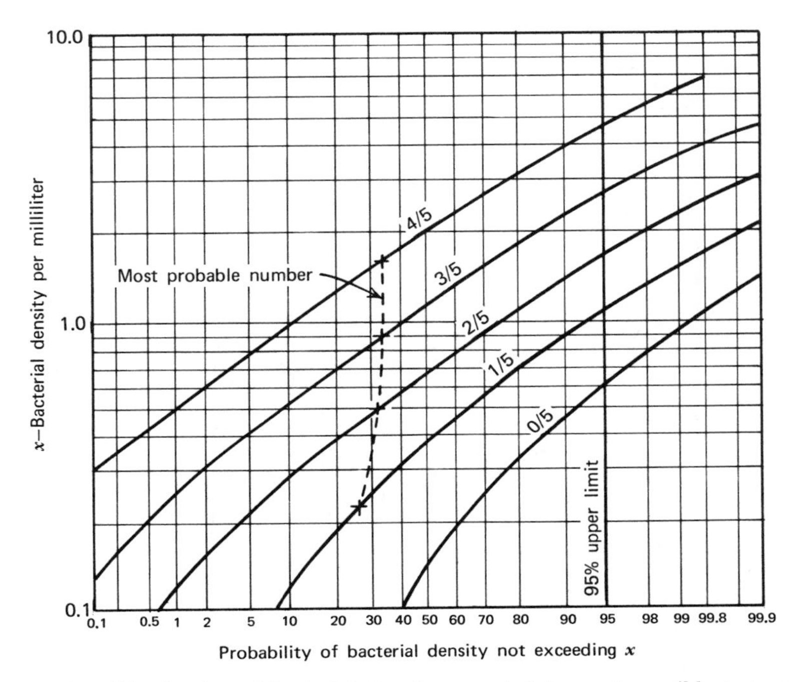

Figure B–1*a* Distribution of bacterial density expected for each possible test result for single-dilution five portion sample. (Undiluted sample, test portion volume 1 ml. Adjust x for other dilution or portion volume.)

from a wide range of bacterial densities as defined by the (4/5) distribution curve with the upper 95 percent limit (95 percent $\overline{<}$) of 4.6 per millimeter. For a 30-portion sample a test result of the same proportion of positives (24/30) has the same most probable number, 1.6 per milliliter, but at a slightly higher position on the probability scale (42 percent), and a less steep distribution curve, with a substantially lower density of 2.4 per milliliter at the 95-percent upper probability limit.

It is evident that a more sensitive reflection of reliability of the most probable number of a single sample of multiple portions than the probability of the most probable number is the density expected for the test result at some upper limit on its probability distribution curve, such as 95 percent equal to or less than. The most probable number is obtained from equation 3–6 or the distribution curve, and the 95-percent density is readily derived from the 95-percent probability intercept. Hence in reporting the test results of a single sample of multiple portions as the most probable number the 95 percent equal to or less than density

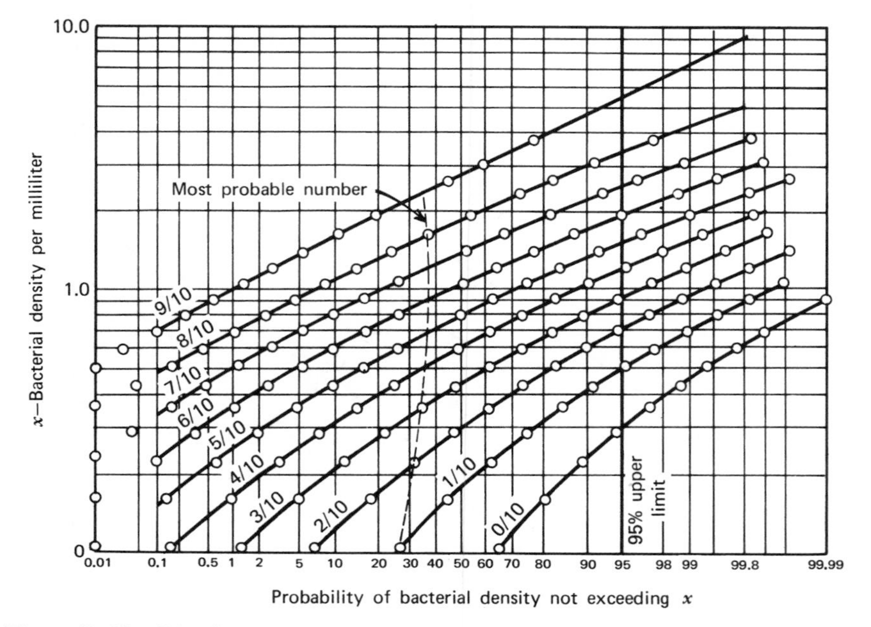

Figure B–1*b* Distribution of bacterial density expected for each possible test result for single-dilution 10-portion sample. (Undiluted sample, test portion volume 1 ml. Adjust x for other dilution or portion volume.)

should also be given to reflect the reliability of the most probable number.*

Evaluation of Series of Samples of Multiple Portions of a Single Dilution

The mean density is obtained only from evaluation of a series of samples. For this it is not necessary to resort to the determination of most probable numbers, since the mean density is obtained directly from an analysis of the percentage of the samples that are positive. For example, for the mean density x of 1.05 per milliliter for the 10-portion sample (Figure B–1*b*) the probability of not exceeding the number of portions that are positive is obtained where the respective distribution curves 9/10 to 0/10 intersect the horizontal line drawn at x = 1.05 per milliliter and reading probabilities on the upper reverse scale. Transfer of these

* *Standard Methods* recommends the 95-percent confidence limits; that is, the lower limit that 2.5 percent are equal to or less than and the upper limit that 97.5 percent are equal to or less than.

readings to normal probability paper develops a straight line distribution for the central portion, as shown in Figure B–2.

A nest of such nearly normal distributions is obtained by similar intercepts on Figure B–1*b* for the effective central range of mean densities 0.4 to 1.1 per milliliter, as shown in Figure B–3 for 10-portion samples. (Similar nests can be developed for the 5-, 15-, 20-, and 30-portion samples.) This range in mean density and the blocked-in central portion of the distributions that are normal provide a reference base, or *control frame,* for efficient evaluation of bacterial density.

In the central portion the distribution lines are parallel, with a standard deviation σ of 1.12, 1.57, 1.95, 2.23, and 2.74 for samples of 5, 10, 15, 20, and 30 portions, respectively. From these the general relation of the standard deviation to the number of portions in a sample may be expressed as

$$\sigma = \tfrac{1}{2}\sqrt{n}. \tag{B–9}$$

Thus the standard deviation of 1, 2, 3, 4, and 5 portions that are positive is expected with samples of 4, 16, 36, 64, and 100 portions. The standard deviation σ of portions that are positive increases as the number of portions n in a sample increases; however, the proportionate increase σ/n decreases as n increases, and hence reliability increases with increasing number of portions in a sample. A single sample of 5 or 10 portions is a rather imprecise measure. Improved reliability can be obtained either

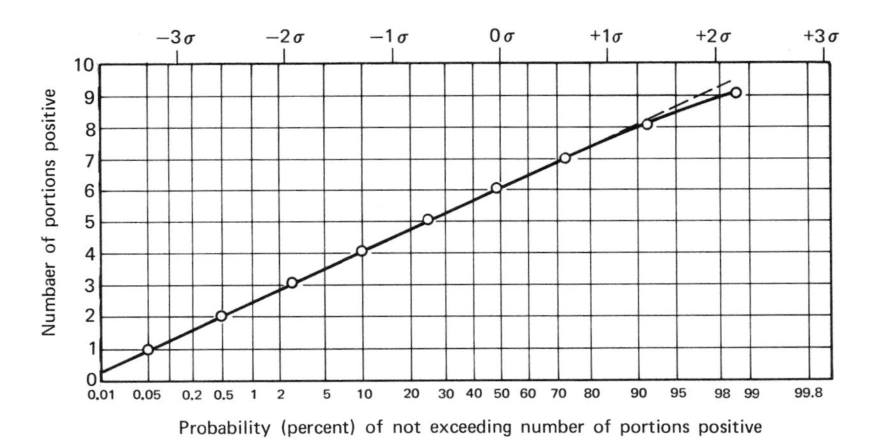

Figure B–2 Distribution of number of portions positive for single-dilution 10-portion sample for mean density of 1.05 per milliliter (developed from Figure B–1*b*).

by increasing the number of portions in a sample or by taking a series of samples.

From the control frames of normal distributions and the standard deviations, evaluation of series of multiple portion samples can be made directly from the portions that are positive in each sample without the necessity of reducing each sample result to most probable numbers. The distribution of such series referred to the control frames provides directly the mean density and also indicates whether the water quality remained stable or varied during the sampling period and, if so, within what ranges in mean density. This is illustrated by the following examples of analyses of two sets of 30 samples each.

The first set of 30 samples was taken in a single day from water considered to be reasonably stable in bacterial quality; 10 portions of 1 ml each were examined, and results reported as number of portions positive in each sample ranging from 1/10 to 7/10. The 30-sample results arranged in order of number of positives in each are plotted on normal probability paper on which the control frame for a 10-portion sample is located as a background, as shown in Figure B–4. The plotted results form a reasonably straight line (curve A), which is practically parallel to the distribution expected of 10-portion samples with standard devia-

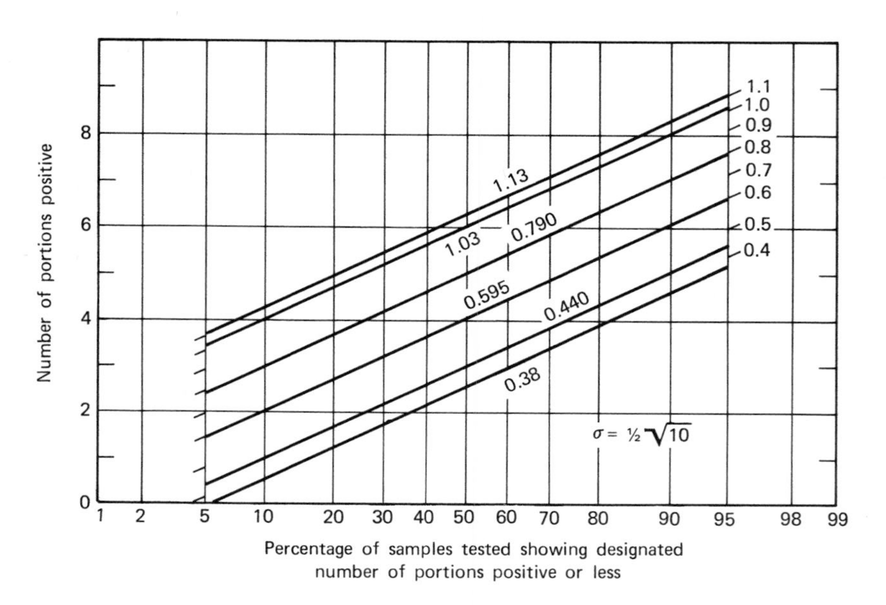

Figure B–3 Distribution of number of portions positive for single-dilution 10-portion sample for mean densities of 0.38 to 1.13 per milliliter.

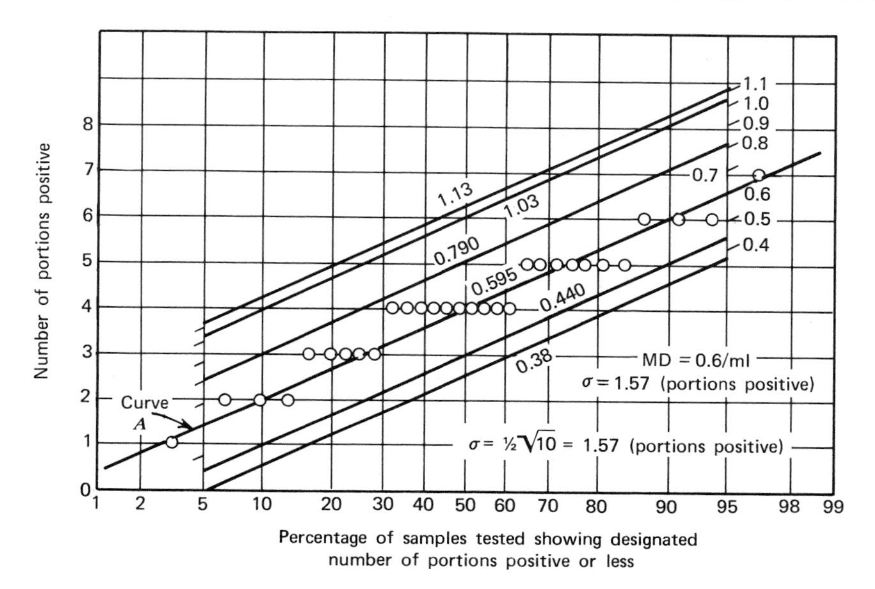

Figure B–4 Evaluation of mean density (MD) from series of single-dilution 10-portion samples (stable water quality).

tion σ of 1.57 and coincides with the distribution expected from sampling water of a true mean density of 0.6 per milliliter. Thus the mean density of the sampled water is 0.6 per milliliter, and, since the distribution is that expected of 10-portion samples, the quality remained stable during sampling.

The second set of 30 samples was taken over a period of 30 days, one each day, and was examined in samples of 10 portions of each with results plotted as in the first set, as shown in Figure B–5. A reasonably straight line distribution (curve A) develops, but it is steeper than that expected of 10-portion samples. From curve A the most probable mean density at a probability of 50 percent in the control frame is 0.7 per milliliter; the standard deviation is 2.8 portions positive, as compared with 1.57 expected for 10-portion samples taken from water of stable quality. Hence the bacterial density of the sampled water varied during the 30-day sampling period. The extent of this variation in mean density for any confidence range, such as 80 percent, is readily obtained from Figure B–5: the lower limit at 10 percent equal to or less than is 0.44 per milliliter, and the upper limit at 90 percent equal to or less than is 1.0 per milliliter.

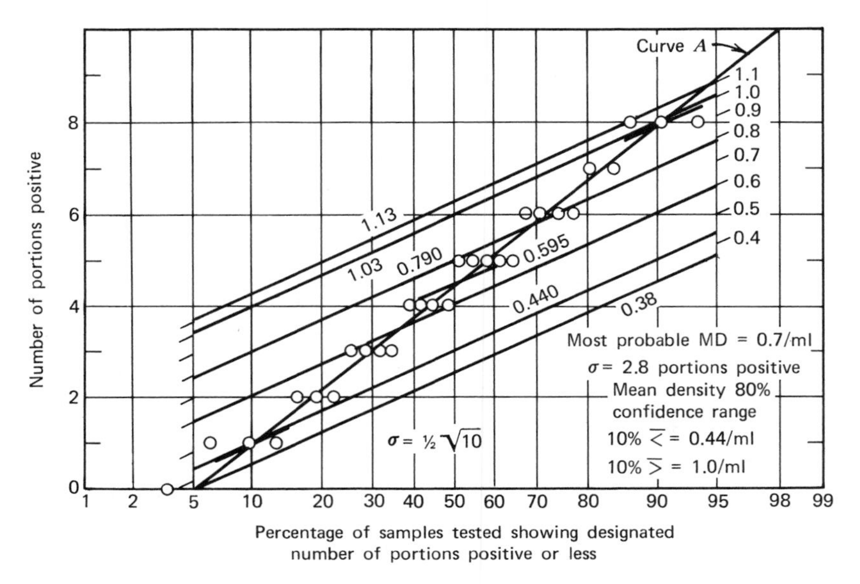

Figure B–5 Evaluation of mean density (MD) from a series of single-dilution 10-portion samples (unstable water quality).

Multiple Portions in Each of Three Decimal Dilutions

Bacteriological examinations of the more polluted waters and treated wastewater effluents require samples of multiple portions in each of several decimal dilutions of which three dilutions providing positive and negative results are selected. Any individual sample result can be reported as the most probable number with a 95-percent upper limit, as with the sample of a single dilution of multiple portions. Series of multiple portion three decimal dilution samples likewise can be reduced to mean density and evaluated as to stability of the water sampled.

The probability of any test result can be determined from equation B–4, and by successive approximations the MPN value also can be obtained. Hoskins' [2] tables provide most probable numbers for all possible test results up to five portions in each of three decimal dilutions. A table of most probable numbers for all possible combinations employing 10 portions in each of three decimal dilutions has been provided by Halverson and Ziegler [3].

The probability summation of bacterial density not exceeding a certain value is obtained from summations of Halverson's and Ziegler's data [4]. Plotting the densities against corresponding probability summations on log-normal probability paper produces a straight line, with the midpoint of the distribution coinciding with the assigned initial true mean density. Hence the distribution of the log of density is normal, and the central value at probability 0.50 represents the true mean density. The slope of this line provides the standard deviation.

Figure B–6 represents the distribution curve for three decimal dilutions employing 10 portions each for assigned true mean densities of 0.15, 0.25, 0.5, and 1.5 per milliliter. Data are reduced to a comparable basis by plotting ratios to the assigned true densities. It is observed that one distribution fits the four sets of data with a common slope of $\sigma_{\log}$ of 0.1735. The slope, $\sigma_{\log}$, is therefore independent of bacterial density.

Similarly linear distributions are obtained for samples of 5, 20, and 40 portions in each of three decimal dilutions for an assigned true mean density of 1.5 per milliliter, with slopes of $\sigma_{\log}$ of 0.2454, 0.1227, and 0.0868, respectively.

The slopes as represented by $\sigma_{\log}$ are related to the number of portions n employed and may be expressed as

$$\sigma_{\log} = \frac{0.5487}{\sqrt{n}} \tag{B–10}$$

Equation B–10 developed from Halverson's and Ziegler's data is in agreement with the equation proposed by Fisher and Yates [5]:

$$\sigma_{\log} = \frac{\log 2}{\sqrt{n}}.$$

Figure B–7 illustrates how the slopes of the distributions flatten as n increases. With $\sigma_{\log}$ thus determined, the expected variation in test results can be defined. For example, in a test involving five portions in each dilution the result (5/5 in 10^{-3},2/5 in 10^{-4}, 0/5 in 10^{-5}) gives a most probable number of 4900 per milliliter. The standard deviation for five portions from equation B–10 is 0.245. Since the distribution is logarithmically normal, any desired confidence range can be established as an indication of reliability, such as:

50-percent confidence: log 4900 $\pm$ (0.6745 $\times$ 0.245), or 3350 to 7170.

95-percent confidence: log 4900 $\pm$ (2 $\times$ 0.245), or 1590 to 15150.

Likewise the 95-percent upper limit is log 4900 + (1.642 $\times$ 0.245), or 12,400 per milliliter.

Although the standard deviation $\sigma_{\log}$ is the conventional measure of variation, it can be stated only in terms of a logarithm and hence does not directly reflect the extent of accuracy of the most probable number. Hence it is recommended again that in reporting bacterial den-

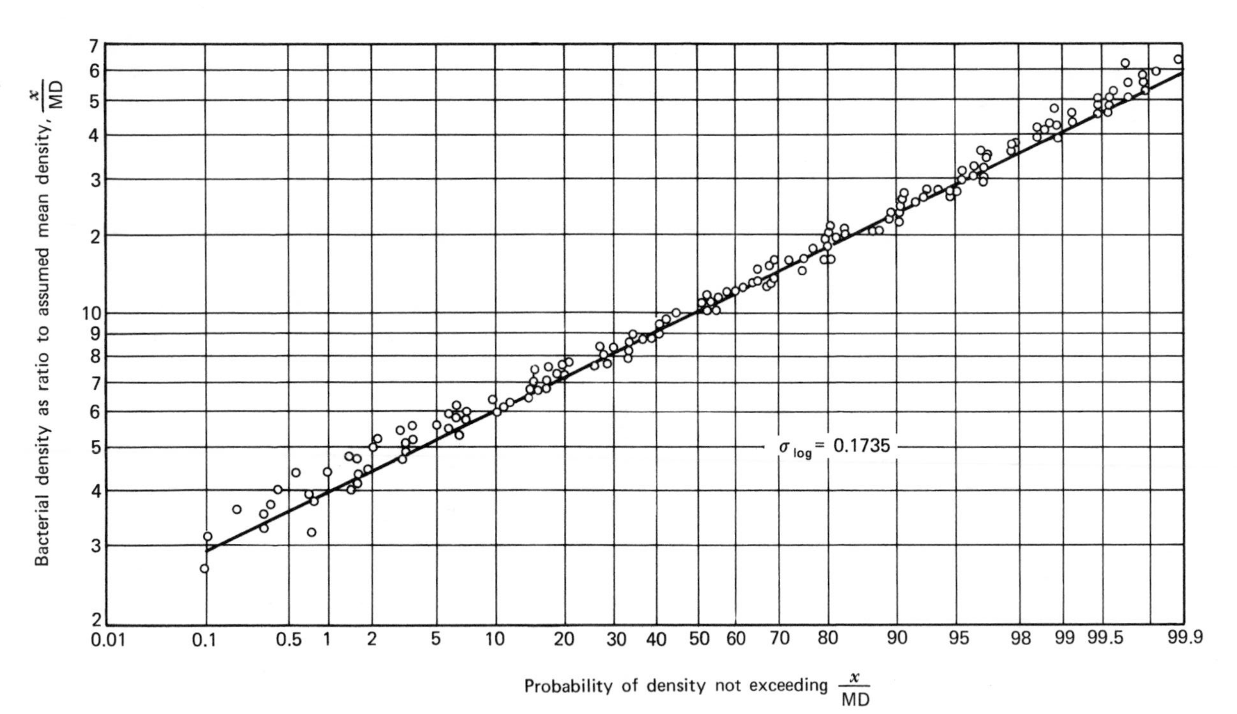

Figure B–6 Distribution of bacterial density expected for samples of 10 portions each of three decimal dilutions at true mean densities of 0.05, 0.25, 0.50, and 1.5 per milliliter (data after Halverson and Ziegler [3]).

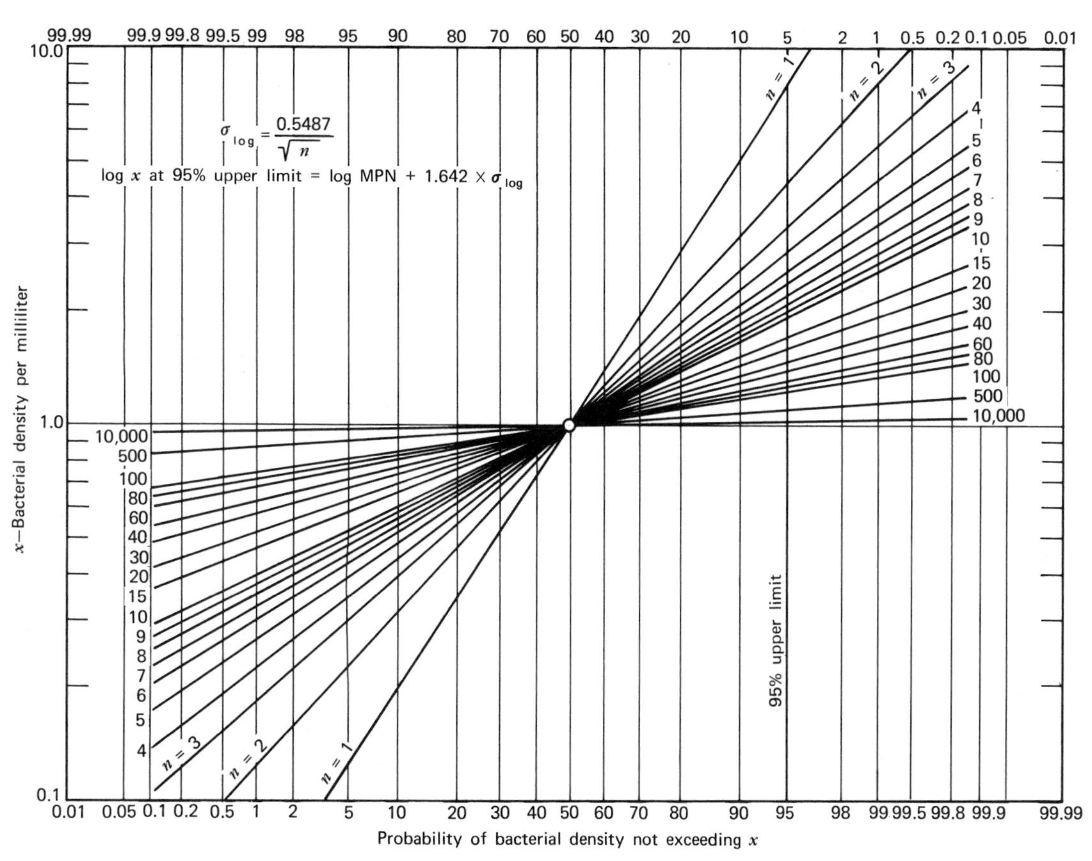

Figure B–7 Standard distribution slopes for various numbers of portions in each of three decimal dilutions.

sity from a single sample test result of multiple portions in each of three decimal dilutions both the most probable number and the 95-percent upper limit be stated; for example, MPN = 4900 per milliliter, 95 percent $\lesssim$ 12,400 per milliliter.

Evaluation of Series of Samples of Multiple Portions in Three Decimal Dilutions

Figure B–8 is the log-normal distribution of a *series* of 16 samples (taken under stable conditions) of 10 portions in each of three decimal dilutions. A straight line forms, and the mean density is obtained directly where the line of the distribution intersects the probability of 50 percent. As already shown, this coincides with the theoretical true mean density. Thus the mean density by curve *A* is 0.486 per milliliter; the distribution is practically parallel to the standard slope expected for 10-portion samples, with $\sigma_{\log}$ of 0.174; hence the mean density did not change during sampling.

There remains, however, a further question: How would other mean densities based on repeated series of 16 samples vary? This is defined by the standard error of the mean density (see Appendix A), which is given by $S.E._{\log} = \sigma_{\log}/\sqrt{N}$, or

$$S.E._{\log} = \frac{0.174}{\sqrt{16}} = 0.0434.$$

From this we are 95-percent confident that other mean densities based on 16 samples each taken under the same conditions of stability will fall within the range $\log 0.486 \pm (2 \times 0.0434)$, or in terms of mean density 0.398 to 0.594 per milliliter. Since the repeated series of samples are taken under the same stable conditions, we would expect the slope of the distribution of each series to approximate that expected for 10 portions. These variations in mean density do not represent a real change in the quality of the sampled water; they simply are associated with the small number of samples in the series. As the number of samples is increased, the $S.E._{\log}$ would decrease, narrowing the range toward the true mean density of the sampled water.

In most instances sampling polluted waters involves instability of quality, real variation in the mean density of the water sampled, and therefore the distributions of series of most probable numbers develop slopes that are steeper than those expected for samples of the given number of portions employed. The range of such a real change in mean density is readily obtained graphically from the probability plot of the sampling results, as illustrated in Figure B–9. In this example 92 samples

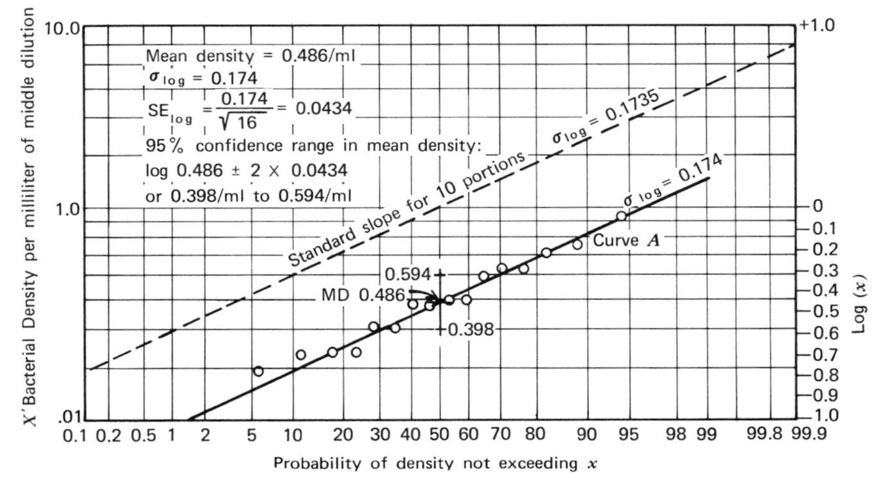

Figure B–8 Evaluation of mean density from a series of samples of 10 portions in each of three decimal dilutions, density stable during sampling period. The following test result for 16 samples are from the experimental data of Halverson and Ziegler [6] (most probable number per milliliter in middle dilution, 10 portions each 10^6, 10^7, and 10^8 dilutions):

Most Probable Number per Milliliter: Order of Magnitude (x)	Probability of Not Exceeding x $\left(\frac{100m}{n+1}\right)$
0.288	5.88
0.329	11.8
0.329	17.6
0.334	23.5
0.399	29.4
0.399	35.3
0.474	41.2
0.474	47.0
0.493	52.8
0.493	58.8
0.589	64.6
0.622	70.5
0.622	76.5
0.700	82.3
0.742	88.2
0.933	94.1

of three portions in each of three decimal dilutions were taken from the Detroit River over a period of 3 months at a location suspected of variation in quality. The distribution line developed from the plotting of the 92 MPN values, line A, gives a mean density of 22,700 per 100 ml and a $\sigma_{\log}$ of 0.535. The standard slope expected for a three-portion

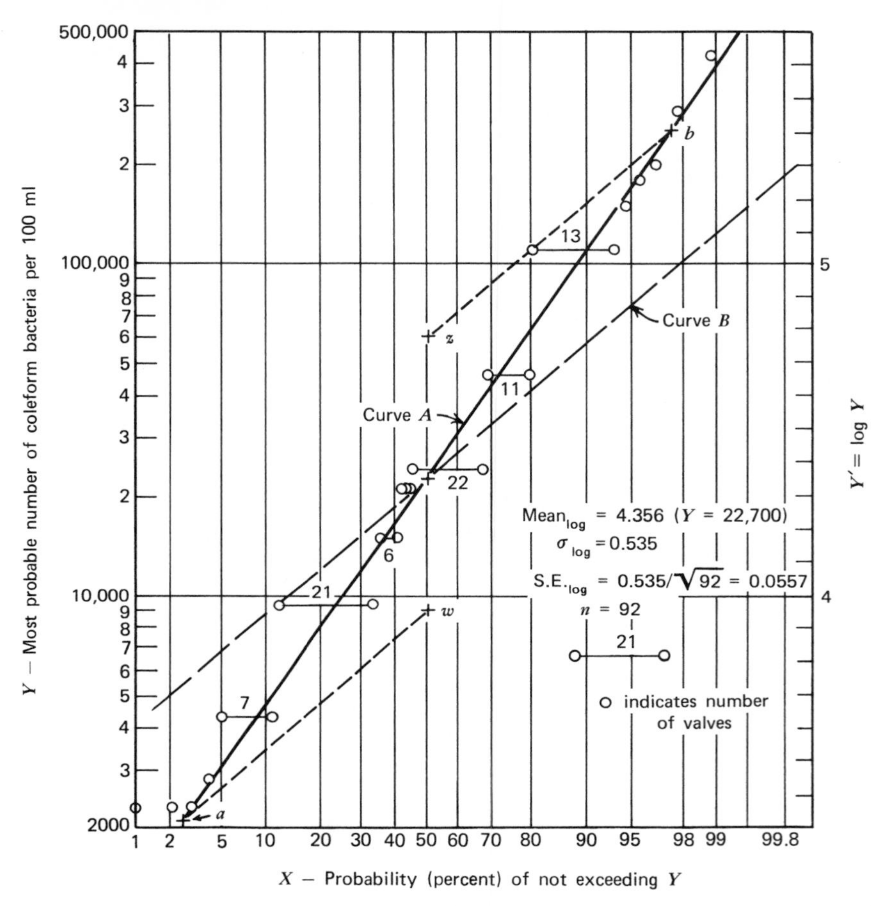

Figure B–9 Distribution of 92 most probable numbers (three portions) in each of three decimal dilutions.

test procedure $(0.5487/\sqrt{3} = 0.317)$ is drawn through the mean as line *B*. The actual distribution, line *A*, is steeper than the standard slope expected for a three-portion test; this indicates at once that changes in mean density, and therefore real changes in quality, took place during the sampling period.

The question now arises, Within what range did the level of mean density shift? The answer to this depends on the confidence we wish to have in the range defined. A range of 95-percent confidence is obtained by projecting *a* and *b*, the 2.5 and 97.5 percent intercepts with line *A*, to the 50-percent probability, locating *W* and *Z*, the mean density range of 9000 to 60,000 per 100 ml. It is reasoned that if 2.5 percent of the MPN values do not exceed 2100 (a) and the quality remained stable, the distribution for a three-portion test would arise from a mean density of 9000 (W); similarly, if 97.5 percent of the MPN values do not exceed 250,000 (b), the distribution would arise from a mean density of 60,000 (Z).

In a similar manner the range in shift of mean density can be defined for any desired confidence limits.

It is of special significance to note that this range in mean density (W to Z) cannot be accounted for in normal variation in the mean due to the number of samples in the series. If 22,700 per 100 ml is accepted as the true mean density and 0.317 as the true $\sigma_{\log}$ for a three-portion test, then the standard error of the mean (S.E.$_{\log}$) for a series of 92 samples is $0.317/\sqrt{92}$, or 0.033. The 95-percent confidence range in other means about 22,700 would be log $22{,}700 \pm 2$ S.E.$_{\log}$, or 19,500 to 26,400 per 100 ml. This range is small relative to the range as defined between W and Z. Similarly, if the observed $\sigma_{\log}$, 0.535, is substituted for 0.317, the 95-percent confidence range would extend from 17,600 to 29,300 per 100 ml, still well within the range W to Z, 9000 to 60,000 per 100 ml. The observed distribution of MPN values (curve A) cannot be accounted for by variations associated with the number of samples employed nor with the three-portion test procedure; the distribution, clearly, reflects a real shift in the quality of the river water during the sampling period.

Reliability of the Membrane Filter Count (MFC)

As with the dilution method, the law of mixing, the binomial and Poisson series apply to determination of reliability of membrane filter counts. The standard deviation of the Poisson series is readily derived from the basic binomial. If the binomial

$$(0.001 + 0.999)^{1000x}$$

represents the distribution of bacteria in 1000 unit volumes of water, the mean density being x per unit volume, the standard deviation of the distribution is

$$\sigma_c = \sqrt{pqn} = \sqrt{0.999x}.$$

This expression approaches $\sqrt{x}$ as the size of the supply is increased; and since there is no necessary limit to this size, we may write

$$\sigma_c = \sqrt{x}.$$

In the Poisson series x is the true mean density; since $\sigma_c = \sqrt{x}$, an estimate of x provides an estimate of its own standard deviation, and the actual trial count T of a sample constitutes the most probable estimate of the true mean density of the water sampled, and σ_c may be derived as

$$\sigma_c = \sqrt{T}.$$

From this it is evident that for samples producing large counts T the standard deviation of the counts also increases; however, proportionately the deviation decreases.

$$\sigma_P = \frac{\sigma_c}{T} = \frac{\sqrt{T}}{T}.$$

Hence counts of samples producing a small number of colonies are less reliable than those of a large number of colonies, as illustrated by the following tabulation:

Colony Count	Standard Deviation as a Count σ_c	Standard Deviation as a Percentage of the Count $\frac{100\,\sigma_c}{T}$
4	2	50.0
9	3	33.3
16	4	25.0
25	5	20.0
49	7	14.3
100	10	10.0
196	14	7.1

The practical limit on the number of colonies is approximately 200 as crowding and overgrowth beyond this density is a limitation; hence the recommendation in *Standard Methods* specifies colony density in the range 50 to 200.

In general the maximum possible reliability of a membrane filter count, or any similar plate count is given by the standard deviation σ_c, which is the square root of the value found, increasing in reliability as the count increases. However, this is the *residual, unavoidable error of sampling.*

There are other errors, personal and mechanical, involved in the laboratory procedure that add to variance, such as measurement, dilution, identification of colonies, and counting. Consequently deviations in membrane filter counts usually exceed that of maximum theoretical reliability defined by σ_c, the square root of the count. The actual deviation of an MFC σ_a, although not well defined from extensive experience, is generally considered not to exceed that of a most probable number determined by the dilution method and has ultimate possibility of being much smaller as other errors are eliminated and the minimum of its sampling error is approached.

An analysis made of duplicate membrane filter counts of Detroit River water sampled in the interval April 1956–March 1957 illustrates one order of magnitude of the *actual* standard deviation. Samples were limited to the larger of duplicates within the range 50 to 200 colonies, and the difference in count on each set of duplicates was expressed in terms of ratio to the standard deviation based on the square root of the larger count. On this basis the distribution of differences expressed in terms of standard deviation units should plot as a normal distribution on normal probability paper, with a mean of 0.0 at 50 percent and a slope of σ equal to 1.0 if the variance of the membrane filter counts approached that expected solely from sampling error. The actual distribution of 85 sets of duplicates disclosed a quite normal plot on normal probability paper, with a mean of 0.0, but the slope of the distribution produced a standard deviation of 2.85. This indicates that the *actual* standard deviation σ_a was for these data 2.85 times that of the standard deviation σ_c expected for sampling error alone based on the Poisson distribution.

Evaluation of Series of Samples, and Comparison of MPN and MFC Enumeration

It is expected that if the values obtained by the MFC method are equivalent to those obtained by the dilution method the distributions of membrane filter counts of a series of samples should be logarithmically normal with true mean density at the 50-percent probability.

This has been repeatedly demonstrated in parallel series of MPN and MFC enumerations. For example, series of parallel samples of Detroit River* water examined for coliform bacteria by MPN and MFC procedure show log-normal distributions. The series involved 239 samples. The MPN determinations were made by the decimal dilution method (confirmed), employing five portions in each dilution; the parallel MFC determinations were made by the immediate and by the delayed procedure. The following summary statistics were developed from the probability plots:

* Data courtesy of George Hazey, Chief Operator, Wyandotte, Michigan, Water Treatment Plant.

Parameter	Most Probable Number	Membrane Filter Count	
		Immediate	Delayed
Number of samples	239	239	239
Mean density per 100 ml	4700	2650	3150
95-Percent upper limit (per 100 ml)	33,000	21,000	21,000
Log mean density	3.672	3.423	3.498
$\sigma_{\log}$	0.52	0.55	0.49
$S.E._{\log}$	0.034	0.036	0.032
Mean density ± 2 S.E. (per 100 ml)	4010–5500	2240–3120	2720–3650

The mean density of the MFC distribution of the delayed procedure is 67 percent of the mean density of the MPN distribution. The standard deviations ($\sigma_{\log}$) of the MPN and MFC distributions are approximately the same. The standard deviation of the MPN distribution, however, is larger than that expected for tests employing five portions, which indicates that a real change in mean density took place during the sampling period.

McCarthy and associates [7] who took samples of rivers, lakes, and ponds and examined them in parallel for coliform bacteria by MPN and MFC procedures report log-normal distribution by plotting ratios of MPN to MFC for 49 sets of samples. For agreement between mean densities of both methods of examination the distribution of MPN/MFC should intersect the 50-percent probability at a mean of 1.0. They employed the usual one-step membrane filter method with M-Endo broth and found that the MPN/MFC distribution developed a mean intercept of 1.53, which indicates that this membrane filter procedure produced a mean density that was approximately 65 percent of that obtained by the MPN dilution method. This is in close agreement with the 67-percent result obtained for the Detroit River series. However, when the E and A procedure was used in the membrane filter counts, the distribution of MPN/MFC values developed a mean intercept of 0.97, which indicates that the mean densities obtained by the MPN and MFC procedures (E and A) were in very close agreement.

Similarly in examining samples of biologically treated sewage and seawaters the distributions of MPN/MFC (employing the E and A procedure in the membrane filter count) produced mean intercepts of 1.10 and 0.96, resppectively, which indicates close agreement in mean density between the MPN dilution and the MFC (E and A) methods.

The standard deviations $\sigma_{\log}$ estimated from the MPN/MFC plots of McCarthy et al. [7] were 0.275 and 0.30 for samples of rivers, and lakes and ponds, respectively, and approached the $\sigma_{\log}$ of 0.245 expected in five-portion MPN samples; the standard deviation $\sigma_{\log}$ for the treated sewage samples, however, was larger, estimated as 0.42. For the seawater samples the standard deviation of the MPN/MFC distribution was estimated to be 0.42, as compared with the expected $\sigma_{\log}$ of 0.317 for three portion samples employed in the MPN procedure.

It is not definitely established that the membrane filter enumeration identifies entirely the same organisms that are enumerated by the dilution method. The membrane filter procedure has advantages in dealing with relatively uncontaminated waters. As procedures are perfected, accuracy and precision should improve. Both methods have wide application and are in common use.

REFERENCES

[1] *Report on Biological Standards,* Special Report Series No. 128, Privy Councils, Medical Research Council, London, 1929.
[2] Reprint No. 1621, *Public Health Reports* (revised 1940).
[3] *J. Bacteriology,* **25,** 101 (1933).
[4] *J. Bacteriology,* **26,** 559 (1933).
[5] Fisher, R. A., and Yates, F., *Biological, Agricultural and Medical Research,* Oliver and Boyd, London, 1943.
[6] *J. Bacteriology,* **29,** No. 6 (1935).
[7] McCarthy, J. A., Delaney, J. E., and Grasso, R. J., *Water Sew. Works,* **238** June (1961).

Appendix C

COMPUTER ADAPTATION*

There is scarcely a field today in which the high speed computer does not play a significant role. Many evaluations of practical problems, as well as research investigations, are greatly aided by the refinement and extension of analyses that were previously impossible either because of complexity or the sheer time and labor involved. The astounding speed, great versatility, and accuracy of each new generation of computers seem to offer inexhaustible potential.

A word of caution. Contrary to the popular misconception, computers are not brains. Actually they are only machines and will do only what they are instructed to do; if the input is nonsense, the output is also inevitably nonsense. There is a tendency to include so much in a single processing that irrelevant or overgeneralized factors may produce suprious solutions. It is well to remember that the raw data with which we must work and the assumptions that must be made are such that it is desirable to break the problem into rational parts to permit evaluation of partial outputs before proceeding with subsequent steps. The computer is not a substitute for reasoning and professional judgment—it is only a tool, albeit a powerful one.

ANALOG AND DIGITAL COMPUTATION

Two mathematical principles distinguish between types of computers. These are the principle of *correspondence* and the principle of *succession.*

* In the development and testing of programs for adaptation of computers in the solution of various problems encountered in stream analysis the author is indebted to members of his staff, particularly Professors John J. Gannon and Rolf Deininger, to staff members of the University of Michigan Computing Center, and to many of his graduate students, particularly Thomas D. Downs, John Z. Reynolds, and James Westfield. In the presentation of this section he has drawn liberally on their contributions and upon the excellent pamphlets of the University of Michigan Computing Center.

In numbering the principle of correspondence implies a one-to-one relationship between the things numbered and the units of the numbering system. The count is made one by one, without regard to order. This count constitutes a cardinal number.

In the principle of succesion the order in which the count is made is all important. The position in the succession of the count is identified. This number is an ordinal number and is operational in nature and relies on an intellectual process.

The analog computer, essentially a measuring device, is based on the principle of correspondence. There is a one-to-one relationship between the computer's reading and some selected physical property, to which the reading can be referred at any moment of operation. The analog computer operates by representing variables as physical quantities of a *continuous* nature, generally either mechanical or electrical. In the electronic analog computer there is a direct correspondence between a voltage reading and the magnitude of a particular variable. Analog computers are usually special purpose devices, uniquely suited for such purpose as spatial simulation, industrial process control, or differential analyzer for solving (analogizing) systems of simultaneous differential equations. Inexpensive analog computers can be constructed to solve equations (e.g., the simplified dissolved oxygen sag equation) and read out on an oscilloscope. The analog computer, based as it is on the principle of correspondence, is incapable of the speed, accuracy, and versatility that characterize the digital computer.

The digital computer, essentially a counter, is based on the principle of succession. It deals with discrete entities and therefore treats a continuous function by representing it as a *succession* of entities in ordinal counting. These entities do not vary in magnitude and consequently are not subject to error in measurement; hence accuracy is high (one part in a billion). The digital computer makes no direct or indirect reference to any physical property except by arbitrary designation. This abstraction, combined with simplicity of operation, endows the digital computer as one of the most revolutionary forces in human history.

THE BASIS OF THE SYSTEM

The Binary Number System

The digital computer employs the binary number system. Through this two-entity method of reckoning, the most complex operations are made possible by the simplest elements. In the decimal system in common everyday use there are 10 symbols (0 to 9) to work with in numbering. In the binary system there are only two symbols, 0 and 1. This necessitates many more positions than in the decimal system to represent the same magnitude, and hence is confusing for ordinary usage. However, this is not confusing to a high speed computer, which takes advantage of the two-symbol simplicity. Further, in the binary system subtraction, multiplication, and division are variations on the one basic additive process. Thus the four basic arithmetic processes are reduced to one simple process, addition.

The two-symbol binary system has a still greater advantage in that it is compatible with and can be used as interchangeable parts of two other situations, the *logical* situation and the *physical* situation.

Logic, the science of formal principles of reasoning, has its basis of distinction in the two opposing entities, *true* or *false,* governed by precise rules for manipulating concepts. The interchangeability between the binary 0/1 and true/false of logical reasoning was well known in early centuries but had no far-reaching effect until a third compatible pair of entities was developed by which the other two pairs could be put to work.

The possibility of *physical* linkage between the binary 0/1 and the true/false logical distinction was only recently recognized in the on/off states of the electrical switch. This gave birth to the modern computer. Although computer circuitry has advanced from electrical relays to diodes and transistors, the simple switch illustrates the two opposite entites on/off. The two possible physical states may be accepted as absolutely *discrete.* As pure, successive entities, they can be manipulated without the error that tends to accumulate in physical continuous measuremets, such as those characteristic of analog computation.

The digital computer, which combines the three basic systems—the physical, the logical, and the arithmetical—is an instrument of great power. Ever more elaborate patterns of combinations of 0/1, true/false, and on/off can be put to effective use.

DIGITAL COMPUTER FUNCTIONS

The gap between simplicity of basic theory and complexity of operation of computer hardware (mechanical, magnetic, electrical, and electronic devices) widens with each extension of the system to a point where it well may be the limiting factor in future growth. A rather forbidding array of apparatus is involved, and is increasingly intricate. This so-called hardware operates in the performance of the basic functions of the system. These operations are controlled by the so-called software, the principal item of which is the computer program, which is a plan for implementing and solving a problem by specifying a precise and detailed sequence of instructions to be followed by the computer. In the operations received from the program the computer performs four basic functions referred to as input, storage, processing, and output.

Input

The input refers to any information or instructions transferred or transmitted to the computer. The nature of the digital computation, as we have seen, demands that all information be reduced to some set of two-state symbols. The punched card is usually the first step in such reduction. A card contains 960 punching locations. A punch at a certain location can represent a 1, on, or true state; when it is not punched, a 0, off, or false state. A substantial body of information can be transferred by a file of such punched cards. The punched card recording of data is the most flexible and provides a simple means of verifying the data and correcting errors. In all large computer systems the high speed of operation necessitates transfer of data to magnetic tape as the final input medium. Tapes are a convenient, compact means of external storage of data for future reuse, or they may be erased and used again for other data.

Storage

The storage component, usually a magnetic drum, disk, or core, is an internal device of the computer in which any type of information can be stored, ranging from raw data intended for processing to full programs intended for execution under prescribed conditions. Again the data are stored in the two-state symbols; a state of magnetization of a spot on the storage surface indicates a 1, on, or true; a state of nonmagnetization indicates a 0, off, or false. These data can be taken out of their place in storage, used, and returned unaltered.

The feature of modern storage is the capacity to store instructions (programs) as well as numbers. This gives digital computers the unique properties of *self-control.* The investigator can specify his entire program in advance, and the execution can be carried out by an attendant. Also, arrangements can be made in storage for the computer to make changes in a program in progress, such as the determination of choice between alternative instructions or the order in which instructions are carried out. This process permits efficient use of computer time.

Processing

The central processing component of the digital computer, the unit that actually manipulates the data, is composed of two parts, the arithmetic section and the control section. The arithmetic section is analogous to an adding machine, capable also of subtraction, multiplication, and division. (By virtue of the employment of the binary number system, the arithmetic section also serves for the logical situations.) It works directly on the problems to be solved, performing the calculations and transferring the results of calculations to and from storage.

The control section coordinates and implements the arithmetic processes and serves as an intermediary as instructions are issued through it to the arithmetic section. The basic operations that can be initiated by the control section are built into the machine (not stored) and are fixed in number, but they can be at almost any level of specificity, such as Start; Add; Store; Divide; Proceed; Print.

Output

After the processing unit has completed its operations, the results are fed into devices similar to input devices except that they emit rather than receive the media containing information, which may be punched cards, perforated or magnetic tapes, or sheets of paper. Intermediary devices are usually required in producing the printout in a form prescribed by the investigator in his program, in the form of symbols, numbers, words, or plotted graphs.

COMPUTER PROGRAMMING

Computer Language and Grammar

A program for a digital computer is the stepwise, precise statement of a method solving a problem (an algorithm) expressed in a language that the computer can interpret. The language for a given computer is the means of communication between the programmer and the computer. The language contains symbols that can be converted by a combined mechanical and electronic process to electronic impulses used to represent symbols inside the machine. Each computer language has its own set of symbols (alphabet) and rules for assembling these symbols into meaningful instructions (grammar). Whether the information consists of data intended for processing or instructions intended for execution, it has undergone all necessary conversions from a verbal or mathematical language and is ready for the actual computing process.

The storage properties of digital computers have been exploited to make available a whole series of subroutine instructions that direct the computer to carry out some well defined operation. Such subroutines, as well as routines and complete programs for specialized problems of more common application, are now available in libraries. Such libraries of parts of programs greatly aid programming.

Further aids to programming are in the form of assembly programs and compiler programs. These are designed to relieve the human programmer of the need to write out machine instructions. They can carry out and coordinate such essential tasks as converting numerical information from one base to another, allocating storage locations, or selecting pertinent subroutines.

The language for programming is important. If it is oriented to the machine, the ordinary investigator is quickly lost, as it is not possible for him to know the intracacies of machine design, nor could he keep abreast of the rapidly changing computer hardware. Fortunately compiler programs relieve the investigator of much of this and permit the use of a *problem-oriented* language. Anyone with rudimentary competence in mathematics can learn to write a program or algorithm for solving a problem using a problem-oriented language.

One such problem-oriented language is that developed at the University of Michigan Computing Center; it is called MAD (an acronym for Michigan Algorithm Decoder). It is quite similar in structure to all the better known algorithmic languages, such as FORTRAN (Formula Translator) developed by IBM for large scale computers; familiarity with one permits quickly adapting to the other.

The Flow Diagram

Formulation of an efficient computer program is greatly facilitated by a graphical outline of the algorithm in the form of a flow diagram. A graphical visualization is much easier to perceive than a verbal description. The flow diagram enables the investigator to see at a glance sequences of operations and helps him to keep track of the alternative sequences which are dependent on intermediate results. In the formulation of his problem and model of solution the investigator has identified the relevant variables and assembled his raw data. These are listed and expressed in terms of the language to be employed, for ready designation on the flow diagram.

Figure C–1 illustrates a flow diagram for computing by the rational method (Chapter 4) the dissolved oxygen profile along the course of a river, prepared for MAD language program. The model, variables, and raw data to be incorporated into the program are listed below.

ADAPTATIONS IN APPLIED STREAM SANITATION

Where and when to resort to computer adaptation in the solution of applied stream sanitation problems depends on the nature of the prob-

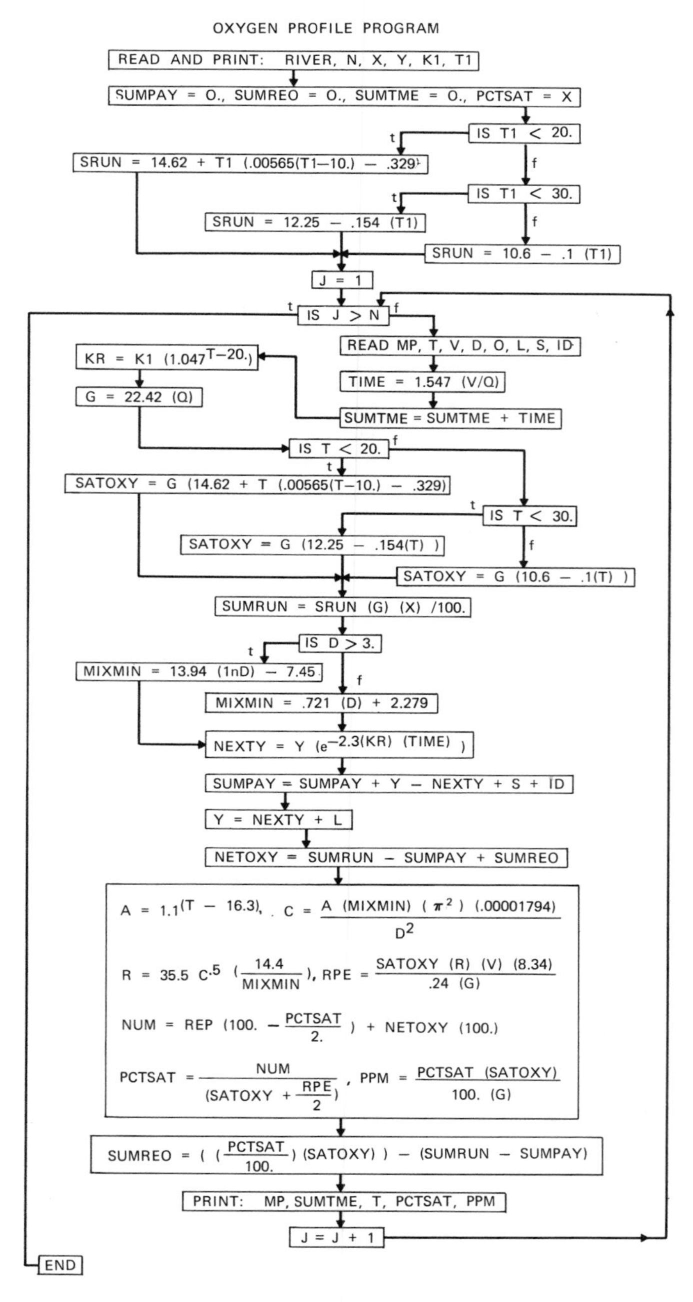

Figure C–1 Flow diagram for dissolved oxygen profile program. Note: in question statements t is followed if statement is true and f is followed if statement is false.

MAD Language Program for Computing a Dissolved Oxygen Profile by the Rational Method

Model and Variables

The reach of river under consideration is partitioned into N sections (J equals I through N). The parameters are designated for each section as follows:

MP	mile point at end of section
T	temperature of water in section (°C)
V	volume in section (million gallons)
D	average depth for section (feet)
Q	runoff through section (cfs)
L	amount of oxygen demanding waste added at end of section (population equivalents)
S	amount of oxygen demand of sludge accumulation in section (population equivalents)
I.D.	amount of immediate oxygen demand exercised in section (population equivalents)

Initial Data

RIVER	identification number to designate problem
N	number of stations
X	oxygen content of runoff before entering stream (percent of saturation)
Y	initial waste load (population equivalents)
Kl	BOD rate of amortization at 20°C per day
Tl	temperature of runoff before entering stream (°C)

Other Physical Parameters

KR	BOD rate of amortization per day at river temperature
MIXMIN	time of mix (minutes)
NETOXY	summation of oxygen budget (population equivalents)
NEXTY	BOD after amortization in section (population equivalents)
PCTSAT	oxygen content of stream (percent of saturation)
PPM	oxygen content of stream (ppm)
SATOXY	saturation level of oxygen for stream (population equivalents)
SRUN	oxygen concentration of inflow (ppm)
SUMPAY	summation of amount of BOD satisfied (population equivalents)
SUMREA	summation of amount of reoxygenation (population equivalents)
SUMRUN	oxygen contributed from inflow (population equivalents)
SUMTME	summation of time of passage (days)
TIME	time of passage in section (days)

A, C, R, and RPE are intermediate variables in the computations of reoxygenation according to the procedure outlined for the rational method in Chapter 4 based on curve *B* of Figure 4–8. (This can be simplified if Deininger's linear relation is employed.) NUM is an intermediate variable in the solution for PCSAT at the end of each section, and G is an intermediate variable that converts ppm to population equivalents.

To save computer storage space dissolved oxygen saturation values (ppm) are

MAD (Continued)

given in equation form for close approximation from the following relations:

$$14.62 + T[0.00565(T - 10) - 0.329] \quad \text{if} \quad T < 20;$$

$$(12.25 - 0.154T) \quad \text{if} \quad T \geqq 20 < 30;$$

$$(10.6 - 0.1T) \quad \text{if} \quad T > 30.$$

The empirical graphical relation between stream depth D and mix interval (Figure 4–13, curves *A* and *C*) for nontidal streams is closely approximated by the following relations:

$$\text{MIXMIN} = 13.94 \log_e D - 7.45 \quad \text{if} \quad D > 3;$$

$$\text{MIXMIN} = 0.721D + 2.279 \quad \text{if} \quad D < 3.$$

lem, the scope, the magnitude, and complexity of the computations involved, and the extent and degree to which solutions are sought. Time and cost are also considerations. In some instances the development and pretesting of a computer program involves more time and effort than solution by ordinary computation or by desk computer. Elaborate programs for solutions of minor consequence are in the category of "much ado about nothing," or "using a missile on a gnat."

The most advantageous use of computers is in the repeated use of a program, such as the routine analysis of river monitoring, or in projecting stream quality as a basis for stream pollution control programs. For example, the Michigan Water Resources Commission in its river basin studies makes extensive use of computer programs based on the rational method, as illustrated in Figure C–1. Several hundred complicated dissolved oxygen profiles forming a frame of reference can be run on the digital computer in a matter of minutes, thus greatly refining the basis for making decisions in design, water use allocations, water quality criteria, and water resource management. The computer is an indispensable tool in the systems analysis approach to comprehensive river basin studies involving evaluation of alternatives in the development, use, and management of water resources.

Computer programs, routines, and subroutines adapted to the digital computer have been developed, tested, and found highly effective tools in the following problem areas:

1. Selection and preparation of stream runoff records for drought and flood flow analyses.
2. Selection and preparation of meterologic records (temperature, wind velocity, vapor pressure, or solar radiation).
3. Reduction of river channel cross-section data along extensive

reaches of a stream and determination of occupied channel volume, time of passage, effective depth, and other parameters under various runoff probabilities. (This may also serve as a subroutine in other programs.)

4. Deoxygenation, reoxygenation, and dissolved oxygen balance profiles under various organic waste loadings and ranges in stream runoff both in freshwater streams and in estuaries.

5. Integration of organic sludge deposit accumulation and demand under various waste loadings and runoff probabilities. (This may also serve as a subroutine in item 4.)

6. Heat dissipation, equilibrium water temperature, and river temperature profile under various waste heat loads and probabilities of steam runoff and meteorologic conditions. (This may also serve as a subroutine or routine in other programs, particularly item 4.)

The final decision with respect to adaptation of computers rests with the investigator. Many adaptations other than those listed above may be profitably exploited. Time and cost are not the only considerations; if a high degree of accuracy is necessary in extensive computations, there is nothing that can match the digital computer. This is true only insofar as the program is not ambiguous. Overall, the reliability of computer output, as with any other method, depends on the quality of the raw data input.

INDEX

Rapide Croche Dam (Wis.), 107, 111
Rating curve, 85, 86, 123, 125